Microbiology for Cleaner Production and Environmental Sustainability

Growth of populations, increasing urbanization, and rising standards of living due to technological innovations demand not only the meticulous use of shrinking resources but also sustainable ways of producing materials for human welfare. Cleaner production involves preventive and protective initiatives which are intended to minimize waste and emissions and maximize product output. These novel microbiological techniques are a practical option for achieving environmental sustainability. *Microbiology for Cleaner Production and Environmental Sustainability* serves as a valuable source of information about microbiological advancements for a sustainability in diversified areas such as energy resources, food industries, agricultural production, and environmental remediation of pollution.

Features:

- Covers key issues on the role of microbiology in the low-cost production of bioenergy
- Provides comprehensive information on microorganisms for maximizing productivity in agriculture
- Examines green pharmaceutical production
- Provides the latest research on microbiological advancements in the restoration of contaminated sites

Microbiology for Cleaner Production and Environmental Sustainability

Edited by
Naga Raju Maddela,
Lizziane Kretli Winkelströter Eller, and Ram Prasad

CRC Press
Taylor & Francis Group
Boca Raton London New York

CRC Press is an imprint of the
Taylor & Francis Group, an **informa** business

Designed cover image: Shutterstock

First edition published 2023
by CRC Press
2385 NW Executive Center Drive, Suite 320, Boca Raton FL 33431

and by CRC Press
4 Park Square, Milton Park, Abingdon, Oxon, OX14 4RN

CRC Press is an imprint of Taylor & Francis Group, LLC

© 2023 selection and editorial matter, Naga Raju Maddela, Lizziane Kretli Winkelströter Eller, and Ram Prasad; individual chapters, the contributors

Library of Congress Cataloging-in-Publication Data
Names: Maddela, Naga Raju, editor.
Title: Microbiology for cleaner production and environmental sustainability /
Naga Raju Maddela, Lizziane Kretli Winkelströter Eller, and Ram Prasad, [editors].
Description: First edition. | Boca Raton, FL: CRC Press, 2023. |
Includes bibliographical references and index.
Identifiers: LCCN 2023002735 (print) | LCCN 2023002736 (ebook) |
ISBN 9781032496061 (hbk) | ISBN 9781032496085 (pbk) | ISBN 9781003394600 (ebk)
Subjects: LCSH: Microbial biotechnology. | Industrial microbiology. |
Green chemistry. | Environmental protection.
Classification: LCC TP248.27.M53 M56 2023 (print) |
LCC TP248.27.M53 (ebook) | DDC 660.028/6—dc23/eng/20230216
LC record available at https://lccn.loc.gov/2023002735
LC ebook record available at https://lccn.loc.gov/2023002736

ISBN: 978-1-032-49606-1 (hbk)
ISBN: 978-1-032-49608-5 (pbk)
ISBN: 978-1-003-39460-0 (ebk)

DOI: 10.1201/9781003394600

Typeset in Times
by codeMantra

Contents

SECTION I   Microorganisms in cleaner production

SECTION II   Understanding microbiology
for environmental sustainability

SECTION III   Microbial remediation

Preface

'Cleaner production' is a preventive and protective initiative which is intended to minimize waste and emissions and maximize product output. Growth of population, increasing urbanization, and rising standards of living due to technological innovations demand not only the meticulous use of shrinking resources but also sustainable ways of producing materials for human welfare. One of the practically feasible options in the 'cleaner production' is the use of microbiological techniques for achieving sustainability in areas such as Industries, Agriculture, Medicine, and Wastewater Treatment facilities. For example, microbial cells play a vital role in the production of biofuels and bioenergy, which is attributed to an efficient consumption of organic substrates and further production of useful products (to be used as biofuels) by novel processes operated by microorganisms. Diverse metabolic activities of microorganisms enable us to reuse organic wastes in the production of sustainable produce. Similarly, microbial solutions enable farmers in an economical way through sustainable yield and productivity. Various naturally occurring microorganisms have potential in the crop protection from pests and diseases and enhance productivity and fertility. However, implications of microbiology for the cleaner production are not in the encouraging way; this situation is mainly due to the following factors: there is no proper understanding about what is cleaner production, what is the functional relevance of microbiology in the cleaner production, why it is needed, what are the benefits, and what does it mean in practice? This book typically aims to establish advances made by the allied field of microbiology till date. This book in its present form has been designed in such a manner so that any enthusiast can be well aware about the very recent works that have been commenced since a long time. It can serve as a 'handbook' dealing with modern microbial technologies that evolved recently.

To reflect the title of the book, there are 20 chapters in three sections – (i) Microorganisms in Cleaner Production, (ii) Understanding Microbiology for Environmental Sustainability, and (iii) Microbial Remediation. Topics ranging from microscale studies to macro, it covers up a huge domain of microbiology. The chapters of Section I are related to production and commercial importance of biosurfactants, microalgal proteins as food, microbial production of acetic acid and citric acid, green pharmaceutical products and their benefits, anaerobic microbial communities for bioenergy production, and microbially synthesized nanoparticles for environmental sustainability. In the second section, there is emphasis on soil microbiome, bacterial sensory mechanisms for xenobiotics utilization, biofilms for bioremediation, bacterial enzymes in aquatic systems, and microbial responses to trace elements. The third section is exclusively designed to include the chapters related to microbial remediation of sites contaminated by dyes, heavy metals, antibiotics, plastics, and agrochemicals.

A special attention was paid in the selection of chapter contributors of this volume. We invited around 55 subject experts (such as Researchers, Academicians, and Industrialists) with sufficient knowledge about basic and applied microbiology from 12 different countries (i.e. Australia, Bangladesh, Brazil, China, Ecuador, India,

Nigeria, Spain, Pakistan, Turkey, UK, and USA) so that this initiative could have a positive impact on the quality of the volume, as well as this volume can reach a wider audience. Furthermore, all the listed editors of this volume have sound knowledge in Microbiology and its applications. NRM has been working in the area of Environmental Microbiology since 2003 and his research activities were related to soil enzyme activities, microbial remediation of crude oil contaminated soils (lab- and field-level), bacterial biofilms, characterization of bacterial exopolysaccharides, quorum quenching for biofouling control in membrane bioreactors, etc. During 2018–2020, the work of NRM was published in internationally reputed journals such as *Applied and Environmental Microbiology, Chemosphere, Science of the Total Environment, Environmental Pollution*, whereas LKWE has hands on experience with the research related to pure-species bacteria derived biofilms, their virulence factors, evaluation of probiotic potentials, mycology, fungal enzymes, biofilm biology under stress, antimicrobial activities, etc. She has published her data in internationally reputed and high impact factor journals such as *LWT, Journal of Food Science and Technology, Advances in Microbiology, International Journal of Dairy Technology*, and *American Journal of Infection Control*. Last but not least, RP, another editor of this volume, is an internationally renowned and high cited (~11,000 citations) researcher who has published >200 articles, and his main research areas include, but not limited to, microbiology, nanotechnology, plant-microbe interaction, and nanobiotechnology. Overall, the book portrays a very clear idea about the modern technologies evolving and also directs young minds in the same path. This book has been designed to serve as a kind of information hub about modern microbiology toward cleaner production for the sustainability of the environment. It will also serve as a ready reference for practicing students and researchers of biotechnology, environmental engineering, chemical engineering, and other allied fields likewise.

Naga Raju Maddela, Ph.D.,
Universidad Técnica de Manabí,
Portoviejo, Ecuador

Lizziane Kretli Winkelströter Eller, Ph.D.,
University of Western Sao Paulo,
Sao Paulo, Brazil

Ram Prasad, Ph.D.,
Mahatma Gandhi Central University,
Motihari, Bihar, India

Foreword

Minimizing the waste and emissions and maximizing the product output is an important initiative for environmental protection. To accomplish this, analysis of flow of materials and energy during the production process and identification of options for minimizing the waste and emissions out of industrial processes through source reduction strategies are greatly warranted. Use of microbial science in the production of high-value products with less waste and remediation of polluted sites without leaving toxic metabolites are along the lines of *cleaner production* and *environmental sustainability.*

Centring around the above issue, this volume has been well designed to address the latest issues on microbial production of high value-added products, microbial activities in the environment, and microbial remediation of contaminated sites. This edited volume has three sections – (i) Microorganisms in Cleaner Production, (ii) Understanding Microbiology for Environmental Sustainability, and (iii) Microbial Remediation. This volume has been edited by three microbiology experts from Ecuador, Brazil, and India. There are 20 chapters contributed by 55 researchers from 12 different countries of Asia-Pacific, Africa, Americas, and Europe. Therefore, this volume has an immense global appeal in terms of editors and contributors. There is comprehensive revision of literature on diverse topics such as production of biosurfactants, microbial production of value-added products, microbially synthesized nanoparticles, soil microbiome, microbial sensory mechanisms for pollutants, biofilms, microbial enzyme activities, and microbial remediation of contaminated sites. I truly believe that this volume will have a wider readership and will serve researchers, environmental policy makers, industrialists, technicians, and students for a considerable length of time.

Santiago Quiroz Fernández, Ph.D.,
Rector, Universidad Técnica de Manabí
Portoviejo, Ecuador
09 September 2022

Acknowledgements

It is our great honour to acknowledge the support of the chapter's contributors for their valuable contribution and timely responses for the success of this project. All contributors are immensely appreciated for their eagerness and inordinate support for this volume to be ready in the scheduled time; we therefore appreciate their team-work and partnership. We also thank anonymous reviewers for their constructive criticism, which had helped us in improving the quality of this book by inviting the experts to contribute the additional chapters. We greatly acknowledge the CRC Press Editorial and Production team for their respected support; without their guidelines, this project would not be finished in such a very short time. It is an honest honour and privilege to work with them all. Finally, yet importantly, we are very much thankful to the colleagues at Universidad Técnica de Manabí (Ecuador), University of Western Sao Paulo (Brazil), and Mahatma Gandhi Central University (India) for their unrestricted backing and for the establishment of treasured propositions at the time of book proposal and final book preparation.

Naga Raju Maddela, Ph.D.,
Universidad Técnica de Manabí,
Portoviejo, Ecuador

Lizziane Kretli Winkelströter Eller, Ph.D.,
University of Western Sao Paulo,
Sao Paulo, Brazil

Ram Prasad, Ph.D.,
Mahatma Gandhi Central University,
Motihari, Bihar, India

About the Editors

Naga Raju Maddela received his M.Sc. (1996–1998) and Ph.D. (2012) in Microbiology from Sri Krishnadevaraya University, Anantapur, India. During his doctoral program in the area of environmental microbiology, he investigated the effects of industrial effluents/insecticides on soil microorganisms and their biological activities and worked as a faculty in microbiology for 17 years, teaching undergraduate and postgraduate students. He received 'Prometeo Investigator Fellowship' (2013–2015) from Secretaría de Educación Superior, Ciencia, Tecnología e Innovación (SENESCYT), Ecuador, and 'Postdoctoral Fellowship' (2016–2018) from Sun Yat-sen University, China. He also received external funding from 'China Postdoctoral Science Foundation' in 2017, internal funding from 'Universidad Técnica de Manabí' in 2020, worked in the area of Environmental Biotechnology, participated in 20 national/international conferences, and presented research data in China, Cuba, Ecuador, India, and Singapore. Currently, he is working as a full-time Professor at the Facultad de Ciencias de la Salud, Universidad Técnica de Manabí, Portoviejo, Ecuador. He has published eight books, 40 chapters, and 60 research papers.

Lizziane Kretli Winkelströter Eller graduated in Biochemical Pharmacy from the Federal University of Alfenas – Brazil (2006). She earned her Master's degree in Pharmacy from the Faculty of Pharmaceutical Sciences at USP of Ribeirão Preto – Brazil (2009), doctorate in Pharmacy from the Faculty of Pharmaceutical Sciences of Ribeirão Preto – Brazil (2012), and completed postdoc at the Faculty of Pharmaceutical Sciences at USP in Ribeirão Preto – Brazil (2014). She has published 28 manuscripts and four book chapters and is an editor of a research topic in *Frontiers in Microbiology*. She is currently a Professor at the University of Western of São Paulo-Brazil. She has experience in the field of Pharmacy, acting mainly on the following topics: lactic acid bacteria, foodborne pathogens, microbiology, molecular biology, biofilms, probiotics, and cell culture.

Ram Prasad, Ph.D., is associated with the Department of Botany, Mahatma Gandhi Central University, Motihari, Bihar, India. His research interest includes applied and environmental microbiology, plant–microbe interactions, sustainable agriculture, and nanobiotechnology. Dr. Prasad has more than two hundred fifty publications (Total citations 12505 with an h-index 56, i10-index 176) to his credit, including research papers, review articles & book chapters and seven patents issued or pending, and edited or authored several books. Dr. Prasad has 14 years of teaching experience and has been awarded the Young Scientist Award & Prof. J.S. Datta Munshi Gold Medal by the International Society for Ecological Communications; Fellow of Biotechnology Research society of India; Fellow of the Society for Applied Biotechnology; Fellowship by the Society for Applied Biotechnology; the American Cancer Society UICC International Fellowship for Beginning Investigators, USA; Outstanding Scientist Award in the field of Microbiology; BRICPL Science Investigator Award and Research Excellence Award, etc. He has been serving as editorial board member of *Nanotechnology Reviews; Green Processing and Synthesis; BMC Microbiology; BMC Biotechnology; Current Microbiology; Archives of Microbiology; Annals of Microbiology; Nanotechnology for Environmental Engineering; Journal of Renewable Material; Journal of Agriculture and Food Research; SN Applied Sciences; Agriculture etc., including Series editor of Nanotechnology in the Life Sciences, Springer Nature, USA.* Previously, Dr. Prasad served as an Assistant Professor, Amity University Uttar Pradesh, India; a Visiting Assistant Professor, Whiting School of Engineering, Department of Mechanical Engineering at Johns Hopkins University, Baltimore, USA; and a Research Associate Professor at School of Environmental Science and Engineering, Sun Yat-sen University, Guangzhou, China.

Contributors

Binta Buba Adamu
National Biotechnology Development
  Agency
Abuja, Nigeria

Beauty Akter
Jahangirnagar University
Savar, Dhaka, Bangladesh

Sesan Abiodun Aransiola
Bioresources Development Centre,
  National Biotechnology Development
  Agency
Ogbomoso, Oyo State, Nigeria

Abraham O. Ayanwale
Federal University of Technology
  Minna
Minna, Niger State, Nigeria

Shireen Aziz
University of Faisalabad
Faisalabad, Punjab, Pakistan

Babafemi Raphael Babaniyi
Bioresources Development Centre,
  National Biotechnology Development
  Agency
Ogbomoso, Oyo State, Nigeria

Ebunoluwa Elizabeth Babaniyi
Obafemi Awolowo University
Ile-Ife, Osun State, Nigeria

Gabriel Gbenga Babaniyi
Agricultural and Rural Management
  Training Institute (ARMTI)
Ilorin, Kwara State, Nigeria

Iqra Bano
Shaheed Benazir Bhutto University of
  Veterinary & Animal Sciences
Sakrand, Sindh, Pakistan

Ademola Bisi-omotosho
University of Westminster
London, United Kingdom

Ahmet Çabuk
Eskişehir Osmangazi University
Eskişehir, Turkey

Patrícia Caetano
Federal University of Santa Maria
  (UFSM)
Santa Maria, RS, Brazil

Pınar Aytar Çelik
Eskişehir Osmangazi University
Eskişehir, Turkey

M. Subhosh Chandra
Yogi Vemana University
Kadapa, Andhra Pradesh, India

Rebeca Díez-Antolínez
Instituto Tecnológico Agrario de
  Castilla y León (ITACyL)
León, Spain

Evans C. Egwim
Federal University of Technology
  Minna
Minna, Niger State, Nigeria

Sharmin Zaman Emon
University of Dhaka
Dhaka, Bangladesh

Andressa Fernandes
Federal University of Santa Maria
   (UFSM)
Santa Maria, RS, Brazil

Daniel Gana
Federal University of Technology
   Minna
Minna, Niger State, Nigeria

Mordecai Gana
Federal University of Technology
   Minna
Minna, Niger State, Nigeria

Xiomar Gómez
University of León
León, Spain

Rubén González
University of León
León, Spain

Syed Shams ul Hassan
Shanghai Jiao Tong University
Shanghai, China

Eduardo Jacob-Lopes
Federal University of Santa Maria
Santa Maria, RS, Brazil

Zobaidul Kabir
University of Newcastle
NSW, Australia

Srinivasan Kameswaran
Vikrama Simhapuri University College
Kavali, Andhra Pradesh, India

SMA Karim
Ahsanullah University of Science and
   Technology
Dhaka, Bangladesh

Vijaya Sudhakara Rao Kola
University of Nebraska Medical Center
Omaha, Nebraska, United States

B. Lakshmanna
Yogi Vemana University
Kadapa, Andhra Pradesh, India

M. Madakka
Yogi Vemana University
Kadapa, Andhra Pradesh, India

D. Mallaiah
Yogi Vemana University
Kadapa, Andhra Pradesh, India

Enuh Blaise Manga
Eskişehir Osmangazi University
Eskişehir, Turkey

G. Jaffer Mohiddin
Universidad de las Fuerzas Armadas
   – ESPE
Quito, Ecuador

G. Narasimha
Sri Venkateswara University
Tirupati, Andhra Pradesh, India

Tatiele Nascimento
Federal University of Santa Maria
   (UFSM)
Santa Maria, RS, Brazil

Pricila Nass
Federal University of Santa Maria
   (UFSM)
Santa Maria, RS, Brazil

Olusola David Ogundele
Achievers University
Owo, Ondo State, Nigeria

Olaniran Victor Olagoke
Agricultural and Rural Management
Training Institute (ARMTI)
Ilorin, Kwara State, Nigeria

Oluwafemi Adebayo Oyewole
Federal University of Technology
Minna
Minna, Niger State, Nigeria

Daniela Peña-Carrilloa
University of León
León, Spain

Gopi Krishna Pitchika
Vikrama Simhapuri University College
Kavali, Andhra Pradesh, India

M. Ramakrishna
Vikrama Simhapuri University College
Kavali, Andhra Pradesh, India

Bellamkonda Ramesh
University of Nebraska Medical Center
Omaha, Nebraska, United States

Muhammad Ahmer Raza
University of Faisalabad
Faisalabad, Punjab, Pakistan

P. Sudhakar Reddy
Vikrama Simhapuri University College
Kavali, Andhra Pradesh, India

Muhammed Muhammed Saidu
Federal University of Technology
Minna
Minna, Niger State, Nigeria

G. Mary Sandeepa
Vikrama Simhapuri University
Nellore, Andhra Pradesh, India

Luisa Schetinger
Federal University of Santa Maria
(UFSM)
Santa Maria, RS, Brazil

Mashura Shammi
Jahangirnagar University
Savar, Dhaka, Bangladesh

M. Srinivasulu
Yogi Vemana University
Kadapa, Andhra Pradesh, India

Gujjala Sudhakara
Sri Krishnadevaraya University
Anantapur, Andhra Pradesh, India

B. Swapna
Vikrama Simhapuri University College
Kavali, Andhra Pradesh, India

Japhet Gaius Yakubu
Federal University of Technology
Minna
Minna, Niger State, Nigeria

Belma Nural Yaman
Eskişehir Osmangazi University
Eskişehir, Turkey

Leila Q. Zepka
Federal University of Santa Maria
(UFSM)
Santa Maria, RS, Brazil

Section I

Microorganisms in cleaner production

1 Production and commercial significance of biosurfactants

Pınar Aytar Çelik, Belma Nural Yaman,
Enuh Blaise Manga, and Ahmet Çabuk

CONTENTS

DOI: 10.1201/9781003394600-2

1.1 INTRODUCTION

Biosurfactants are amphiphilic compounds that can lower the surface tension between two liquids. Since the first report on biosurfactants in 1968, a lot of research has been carried out characterizing the structures of various biosurfactants produced including the microorganisms that produce them. Various properties of their structures have guided different classification schemes over time. The differences in their structures also confer different properties to biosurfactants enabling them to suit various processes across a wide range of fields. Biosurfactants are eco-friendly, highly biodegradable, have low toxicity, reduce surface tension, have good foaming properties, and have antimicrobial activity. These properties make them more desirable for use in the industry over synthetic surfactants. The applications of biosurfactants in cosmetics, pharmaceutical, food, petroleum, mining, wastewater treatment, agriculture, textile, bioremediation, and a few miscellaneous industries have shown the high importance of these molecules and their potential to impact significantly important industrial processes (Manga et al., 2021). Biosurfactants are synthesized naturally by many fungi and bacteria species. The production yields in nature are very small. Various approaches namely the use of specific media formulations, low-cost substrates, an improvement in fermentation processes, and genetic engineering of producer strains have been used to improve production yields and production cost-effectiveness. These approaches though have yielded significant progress; research is still needed to continue improving production to make biosurfactants competitive with synthetic surfactants in the markets. However, amid high production costs, companies are still involved in biosurfactant production. The high demands driving the biosurfactant markets are a result of awareness of consumers on current global environmental challenges and the need to use green products. With the increase in demand for biosurfactants globally and the increase in patents of new technologies for production, in the future, biosurfactants could replace synthetic surfactants in many industrial processes.

1.2 DISCOVERY OF BIOSURFACTANTS

The first biosurfactant to be produced, purified, and characterized was surfactin (Arima et al., 1968). The "greenness" and surface-active properties showed their versatile industrial applicabilities, which have developed research interests in many fields over the years enabling them to occupy a niche of their own in the world markets over time (Sekhon Randhawa and Rahman, 2014). Today, biosurfactants play a huge role in various industries and are a significant component in the various consumables that we use each day of our lives.

1.3 PROPERTIES OF BIOSURFACTANTS

Surfactants are amphiphilic compounds that lower the surface tension between liquids. Biosurfactants are surfactants synthesized from bacteria, fungi, or plants (Sobrinho et al., 2013). Their eco-friendliness shows a better promise in industrial applications compared to synthetic ones. Their attractive physicochemical properties include high stability (Ben Ayed et al., 2014), high biodegradability (Lima et al., 2011), low toxicity, surface tension reduction, foaming capacity, and antimicrobial activity (Akbari et al., 2018; Liu et al., 2015; Morais et al., 2017).

1.4 TYPES OF BIOSURFACTANTS

The classification of surfactants follows many schemes, for instance, the charge on the polar component is used for the classification of surfactants. Sulfur group or sulfonate group containing surfactants are negatively charged termed anionic, those containing the positively charged quaternary ammonium groups are cationic, both positive and negative charge groups containing ones are amphoteric, and those completely lacking the ionic moiety are non-ionic (Rahman and Gakpe, 2008). Biosurfactants obtained from microbes are generally anionic or neutral, while a few are cationic (Sobrinho et al., 2013). Long-chain fatty acids generally characterize the hydrophobic moiety while organic acid, alcohol, amino acid, or carbohydrate functional groups characterize the hydrophilic moiety (Sobrinho et al., 2013). They are also grouped concerning their chemical structures, which can be broadly referred to as high molecular weight and low molecular weight molecules. Low molecular weight molecules are the glycolipids, lipopeptides, and phospholipids while the high molecular weight groups are the polymeric and particulate biosurfactants (Shoeb et al., 2013). We will examine the five-group classification based on their structures.

1.4.1 GLYCOLIPID BIOSURFACTANTS

Most of the known biosurfactants are glycolipids (Rahman and Gakpe, 2008; Shoeb et al., 2013). Often, they are carbohydrates bonded to either long-chain aliphatic acids or fatty acids by an ester or ether linkage. Within this group, rhamnolipids, trehalolipids, and sophorolipids are the best characterized (Shoeb et al., 2013). They contain mono-, di-, tri-, and tetrasaccharides, which could be glucose, mannose, galactose, glucuronic acid, rhamnose, and galactose sulfate (Rahman and Gakpe, 2008). They are generally produced as secondary metabolites mixtures of glycolipids. A mixture of carbohydrate and vegetable oil is usually required in culture to obtain large yields of glycolipids (Cooper and Paddock, 1984).

1.4.1.1 Rhamnolipids

Rhamnolipids contain a rhamnose sugar moiety bonded to one or two molecules of β-hydroxydecanoic acid. One of the hydroxydecanoic acid's hydroxyl groups is bonded by glycosidic bond to the rhamnose sugar's reducing end; the second hydroxydecanoic acid's OH group is used for ester bonding (Shoeb et al., 2013). The genus *Pseudomonas* comprises the commonly known bacteria producers of rhamnolipids (Edwards and Hayashi, 1965).

1.4.1.2 Trehalose lipids

Trehalose is a disaccharide containing two glucose moieties linked by α,α-1,1-glycosidic bonds (Shoeb et al., 2013). The α-d-trehalose is a non-reducing disaccharide generally produced by bacteria, fungi, and algae. However, just the α,α'-isomer has been observed to occur naturally, except for some biosurfactants produced by *Streptococcus faecalis* (Asselineau and Asselineau, 1978).

1.4.1.3 Sophorolipids

Sophorolipids are extracellular glycolipids produced by ascosporous yeasts, which consist of sophorose, the carbohydrate, which is two glucose bonded by β-1,2' bond, and a fatty acid. More than one unsaturated bond can form in the hydroxyl group of the fatty acid while the carboxylic group can be free (Open) or form ester bonds (Lactones) (Kulakovskaya and Kulakovskaya, 2014). Six to nine different sophorose molecules can be found in a mixture. They have shown the greatest promise in bioremediation and many industrial processes some have also shown antifungal activities (Sen et al., 2017).

1.4.2 LIPOPEPTIDE AND LIPOPROTEIN BIOSURFACTANTS

Lipopeptide and lipoprotein biosurfactants are synthesized by many bacteria species, fungi, and yeasts. Surfactin produced by *Bacillus subtilis* is a well-known lipopeptide. They are antibiotic-like compounds that consist of a fatty acid chain linked to a cyclic peptide moiety. The changes in the peptide moiety, the fatty acid chain length, and the linkage between these two give rise to different isoforms. Several isoforms can be produced by one strain in the same mixture (Mnif and Ghribi, 2015). They can be produced by extreme acidophiles, which show another desirable industrial property (Paranji et al., 2014).

1.4.3 FATTY ACID, PHOSPHOLIPID, AND NEUTRAL LIPID BIOSURFACTANTS

This group of biosurfactants is synthesized by many species of bacteria and yeasts when grown on *n*-alkanes containing media. They are also components of the outer membranes of extracellular vesicles of some *Acinetobacter* species used to degrade hydrocarbons (Käppeli and Finnerty, 1979).

1.4.4 POLYMERIC BIOSURFACTANTS

Polymeric biosurfactants occur generally in nature in polymeric forms and those well-characterized include emulsan, liposan, lipomanan, alasan, mannoprotein, and polysaccharide-protein compounds (Santos et al., 2016; Vijayakumar and Saravanan, 2015). *Acinetobacter calcoaceticus* is known to synthesize emulsan containing a heteropolysaccharide backbone linked to fatty acids (Rosenberg et al., 1988).

1.4.5 PARTICULATE BIOSURFACTANTS

Particulate biological surfactants form extracellular membrane vesicles that facilitate the uptake of alkane substrates by forming small emulsions (Vijayakumar and Saravanan, 2015). Their production was found to be independent of the substrate and hydrocarbon used in *Pseudomonas marginalis* but proportional to bacterial biomass (Burd and Ward, 1997).

1.5 USES OF BIOSURFACTANTS

Their multifunctional and numerous applications are due to their inherent properties which are: surface and interfacial tension reduction, low critical micelle concentration (CMC), wettability, specificity, antimicrobial, and their advantages over synthetic counterparts such as eco-friendliness, biodegradability, biocompatibility, low toxicity, easy production, chemical diversity, effective at high temperatures and pH, and sometimes cost-effectiveness. The CMC is the smallest biosurfactant concentration that produces the highest reduction of surface tension. These exclusive properties of biosurfactants have made them economically attractive, and in the future, they may take the place of synthetic biosurfactants in various industrial operations (Akbari et al., 2018; Banat et al., 2000; Fenibo et al., 2019). Biosurfactants have applications in many industrial areas such as cosmetics, pharmaceutics, food, petroleum, wastewater treatment, agriculture, textile, painting, and many others (Banat et al., 2000).

1.5.1 COSMETICS INDUSTRY

Cosmetic products of everyday use such as shampoos, makeup, body creams, and soaps contain biosurfactants. The properties of biosurfactants that make them desirable in the cosmetic industry are those related to their CMC, ionic performance, and the balance between hydrophilicity and lipophilicity. These measurements determine their applications in cosmetic formulations. The great demand for green products gives an advantage to the use of biosurfactants in cosmetic formulations (Vecino et al., 2017). Furthermore, they have other benefits such as low cytotoxicity and easy biodegradability even after use. Up to 97% proliferation has been obtained on mouse cell lines as compared to inhibition for SDS on the application of 0.5 g/l each of both SDS and biosurfactant (Ferreira et al., 2017). Better properties have also been reported in some formulated biosurfactant-based toothpaste (Das et al., 2018). In cosmetic formulations, they serve as detergents, emulsifiers, wetting, foaming solubilizing, and dispersing agents. Their most important function is being emulsifying agents. They are easy to use, cheaper, and can be used at the same time with other liposoluble and hydrosoluble compounds. Safety regulations are not yet available for the use of biosurfactants (Vecino et al., 2017). Synthetic surfactants are capable of causing various skin irritations because they can bind skin proteins and destroy lipid structures both on the skin surface and intercellularly (Bujak et al., 2015). Biosurfactants have been reported to have antimicrobial and anti-adhesive properties that are desirable for use in cosmetic products for application on the skin microflora. Antimicrobial activity has been shown for common skin pathogens such as *Staphylococcus aureus, Streptococcus pyogenes,* and *Streptococcus agalactiae* (Vecino et al., 2018). They can be used in formulations for sanitizers to protect against bacterial and viral infections (Çelik et al., 2021). Their low toxicity also enables them to be used as permeability enhancers (Okasaka et al., 2019). In addition, prebiotic characteristics have been described; the mechanism is preventing the proliferation of skin pathogens while favoring the proliferation of beneficial ones (Vecino et al., 2017).

1.5.2 PHARMACEUTICAL INDUSTRY

They are one of the promising biomolecules in the pharmaceutical industry due to their structural novelty, adaptability, and diverse properties that have potential applications in therapy. Their surface activity is also of interest in the pharmaceutical industry. Because of their surface activity, they can interact with cell membranes and their surrounding environments making them suitable for use in the delivery of drugs for diseases like cancer or enhancing their solubility and bioavailability (Bhadoriya and Madoriya, 2013; Gudiña et al., 2013). Biosurfactants have promising applications in numerous areas within the pharmaceutical industry. There is demonstrated antimicrobial activity against several pathogens (Bucci et al., 2018; Rodrigues et al., 2006), antiviral activity against enveloped viruses such as herpes and retroviruses (Vollenbroich et al., 1997), antitumor activity (Kameda et al., 1974), anti-adhesive agent and drug adjuvants (Bhadoriya and Madoriya, 2013), and gene transfection (Kitamoto et al., 2002). Other relevant applications are in immunoglobulin binding ligands, inhibition of clotting by activating the enzyme fibrin, and coating for medical instruments. They may also be used as probiotics to combat lung infections (Gharaei, 2011; Gudiña et al., 2013; Rodrigues et al., 2006). Furthermore, the properties of biosurfactants have been evaluated to be desirable in the pharmaceutical industry for producing drugs, sanitizers, vaccines, etc., that can be very effective in combatting the COVID-19 pandemic. In addition, these novel approaches can be applied in the management of future threats of microbial origin (Çelik et al., 2021).

1.5.3 FOOD INDUSTRY

The antimicrobial, antioxidant, emulsification, and anti-biofilm activities against food pathogens are most beneficial to the food industry (Karlo et al., 2023). Very important are their surface treatment benefits; as anti-adhesives, they prevent the formation of biofilms on biological surfaces by preventing colonization of pathogens (Nitschke and Silva, 2018). Delayed adhesion to stainless steel surfaces was shown for foodborne pathogenic bacteria such as *Listeria, Enterobacter,* and *Salmonella* showing applicability in extending shelf life during food storage (Nitschke et al., 2009). Furthermore, biosurfactants have shown antagonistic activity against important foodborne pathogens of the genus *Staphylococcus* (Moore et al., 2013). The growing demand for probiotics and the use of lactic acid bacteria in probiotics have also attracted interest. In addition to their probiotic properties, lactic acid bacteria were found to produce biosurfactants simultaneously making them suitable food ingredients (Sharma and Singh Saharan, 2014).

1.5.4 PETROLEUM INDUSTRY

The degradation of petroleum hydrocarbons is possible only after it has been made soluble. Surfactants can widen the surface area of hydrophobic hydrocarbons from petroleum oil spills on land or water facilitating their uptake by microorganisms. Therefore, asphaltenes, resins, or other molecularly complex hydrocarbons degradation can be rendered feasible (Mazaheri Assadi and Tabatabaee, 2010). The petroleum industry applies surfactants in many of its processes. Petrochemically derived synthetic surfactants are used to enhance the recovery of petroleum, extraction, and treatments, cleaning of storage tanks, and facilitate oil transport within pipelines. Further uses include emulsification agents, anticorrosive, control of sulfate-reducing bacteria, and new formulations. The demand for sustainable technologies globally which can improve petrochemical processes and be environmentally friendly has also turned interests toward the use of biosurfactants (Almeida et al., 2016).

1.5.4.1 Microbial enhanced oil recovery (MEOR)

Oil production goes through three stages: the primary and secondary methods employ pressures for the extraction of oil recovering an estimated 40% of the oil while the tertiary process, also known as the enhanced recovery, is then used to recover the 60% still left in the reservoir. Temperature-dependent and -independent methods are used for recovery at the tertiary stage. The latter uses chemicals and biological methods to achieve recovery. The biological methods are termed MEOR (Fenibo et al., 2019). Using biosurfactant, 16% oil recovery was achieved with *Bacillus licheniformis* and biocide activity against sulfate-reducing bacteria using 1.0% biosurfactant after 3 hours (El-Sheshtawy et al., 2015).

1.5.4.2 Emulsified fuel formulations

Much interest has not been shown in this area of application. These are mixtures of surfactant and fuel that lead to the formation of a stable emulsion of other liquids in the fuel phase (Almeida et al., 2016).

1.5.4.3 Biocide and anticorrosive

Microorganisms that reduce sulfate to sulfide when present in petroleum reservoirs can cause a process known as souring. During secondary treatment of petroleum, water is flooded in the reservoir promoting the souring process. The hydrogen sulfide produced at high concentrations can be very toxic or cause an explosion (Korenblum et al., 2012). The need for effective control of sulfide producers therefore arises. Biosurfactants have been shown to have biocide and anticorrosion properties. By interacting with the metal surfaces, they inhibit the formation of corrosions and bacteria biofilms (Fenibo et al., 2019). The mechanism of biocide activity as seen with *Bacillus* sp. lipopeptide is by forming cytoplasmic inclusions, osmotic lysis of cells, and reducing the hydrophobicity of surfaces (Korenblum et al., 2012). In another study, seven isolates were screened for bioemulsifier characteristics. It showed good surface activity after 48 hours of incubation and at 35°C than that under 10°C (Yu and Huang, 2011). *Acinetobacter calcoaceticus* PTCC1318 produced emulsan which reduced water surface tensions to 24 mN/m and interfacial tension to 3 mN/m. A 98% crude oil emulsification was achieved at a water and oil dilution of 1:2. From the washing time and flow rates tested, emulsan showed usefulness in cleaning tubes with 100% removal at room temperatures and relying on cleaning conditions (Amani and Kariminezhad, 2016). They can be applied in cleaning petroleum tanks or conditioning oil in long petroleum pipelines.

1.5.5 BIOMINING

Biosurfactants can be used in biomining during the leaching process to remove organic matter allowing the accumulation of metals. Organic matter is usually made up of aromatic compounds. The role of biosurfactants is to bind the compounds and increase their solubility hence bioavailability for other microorganisms to use as carbon sources (Czaplicka and Chmielarz, 2009). Biosurfactants have been shown to remove 20–30% of slime from copper during smelting (Czaplicka and Chmielarz, 2009). Biomining has also been utilized to enhance bioleaching with bacteria consortia. Biosurfactant-producing bacteria and sulfur-oxidizing bacteria were mixed to bleach Cu, Pb, Zn, Cd, Ni, and Cr. By removing the organic compounds in the combustion waste, metal bioleaching efficiencies of up to 90% were achieved (Karwowska et al., 2015). Using a similar consortium, metals were bioleached from waste circuit boards and the best removal efficiency of 93% was achieved with cadmium. Increasing aeration and temperature also increased the release of metals (Karwowska et al., 2014). Optimizing conditions for each bioleaching experiment promise to provide higher amounts of metals extracted from wastes.

1.5.5.1 Biodesulfurization

Coal contains many unwanted materials some of which are released into the atmosphere during combustion. Organic and inorganic sulfur compounds when burned generate SO_2, which is air pollution causing gas. SO_2 in the atmosphere can form acid rain which causes respiratory and pulmonary diseases alongside, skin diseases and cancer (Çelik et al., 2019). The demand for coal, especially with current applications of clean coal technologies, continues to increase (Aytar et al., 2014; Çelik et al., 2019). The prevention of SO_2 gas production can be achieved by using biosurfactants to increase the bioavailability of sulfur compounds to microbes within the treatment process. Biosurfactants are more advantageous than synthetic surfactants in this process mainly because they are easy to produce and cost-effective among other minor advantages (Fenibo et al., 2019). Bacillus species and *R. erythropolis* were grown and optimized to produce surfactin in a vertical rotating immobilized cell reaction and surfactant was evaluated for the biodesulfurization of dibenzothiophene. The desulfurization rate and volumetric reactor productivity both increased to 210.5 and 16.6 mM 2 HBP/l/h, respectively. The concentration of sulfur decreased to <5 ppm from 398 and surfactin was produced at 4.8 g/l measuring at 720 mg/l/h. This demonstrates the value of surfactin in potentially decreasing operation costs (Amin et al., 2013).

1.5.5.2 Bioflotation

The increasing amounts of low-grade mineral ores as higher grades are depleted warrant better methods for efficient extraction. The interaction of bacteria and minerals can be exploited in a technique known as bioflotation. Bacteria can change the surface properties of minerals making them prone to redox reactions of their constituents. The microbial activity promotes flotation depending on the characteristics of the minerals. Biosurfactants' low CMC and foaming abilities are beneficial for use as biofrothers, biodepressants, or biocollectors in mineral flotation (Aytar Çelik et al., 2021; Behera and Mulaba-Bafubiandi, 2017; Sanwani et al., 2016). Due to the high compatibility of biosurfactants with many processes, bioflotation can be beneficial in the various mineral processes; biomodification of serpentinite and quartz (Didyk and Sadowski, 2012), frother in coal flotation recovering 72–79% combustible fractions (Fazaelipoor et al., 2010), the precipitation of calcium carbonate and simultaneously favoring the formation of good surface characteristics which can enable their application in polymeric capsules (Bastrzyk et al., 2019) and increasing flotation of minerals from copper ores (Khoshdast et al., 2012).

The usability of bioreagents such as biosurfactants was utilized as collector, frothing agent, depressant, and activator with high selectivity and specificity in the literature. Fazaelipoor and colleagues examined biosurfactants extracted from *Pseudomonas aeruginosa* as frother agents for flotation of coal. 72–79% of combustible matter recovery with 10–15.5% ash content was obtained in the flotation procedure. Furthermore, the biosurfactant provided was better than methyl isobutyl carbinol (MIBC) as a conventional frother in terms of froth height and stability (Fazaelipoor et al., 2010). In another study, the flotation of Cr (III) was carried out using rhamnolipid as a collector. Within five minutes, about 95% chromium was removed at an initial concentration of 40 ppm (Abyaneh and Fazaelipoor, 2016). Similarly, surfactin was shown to float calcite efficiently and shown to have potential to compete with oleate at an industrial scale (Aytar Çelik et al., 2021). *Acidithiobacillus ferrooxidans, Alicyclobacillus ferrooxidans* (Sanwani et al., 2016), and *Halomonas* biosurfactants (Consuegra et al., 2020) have also been used as biodepressant for pyrite depression. Nevertheless, bioflotation is not yet fully extended to mineral processes in industries despite its wide scope and eco-friendliness. Besides, challenges in obtaining the bioreagent, and the time for the bioreagent to react with the mineral still have to be overcome (Aksoy et al., 2022).

1.5.6 WASTEWATER INDUSTRY

The use of many chemical products is causing more complex sewage problems over the world. Industrial wastes also often contain high contents of waste oils and heavy metals. This is an important global challenge that is causing significant amounts of pollution. Biosurfactants provide an alternative advantageous method for the treatment of these wastewaters. Biosurfactant-producing bacteria are found ubiquitously in wastewater treatment plants already providing some basic favorable conditions for wastewater treatment (Ndlovu et al., 2016). *Citrobacter freundii* MG812314.1 produced biosurfactant in a study removing high percentages of cadmium (55), copper (44), zinc (66), aluminum (80), iron (45), lead (67), and manganese (41) from wastewater (Gomaa and El-Meihy, 2019). Biosurfactants used as enhancers with lignocellulose biocomposite achieved an additional increase in the removal of dyes (10%) and sulfates (62%) attributed to their good emulsifying ability and formation of better structures with lignocellulose favoring removal of more waste substances (Perez-Ameneiro et al., 2015). The rich oily content of industrial wastewaters can also be dispersed using biosurfactants. Rhamnolipids promoted the removal of up to 92% of lubricating oil after 24 hours at 20°C from wastewater. In combination with enzyme pools, biosurfactants have achieved higher removal of fats from wastewater and be more cost-effective (Damasceno et al., 2018).

1.5.7 AGRICULTURE INDUSTRY

The growing need for sustainable agricultural methodologies and demand for food has also put biosurfactants in the spotlight of the agricultural industry. Biosurfactants are environmentally compatible molecules with diverse uses in the agricultural industry improving plant-microbe interactions, promoting the availability of nutrients to plants, bioremediating the soil, and controlling plant pathogens (Naughton et al., 2019; Sachdev and Cameotra, 2013; Sonowal et al. 2022). Biosurfactant was produced from agricultural waste and then used to promote cabbage seed germination (Araújo et al., 2019). The availability to plants of micronutrients that are often found naturally or in wastes can be improved by using biosurfactants. Zinc, manganese, iron, and copper can be chelated by biosurfactants increasing their bioavailability in trace element fertilizers (Stacey et al., 2007). Improvement has occurred in plant growth and bioaccumulation of cadmium when soils are inoculated with biosurfactant produced by Bacillus species (Sheng et al., 2008). Increasing soil hydrophilization to achieve uniform distribution of fertilizers also better pesticide formulations to increase their solubility (Banat et al., 2000). The third highest selling insecticide DDT after application leaves behind recalcitrant compounds such as endosulfan and chlorinated cyclodiene causing many environmental problems. Their bioavailability can be improved by adding biosurfactants, which will enhance biodegradation. Biosurfactant-producing bacteria have been shown to degrade 99% endosulfan while reducing surface tension to 37 dynes/cm (Odukkathil and Vasudevan, 2013).

1.5.8 TEXTILE INDUSTRY

Textile processes at various steps employ the use of surfactants. Scouring (removing impurities), lubrication and softening (to remove fats), dying (wetting to increase dye penetration), and finishing processes employ surfactants. During dye pre-treatment, various unwanted fiber mixtures are removed with the aid of lubricants. During the removal of these lubricants, emulsification problems leaving some oil deposits on the fiber surface occur. Considering ecological problems arising from the use of synthetic surfactants for washing off these oils, biosurfactants provide attractive alternative approaches for washing (Kesting et al., 1996). The broader range of bio-surfactants with their versatile properties provides larger avenues for application in the textile industry (Kesting et al., 1996). Rhamnolipid was used to treat Ryeland and Shetland wool to remove surface keratinous contaminants. Two days later, a smooth outer layer was obtained after the surface contaminants were displaced. It showed promise to be a faster and easier methodology for wool scouring during pre-treatment in textile companies (Jibia et al., 2018). Using biomass residue from urban wastes, six biosurfactants were isolated from aerobic digestion culture and tested for auxiliary function during dyeing of cellulose fabrics to enhance dye solubility in water. No significant difference in performance was evident using biosurfactant and synthetic surfactants. The generation of biosurfactants from recyclable waste offered it a cost advantage over the synthetic surfactant (Savarino et al., 2009).

1.5.9 ENVIRONMENTAL REMEDIATION

Environmental pollution is one of the major issues on global agendas in the 21st century. Chemical methods have been used to mitigate pollution, but their limitations are still causing concerns. The native properties of biosurfactants enable them to offer better environmentally friendly solutions to environmental remediation (Patel et al., 2019).

1.5.9.1 Oil spill bioremediation

Biosurfactants can increase the degradation of oils by dispersion and increasing their uptake as nutrients by microorganisms. *Gordonia* sp. strain JE-1058 was prepared as a remediation agent alongside a biosurfactant and it showed strong oil spill dispersant properties without further need for any solvent. In seawater, weathered crude oil was degraded effectively by the natural marine bacteria population with the same results observed for sea sand (Saeki et al., 2009). Many factors affect the efficient use of biosurfactants as bioremediation agents. Biosurfactants are strain specific and often show inhibition of the growth of other species (Uttlová et al. 2016). Consortia are generally required for oil spill remediation therefore care should be taken that the biosurfactant being employed does not destroy other bacteria supposed to use the available hydrocarbons as substrates. Biosurfactant production is highly regulated making its use very complex. The reason for the production of biosurfactants is not fully understood. Therefore, intended effects may not be observed (Patel et al., 2019). Cases of no degradation have been reported even with supplementation of biosurfactants (Chrzanowski et al., 2012). Therefore, fully understanding the reason for rhamnolipid production and their interactions with other microorganisms and pollutants are important for more effective implementations within this area.

1.5.9.2 Metal bioremediation

Biosurfactants have a wide range of structures enabling them to form many complexes with many metals. They thus provide alternative environmentally friendly methods for the bioremediation of metal-contaminated soils. Given that metals are not biodegradable, they influence their mobility enabling them to be picked up by bioaccumulators or immobilized in the soil (Ayangbenro and Babalola, 2018; Fenibo et al., 2019; Miller, 1995). Co-contamination sites having metals with other organic compounds have also shown promising remediation when biosurfactants or biosurfactant producers are used. This is a more economical approach compared to using single chelators (Pacwa-Płociniczak et al., 2011).

1.5.9.3 Degradation of antibiotics

Large antibiotic discharge in the soil influencing microbial community compositions and increasing antimicrobial resistance are major global concerns. Biosurfactants offer safer approaches for the biotransformation of these antibiotics to serve as an energy source of carbon (Karlapudi et al., 2018). Biosurfactant-producing *Bacillus subtilis* strains have been shown to degrade antibiotics in the soil (Jałowiecki et al., 2017). Consortia of biosurfactant-mediated antibiotic degraders have also been identified (Jałowiecki et al., 2018). Apart from degradation, the extraction of antibiotics with sophorolipids has been demonstrated and their potential, environmental friendliness, and sustainability appreciated (Chuo et al., 2018).

1.5.9.4 Soil washing

The extensive environmental damage from soil pollution has increased interest to use biosurfactants to the improvement of various contaminated soils worldwide. Surfactants can promote the desorption of soil pollutants and the bioremediation of organic matter by making the pollutants more available as substrates to other microorganisms. The extraction from the soil of heavy metals and radionuclides includes dissolution, complexation with surfactants, and ion exchange mechanisms. Mixing surfactants or combining them with other additives enhances their performance as soil washing agents (Mao et al., 2015). The successful removal of malathion from the soil by surfactant washing as a process for the remediation of nonaqueous phase liquid-contaminated soil has been reported (Chu et al., 2005). Furthermore, microbial surfactants mixed with plant surfactants showed great potential for application in sequential washing of soils of various properties and multi-metal contamination (Gusiatin et al., 2019).

1.5.10 OTHER INDUSTRIES

Paper companies produce large amounts of sludge, which requires a lot of money for management. Using paper sludge as a substrate for biosurfactant production provides a cheaper method for management (Pervaiz and Sain, 2010), low-cost degreasing of leather (Kiliç, 2013), and biosynthesis of selected metallic nanoparticles (Płaza et al.,

2014). Biosurfactants are used in the paint industry, dispersion of inorganic minerals in mining and manufacturing processes (Banat et al., 2000).

1.6 PRODUCERS AND PRODUCTION METHODS

Microorganisms produce biosurfactants under different conditions. These conditions often help to determine the type of production methods that will produce optimal biosurfactants in culture. The various producer microbes and methods of production will be examined in this section.

1.6.1 PRODUCER MICROBES OF BIOSURFACTANTS

Biosurfactants are synthesized naturally by a variety of bacteria, fungi, and yeasts. Most of these biosurfactant-producing microbes are isolated from industrial waste effluent contaminated samples which do not permit the growth of other microorganisms. The major bacteria producing biosurfactants are the species *Pseudomonas*, *Bacillus*, *Acinetobacter*, *Halomonas*, *Myroide*, and *Rhodococcus*. Fungal producer species are mainly *Candida* and yeasts *Pseudozyma*, *Torulopsis*, *Kurtzmanomyces*, and *Saccharomyces* (Shekhar et al., 2015).

1.6.2 CONVENTIONAL METHODS OF PRODUCTION

Conventional methods of producing biosurfactants relied mainly on the optimization of major factors that affect microbial production in cultures such as the media requirements of carbon and nitrogen or environmental factors such as temperature and pH. Alternative low-cost substrates such as industrial wastes have also been used as a means to reduce the costs of production.

1.6.2.1 Media formulation in the production of biosurfactants

Bioprocesses involving the use of single cells are very sensitive to many parameters, which makes their monitoring tedious, time-consuming, and more expensive. To multi-task with different data simultaneously, response surface methodology (RSM) such as the central composite design and Box–Behnken design and factorial designs based on statistical methods like Taguchi and Plackett–Burman design are used often to optimize growth media to increase biosurfactant output. In addition to statistical methods, another method known as artificial neural intelligence coupled with genetic algorithm (ANN-GA) is also being used for optimal media formulation (Bertrand et al., 2018; Singh et al., 2019). It is been advanced that the use of more than one of these approaches produces more reliable results considering the limitations of each. These are referred to as mixed strategies. The effect on biosurfactant production of nutritional and physicochemical factors is commonly evaluated. The main nutritional factors are the concentrations and proportions of C and N sources meanwhile temperature, shaking, oxygenation, and pH are the physicochemical factors. Upon choosing a particular method there is a need for proper justification on its use following the intended study (Bertrand et al., 2018).

The Plackett–Burman design was used to investigate factors affecting biosurfactant production by *Pseudomonas putida* CB-100 using corn oil and glucose as a carbon source. Using a regression analysis it was shown the reduction in the surface tension was being affected by KH_2PO_4 ($p < 0.0001$), K_2HPO_4 ($p < 0.0006$), yeast extract ($p < 0.0006$), and glucose ($p < 0.0008$), whereas cell growth was affected by $FeSO_4 \cdot 7H_2O$ ($p < 0.0001$), NH_4Cl ($p < 0.0001$), glucose ($p < 0.0001$), and temperature ($p < 0.0001$) (Martinez-Toledo et al., 2015). Using Box–Behnken design with regression analysis, reports of culture medium formulations have also shown that biosurfactant production was greatly improved significantly by glucose and ammonium chloride interactions ($P < 0.004$) (Martínez-Toledo and Rodríguez-Vázquez, 2011). When doing media formulations and determining optimum requirements, it becomes necessary to evaluate the possible different media sources, which can provide the necessary nutritional components while considering its availability and cost. We will examine the various nutritional factors to consider when formulating media for biosurfactant production.

1.6.2.1.1 Carbon and nitrogen sources for biosurfactant production

Carbon and nitrogen sources are important process parameters in the production of biosurfactants. Different carbon and nitrogen formulations have been evaluated for rhamnolipid produced by *Pseudomonas nitroreducens*. The higher yields of rhamnolipids were obtained when glucose and sodium nitrate were used producing 5.28 and 4.38 g/l, respectively. The best C/N ratio of 22 for both sources also produced the highest rhamnolipid concentrations of 5.46 g/l (Onwosi and Odibo, 2012). Using olive oil as carbon source and ammonium nitrate as nitrogen source, *Pseudomonas fluorescens* Migula 1895-DSMZ biosurfactant amount obtained was highest at a C/N ratio of 10 (Abouseoud et al., 2008). Same results were obtained in a similar study evaluating different sources of carbon (glucose, olive oil, and hexadecane) and nitrogen (urea, NH_4NO_3, KNO_3, and NH_4Cl). Olive oil and ammonium nitrate at a C:N of 10:1 produced an optimal biosurfactant yield of 9 g/l (Yataghene et al., 2008).

1.6.2.1.2 Critical environmental parameters for biosurfactant production

Different environmental parameters such as temperature, pH, and salinity have different effects on the production of rhamnolipids. In cultures of *Bacillus mycoides*, glucose and temperature were found to be the most important factors affecting biosurfactant synthesis. Meanwhile, pH and salinity did not affect production within the investigated range. Surface tension was reduced highest under the conditions of 16.55 g/l glucose concentration, 39.03°C, 55.05 g/l total salt concentration, and medium pH of 7.37 (Najafi et al., 2010). Biosurfactant production of *Bacillus thuringiensis* (JB) has been investigated at various environmental parameters such as pH (6–9), NaCl concentrations (1–4%), and water-soluble fraction concentrations (WSF 1–4%). It showed maximum activity and efficiency at pH 7 and 3 and 4% NaCl. Three percent WSF showed good activity while 4% WSF provided good biosurfactant yield (Jaysree et al., 2015).

1.6.2.2 Alternative eco-friendly and low-cost substrates

The search for more sustainable production and the need for cost-effectiveness has driven research interests toward finding alternative cheap substrates. In this light, a good number of industrial wastes suit the purpose of producing biosurfactants reducing both production costs and greenhouse emissions during production (Manga et al., 2021). Various properties of the substrate such as form, size, purity, stability, and the amounts that are to be used form the basis for the selection and formulation of substrate for production (Singh et al., 2019). Wastewater from olive oil mills was used as substrate for rhamnolipid production. Oil mill wastewater (OMW), a residue generated during olive oil extraction, was evaluated as an inducer of rhamnolipid production. The main media contents were corn steep liquor, sugarcane molasses and oil mill wastewater. *P. aeruginosa* #112 strain produced 4.5 and 5.1 g of rhamnolipid per liter in flasks and reactor, respectively, with a low CMC of 13 mg/l (Gudiña et al., 2016). Other substrates such as agro-industrial waste (Ebadipour et al., 2016; Gudiña et al., 2016; Pele et al., 2019), glucose syrup (Lima et al., 2020), soybean oil and glycerol (Accorsini et al., 2012), coconut, soybean, and sesame waste (Waghmode et al., 2014), waste canola oil (Pérez-Armendáriz et al., 2019), and fish waste (Kazemi, 2017) among others have been used as substrate to produce biosurfactants. These substrates are not only cheap but also readily available, renewable, even profitable waste management alternatives.

1.6.3 ALTERNATIVE FAVORABLE STRATEGIES FOR BIOSURFACTANT PRODUCTION

Approaches to improve the yield of biosurfactants have also gone beyond improving the characteristics of the media to providing favorable fermentation conditions, introducing other organisms to coproduce other beneficial products, or introducing agents that can improve downstream processing of biosurfactants. Some of these approaches are discussed below.

1.6.3.1 Solid-state fermentation process

Solid-state fermentation (SSF) defines processes in which microbes are cultured on or within solid support without water (Pandey et al., 2000). It is less capital and energy-intensive while providing at the same time a natural environment for fermentation (Zhu et al., 2013). This alternative methodology brings two key advantages to the fermentation process namely its use of cheaper substrates and the absence of foam in the fermentation process which is a huge problem during submerged fermentation. Current challenges in the use of SSF include the selection of the appropriate bioreactor, substrates with adequate physicochemical properties, and the design of efficient downstream processes (Costa et al., 2018; Krieger et al., 2010). The compatibility of most industrial wastes as growth media in SSF contributes to cost reduction. Biosurfactant production conditions were optimized to grow *Pleurotus djamor*, on sunflower seed shells, grape wastes, or potato peels as substrates in SSF. The substrate amount and volume were optimized as well as other environmental factors such as temperature, pH, and Fe^{2+}. In optimum conditions, the biosurfactant produced reduced water surface tension to 28.82 ± 0.3 mN/m with an oil displacement diameter of 3.9 ± 0.3 cm, 10.205 ± 0.5 g/l. The process was suggested to be economical and have promising applicability in large-scale production (Velioglu and Ozturk Urek, 2015). Generally, it is observed that up to about 60% of production costs are due to downstream processing. It is important to obtain the product as pure as possible because the risks of contamination are very high during this process (Costa et al., 2018). Improving on this part of the production could also increase the quality of the product and/or make the process much cheaper.

1.6.3.2 Biosurfactant coproduction

In biosurfactant coproduction, a biosurfactant is produced simultaneously with another product of economic importance in the same bioprocess. The ability to achieve this makes the process very economical (Singh et al., 2019). Numerous applications of biosurfactant coproduction have been demonstrated. Enhanced biosurfactant prod has been shown to increase power output in microbial fuel cells which are one of the most promising green energy technologies. A methodology to increase the current flow and power output of microbial fuel cells was developed. The rhamnolipid production enzyme gene rhlA was overexpressed in *Pseudomonas aeruginosa* PAO1 to favor its rhamnolipid overproduction which simultaneously increased electron transfer by increasing the electron shuttle pyocyanin synthesis and favoring attachment to the anode of bacteria. Comparatively, the enhanced strains showed 2.5 times more power production than the wild type (Zheng et al., 2015). Another study demonstrating coproduction saw the production of the phenazine dye pyocyanine alongside rhamnolipids when *Pseudomonas* BOP100 was grown on ethanol as the only C source. The products were obtained at different levels of the growth process. Pyocyanine was obtained during the exponential phase while rhamnolipids were obtained during the stationary phase. The coproduction was optimal at 3% ethanol yielding 3 g/l of rhamnolipids and 0.2 g/l of dye (Osman et al., 1996).

1.6.3.3 Immobilization process

The continuous culturing of cells and required processing to obtain products has many disadvantages such as the cells being washed out of the reactor, the need to change reactor conditions, management of the effects of unwanted metabolites on cell growth, and the formation of foam by free cells. Immobilized cells provide an alternative to the foam-producing free cells used in the bioproduction process (Singh et al., 2019). Alginate beads have commonly been used as immobilization material for bioproduction. *P. aeruginosa* was immobilized in alginate beads using the calcium alginate encapsulation technique to entrap the cells. The cells remained viable and produced biosurfactants effectively. The parameters were well adjusted to provide stability of alginate beads within the bioprocess (Dehghannoudeh et al., 2019). Another biosurfactant producer, *B. licheniformis* PTCC 1320 was immobilized on alginate beads and its rhamnolipid production capacity was investigated. Cell viability and rhamnolipid production were demonstrated alongside biosurfactant production. Optimal biosurfactant production was observed in the first 24 hours (Ohadi et al., 2014). Other immobilization matrixes poly (ethylene oxide) (PEO) and polyacrylamide (PAAm) cryogels were used to immobilize *Pseudomonas aeruginosa* strain BN10 to produce rhamnolipids. The biosurfactant production output was compared with that of free cells all grown under shake flask conditions. After nine cycles, the highest yields of rhamnolipids (4.6 g/l) were obtained at the sixth cycle for immobilized cells at 3 g of PEO in contrast to 4.2 g/l for free cells. The cells structures were well preserved and well distributed in the matrix. The matrix showed chemical, mechanical, and biological stability, indicating that it will be long-lasting material in the production process for rhamnolipids (Christova et al., 2013).

1.6.3.4 The use of nanotechnology

Nanoparticles have numerous antimicrobial properties and are widely used in the medical and cosmetic sectors. Biochemical methods are attractive alternatives to physical and chemical methods in the synthesis of nanoparticles due to their lesser toxicity and energy efficiency (Eswari et al., 2018; Prasad et al. 2016). The biological synthesis of nanoparticles has been effective so far but the process to obtain nanoparticles is very slow compared to other methods that employ reducing agents. The long time for reduction and complicated processing to obtain the particles from bacterial biomass is a major setback for biological production (Kiran et al., 2011). Biosurfactant-mediated nanoparticle synthesis offers the advantage of reducing aggregates formed as a result of electrostatic forces of attraction facilitating the uniformity in the structure of nanoparticles (Kiran et al., 2011). Biosurfactant from *Bacillus subtilis* was used to produce silver nanoparticles showing a potential to be used in drug formulations (Eswari et al., 2018). Biosurfactants can also be used as stabilizing agents of nanoparticles in liquid phases. In addition to the production from industrial waste, the process was simpler than using whole organisms, or their extracts for nanoparticle synthesis (Farias et al., 2014). Similar results have been obtained by synthesizing nickel oxide nanoparticles using a microemulsion technique giving an alternative eco-friendly alternative to using organic surfactants in microemulsion (Palanisamy and Raichur, 2009). Overall, it can be said that biosurfactants have a promising future in the production of nanoparticles. More research integrating other processes and cost-effectiveness is recommended.

1.6.3.5 Enzymatic synthesis of biosurfactants

An approach to reduce production costs that are being researched a lot is the modification of the genes of industrial microbe strain biosurfactant producers to increase their enzyme productivity. Protein engineering and directed protein evolution techniques have enabled the modification of enzymes enhancing their catalytic efficiency and improving their activity over a broad range of conditions that are similar to those in the industrial fermentations (Helmy et al., 2011). A lot of attention has been directed toward the enzymatic synthesis of lipoamino acid/peptides surfactant analogs due to their functional versatility; highly efficient, good aggregation, and wide biological activity. They enable simpler formulations and reduce costs (Clapés and Rosa Infante, 2002). There are tailor-made glycolipid surfactants lipases to synthesize amino acid-based surfactants (Valivety et al., 1997) and catalyze the esterification of acid and glycerol including the development of a modeling framework for the reaction (Martínez et al., 2011).

1.6.4 OVERPRODUCTION STRATEGIES FOR BIOSURFACTANT PRODUCTION

The production of biosurfactants is still less competitive compared to synthetic surfactants. To meet up, various strategies directed toward cutting production costs have been employed. These include: using cheaper media, optimization for efficient bioprocesses, and genetic engineering approaches (Manga et al., 2021; Mukherjee et al., 2019).

1.6.4.1 Modifying of media to increase specific yield

Several approaches intending to achieve overproduction of biosurfactants have been researched showing promising results for production cost reduction. Cheap porous carriers such as activated carbon or expanded clay have been introduced into fermentation broths to increase biosurfactant production. A 36-fold increase in the surfactin produced by *Bacillus subtilis* ATCC 21332 in a culture with 25 g/l of porous carriers was observed. An increase in surfactin production was attributed to the presence of activated carbon carriers though agitation rates also had a little effect. A high purity of biosurfactant (90%) and recovery efficiency of 72% was obtained (Yeh et al., 2005). Affecting the availability of certain nutrients in the culture media has also been shown to affect various pathways in biosynthetic pathways. Limiting nitrogen in fed-batch operation has shown to increase rhamnolipid production by 3.8 fold in *Pseudomonas aeruginosa* strain PA1 (Soares dos Santos et al., 2016). The availability of other nutrients such as phosphorus has been evaluated in excess of nitrogen. Nevertheless, it still suggests that there could be a non-specificity of limiting nutrients, but production is dependent on the provision of carbon in excess of metabolic capacity (Clarke et al., 2010). Other metal ions such as Zn^{2+} and Mn^{4+} have been shown to contribute to the enhancement of biosurfactant production (Manna Sahoo et al., 2011). In contrast, Fe which served as an external signal for the initiation of rhamnolipid synthesis was found to be an enhancer for rhamnolipid synthesis under limiting conditions. When iron was limited, the synthesis of rhamnolipids occurred much earlier during biofilm development as compared to later during normal iron concentrations (Glick et al., 2010).

1.6.4.2   Use different fermentation modes

Levels of dissolved oxygen have an important effect on rhamnolipid produced in fed-batch cultivation. A comparative study to investigate the effect of various concentrations of dissolved oxygen on *P. aeruginosa* biosurfactant production in batch and fed-batch control fermentation was carried out. Optimal rhamnolipid production was 22.5 g/l at 40% dissolved oxygen in the batch fermentation. For higher rhamnolipid production, a fed-batch system under 40% tight dissolved oxygen control was carried out. Production was 10.7- and 4.8-fold higher than batch experiments. This high level of biosurfactant production was suggested to be due to the dissolved oxygen concentration and the feeding method (Bazsefidpar et al., 2019). Another study established the importance of operational modes, the supply of oxygen, and feeding strategy in the biosurfactant synthesis process. Using *Aneurinibacillus thermoaerophilus* HAK01, lipopeptide biosurfactant was produced at 4.9 g/l, at 45°C. Culture components were optimized to obtain the best fed-batch strategy to overproduce the lipopeptide by the microorganism. To achieve this, the fed-batch strategies pH-stat mode, constant feeding rate strategy, dissolved oxygen-stat mode, and combined feeding strategy were implemented. The biosurfactant produced increased from 4.9 g/l in batch mode to 5.9, 7.1, 8.8, and 11.2 g/l in each fed-batch mode, respectively. Better biosurfactant production was observed in the dissolved oxygen stat mode than in the pH-stat mode (Hajfarajollah et al., 2019).

1.6.4.3   Genetic engineering strategies

Genes of various biosurfactant producers have been modified to enhance biosurfactant production. Precursors of rhamnolipids are shared with polyhydroxyalkanoates (PHA) in a competitive fashion. Many studies have tested the hypothesis of inhibiting the PHA synthesis pathway at various steps. Blocking the PHA synthesis genes in *Burkholderia thailandensis* by knocking out the *phbA*, *phbB*, and *phbC* genes for PHA synthesis increased rhamnolipid production (Funston et al., 2017). In another approach, *Pseudomonas aeruginosa* PA14 mutants were developed but in addition to blocking the synthesis of the carbon-storage polymer PHA, the rhlAB-R operon that encodes for enzymes of rhamnolipid synthesis and the RhlR transcriptional regulator were overexpressed. Production of rhamnolipids was found to depend on many factors and on the strain used (Gutiérrez-Gómez et al., 2018). Other approaches to *Bacillus* sp. to enhance rhamnolipid have to be implemented. Mutagenesis was done on *Bacillus subtilis* SPB1 using a combination of UV irradiation and nitrous acid treatment to improve biosurfactant production. The result was a production of double the amount of biosurfactant than the wild type (Bouassida et al., 2018). In a more specific genetic engineering approach, the branched fatty acid metabolic pathway of *Bacillus subtilis* was engineered to overproduce surfactin C14 isoform. The gene *codY*, encoding the global transcriptional regulator, and the gene *lpdV*, important in branched-chain amino acid degradation in acyl-CoA were deleted. The mutants showed an increase in surfactin production by 5.8-fold for the *codY* mutant and 1.4-fold for the *lpdV* mutant (Dhali et al., 2017).

An in silico approach investigating the metabolic abilities of *P. putida* to produce rhamnolipids combined statistical, metabolic, and synthetic engineering methodologies after the introduction of RhlA and RhlB genes from *P. aeruginosa* into a genome-scale model of *P. putida*. There was a significant increase in the biosurfactant produced by the engineered model compared to the control model (Occhipinti et al., 2018). A computational approach to increase biosurfactant production in an already enhanced strain of *Bacillus subtilis* aimed at increasing the production of leucine precursor showed that increasing the synthesis of biosurfactant precursor can increase the synthesis of biosurfactant (Coutte et al., 2015). The genome and transcriptional profile of *Bacillus amyloliquefaciens* MT45 were analyzed and overexpression targets were identified. The central carbon metabolism and fatty acid biosynthesis revealed possible precursor enrichment targets that can lead to the enhancement of biosurfactant production as well as many regulatory genes (Zhi et al., 2017). More recently, a systematic approach to engineering various genes in *Bacillus subtilis* 168 achieved an increase in biosurfactant to 12.8 g/l (Wu et al., 2019) and 10% genome reduction showing better glucose to biomass conversion but inferior in biosurfactant production at the experimental temperatures and oxygen concentrations (Geissler et al., 2019). Genetic engineering strategies are getting better every day and there are high hopes that novel strains will be developed that will produce biosurfactants at very minimal costs.

1.7   DISCOVERY OF NOVEL BIOSURFACTANTS

The possibility of discovering novel biosurfactants in uncultured microorganisms could lead us to find cost-effective sources or even more useful biosurfactants. Many new biosurfactants have been discovered to date using metagenomics, thanks to the availability of high-throughput screening methods and heterologous hosts for gene expression (Jackson et al., 2015). Metagenomic methods employing the development of genomic libraries from samples and expression in heterologous hosts have been described and employed for the discovery of novel biosurfactants (Kennedy et al., 2011). Probably the first application of high-throughput methods in metagenomics to screen novel biosurfactants was described by Thies et al. (2016). N-acyltyrosines with N-myristoyltyrosine biosurfactants described showed good CMC, surface activity, and antimicrobial properties (Thies et al., 2016). More recently, a lyso-ornithine lipid from functional metagenomic screening has been described but biosurfactant potential is still unexplored (Williams et al., 2019).

1.8 INDUSTRIAL-SCALE PRODUCTION AND CHALLENGES FROM LAB TO MARKET

The green nature of biosurfactants has already been established, their properties and areas of applications are well established, the metabolism and physiology of microorganisms producing biosurfactants are well known, a wide range of substrates both water-soluble and insoluble have been used to produce biosurfactants including use of low-cost substrates or wastes. Improved industrial fermentation methods have been described. Prospects for large-scale industrial production are promising (Fenibo et al., 2019). However, producing biosurfactants at an industrial scale usually requires overcoming the effects of foaming in bioreactors and the use of an ideal substrate. Foam control strategies such as dispersion with foam stirrers and disruption using antifoaming agents or mechanically are being used to minimize the effects during fermentation (Chen et al., 2006; Jiang et al., 2020; Kosaric and Sukan, 2014). In addition, foam avoidance strategies have also been developed including using a rotating disk bioreactor, a bubble-free membrane bioreactor, using a solid-state cultivation method, and foam fractionation. These methodologies vary in use across less foaming biosurfactants such as sophorolipids to more foaming ones such as rhamnolipids. When considering the mitigation of the foaming effects, it is also important to consider the effect of the strategy to be used on oxygen distribution as oxygen distribution could affect growth conditions within the bioreactor. There is still a need for the development of adequate foam management strategies though which requires a better understanding of foam-forming mechanisms (Kosaric and Sukan, 2014).

An improvement in downstream processing (foam separation, phase separation, cell separation, ultra-filtration, extraction, precipitation, crystallization, and chromatography) costs is also very important given that 60–80% of production costs are usually incurred during this phase. A significant improvement in these strategies will certainly lead to cost-effective and competitive production (Anic et al., 2018; Kosaric and Sukan, 2014). A current check on product prices for 90% pure rhamnolipid is 84.9 euros for 10 g, 1 g of sophorolipid 2-(p-tolyl)ethylamine at 41.40 euros against 1 kg of 99% SDS dust-free pellets at 88.5 euros or Tween 20 1 kg at 85.6 euros on Sigma Aldrich website shows a huge difference in cost between biosurfactants and synthetic surfactants (Sigma Aldrich, 2020). Presently, commercial production of biosurfactants is carried out by only a few companies globally. The introduction of biosurfactants in more products in industries would increase the demand for biosurfactants and thus increase the global market (Fenibo et al., 2019).

1.8.1 MARKET AND FORECAST

In 2018, the market for biosurfactants worldwide was estimated to be USD 4.70 Billion, growing at a CAGR of 5.6% it is expected to reach a USD 7.25 Billion coming 2026 (Maximize Market Research, 2019). Growth is driven by high demands from the home and personal care industries. Europe, North America and Asia Pacific are expected to have the highest demands by 2022. Demand in Asia Pacific exceeded that of Europe in 2017 and it is projected to increase steadily over the forecasted period (Maximize Market Research, 2019). An increase in demand for biosurfactants in the personal care industry in Europe was explained to be mainly driven by consciousness toward environmental protection (Maximize Market Research, 2019).

1.8.2 PATENTS AND COMPANIES FOR BIOSURFACTANT PRODUCTION

Despite the costs associated with biosurfactant production, there is still a considerable level of industrial production worldwide. Despite the limitation of the high cost of production, some key player companies such as Belgium-based Ecover, US-based Jeneil Biotech, and Germany-based Evonik, Givaudan, Henkel, AGAEm, AkzoNobel N.V., and BASF are still producing biosurfactants (Global Market Insights, 2019; Markets and Markets, 2017). A list of companies involved in biosurfactant production and the types produced are shown in Table 1.1. The table was adapted from Jimoh and Lin (2019).

The need to make biosurfactants sustainable has also driven research toward producing novel methods to increase production. A search of patents with the keyword biosurfactant in the Espacenet database returned 3,868 results spanning from 1978 to April 2022 showing a steep rise in patents from 1978 to April 2022. A sharp drop is also observed in 2020 which could have been due to the COVID-19 pandemic (Figure 1.1). At the country level, China, the United States of America, and Japan are leading in the patents filed (Figure 1.2) while within the industry, TOYO BOSEKI is leading with 85 patents filed (Figure 1.3).

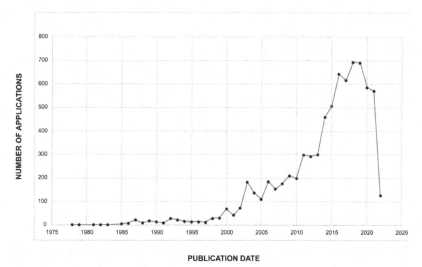

FIGURE 1.1 Number of patents applications for biosurfactants from 1978 to April 2022.

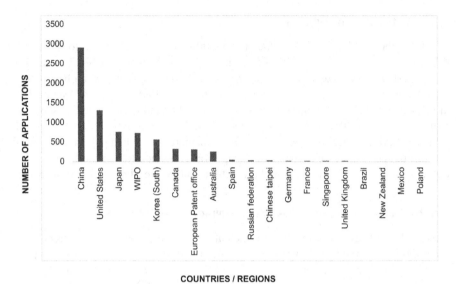

FIGURE 1.2 Number of biosurfactant related patents filed from 1978 to April 2022 according to country/region.

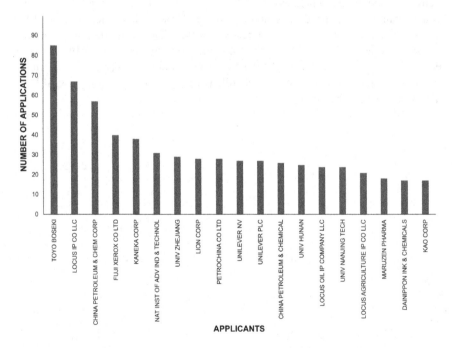

FIGURE 1.3 Number of patents for biosurfactants filed by institutions from 1978 to April 2022.

TABLE 1.1

Companies producing biosurfactants worldwide

Product	Company name	Web address/contact	Country
Rhamnolipids	Jeneil biosurfactant	http://www.jeneilbiotech.com/	USA
Rhamnolipids	Paradigm Biomedical Inc	http://www.akama.com/company/Paradigm_Biomedical_Inc_a7bcb2680775.html	USA
Rhamnolipid	AGAE Technologies Ltd	http://www.agaetech.com/	USA
Sophorolipid	Ecover Eco-Surfactant	https://www.ecover.com/	Belgium
Green surfactant alkyl polyglucoside (APG)	Cognis (BASF)	http://saifuusa.com/portfolio-item/mildsurfactants/	Germany, USA
Sopholiance S (sophorolipids)	Groupe Soliance	http://www.soliance.com/dtproduit.php?id=42	France
Rhamnolipid	Biotensidon	https://biotensidon.com/en/home/	Germany
Sophrololipid	Saraya	http://worldwide.saraya.com/	Japan
Green surfactant alkyl polyglucoside (APG)	Cognis	https://www.globalsources.com/si	China
Rhamnolipids(chemically synthesized)	GlycoSurf LLC.	http://glycosurf.com/	USA
NatSurFact (Rhamnolipids)	Logos Technologies, LLC.	https://www.natsurfact.com/	USA
Mannosylerythritol lipid B (MEL-B)	Kanebo Cosmetics Inc.	http://www.kanebo.com/science/skincare/biosurfactants.html	Japan
BERO biosurfactant	ZFA Technologies Inc.	http://www.zfatech.com/index.php?lang=en	China
Lipopeptides	Lipofabrik	http://www.lipofabrik.com/	France
Sopholine (Sophorolipids)	MG Intobio Co., Ltd.	https://mgintobio.fm.alibaba.com/productlist.html	South Korea
KANEKA Surfactin (Sodium Surfactin)	Kaneka Corporation	http://www.kaneka.co.jp/kaneka-e/	Japan
Rhamnolipids and Lipopeptides	TeeGene Biotech LTD.	http://www.teegene.co.uk/	UK
Rhamnolipids	Urumqi Unite Bio-Technology Co., Ltd.	https://unite-xj.en.alibaba.com/productlist.html	China
Surfactin	Soft Chemical Laboratories	www.probac.co.za	South Africa
Rhamnolipids	Rhamnolipids Companies, Inc.	http://rhamnolipid.com/	USA
Sophorolipids	SyntheZyme LLC.	http://www.synthezyme.com/index.html	USA
BioSurfactantsACS-Sophor®(sophorolipid)	Allied Carbon Solutions Co., Ltd.	http://www.allied-c-s.co.jp/english.php	Japan
Blends of biosurfactant producing bacteria	BioFuture Ltd.	http://www.biofuture.ie/	Ireland
Rhamnolipids, Lipopeptides, Trehalolipids	TensioGreen Corporation	http://www.tensiogreen.com/index.php	USA
Sophorolipid	Evonik	https://corporate.evonik.com/misc/micro/biosurfactants/index.en.html	Germany

1.9 FUTURE TRENDS

We have discussed many novel technologies and recent areas of applications for bio-surfactants. Novel areas of applications continue to emerge but there is still a need to increase the cost-effectiveness of the production process. Research for the discovery of better over producer strains or engineering to obtain synthetic ones needs to continue to decrease fermentation costs. The optimization of production and purification processes must continue to increase the feasibilities for large-scale production. As high-throughput technologies continue to develop at a speedy rate, with their applications in biosurfactant research we expect to overcome the bottlenecks of production in the future.

1.10 CONCLUSIONS

Biosurfactants are indeed sustainable multifunctional "products" worthy of our interest. The versatility in their properties promoting their use across various industries keeps attracting broad interest. There is a need for continuous research to come up with better biosurfactant overproducer strains, improvement and standardization of fermentation procedures, betterment of the downstream purification processes, and galvanization of the biosurfactant market to enhance the sustainability of bio-surfactants (Singh et al., 2019). The sustainability of bioproducts can only be valid if the production process is cost-effective and commercialization is profitable. In order to achieve this, all research aimed at improving the production and promoting the commercialization of the product must continue relentlessly. Furthermore, as more and more companies producing the "green" products increase, important industrial processes requiring surfactants will gradually replace synthetic surfactants with bio-surfactants. Biosurfactants can be said to be one of the most promising future green technologies.

REFERENCES

Abouseoud, M., Maachi, R., Amrane, A., Boudergua, S., Nabi, A., 2008. Evaluation of different carbon and nitrogen sources in production of biosurfactant by Pseudomonas fluorescens. Desalination, European Desalination Society and Center for Research and Technology Hellas (CERTH), Sani Resort 22–25 April 2007, Halkidiki, Greece 223, 143–151. https://doi.org/10.1016/j.desal.2007.01.198

Abyaneh, A.S., Fazaelipoor, M.H., 2016. Evaluation of rhamnolipid (RL) as a biosurfactant for the removal of chromium from aqueous solutions by precipitate flotation. *Journal of Environmental Management* 165, 184–187.

Accorsini, F.R., Mutton, M.J.R., Lemos, E.G.M., Benincasa, M., 2012. Biosurfactants production by yeasts using soybean oil and glycerol as low cost substrate. *Brazilian Journal of Microbiology* 43, 116–125. https://doi.org/10.1590/S1517-83822012000100013

Akbari, S., Abdurahman, N.H., Yunus, R.M., Fayaz, F., Alara, O.R., 2018. Biosurfactants—a new frontier for social and environmental safety: a mini review. *Biotechnology Research and Innovation* 2, 81–90. https://doi.org/10.1016/j.biori.2018.09.001

Aksoy, D.Ö., Özdemir, S., Aytar Çelik, P., Koca, S., Çabuk, A., Koca, H., Brito-Parada, P., 2022. Fusion of the microbial world into the flotation process. *Mineral Processing and Extractive Metallurgy Review*, 1–15. https://doi.org/10.1080/08827508.2021.2023518

Almeida, D.G.D., Silva, R. de C.F.S.D., Luna, J.M., Rufino, R.D., Santos, V.A., Banat, I.M., Sarubbo, L.A., 2016. Biosurfactants: promising molecules for petroleum biotechnology advances. *Frontiers in Microbiology* 7. https://doi.org/10.3389/fmicb.2016.01718

Amani, H., Kariminezhad, H., 2016. Study on emulsification of crude oil in water using emulsan biosurfactant for pipeline transportation. *Petroleum Science and Technology* 34, 216–222. https://doi.org/10.1080/10916466.2015.1118500

Amin, G.A., Bazaid, S.A., El-Halim, M.A., 2013. A two-stage immobilized cell bioreactor with Bacillus subtilis and Rhodococcus erythropolis for the simultaneous production of biosurfactant and biodesulfurization of model oil. *Petroleum Science and Technology* 31, 2250–2257. https://doi.org/10.1080/10916466.2011.565292

Anic, I., Apolonia, I., Franco, P., Wichmann, R., 2018. Production of rhamnolipids by integrated foam adsorption in a bioreactor system. *AMB Express* 8. https://doi.org/10.1186/s13568-018-0651-y

Araújo, H.W.C., Andrade, R.F.S., Montero-Rodríguez, D., Rubio-Ribeaux, D., Alves da Silva, C.A., Campos-Takaki, G.M., 2019. Sustainable biosurfactant produced by Serratia marcescens UCP 1549 and its suitability for agricultural and marine bioremediation applications. *Microbial Cell Factories* 18, 2. https://doi.org/10.1186/s12934-018-1046-0

Arima, K., Kakinuma, A., Tamura, G., 1968. Surfactin, a crystalline peptidelipid surfactant produced by Bacillussubtilis: isolation, characterization and its inhibition of fibrin clot formation. *Biochemical and Biophysical Research Communications* 31, 488–494. https://doi.org/10.1016/0006-291X(68)90503-2

Asselineau, C., Asselineau, J., 1978. Trehalose-containing glycolipids. *Progress in the Chemistry of Fats and Other Lipids* 16, 59–99. https://doi.org/10.1016/0079-6832(78)90037-X

Ayangbenro, A.S., Babalola, O.O., 2018. Metal (loid) bioremediation: strategies employed by microbial polymers. *Sustainability* 10, 3028.

Aytar, P., Aksoy, D., Toptaş, Y., Çabuk, A., Koca, S., Koca, H., 2014. Isolation and characterization of native microorganism from Turkish lignite and usability at fungal desulphurization. *Fuel* 116, 634–641. https://doi.org/10.1016/j.fuel.2013.08.077

Aytar Çelik, P., Çakmak, H., Öz Aksoy, D., 2021. Green bioflotation of calcite using surfactin as a collector. *Journal of Dispersion Science and Technology* 0, 1–11. https://doi.org/10.1080/01932691.2021.1979999Banat, I.M., Makkar, R.S., Cameotra, S.S., 2000. Potential commercial applications of microbial surfactants. *Applied Microbiology and Biotechnology* 53, 495–508. https://doi.org/10.1007/s002530051648

Bastrzyk, A., Fiedot-Toboła, M., Polowczyk, I., Legawiec, K., Płaza, G., 2019. Effect of a lipopeptide biosurfactant on the precipitation of calcium carbonate. *Colloids and Surfaces B: Biointerfaces* 174, 145–152. https://doi.org/10.1016/j.colsurfb.2018.11.009

Bazsefidpar, S., Mokhtarani, B., Panahi, R., Hajfarajollah, H., 2019. Overproduction of rhamnolipid by fed-batch cultivation of Pseudomonas aeruginosa in a lab-scale fermenter under tight DO control. *Biodegradation* 30, 59–69. https://doi.org/10.1007/s10532-018-09866-3

Behera, S.K., Mulaba-Bafubiandi, A.F., 2017. Microbes assisted mineral flotation a future prospective for mineral processing industries: a review. *Mineral Processing and Extractive Metallurgy Review* 38, 96–105. https://doi.org/10.1080/08827508.2016.1262861

Ben Ayed, H., Jridi, M., Maalej, H., Nasri, M., Hmidet, N., 2014. Characterization and stability of biosurfactant produced by Bacillus mojavensis A21 and its application in enhancing solubility of hydrocarbon. *Journal of Chemical Technology & Biotechnology* 89, 1007–1014.

Bertrand, B., Martínez-Morales, F., Rosas-Galván, N., Morales-Guzmán, D., Trejo-Hernández, M., 2018. Statistical design, a powerful tool for optimizing biosurfactant production: a review. *Colloids and Interfaces* 2, 36.

Bhadoriya, S.S., Madoriya, N., 2013. Biosurfactants: a new pharmaceutical additive for solubility enhancement and pharmaceutical development. *Biochemical Pharmacology* 02. https://doi.org/10.4172/2167-0501.1000113

Bouassida, M., Ghazala, I., Ellouze-Chaabouni, S., Ghribi, D., 2018. Improved biosurfactant production by Bacillus subtilis SPB1 mutant obtained by random mutagenesis and its application in enhanced oil recovery in a sand system. *Journal of Microbiology and Biotechnology* 28, 95–104.

Bucci, A.R., Marcelino, L., Mendes, R.K., Etchegaray, A., 2018. The antimicrobial and antiadhesion activities of micellar solutions of surfactin, CTAB and CPCl with terpinen-4-ol: applications to control oral pathogens. *World Journal of Microbiology and Biotechnology* 34, 86. https://doi.org/10.1007/s11274-018-2472-1

Bujak, T., Wasilewski, T., Nizioł-Łukaszewska, Z., 2015. Role of macromolecules in the safety of use of body wash cosmetics. *Colloids and Surfaces B: Biointerfaces* 135, 497–503. https://doi.org/10.1016/j.colsurfb.2015.07.051

Burd, G., Ward, O.P., 1997. Energy-dependent accumulation of particulate biosurfactant by Pseudomonas marginalis. *Canadian Journal of Microbiology* 43, 391–394. https://doi.org/10.1139/m97-054

Çelik, P.A., Aksoy, D.Ö., Koca, S., Koca, H., Çabuk, A., 2019. The approach of biodesulfurization for clean coal technologies: a review. *International Journal of Environmental Science and Technology* 16, 2115–2132. https://doi.org/10.1007/s13762-019-02232-7

Çelik, P.A., Manga, E.B., Çabuk, A., Banat, I.M., 2021. Biosurfactants' potential role in combating COVID-19 and similar future microbial threats. *Applied Sciences* 11, 334. https://doi.org/10.3390/app11010334

Chen, C.-Y., Baker, S.C., Darton, R.C., 2006. Batch production of biosurfactant with foam fractionation. *Journal of Chemical Technology & Biotechnology* 81, 1923–1931. https://doi.org/10.1002/jctb.1625

Christova, N., Petrov, P., Kabaivanova, L., 2013. Biosurfactant production by Pseudomonas aeruginosa BN10 cells entrapped in cryogels. *Zeitschrift für Naturforschung C* 68, 47–52.

Chrzanowski, Ł., Dziadas, M., Ławniczak, Ł., Cyplik, P., Białas, W., Szulc, A., Lisiecki, P., Jeleń, H., 2012. Biodegradation of rhamnolipids in liquid cultures: effect of biosurfactant dissipation on diesel fuel/B20 blend biodegradation efficiency and bacterial community composition. *Bioresource Technology* 111, 328–335. https://doi.org/10.1016/j.biortech.2012.01.181

Chu, W., Choy, W.K., Hunt, J.R., 2005. Effects of nonaqueous phase liquids on the washing of soil in the presence of nonionic surfactants. *Water Research* 39, 340–348. https://doi.org/10.1016/j.watres.2004.09.028

Chuo, S.C., Abd-Talib, N., Mohd-Setapar, S.H., Hassan, H., Nasir, H.M., Ahmad, A., Lokhat, D., Ashraf, G.M., 2018. Reverse micelle extraction of antibiotics using an eco-friendly sophorolipids biosurfactant. *Scientific Reports* 8, 1–13. https://doi.org/10.1038/s41598-017-18279-w

Clapés, P., Rosa Infante, M., 2002. Amino acid-based surfactants: enzymatic synthesis, properties and potential applications. *Biocatalysis and Biotransformation* 20, 215–233. https://doi.org/10.1080/10242420290004947

Clarke, K., Ballot, F., Reid, S., 2010. Enhanced rhamnolipid production by Pseudomonas aeruginosa under phosphate limitation. *World Journal of Microbiology and Biotechnology* 26, 2179–2184. https://doi.org/10.1007/s11274-010-0402-y

Consuegra, G.L., Kutschke, S., Rudolph, M., Pollmann, K., 2020. Halophilic bacteria as potential pyrite bio-depressants in Cu-Mo bioflotation. *Minerals Engineering* 145, 106062.

Cooper, D.G., Paddock, D.A., 1984. Production of a biosurfactant from Torulopsis bombicola. *Applied and Environmental Microbiology* 47, 173–176.

Costa, J.A.V., Treichel, H., Santos, L.O., Martins, V.G., 2018. Chapter 16 - Solid-state fermentation for the production of biosurfactants and their applications, in: Pandey, A., Larroche, C., Soccol, C.R. (Eds.), *Current Developments in Biotechnology and Bioengineering*. Elsevier, pp. 357–372. https://doi.org/10.1016/B978-0-444-63990-5.00016-5

Coutte, F., Niehren, J., Dhali, D., John, M., Versari, C., Jacques, P., 2015. Modeling leucine's metabolic pathway and knockout prediction improving the production of surfactin, a biosurfactant from Bacillus subtilis. *Biotechnology Journal* 10, 1216–1234. https://doi.org/10.1002/biot.201400541

Czaplicka, M., Chmielarz, A., 2009. Application of biosurfactants and non-ionic surfactants for removal of organic matter from metallurgical lead-bearing slime. *Journal of Hazardous Materials* 163, 645–649. https://doi.org/10.1016/j.jhazmat.2008.07.010

Damasceno, F.R.C., Cavalcanti-Oliveira, E.D., Kookos, I.K., Koutinas, A.A., Cammarota, M.C., Freire, D.M.G., Damasceno, F.R.C., Cavalcanti-Oliveira, E.D., Kookos, I.K., Koutinas, A.A., Cammarota, M.C., Freire, D.M.G., 2018. Treatment of wastewater with high fat content employing an enzyme pool and biosurfactant: technical and economic feasibility. *Brazilian Journal of Chemical Engineering* 35, 531–542. https://doi.org/10.1590/0104-6632.20180352s20160711

Das, A., Ambust, S., Kumar, R., 2018. Development of biosurfactant based cosmetic formulation of toothpaste and exploring its efficacy. *Advances in Industrial Biotechnology* 1. https://doi.org/10.24966/AIB-5665/100005

Dehghannoudeh, G., Kiani, K., Moshafi, M.H., Dehghannoudeh, N., Rajaee, M., Salarpour, S., Ohadi, M., 2019. Optimizing the immobilization of biosurfactant-producing Pseudomonas aeruginosa in alginate beads. *Journal of Pharmacy & Pharmacognosy Research* 7, 413–420.

Dhali, D., Coutte, F., Arias, A.A., Auger, S., Bidnenko, V., Chataigné, G., Lalk, M., Niehren, J., de Sousa, J., Versari, C., Jacques, P., 2017. Genetic engineering of the branched fatty acid metabolic pathway of Bacillus subtilis for the overproduction of surfactin C14 isoform. *Biotechnology Journal* 12, 1600574. https://doi.org/10.1002/biot.201600574

Didyk, A.M., Sadowski, Z., 2012. Flotation of serpentinite and quartz using biosurfactants. *Physicochemical Problems of Mineral Processing* 48, 607–618.

Ebadipour, N., Lotfabad, T.B., Yaghmaei, S., RoostaAzad, R., 2016. Optimization of low-cost biosurfactant production from agricultural residues through response surface methodology. *Preparative Biochemistry & Biotechnology* 46, 30–38. https://doi.org/10.1080/10826068.2014.979204

Edwards, J.R., Hayashi, J.A., 1965. Structure of a rhamnolipid from Pseudomonas aeruginosa. *Archives of Biochemistry and Biophysics* 111, 415–421. https://doi.org/10.1016/0003-9861(65)90204-3

El-Sheshtawy, H.S., Aiad, I., Osman, M.E., Abo-ELnasr, A.A., Kobisy, A.S., 2015. Production of biosurfactant from Bacillus licheniformis for microbial enhanced oil recovery and inhibition the growth of sulfate reducing bacteria. *Egyptian Journal of Petroleum* 24, 155–162. https://doi.org/10.1016/j.ejpe.2015.05.005

Eswari, J.S., Dhagat, S., Mishra, P., 2018. Biosurfactant assisted silver nanoparticle synthesis: a critical analysis of its drug design aspects. *Advances in Natural Sciences: Nanoscience and Nanotechnology* 9, 045007.

Farias, C.B.B., Ferreira Silva, A., Diniz Rufino, R., Moura Luna, J., Gomes Souza, J.E., Sarubbo, L.A., 2014. Synthesis of silver nanoparticles using a biosurfactant produced in low-cost medium as stabilizing agent. *Electronic Journal of Biotechnology* 17, 122–125. https://doi.org/10.1016/j.ejbt.2014.04.003

Fazaelipoor, M.H., Khoshdast, H., Ranjbar, M., 2010. Coal flotation using a biosurfactant from Pseudomonas aeruginosa as a frother. *Korean Journal of Chemical Engineering* 27, 1527–1531. https://doi.org/10.1007/s11814-010-0223-6

Fenibo, E.O., Ijoma, G.N., Selvarajan, R., Chikere, C.B., 2019. Microbial surfactants: the next generation multifunctional biomolecules for applications in the petroleum industry and its associated environmental remediation. *Microorganisms* 7. https://doi.org/10.3390/microorganisms7110581

Ferreira, A., Vecino, X., Ferreira, D., Cruz, J., Moldes, A., Rodrigues, L., 2017. Novel cosmetic formulations containing a biosurfactant from Lactobacillus paracasei. *Colloids and Surfaces B: Biointerfaces* 155, 522–529.

Funston, S.J., Tsaousi, K., Smyth, T.J., Twigg, M.S., Marchant, R., Banat, I.M., 2017. Enhanced rhamnolipid production in *Burkholderia thailandensis* transposon knockout strains deficient in polyhydroxyalkanoate (PHA) synthesis. *Applied Microbiology and Biotechnology* 101, 8443–8454. https://doi.org/10.1007/s00253-017-8540-x

Geissler, M., Kühle, I., Morabbi Heravi, K., Altenbuchner, J., Henkel, M., Hausmann, R., 2019. Evaluation of surfactin synthesis in a genome reduced Bacillus subtilis strain. *AMB Express* 9, 84. https://doi.org/10.1186/s13568-019-0806-5

Gharaei, E., 2011. Biosurfactants in pharmaceutical industry (a mini-review). *American Journal of Drug Discovery and Development* 1, 58–69. https://doi.org/10.3923/ajdd.2011.58.69

Glick, R., Gilmour, C., Tremblay, J., Satanower, S., Avidan, O., Déziel, E., Greenberg, E.P., Poole, K., Banin, E., 2010. Increase in rhamnolipid synthesis under iron-limiting conditions influences surface motility and biofilm formation in Pseudomonas aeruginosa. *Journal of Bacteriology* 192, 2973–2980. https://doi.org/10.1128/JB.01601-09

Global Market Insights, 2019. Biosurfactants Market Size, Value | Industry Share Report 2025 [WWW Document]. Global Market Insights, Inc. https://www.gminsights.com/industry-analysis/biosurfactants-market-report (accessed 1.26.20).

Gomaa, E.Z., El-Meihy, R.M., 2019. Bacterial biosurfactant from Citrobacter freundii MG812314.1 as a bioremoval tool of heavy metals from wastewater. *Bulletin of the National Research Centre* 43, 69. https://doi.org/10.1186/s42269-019-0088-8

Gudiña, E.J., Rangarajan, V., Sen, R., Rodrigues, L.R., 2013. Potential therapeutic applications of biosurfactants. *Trends in Pharmacological Sciences* 34, 667–675. https://doi.org/10.1016/j.tips.2013.10.002

Gudiña, E.J., Rodrigues, A., Freitas, V., Azevedo, Z., Teixeira, J., Rodrigues, L., 2016. Valorization of agro-industrial wastes towards the production of rhamnolipids. *Bioresource Technology* 212. https://doi.org/10.1016/j.biortech.2016.04.027

Gusiatin, Z.M., Radziemska, M., Żochowska, A., 2019. Sequential soil washing with mixed biosurfactants is suitable for simultaneous removal of multi-metals from soils with different properties, pollution levels and ages. *Environmental Earth Sciences* 78, 529. https://doi.org/10.1007/s12665-019-8542-3

Gutiérrez-Gómez, U., Soto-Aceves, M.P., Servín-González, L., Soberón-Chávez, G., 2018. Overproduction of rhamnolipids in Pseudomonas aeruginosa PA14 by redirection of the carbon flux from polyhydroxyalkanoate synthesis and overexpression of the rhlAB-R operon. *Biotechnology Letters* 40, 1561–1566. https://doi.org/10.1007/s10529-018-2610-8

Hajfarajollah, H., Mokhtarani, B., Tohidi, A., Bazsefidpar, S., Noghabi, K.A., 2019. Overproduction of lipopeptide biosurfactant by Aneurinibacillus thermoaerophilus HAK01 in various fed-batch modes under thermophilic conditions. *RSC Advances* 9, 30419–30427. https://doi.org/10.1039/C9RA02645B

Helmy, Q., Kardena, E., Funamizu, N., Wisjnuprapto, 2011. Strategies toward commercial scale of biosurfactant production as potential substitute for its chemically counterparts. *International Journal of Biotechnology* 12, 66–86.

Jackson, S.A., Borchert, E., O'Gara, F., Dobson, A.D., 2015. Metagenomics for the discovery of novel biosurfactants of environmental interest from marine ecosystems. *Current Opinion in Biotechnology* 33, 176–182. https://doi.org/10.1016/j.copbio.2015.03.004

Jałowiecki, Ł., Żur, J., Chojniak, J., Ejhed, H., Płaza, G., 2018. Properties of antibiotic-resistant bacteria isolated from onsite wastewater treatment plant in relation to biofilm formation. *Current Microbiology* 75, 639–649. https://doi.org/10.1007/s00284-017-1428-2

Jałowiecki, Ł., Żur, J., Płaza, G., 2017. Norfloxacin degradation by Bacillus subtilis strains able to produce biosurfactants on a bioreactor scale. E3S Web of Conferences 17, 00033, 9th Conference on Interdisciplinary Problems in Environmental Protection and Engineering EKO-DOK 2017. https://doi.org/10.1051/e3sconf/20171700033

Jaysree, R.C., Patel, J., Na, R., 2015. Optimization of environmental parameters for biosurfactant production by Bacillus thuringiensis. *Journal of Pure and Applied Microbiology* 9, 459–465.

Jiang, J., Zu, Y., Li, X., Meng, Q., Long, X., 2020. Recent progress towards industrial rhamnolipids fermentation: process optimization and foam control. *Bioresource Technology* 298, 122394. https://doi.org/10.1016/j.biortech.2019.122394

Jibia, S.A., Mohanty, S., Dondapati, J.S., O'hare, S., Rahman, P.K.S.M., 2018. Biodegradation of wool by bacteria and fungi and enhancement of wool quality by biosurfactant washing. *Journal of Natural Fibers* 15, 287–295. https://doi.org/10.1080/15440478.2017.1325430

Jimoh, A.A., Lin, J., 2019. Biosurfactant: a new frontier for greener technology and environmental sustainability. *Ecotoxicology and Environmental Safety* 184, 109607.

Kameda, Y., Ouhira, S., Matsui, K., Kanatomo, S., Hase, T., Atsusaka, T., 1974. Antitumor activity of Bacillus natto. V. Isolation and characterization of surfactin in the culture medium of Bacillus natto KMD 2311. *Chemical and Pharmaceutical Bulletin* 22, 938–944.

Käppeli, O., Finnerty, W.R., 1979. Partition of alkane by an extracellular vesicle derived from hexadecane-grown Acinetobacter. *Journal of Bacteriology* 140, 707–712.

Karlapudi, A.P., Venkateswarulu, T.C., Tammineedi, J., Kanumuri, L., Ravuru, B.K., Dirisala, V.r., Kodali, V.P., 2018. Role of biosurfactants in bioremediation of oil pollution-a review. *Petroleum* 4, 241–249. https://doi.org/10.1016/j.petlm.2018.03.007

Karlo, J., Prasad, R., Singh, S.P., 2023. Biophotonics in food technology: Quo vadis ? *Journal of Agriculture and Food Research.* https://doi.org/10.1016/j.jafr.2022.100482

Karwowska, E., Andrzejewska-Morzuch, D., Łebkowska, M., Tabernacka, A., Wojtkowska, M., Telepko, A., Konarzewska, A., 2014. Bioleaching of metals from printed circuit boards supported with surfactant-producing bacteria. *Journal of Hazardous Materials* 264, 203–210. https://doi.org/10.1016/j.jhazmat.2013.11.018

Karwowska, E., Wojtkowska, M., Andrzejewska, D., 2015. The influence of metal speciation in combustion waste on the efficiency of Cu, Pb, Zn, Cd, Ni and Cr bioleaching in a mixed culture of sulfur-oxidizing and biosurfactant-producing bacteria. *Journal of Hazardous Materials* 299, 35–41. https://doi.org/10.1016/j.jhazmat.2015.06.006

Kazemi, K., 2017. Solid waste composting and the application of compost for biosurfactant production. Doctoral thesis, Memorial University of Newfoundland, St. John's, Canada. http://research.library.mun.ca/id/eprint/12562

Kennedy, J., O'Leary, N.D., Kiran, G.S., Morrissey, J.P., O'Gara, F., Selvin, J., Dobson, A.D.W., 2011. Functional metagenomic strategies for the discovery of novel enzymes and biosurfactants with biotechnological applications from marine ecosystems. *Journal of Applied Microbiology* 111, 787–799. https://doi.org/10.1111/j.1365-2672.2011.05106.x

Kesting, W., Tummuscheit, M., Schacht, H., Schollmeyer, E., 1996. Ecological washing of textiles with microbial surfactants, in: Jacobasch, H.-J. (Ed.), *Interfaces, Surfactants and Colloids in Engineering, Progress in Colloid & Polymer Science.* Steinkopff, Darmstadt, pp. 125–130. https://doi.org/10.1007/BFb0114456

Khoshdast, H., Sam, A., Manafi, Z., 2012. The use of rhamnolipid biosurfactants as a frothing agent and a sample copper ore response. *Minerals Engineering* 26, 41–49. https://doi.org/10.1016/j.mineng.2011.10.010

Kiliç, E., 2013. Evaluation of degreasing process with plant derived biosurfactant for leather making: an ecological approach. *Tekstil ve Konfeksiyon* 23, 181–187.

Kiran, S., Selvin, J., Manilal, A., Sugathan, S., 2011. Biosurfactants as green stabilizer for the biological synthesis of nanoparticles. *Critical Reviews in Biotechnology* 31, 354–364. https://doi.org/10.3109/07388551.2010.539971

Kitamoto, D., Isoda, H., Nakahara, T., 2002. Functions and potential applications of glycolipid biosurfactants--from energy-saving materials to gene delivery carriers. *Journal of Bioscience and Bioengineering* 94, 187–201. https://doi.org/10.1263/jbb.94.187

Korenblum, E., de Araujo, L.V., Guimarães, C.R., De Souza, L.M., Sassaki, G., Abreu, F., Nitschke, M., Lins, U., Freire, D.M.G., Barreto-Bergter, E., 2012. Purification and characterization of a surfactin-like molecule produced by Bacillus sp. H2O-1 and its antagonistic effect against sulfate reducing bacteria. *BMC Microbiology* 12, 252.

Kosaric, N., Sukan, F.V., 2014. *Biosurfactants: Production and Utilization-Processes, Technologies, and Economics*. CRC Press, Boca Raton, FL.

Krieger, N., Neto, D.C., Mitchell, D.A., 2010. Production of microbial biosurfactants by solid-state cultivation, in: Sen, R. (Ed.), *Biosurfactants, Advances in Experimental Medicine and Biology*. Springer, New York, pp. 203–210. https://doi.org/10.1007/978-1-4419-5979-9_15

Kulakovskaya, E., Kulakovskaya, T., 2014. Chapter 1 - Structure and occurrence of yeast extracellular glycolipids, in: Kulakovskaya, E., Kulakovskaya, T. (Eds.), *Extracellular Glycolipids of Yeasts*. Academic Press, pp. 1–13. https://doi.org/10.1016/B978-0-12-420069-2.00001-7

Lima, F.A., Santos, O.S., Pomella, A.W.V., Ribeiro, E.J., Resende, M.M. de, 2020. Culture medium evaluation using low-cost substrate for biosurfactants lipopeptides production by Bacillus amyloliquefaciens in pilot bioreactor. *Journal of Surfactants and Detergents* 23, 91–98. https://doi.org/10.1002/jsde.12350

Lima, T.M.S., Procópio, L.C., Brandão, F.D., Carvalho, A.M.X., Tótola, M.R., Borges, A.C., 2011. Biodegradability of bacterial surfactants. *Biodegradation* 22, 585–592. https://doi.org/10.1007/s10532-010-9431-3

Liu, J.-F., Mbadinga, S.M., Yang, S.-Z., Gu, J.-D., Mu, B.-Z., 2015. Chemical structure, property and potential applications of biosurfactants produced by Bacillus subtilis in petroleum recovery and spill mitigation. *International Journal of Molecular Sciences* 16, 4814–4837. https://doi.org/10.3390/ijms16034814

Manga, E.B., Celik, P.A., Cabuk, A., Banat, I.M., 2021. Biosurfactants: opportunities for the development of a sustainable future. *Current Opinion in Colloid & Interface Science* 56, 101514. https://doi.org/10.1016/j.cocis.2021.101514

Manna Sahoo, S., Datta, S., Biswas, D., 2011. Optimization of culture conditions for biosurfactant production from Pseudomonas aeruginosa OCD1. *Journal of Advanced Scientific Research* 2, 32–36.

Mao, X., Jiang, R., Xiao, W., Yu, J., 2015. Use of surfactants for the remediation of contaminated soils: a review. *Journal of Hazardous Materials* 285, 419–435. https://doi.org/10.1016/j.jhazmat.2014.12.009

Markets and Markets, 2017. Biosurfactants Market Analysis | Recent Market Developments | Industry Forecast to 2016–2022 [WWW Document]. URL https://www.marketsandmarkets.com/Market-Reports/biosurfactant-market-163644922.html?gclid=CjwKCAiAjrXxBRAPEiwAiM3DQmdr778lPDUjK3XChKXZDoTAKDlQHs7eTX44QDmAHOOo_t91YemOSRoC0n8QAvD_BwE (accessed 1.26.20).

Martínez, M., Oliveros, R., José, A., 2011. Synthesis of biosurfactants: enzymatic esterification of diglycerol and oleic acid. 1. Kinetic modeling. *Industrial & Engineering Chemistry Research* 50, 6609–6614. https://doi.org/10.1021/ie102560b

Martínez-Toledo, A., Rodríguez-Vázquez, R., 2011. Response surface methodology (Box-Behnken) to improve a liquid media formulation to produce biosurfactant and phenanthrene removal by Pseudomonas putida. *Annals of Microbiology* 61, 605–613.

Martinez-Toledo, A., Rodriguez Vazquez, R., Ilizaliturri-Hernández, C., 2015. Culture media evaluation for biosurfactant production by Pseudomonas putida CB-100 using Plackett-Burman experimental design. *African Journal of Microbiology Research* 9, 161–170. https://doi.org/10.5897/AJMR2014.7184

Maximize Market Research, 2019. Global biosurfactants market - industry analysis and forecast 2019–2026. Maximize Market Research. URL https://www.maximizemarketresearch.com/market-report/global-biosurfactants-market/433/ (accessed 1.26.20).

Mazaheri Assadi, M., Tabatabaee, M.S., 2010. Biosurfactants and their use in upgrading petroleum vacuum distillation residue: a review. *International Journal of Environmental Research* 4, 549–572. https://doi.org/10.22059/ijer.2010.242

Miller, R.M., 1995. Biosurfactant-facilitated remediation of metal-contaminated soils. *Environmental Health Perspectives* 103, 59–62. https://doi.org/10.2307/3432014

Mnif, I., Ghribi, D., 2015. Lipopeptides biosurfactants, main classes and new insights for industrial; biomedical and environmental applications. *Biopolymers* 104. https://doi.org/10.1002/bip.22630

Moore, T., Globa, L., Barbaree, J., Vodyanoy, V., Sorokulova, I., 2013. Antagonistic activity of Bacillus bacteria against food-borne pathogens. *Journal of Probiotics & Health* 1, 1–6.

Morais, I.M.C., Cordeiro, A.L., Teixeira, G.S., Domingues, V.S., Nardi, R.M.D., Monteiro, A.S., Alves, R.J., Siqueira, E.P., Santos, V.L., 2017. Biological and physicochemical properties of biosurfactants produced by Lactobacillus jensenii P6A and Lactobacillus gasseri P65. *Microbial Cell Factories* 16, 155. https://doi.org/10.1186/s12934-017-0769-7

Mukherjee, S., Stamatis, D., Bertsch, J., Ovchinnikova, G., Katta, H.Y., Mojica, A., Chen, I.-M.A., Kyrpides, N.C., Reddy, T.B.K., 2019. Genomes OnLine Database (GOLD) v.7: updates and new features. *Nucleic Acids Research* 47, D649–D659. https://doi.org/10.1093/nar/gky977

Najafi, A.R., Rahimpour, M.R., Jahanmiri, A.H., Roostaazad, R., Arabian, D., Ghobadi, Z., 2010. Enhancing biosurfactant production from an indigenous strain of Bacillus mycoides by optimizing the growth conditions using a response surface methodology. *Chemical Engineering Journal* 163, 188–194. https://doi.org/10.1016/j.cej.2010.06.044

Naughton, P., Marchant, R., Naughton, V., Banat, I.M., 2019. Microbial biosurfactants: current trends and applications in agricultural and biomedical industries. *Journal of Applied Microbiology* 127, 12–28. https://doi.org/10.1111/jam.14243

Ndlovu, T., Khan, S., Khan, W., 2016. Distribution and diversity of biosurfactant-producing bacteria in a wastewater treatment plant. *Environmental Science and Pollution Research* 23, 9993–10004. https://doi.org/10.1007/s11356-016-6249-5

Nitschke, M., Araújo, L.V., Costa, S.G.V.A.O, Pires, R.C., Zeraik, A.E., Fernandes, A.C.L.B., Freire, D.M.G., Contiero, J., 2009. Surfactin reduces the adhesion of food-borne pathogenic bacteria to solid surfaces. *Letters in Applied Microbiology* 49, 241–247. https://doi.org/10.1111/j.1472-765X.2009.02646.x

Nitschke, M., e Silva, S.S., 2018. Recent food applications of microbial surfactants. *Critical Reviews in Food Science and Nutrition* 58, 631–638. https://doi.org/10.1080/10408398.2016.1208635

Occhipinti, A., Eyassu, F., Rahman, T.J., Rahman, P.K.S.M., Angione, C., 2018. In silico engineering of Pseudomonas metabolism reveals new biomarkers for increased biosurfactant production. *PeerJ* 6. https://doi.org/10.7717/peerj.6046

Odukkathil, G., Vasudevan, N., 2013. Enhanced biodegradation of endosulfan and its major metabolite endosulfate by a biosurfactant producing bacterium. *Journal of Environmental Science and Health, Part B* 48, 462–469. https://doi.org/10.1080/03601234.2013.761873

Ohadi, M., Amir-Heidari, B., Moshafi, M.H., Mirparizi, A., Basir, M., Dehghan-Noudeh, G., 2014. Encapsulation of biosurfactant-producing Bacillus licheniformis (PTCC 1320) in alginate beads. *Biotechnology* 13, 239–244.

Okasaka, M., Kubota, K., Yamasaki, E., Yang, J., Takata, S., 2019. Evaluation of anionic surfactants effects on the skin barrier function based on skin permeability. *Pharmaceutical Development and Technology* 24, 99–104. https://doi.org/10.1080/10837450.2018.1425885

Onwosi, C.O., Odibo, F.J.C., 2012. Effects of carbon and nitrogen sources on rhamnolipid biosurfactant production by Pseudomonas nitroreducens isolated from soil. *World Journal of Microbiology and Biotechnology* 28, 937–942. https://doi.org/10.1007/s11274-011-0891-3

Osman, M., Ishigami, Y., Someya, J., Jensen, H.B., 1996. The bioconversion of ethanol to biosurfactants and dye by a novel coproduction technique. *Journal of the American Oil Chemists' Society* 73, 851–856. https://doi.org/10.1007/BF02517986

Pacwa-Płociniczak, M., Płaza, G.A., Piotrowska-Seget, Z., Cameotra, S.S., 2011. Environmental applications of biosurfactants: recent advances. *International Journal of Molecular Sciences* 12, 633–654. https://doi.org/10.3390/ijms12010633

Palanisamy, P., Raichur, A.M., 2009. Synthesis of spherical NiO nanoparticles through a novel biosurfactant mediated emulsion technique. *Materials Science and Engineering: C* 29, 199–204.

Paranji, S., Swarnalatha, S., Sekaran, G., 2014. Lipoprotein biosurfactant production from an extreme acidophile using fish oil and its immobilization in nanoporous activated carbon for the removal of Ca2+ and Cr3+ in aqueous solution. *RSC Advances* 4, 34144–34155. https://doi.org/10.1039/C4RA03101F

Patel, S., Homaei, A., Patil, S., Daverey, A., 2019. Microbial biosurfactants for oil spill remediation: pitfalls and potentials. *Applied Microbiology and Biotechnology* 103, 27–37. https://doi.org/10.1007/s00253-018-9434-2

Pele, M.A., Ribeaux, D.R., Vieira, E.R., Souza, A.F., Luna, M.A.C., Rodríguez, D.M., Andrade, R.F.S., Alviano, D.S., Alviano, C.S., Barreto-Bergter, E., Santiago, A.L.C.M.A., Campos-Takaki, G.M., 2019. Conversion of renewable substrates for biosurfactant production by Rhizopus arrhizus UCP 1607 and enhancing the removal of diesel oil from marine soil. *Electronic Journal of Biotechnology* 38, 40–48. https://doi.org/10.1016/j.ejbt.2018.12.003

Perez-Ameneiro, M., Vecino, X., Cruz Freire, J.M., Moldes, A., 2015. Wastewater treatment enhancement by applying a lipopeptide biosurfactant to a lignocellulosic biocomposite. *Carbohydrate Polymers* 131, 186–196. https://doi.org/10.1016/j.carbpol.2015.05.075

Pérez-Armendáriz, B., Cal-y-Mayor-Luna, C., El-Kassis, E.G., Ortega-Martínez, L.D., 2019. Use of waste canola oil as a low-cost substrate for rhamnolipid production using Pseudomonas aeruginosa. AMB Express 9. https://doi.org/10.1186/s13568-019-0784-7

Pervaiz, M., Sain, M., 2010. Extraction and characterization of extracellular polymeric substances (EPS) from waste sludge of pulp and paper mill. *International Review of Chemical Engineering* 2, 550–554.

Płaza, G.A., Chojniak, J., Banat, I.M., 2014. Biosurfactant mediated biosynthesis of selected metallic nanoparticles. *International Journal of Molecular Sciences* 15, 13720–13737. https://doi.org/10.3390/ijms150813720

Prasad, R., Pandey, R., and Barman, I., 2016. Engineering tailored nanoparticles with microbes: quo vadis. *WIREs Nanomedicine & Nanobiotechnology* 8, 316–330. https://doi.org/10.1002/wnan.1363

Rahman, P., Gakpe, E., 2008. Production, characterisation and applications of biosurfactants-review. *Biotechnology* 7, 360–370.

Rodrigues, L., Banat, I.M., Teixeira, J., Oliveira, R., 2006. Biosurfactants: potential applications in medicine. *Journal of Antimicrobial Chemotherapy* 57, 609–618. https://doi.org/10.1093/jac/dkl024

Rosenberg, E., Rubinovitz, C., Legmann, R., Ron, E., 1988. Purification and chemical properties of Acinetobacter calcoaceticus A2 biodispersan. *Applied and Environmental Microbiology* 54, 323–326.

Sachdev, D.P., Cameotra, S.S., 2013. Biosurfactants in agriculture. *Applied Microbiology and Biotechnology* 97, 1005–1016. https://doi.org/10.1007/s00253-012-4641-8

Saeki, H., Sasaki, M., Komatsu, K., Miura, A., Matsuda, H., 2009. Oil spill remediation by using the remediation agent JE1058BS that contains a biosurfactant produced by Gordonia sp. strain JE-1058. *Bioresource Technology* 100, 572–577. https://doi.org/10.1016/j.biortech.2008.06.046

Santos, D.K.F., Rufino, R.D., Luna, J.M., Santos, V.A., Sarubbo, L.A., 2016. Biosurfactants: multifunctional biomolecules of the 21st century. *International Journal of Molecular Sciences* 17. https://doi.org/10.3390/ijms17030401

Sanwani, E., Chaerun, S., Mirahati, R., Wahyuningsih, T., 2016. Bioflotation: bacteria-mineral interaction for eco-friendly and sustainable mineral processing. Procedia Chemistry, 5th International Conference on Recent Advances in Materials, Minerals and Environment (RAMM) & 2nd International Postgraduate Conference on Materials, Mineral and Polymer (MAMIP) 19, 666–672. https://doi.org/10.1016/j.proche.2016.03.068

Savarino, P., Montoneri, E., Bottigliengo, S., Boffa, V., Guizzetti, T., Perrone, D.G., Mendichi, R., 2009. Biosurfactants from urban wastes as auxiliaries for textile dyeing. *Industrial & Engineering Chemistry Research* 48, 3738–3748.

Sekhon Randhawa, K.K., Rahman, P.K.S.M., 2014. Rhamnolipid biosurfactants—past, present, and future scenario of global market. *Frontiers in Microbiology* 5. https://doi.org/10.3389/fmicb.2014.00454

Sen, S., Borah, S.N., Bora, A., Deka, S., 2017. Production, characterization, and antifungal activity of a biosurfactant produced by Rhodotorula babjevae YS3. *Microbial Cell Factories* 16. https://doi.org/10.1186/s12934-017-0711-z

Sharma, D., Singh Saharan, B., 2014. Simultaneous production of biosurfactants and bacteriocins by probiotic Lactobacillus casei MRTL3. *International Journal of Microbiology* 2014. PMID: 24669225.

Shekhar, S., Sundaramanickam, A., Balasubramanian, T., 2015. Biosurfactant producing microbes and their potential applications: a review. *Critical Reviews in Environmental Science and Technology* 45, 1522–1554. https://doi.org/10.1080/10643389.2014.955631

Sheng, X., He, L., Wang, Q., Ye, H., Jiang, C., 2008. Effects of inoculation of biosurfactant-producing Bacillus sp. J119 on plant growth and cadmium uptake in a cadmium-amended soil. *Journal of Hazardous Materials* 155, 17–22. https://doi.org/10.1016/j.jhazmat.2007.10.107

Shoeb, E., Akhlaq, F., Badar, U., Akhter, J., Imtiaz, S., 2013. Classification and industrial applications of biosurfactants 4, 10. Conference Proceedings, Corpus ID: 138593726. https://www.semanticscholar.org/paper/CLASSIFICATION-AND-INDUSTRIAL-APPLICATIONS-OF-Shoeb-Akhlaq/f50a693283179b50722853befb510e82761e8684

Sigma Aldrich, 2020. Sigma Aldrich [WWW Document]. Sigma-Aldrich. URL https://www.sigmaaldrich.com/european-export.html (accessed 2.16.20).

Singh, P., Patil, Y., Rale, V., 2019. Biosurfactant production: emerging trends and promising strategies. *Journal of Applied Microbiology* 126, 2–13. https://doi.org/10.1111/jam.14057

Soares dos Santos, A., Pereira Jr, N., Freire, D.M.G., 2016. Strategies for improved rhamnolipid production by Pseudomonas aeruginosa PA1. *PeerJ* 4. https://doi.org/10.7717/peerj.2078

Sobrinho, H., Luna, J.M., Rufino, R.D., Porto, A., Sarubbo, L.A., 2013. Biosurfactants: classification, properties and environmental applications. *Recent Developments in Biotechnology* 11, 1–29.

Sonowal, S., Nava, A.R., Joshi, S.J., Borah, S.N., Islam, N.F., Pandit, S., Prasad, R., Sarma, H., 2022. Biosurfactants assisted heavy metals phytoremediation: Green technology for the United Nations sustainable development goals. *Pedosphere* 2(1), 198–210. https://doi.org/10.1016/S1002-0160(21)60067-X

Stacey, S., McLaughlin, M., Cakmak, I., Lombi, E., Johnston, C., 2007. Novel chelating agent for trace element fertilizers. Conference paper, Improving crop production and human health, May, 2007, Zinc Crops 2007, Istanbul, Turkey.

Thies, S., Rausch, S.C., Kovacic, F., Schmidt-Thaler, A., Wilhelm, S., Rosenau, F., Daniel, R., Streit, W., Pietruszka, J., Jaeger, K.-E., 2016. Metagenomic discovery of novel enzymes and biosurfactants in a slaughterhouse biofilm microbial community. *Scientific Reports* 6, 27035. https://doi.org/10.1038/srep27035

Valivety, R., Jauregi, P., Gill, I., Vulfson, E., 1997. Chemo-enzymatic synthesis of amino acid-based surfactants. *Journal of the American Oil Chemists' Society* 74, 879–886. https://doi.org/10.1007/s11746-997-0232-8

Vecino, X., Cruz, J.M., Moldes, A.B., Rodrigues, L.R., 2017. Biosurfactants in cosmetic formulations: trends and challenges. *Critical Reviews in Biotechnology* 37, 911–923. https://doi.org/10.1080/07388551.2016.1269053

Vecino, X., Rodríguez-López, L., Ferreira, D., Cruz, J.M., Moldes, A.B., Rodrigues, L.R., 2018. Bioactivity of glycolipopeptide cell-bound biosurfactants against skin pathogens. *International Journal of Biological Macromolecules* 109, 971–979. https://doi.org/10.1016/j.ijbiomac.2017.11.088

Velioglu, Z., Ozturk Urek, R., 2015. Optimization of cultural conditions for biosurfactant production by Pleurotus djamor in solid state fermentation. *Journal of Biosciences and Bioengineering* 120, 526–531. https://doi.org/10.1016/j.jbiosc.2015.03.007

Vijayakumar, S., Saravanan, V., 2015. Biosurfactants-types, sources and applications [WWW Document]. Science Alert. https://doi.org/10.3923/jm.2015.181.192

Vollenbroich, D., Özel, M., Vater, J., Kamp, R.M., Pauli, G., 1997. Mechanism of inactivation of enveloped viruses by the biosurfactant surfactin from Bacillus subtilis. *Biologicals* 25, 289–297.

Waghmode, S., Kulkarni, C., Shukla, S., Sursawant, P., Velhal, C., 2014. Low cost production of biosurfactant from different substrates and their comparative study with commercially available chemical surfactant. *International Journal of Scientific & Technology Research* 3, 146–149.

Williams, W., Kunorozva, L., Klaiber, I., Henkel, M., Pfannstiel, J., Van Zyl, L.J., Hausmann, R., Burger, A., Trindade, M., 2019. Novel metagenome-derived ornithine lipids identified by functional screening for biosurfactants. *Applied Microbiology and Biotechnology* 103, 4429–4441. https://doi.org/10.1007/s00253-019-09768-1

Wu, Q., Zhi, Y., Xu, Y., 2019. Systematically engineering the biosynthesis of a green biosurfactant surfactin by Bacillus subtilis 168. *Metabolic Engineering* 52, 87–97. https://doi.org/10.1016/j.ymben.2018.11.004

Yataghene, A., Abouseoud, M., Maachi, R., Amrane, A., 2008. Effect of the carbon and nitrogen sources on biosurfactant production by Pseudomonas fluorescens – biosurfactant characterization 8. https://www.aidic.it/IBIC2008/webpapers/71Yataghene.pdf

Yeh, M.-S., Wei, Y.-H., Chang, J.-S., 2005. Enhanced production of surfactin from Bacillus subtilis by addition of solid carriers. *Biotechnology Progress* 21, 1329–1334. https://doi.org/10.1021/bp050040c

Yu, H., Huang, G.H., 2011. Isolation and characterization of biosurfactant- and bioemulsifier-producing bacteria from petroleum contaminated sites in Western Canada. *Soil and Sediment Contamination: An International Journal* 20, 274–288. https://doi.org/10.1080/15320383.2011.560981

Zheng, T., Xu, Y.-S., Yong, X.-Y., Li, B., Yin, D., Cheng, Q.-W., Yuan, H.-R., Yong, Y.-C., 2015. Endogenously enhanced biosurfactant production promotes electricity generation from microbial fuel cells. *Bioresource Technology* 197, 416–421. https://doi.org/10.1016/j.biortech.2015.08.136

Zhi, Y., Wu, Q., Xu, Y., 2017. Genome and transcriptome analysis of surfactin biosynthesis in Bacillus amyloliquefaciens MT45. *Scientific Reports* 7, 1–13. https://doi.org/10.1038/srep40976

Zhu, Z., Zhang, F., Wei, Z., Ran, W., Shen, Q., 2013. The usage of rice straw as a major substrate for the production of surfactin by Bacillus amyloliquefaciens XZ-173 in solid-state fermentation. *Journal of Environmental Management* 127, 96–102. https://doi.org/10.1016/j.jenvman.2013.04.017

2 Microalgae proteins as a sustainable food supply

Tatiele Nascimento, Andressa Fernandes,
Pricila Nass, Luisa Schetinger, Patrícia Caetano,
Leila Q. Zepka, and Eduardo Jacob-Lopes

CONTENTS

2.1 INTRODUCTION

Global food consumption is estimated to reach 455 million tons by 2050, representing more than 70% of current production demand. In this global diet increase, meat is one of the primary sources of nutrients (Kell, 2022). However, this massive growth in the food industry will come at a cost. It will pressure the environmental resources as traditional protein production requires lots of land area, water, and energy (Aiking & de Boer, 2020). Furthermore, this industry is responsible for gas emissions that hurt the earth's atmosphere and increase global warming effects (Lopez-Ridaura et al., 2009).

Allied to this, the increase in popularity of the meat-free diet, which at the same time is rich in protein, is a trend, especially among the new generations (Chen & Chaudhary, 2019). These paradigms lead to a new era of ingredients and food products that are increasingly nutritious, functional, innovative, and sustainable.

Considering this scenario is an urgent need to investigate sustainable and clean protein sources to help supply the new global demands (Kumar et al., 2022). Then, microalgae appear as a sustainable nutrient source, with a powerful capacity to convert carbon dioxide and light into biomolecules, contributing to water and land preservation (Mok et al., 2020). Microalgae can thrive in environments where traditional agriculture would not have success. Non-potable water, polluted rivers, and unproductive terrains are not a barrier to microalgae growth (Chisti, 2007).

Conventionally, microalgae are employed as food sources, with their first use dating back over 2000 years to the consumption of *Nostoc* by the Chinese

DOI: 10.1201/9781003394600-3

population. Likewise, other reports point out that centuries ago, the indigenous people also consumed these microorganisms to survive (Wang et al., 2021). Although its use in food dates back a long time, greater interest in microalgae in food was only given in the last century, when these microorganisms became the focus of extensive studies on their potential as alternative sustainable sources of protein for the following centuries (Becker, 2007; Amorim et al., 2021a). In the 1950s, the exploitation of microalgae for food and biochemical applications was suggested at the Algae Mass-Culture Symposium, with the consolidation of initial facilities for the commercial production of *Chlorella* in Japan and only approximately two decades later expanding cultivation of *Arthrospira* in Mexico (Caporgno & Mathys, 2018).

Currently, the number of cataloged microalgae species is increasing. However, it is limited to a few species for use in animal and human food, functional foods, and nutraceuticals. On the other hand, research and development efforts are constantly underway to make other species available for consumption. Although the use of microalgae protein as a sustainable food source shows a promising future, some barriers related to production, digestibility, and palatability (sensory challenges) need to be overcome (Niccolai et al., 2019; Pérez-Lloréns, 2020; Tang et al., 2020; Kumar et al., 2022).

Based on the above, this chapter addresses the main aspects related to the production of microalgae proteins, such as the sustainable approach, the protein quality of microalgae biomass, applications and properties of microalgae proteins, and the challenges and future trends in this area of study.

2.2   MICROALGAE PROTEIN PRODUCTION AS A SUSTAINABLE APPROACH

Achieving a sustainable food system that can deliver a healthy diet with regular protein consumption for a growing population presents challenges. The main matrices for alternative proteins include cereals (including pulses) and oilseeds, which can play a crucial role in replacing animal protein in the diet (Gohara-Beirigo et al., 2022; Oliveira & Bragotto, 2022).

However, there are barriers since their nutritional quality is subpar to animal proteins. Furthermore, geographic availability and seasonality hinder these cultures from consolidating plant-based products. Therefore, emerging matrices arise to solve some of these concerns such as microalgae (Oliveira & Bragotto, 2022).

Microalgae are sources of essential and non-essential amino acids and present a lower carbon and water footprint than beef, whey, soy, and pea. Moreover, microalgae have a higher protein production per hectare than all other sources compared, as shown in the schematic of Figure 2.1 (Klamczynsk & Mooney, 2017).

It is estimated that the market relevance of microalgae as production systems for protein is constantly increasing. Based on this understanding, selecting a microalgae mode of cultivation is vital (Jacob-Lopes et al., 2019).

Historically, circular ponds and open raceway ponds are classic types of photobioreactors that have been widely applied at a large scale for commercial protein production (Lam et al., 2018; Fu et al., 2021).

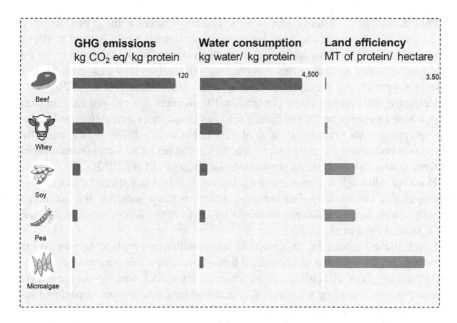

FIGURE 2.1 Illustration of carbon footprint, water consumption, and protein yield per hectare from microalgae versus other protein sources (adapted Klamczynsk & Mooney, 2017).

Systems are characterized because of their cost-effectiveness and simple-to-operate. However, the major bottlenecks are the external agent's contaminations and the vulnerability of seasonal intervention, which directly inhibits their productivity (Mehariya et al., 2021).

Despite the numerous challenges to be overcome open raceway ponds are undeniably well established in the industrial production of microalgae species such as *Chlorella* sp. and *Arthrospira platensis* (Kumar et al., 2021).

On the other hand, closed-loop photobioreactors provide a system with controlled conditions (Fu et al., 2021). In particular, tubular closed-loop photobioreactors have been highlighted for their industrial relevance in microalgae production systems (Lam et al., 2018).

As a result, various tubular photobioreactors have been designed, including horizontal, vertical, conical, near horizontal, and inclined types produced up of plastic or glass tubes in which microalgae cultures were re-circulated either with a pump or airlift technique (Kumar et al., 2021).

The improved design allows a controlled environment, high biomass production (up to 16 times more than the open pond), low evaporation rate, reduced risks of contaminations, higher photosynthetic efficiency, and efficient transfer of nutrients (Kumar et al., 2021; Veerabadhran, 2021).

However, the critical point is the high capital and operating costs that limit the viability of this system. Also, closed photobioreactors face scaling issues. Because enlarging the photobioreactors will affect light availability and trigger microalgae productivity (Veerabadhran, 2021).

Photoautotrophic systems represent a protein source with a production of 50%–70% w/w; however, the need for large areas, limited biomass yield or dependence on artificial light weakens the economic viability of the cultivation. To this end, heterotrophic systems offer a promising alternative: they can grow in light absence supported by an exogenous carbon source (Jacob-Lopes et al., 2020).

Heterotrophic systems allow conventional fermenters (i.e., stirred tank bioreactors), whose operation and maintenance are considered more straightforward, reducing operating costs (Perez-Garcia et al., 2011; Hu et al., 2018). When using this cultivation condition, the productive capacities are superior to the photoautotrophic systems in microalgae biomass production (Jacob-Lopes et al., 2020).

However, although it is an attractive advantage, only a few species can grow heterotrophically successfully. Furthermore, adding organic substrate for microalgae growth can increase capital requirements by up to 50% (Klamczynsk & Mooney, 2017; Jacob-Lopes et al., 2020).

Given these aspects, be important to mention that microalgae cultivation still faces barriers attributed to technological limitations. From this perspective, the path to overcoming these difficulties can be a biorefinery model, seeking out the integration of processes, leading to minimizing nominal production costs, waste, and sustainability (Severo et al., 2019).

In this sense, biorefineries are industrial installation that converts renewable raw materials into several high-value alternative products, such as proteins and peptide hydrolysates, that can be used in food products after the removal of the lipid fraction from biomass through the integration of technologies and processes, being an important scenario of circular economy, closing loops of streams and enriching various outputs (Kumar et al., 2022).

Lastly, the development of multi-product biorefineries has been proactive for valorizing biomass and bioproducts, aiming to sustainably dilute the general costs. However, payback time and incentive policies need to be evaluated (Severo et al., 2019; Banu et al., 2020; Mishra et al., 2020).

2.3   PROTEIN QUALITY OF MICROALGAE BIOMASS

Proteins play an essential role in the structure and metabolism of microalgae cells like cytoplasm, organelles, plastids, cell walls, and nucleus. Thus, the protein concentrations of many microalgae species are different by numerous biotic and abiotic factors and in their genome variation (Amorim et al., 2021b).

According to Table 2.1, the protein content in microalgae biomass range from 23% to 71% on dry biomass basis, depending upon the genera and species. Some species that stand out for their rich protein content are *Arthrospira platensis* (also known as Spirulina), *Chlorella vulgaris, Scenedesmus* sp., *Tetraselmis* sp., *Phaeodactylum* sp., and *Porphyridium* sp. (Kumar et al., 2022).

Among these microorganisms, the most significant amount of protein on a dry weight basis is provided by *A. platensis* (70%) and *C. vulgaris* (58%) (Barkia et al., 2019a). They are world leaders in the segment of microalgae biorefineries (Nass et al., 2022). These two species are marketed as food supplements in dry powder or capsule form and as bioactive ingredients, given their nutritional and functional properties (Lafarga, 2019).

TABLE 2.1
The protein content of different microalgae species and vegetable sources

Microalgae	Protein content (%)[a]	Reference
Chlorella vulgaris	51.0–58.0	Gouveia et al. (2007), Becker (2007), Batista et al. (2011), Kumar et al. (2022)
Chlorella pyrenoidosa	44.0	Safafar et al. (2016), Kumar et al. (2022)
Chlorella sorokiniana	39.0	Kumar et al. (2022)
Chlorella protothecoides	48.2	Grossmann et al. (2020), Kumar et al. (2022)
Dunaliella salina	49.0–57.0	Becker (2007)
Arthospira platensis	46.0–63.0	Tibbetts et al. (2015), Benelhadj et al. (2016), Kumar et al. (2022)
Arthospira maxima	60–71	
Aphanizomenon flos-aquae	62.0	Grossmann et al. (2020), Kumar et al. (2022)
Acutodesmus (Scenedesmus) oblíquo	50.0–56.0	Lum et al. (2013), Kumar et al. (2022)
Tetraselmis chuii	35.0–40.0	Tibbetts et al. (2015), Pereira et al. (2019), Kumar et al. (2022)
Tetraselmis suecia	48.7	Tulli et al. (2012), Kumar et al. (2022)
Phaeodactylum tricornutum	23.0–39.0	Brown (1991), Kumar et al. (2022)
Pavlova salina	26.0	Brown (1991), Kumar et al. (2022)
Nannochloropsis gaditana	44.9	Kumar et al. (2022)
Botryococcus braunii	39.0	Grossmann et al. (2018, 2020), Kumar et al. (2022),
Neochloris oleoabundans	30.1	Grossmann et al. (2018, 2020), Kumar et al. (2022)
Porphyridium cruentum	28.0–39.0	Grossmann et al. (2018, 2020), Kumar et al. (2022)
Porphyridium aerugineum	31.6	Kumar et al. (2022)
Schizochytrium limacinum	24.3	Moaveni et al. (2022), Kumar et al. (2022)
Chickpea	20.5	USDA
Maize	10.0	USDA
Soybean	36.5	USDA

Adapted from Kumar et al. (2022).
[a] Average values.

The nutritional quality of proteins unicellular can be determined by their profile amino acids. Generally, are peptides 2–20 amino acids in length (although most of them are two to four amino acids in length) and a high abundance of hydrophobic amino acids (Lafarga et al., 2021).

Protein is a crucial element in the human diet. Once the human body cannot synthesize it, it needs to be supplied by the diet. Given their importance, essential amino acids are listed: histidine, isoleucine, leucine, lysine, methionine, phenylalanine, tryptophan, and valine. Furthermore, the microalgae profile of amino acids is similar to conventional high-quality protein sources (Moaveni et al., 2022).

Table 2.2 shows the amino acid profile of various microalgae species and compares it with some basic conventional food items. Considering, specifically the essential amino acids profile microalgae such as leucine, arginine, and lysine constituting (7%) each of the total protein while others, such as isoleucine, phenylalanine, threonine, and valine averagely included 4% of each of the complete protein. In general, the microalgae provide a diverse essential amino acid profile of a few fractions (Kumar et al., 2022). It can be seen that the amino acid pattern of almost all microalgae compares favorably with that of the reference recommended by WHO/FAO (1991).

Some proteins of microalgae origin are associated with the reduction of cholesterol and triglyceride levels, with the stimulation of the production of the hormone cholecystokinin, and hypoglycemic properties, which regulate the suppression of appetite and, therefore, have been considered in the formulation of functional foods against obesity (Jacob-Lopes et al., 2019).

Regarding bioactive peptides, especially antioxidant peptides, act in several ways: as inhibitors of lipid peroxidation, metal ion chelators, or scavengers of free radicals (Lafarga et al., 2021). Antioxidants can interfere with oxidation by donating a hydrogen atom or an electron to free radicals or terminating the initiators by inactivating catalysts or metal ions, removing these chain reactions (Markou et al., 2021; Moaveni et al., 2022). Other studies have a relationship between function and structure that was used to explain the antioxidant properties of most amino acids (Nascimento et al., 2019). Some proteins of microalgae origin are associated with the reduction of cholesterol and triglyceride levels, with the stimulation of the production of the hormone cholecystokinin, and hypoglycemic properties, which regulate the suppression of appetite and, therefore, have been considered in the formulation of functional foods against obesity (Jacob-Lopes et al., 2019)

The positive effects depend on protein digestibility and amino acid bioavailability (other parameters that determine the quality of the proteins). Applying is typically defined as the proportion of ingested protein hydrolyzed into amino acids (di- and tri-) peptides available for absorption (Kumar et al., 2022).

Digestibility is an obstacle that needs to be considered, as it varies significantly between species. Some microalgae tend to be less digestible, while others, such as C. *vulgaris* and *A. platensis*, are more digestible (Niccolai et al., 2019). This fact justifies the wide use of these two species as a protein supplement.

Various *in vitro* and *in vivo* research and methods have been recommended to assess such an effect. Nonetheless, the bioavailability of these molecules from the microalgae matrix is usually a limiting step since these compounds are located within the cell, most of the time protected by a robust cell wall (Nass et al., 2022).

Thus, for accurate protein quality and bioactive potential measurement, FAO recommends using *in vivo* methods. This is because the protein quality and its biological efficacy can be determined only by the ability of the ingested protein to enhance the growth and body weight of the test subject (humans or animals) (Adhikari et al., 2022).

Finally, to meet the metabolic demand and to assure proper functioning of the human body, numerous alternative sources have been utilized to generate food proteins. Food supplements are increasingly popular worldwide. The majority of the microalgae biomass currently commercialized for food purposes is sold as a nutritional supplement and is promoted as "superfood", "rich in protein", and/or "rich in

TABLE 2.2

The amino acids profile of a few microalgae species compared with recommended levels

Amino acid	FAO/WHO	Soybean	1. Spirulina	2. Chlorella	3. Scenedesmus	Tetraselmis	Phaeodactylum	Porphyridium
Asparagine	–	1.3	11.8	9.3	3.16	8.9	8.6	15.0
Glutamine	–	19	10.3	13.7	4.38	11.2	11.2	15.6
Serine	–	5.8	5.1	5.8	1.75	4.6	5.9	7.0
Glycine	–	19	5.7	6.3	2.18	5.9	5.8	7.0
Histidine	–	2.6	2.2	2.0	0.71	1.8	1.7	1.9
Arginine	–	7.4	7.3	6.9	1.88	13.2	6.6	8.6
Proline	–	5.3	4.2	5.0	1.46	4.7	6.3	5.0
Valine	5	5.3	7.1	7.0	1.61	5.7	5.9	7.3
Methionine	3.5	1.3	2.5	1.3	0.78	2.3	1.9	3.7
Leucine	7	7.7	9.8	9.5	3.19	8.0	7.7	11.9
Phenylalanine	6	5.0	5.3	5.5	1.87	5.9	6.6	6.3
Lysine	5.5	6.4	4.8	6.4	1.94	6.0	5.6	8.0
Tryptophan	1.0	1.4	0.3	–	0.53	3.8	1.6	3.3

Adapted from FAO (1991) and Kumar et al. (2022).
Values expressed as g/16 g N.

omega-3". Most often, such products contain the biomass of *Arthospira platensis* and for this reason, they are considered healthy and sustainable (Lafarga et al., 2019).

2.4 APPLICATIONS AND MICROALGAE PROTEIN PROPERTIES (RECENT RESEARCH)

Microalgae proteins have versatility in their properties, which strongly corroborates for their possible nutritional and biotechnological applications in various sectors. It is known that protein is an essential nutritional component in the diet, as it contributes to the nutritional contribution of macronutrients for energy gain, metabolic health, maintenance and synthesis of tissues and cells, among other important biological activities.

As mentioned in the previous section, microalgae have a high protein content that allows their exploitation for human and animal nutrition. Since the beginning of studies with microalgae, this class of biomolecules was the most explored bioactive food compounds in these matrices, due to the high concentration (Jacob-Lopes et al., 2019).

According elucidated in Table 2.3, microalgae biomasses are being constantly explored in food formulations to improve the nutritional value or functional properties of several products. Although the addition of isolated microalgae proteins in food formulation has been reported in few works, whole biomass and other biocompounds have been used as food ingredients for different purposes. In general, the vast majority of studies available in the literature explore the incorporation of whole microalgae biomass in the formulation of products, without isolating compounds. The advantage of this approach lies in the use as a whole of all the macro- and micronutrients present in these matrices, as well as the entire contribution of bioactive compounds. Aiming at these aspects, the nutritional and techno-functional properties of microalgae proteins have strongly supported the formulation of nutritious and functional food products.

Bakery products such as breads, cookies, and biscuits are categories widely explored for incorporation of microalgae-based ingredients. Gouveia et al. (2007) observed positive effects with adding *Chlorella vulgaris* biomass in butter cookies. In addition to its significant contribution as a coloring agent, microalgae biomass strongly contributes to the techno-functional properties of the food. The authors suggest that the high protein content of the biomass, in addition to contributing to the nutritional value of the cookies, strongly corroborated with the texture and firmness through the water absorption capacity. Batista et al. (2017) also obtained good results with the addition of the biomass of *Arthrospira platensis, Chlorella vulgaris, Tetraselmis suecica* and *Phaeodactylum tricornutum* as an alternative ingredient in cookies. In this study, it was possible to prove the contribution of the protein content of microalgae biomass to the nutritional protein supply (*A. platensis* and *C. vulgaris*), structuring effect (*A. platensis*) and also enhanced antioxidant activity, related to the phenolic content of phycobiliproteins.

In breads, the incorporation of 5–10% of *Dunaliella salina* biomass improved the physical-chemical parameters of the finished products, such as the porosity of bread

TABLE 2.3

Applications and techno-functional properties of microalgae biomass and proteins in formulation food products for human nutrition

Applications	Microalgae species	Microalgae incorporated (%)	Findings/observations	References
Biscuits/Cookies	*Chlorella vulgaris*	0.5–3%	• Higher textural characteristics • Coloring agents • Nutritional properties	Gouveia et al. (2007)
Pasta	*Chlorella vulgaris* *Spirulina maxima*	0.5–2.0%	• Increase of quality parameters • Enhance the nutritional and sensorial quality • higher protein content	Fradique et al. (2010)
Cookies	*Arthrospira platensis* *Chlorella vulgaris* *Tetraselmis suecica* *Phaeodactylum tricornutum*	2 and 6%	• Structuring effect • Nutritional properties • High protein contents • Antioxidant activity	Batista et al. (2017)
Pasta	*Dunaliella salina*	1 and 3%	• Protein source • Increased the nutritional value • High phytochemicals content	El-Baz et al. (2017)
Snacks	*Arthrospira platensis* (Spirulina)	0.4–3.2%	• High protein contents • Protein content correlated with biomass concentration	Lucas et al. (2017)
Bread	*Dunaliella salina*	5–10%	• Improves physical-chemical parameters • Increased nutritional and biological value • High protein contents	Tertychnaya et al. (2020)
Bread (gluten-free)	*Nannochloropsis gaditana* *Chlamydomonas* sp.	1–3%	• High protein contents • Improvement in techno-functional and nutritional properties	Khemiri et al. (2020)

(Continued)

TABLE 2.3 (Continued)

Applications and techno-functional properties of microalgae biomass and proteins in formulation food products for human nutrition

Applications	Microalgae species	Microalgae incorporated (%)	Findings/observations	References
Bread	*Chlorella vulgaris*	1%	• Improved nutritional and technological properties	Nunes et al. (2020)
Wheat tortillas	*Nannochloropsis* sp. *Tetraselmis* sp.	0.5–3%	• Increased antioxidant capacity • Greater functional properties • Increased protein and fat contents • Nutritional properties • Good sensory properties	Hernández-López et al. (2021)
Corn extrudates	*Arthrospira platensis* (Spirulina)	2–8%	• Enhanced nutritional properties • Higher protein content • Coloring agent	Tańska et al. (2017)
Beef patties	*Chlorella* *Arthrospira platensis* (Spirulina)	1%	• Highest values for the total amino acids content • Improved nutritional content • Acceptable sensory	Žugčić et al. (2018)
Chicken rotti	*Spirulina* *Chlorella* sp.		• Increase of pH and seaweed caused a decrease in some color parameters • Increased protein content • Highest total amino acid content and best ratio of essential and non-essential amino acids	Parniakov et al. (2018)

(Continued)

TABLE 2.3 (Continued)

Applications and techno-functional properties of microalgae biomass and proteins in formulation food products for human nutrition

Applications	Microalgae species	Microalgae incorporated (%)	Findings/observations	References
Sausages fermented "chorizo"	Chlorella Arthrospira platensis (Spirulina)	3%	• Increase protein content • The ratio of essential and nonessential amino acids of Chlorella protein-enriched "chorizo" was higher than the rest of the samples • Color change	Thirumdas et al. (2018)
Fresh pork sausages	Chlorella Arthrospira platensis (Spirulina)	1%	• Increase in the total amino acid content and the ratio of essential/non-essential amino acids	Marti-Quijal et al. (2019)
Meat analogues	Arthrospira platensis (Spirulina)	15.30 and 50%	• Increased total phenolic content, total flavonoid content, and Trolox equivalent antioxidant activity • Dark coloration of meat analogues	Palanisamy et al. (2019)
Meat analogues	Auxenochlorella protothecoides	5–50%	• Protein source • Coloring agent • Nutritional benefits (specifically vitamin E, B1, B3, B6) • Desirable fibrous formation	Caporgno et al. (2020)
Meat analogues	Haematococcus pluvialis	10–40%	• Positive fiber formation • Favorable red coloring • Significantly improved physical properties • Nutritional properties	Xia et al. (2022)
Fermented milks	Arthrospira platensis (Spirulina)	3%	• Increased the essential amino acid and vitamin contents • Bioactive components	Varga et al. (2002)

(Continued)

TABLE 2.3 (Continued)

Applications and techno-functional properties of microalgae biomass and proteins in formulation food products for human nutrition

Applications	Microalgae species	Microalgae incorporated (%)	Findings/observations	References
Yogurt	*Chlorella* sp.	0.25%	• Greater sensory preference • Nutritional properties	Cho et al. (2004)
Cheese	*Chlorella* sp.	0.5 and 1%	• Enhanced in nutritional and functional qualities	Jeon et al. (2006)
Yogurt	*Chlorella vulgaris* *Arthrospira platensis*	0.25–1%	• Increased the viability of probiotic bacteria • Enhanced functional and nutritional properties	Beheshtipour et al. (2012)
Cheese analogue	*Chlorella vulgaris*	1–3%	• Improvement nutritional profile (high protein, carbohydrates, and fiber content) • Enhanced functional properties • Sensory acceptance	Mohamed et al. (2013)
Mayonnaise	*Chlorella vulgaris*	10–20%	• Effect positive on the sensory characteristics • Improvement of properties nutritional, rheological	Abd El-Razik and Mohamed (2013)
Probiotic cheese	*Arthrospira platensis* (*Spirulina*)	0.3–0.8%	• Increased protein and iron content • Color change • Stimulation the growth and survival of probiotic bacteria	Mazinani et al. (2016)
Ayran (yogurt and water)	*Arthrospira platensis* (*Spirulina*)	0.25–1%	• Positive growth of probiotic bacteria • Higher nutritional content • Higher protein content	Çelekli et al. (2019)
Milk and soy fermented	*Arthrospira platensis* (Spirulina)	0.25% and 0.50%	• Increased fermentation performances • Functional effect • Technological advantage	Martelli et al. (2020)

crumb, specific volume, and organoleptic quality (crumb status, taste, and odor). Furthermore, with the inclusion of *D. salina* 10%, a significant increase in the biological value of the proteins was observed (from 66.2 to 76.8%), improving the final product's nutritional and biological value (Tertychnaya et al., 2020). In gluten-free bread, the microalgae *Nannochloropsis gaditana* and *Chlamydomonas* sp. biomass as an additional ingredient contributed to the dough rheology, texture quality and nutritional properties. The authors found that microalgae, mainly *Chlamydomonas* sp., in addition to representing an alternative protein source, showed a significant increase in iron, calcium, and linolenic acid (18:3 ω3) contents. In addition, supplemented breads scored higher for the sensory parameters evaluated. At a technical and functional level, mainly due to the high protein content, the incorporation of microalgae had a positive structuring effect on the texture of gluten-free bread (Khemiri et al., 2020).

Pasta formulations with the addition of microalgae are also a promising field of research and development. These products, consumed worldwide, are potential candidates for incorporating microalgae-based ingredients to improve their structural characteristics and protein supply and also contribute to a more attractive color. Fradique et al. (2010) and El-Baz et al. (2017) explored the incorporation of the microalgae *Chlorella vulgaris, Spirulina maxima*, and *Dunaliella salina* in pasta formulations. They reported that the final products had higher protein content than control formulations. Furthermore, the studies point out that adding microalgae increased quality parameters, phytochemicals content, stable coloring during cooking, and sensory quality.

In dairy formulations, the inclusion of microalgae is a common practice. These foods receive much attention as a good source of protein and probiotics (dairy bacteria). According to literature data, it is observed that microalgae, when incorporated into these products, can strongly contribute to these two biases, even collaborating with the viability of probiotic microorganisms in the development and the gastrointestinal tract (Beheshtipour et al., 2012). The studies by Cho et al. (2004) and Varga et al. (2002) demonstrated that the inclusion of microalgae biomass in yogurts could promote greater sensory preference and nutritional properties, including increased essential amino acid and vitamin contents and bioactive components. Specifically, Beheshtipour et al. (2012) added *Arthrospira platensis* and *Chlorella vulgaris* in yogurt formulations to evaluate the action of these microorganisms' biomass on the viability of probiotic bacteria and their biochemical properties. This study's results revealed that adding microalgae to the product provided the enrichment of milk by proteins and significantly increased and sustained the viability of *Bifidobacterium lactis* and *Lactobacillus acidophilus* at the end of fermentation and during cold storage. In cheese formulations, microalgae have also been added. Mohamed et al. (2013) added *Chlorella* sp. in formulations of cheese analogue at concentrations ranging from 1 to 3% and found that microalgae biomass can contribute to an improved nutritional profile (high protein, carbohydrates, and fiber content), enhanced functional properties, and sensory acceptance.

Currently, protein products or ingredients developed with alternative sources to animal protein are hot points in the area of food formulations. Recently, in a study by Xia et al. (2022), *Haematococcus pluvialis* residue, ranging in concentrations from 10 to 40%, was used for preparing pea protein-based meat analogues by high moisture extrusion. In this study, the authors demonstrated that incorporating biomass residue could significantly improve the protein content and texture properties by contributing to the formation of layers and fibrous structures. The coloring properties of the products were also positive, presenting a reddish color similar to traditional meat products. The addition of *Chlorella* and *Arthrospira platensis* (Spirulina) microalgae biomass in fresh pork sausages was applied to evaluate the influence of the microalgae protein profile on the physicochemical properties and amino acid profile of the products. Other sources of plant proteins and cow's milk whey proteins were also explored in the study (Marti-Quijal et al., 2019).

The data from this research showed that the incorporation of microalgae proteins significantly changed texture of the final products, providing products with desirable textures better than the control, that was soy protein. In addition, a significant change was observed in the color parameters due to the presence of green and blue-green pigments. Specifically, the addition of *A. platensis* proteins increased the essential/non-essential amino acid ratio and the total amino acid content. Overall, microalgae provided favorable nutritional and functional properties compared to other protein sources. Similar studies, with the incorporation of other microalgae species, were reported by Parniakov et al. (2018), Thirumdas et al. (2018), Žugčić et al. (2018), Palanisamy et al. (2019), and Caporgno et al. (2020).

Another relevant category to supply microalgae-based ingredients is the feed sector for ruminant, non-ruminant animals, and aquaculture, mainly aiming at the contribution of macronutrients such as proteins. In addition to having a high protein content with an essential amino acid composition in terms of animal nutrition, a plus of the use of microalgae biomass in these products is the simultaneous addition of bioactive molecules with important biological and functional activities such as structures with antioxidant capabilities (Dineshbabu et al., 2019). Table 2.4 shows the main effects and nutritional and functional properties of supplementation of microalgae biomass with moderate/high protein content in the diet of different animals.

TABLE 2.4

The main effects, nutritional, and functional properties of supplementation of microalgae biomass with moderate/high protein content in the diet of different animals

Animal category	Main effect/benefits
Non-ruminant	
Aquaculture	• Reduced inflammation index
	• Improvement in PUFA, sterol content, pigments as carotenoids
	• Replace fish oil as a PUFA source
	• Live microalgae a preferable feed
	• Increase survival rate and weight gain
	• Enhance coloration
	• Increase disease resistance
	• Increase protein content

(Continued)

TABLE 2.4 (*Continued*)

The main effects, nutritional, and functional properties of supplementation of microalgae biomass with moderate/high protein content in the diet of different animals

Animal category	Main effect/benefits
Swine	• Improved gut health and decrease in intestinal disorders
	• Reduction in plasma and liver triglycerides
	• Improves ovarian development in sows
	• Increase in PUFA content of meat and ham
	• Improves male fertility, increase in sperm mobility, count, and volume
	• Decrease in meat cholesterol
	• Enhanced immune response
	• Increase nutritional value
Poultry	• Decreased cholesterol in meat and egg yolk
	• Eggs with low cholesterol fortified with ômega 3 (EPA,DHA)
	• Superior carotenoid content
	• Increase weight gain
	• Increase in PUFA
	• Increase in egg shell thickness
	• Improved immunity against diseases
	• Increase in egg laying percentage
	• Increase coloration content of egg yolk
	• Intense meat color
Ruminant	
Cattle	• Increase in body weight
	• Reduction in rumen fermentation
	• Increase in milk yield and high ômega 3 (EPA, DHA) and low fat
	• Replace soybean meal as a protein source
	• Alleviates heat stress induced inflammation
	• Enhanced PUFA content of meat and milk
	• Improved reproductive health
	• Reduction in metabolic nitrogen loss
	• Reduction in oxidative stress
Goat/Lamb	• Increased in body weight
	• Meat enriched with essential amino acids and fatty acids (ômega 3, 6)
	• Low cholesterol and improvement of the appearance and the functionality
	• Decrease in meat cholesterol
	• Modulation of gut microflora
	• Increase in PUFA in meat and milk
	• Increased oxidative stability of meat

Source: Adapted from: Yaakob et al. (2015), Dineshbabu et al. (2019), Amorim et al. (2021a), Nagarajan et al. (2021), Saadaoui et al. (2021).

Several beneficial health effects of animals can be obtained by adding microalgae biomass to the feed. Many of these effects may be associated with the protein fraction present in biomass, such as increased body weight and the enrichment of meat, eggs, and milk with essential amino acids. Also, for ruminant feed for example, the use of microalgae biomass as a substitute for soybean meal as a protein source is considered an important alternative (Yaakob et al., 2015; Saadaoui et al., 2021). In summary, microalgae proteins can strongly contribute as an essential nutrient for the growth and increase of farmed animal meat, higher umami and flavor, and softer and tender meat quality (Nagarajan et al., 2021).

In addition to the important applicability of microalgae proteins in food and feed formulations, these compounds have shown important contributions as additives, especially in terms of foaming, emulsifying, gelling, water, and fat absorption (Kumar et al., 2022). However, despite major discoveries regarding the possible technological/functional properties of proteins for these purposes, it is believed that many properties still remain largely unknown. Table 2.5 reports the main technological/functional properties of some proteins and microalgae hydrolysates in different applications.

Gelation is an important technological feature for new protein sources, mainly because this property offers a unique texture to many products. These materials are relevant components in the composition of jellies, desserts, yogurts, jujubes, salad dressings, and also for drug delivery (Banerjee & Bhattacharya, 2012). In this regard, Grossmann et al. (2019) demonstrated that the microalgae protein extract *Chlorella sorokiniana* has important gelling properties. The formation of a stable gel network, with desirable elasticity properties, was achieved with a minimum protein concentration (~9.9%), pH of 5.6, at 61°C. Chronakis (2001), Batista et al. (2012), and Shkolnikov Lozober et al. (2021) also demonstrated the excellent gelling properties of protein fractions from different microalgae species, including the genus *Arthrospira* and *Haematococcus*.

In contrast, microalgae proteins also demonstrate foaming ability. This property is closely related to the amount of interfacial area that can be created during foaming. On the other hand, foam stability refers to the protein's ability to stabilize air bubbles against gravitational stress (Benelhadj et al., 2016). Proteins isolated from microalgae *Chlorella pyrenoidosa, Arthospira platensis*, and *Nannochloropsis oceanica* were evaluated for their physicochemical properties and the effect of pH on their functional properties (Chen et al., 2019). Better oil absorption capacity, higher solubility, foaming capacity, and stability were found for the species *C. pyrenoidosa* and *A. platensis*. The results also indicated that pH and physicochemical properties are determining factors in the functional properties, with the minimum conditions necessary for foam formation and other properties observed at pH close to the isoelectric point. In previous and subsequent work, it has also been shown that foam properties are highly influenced by changes in pH (Schwenzfeier et al., 2013; Benelhadj et al., 2016; Taragjini et al., 2022).

The ability of an emulsifying agent is also very relevant, as it reflects the ability of the compound or sample to rapidly adsorb at oil-water interfaces during the formation of an emulsion, preventing flocculation and coalescence (Benelhadj et al., 2016). Bleakley and Hayes (2021) recently found that protein extracts from *Isochrysis galbana* and *Arthrospira platensis* (Spirulina) showed activities and stabilities emulsifying. The emulsion activity of protein extracts from microalgae was proven, with rates from 10.0 to 22.4%, depending on the oily medium used. However, these values were relatively low when comparing defatted flaxseed protein and whey protein isolate. On the other

TABLE 2.5

Technological/functional properties of microalgae biomass and proteins in different applications

Applications	Microalgae species	Microalgae incorporated (%)	Findings/observations	References
Gel properties	Arthrospira platensis (Spirulina)	2–14%	• Elastic gels in 90°C • Critical concentration between 1.5% and 3% protein content • Higher solubility at pH 9	Chronakis (2001)
Gel properties	Arthrospira platensis (Spirulina) Haematococcus pluvialis	0.75% biomass	• Temperature (70–90°C, 5 min) resulted in more structured gels, while the effect of time (5–30 min, at 90°C) was less pronounced • Haematococcus gels highly structured, less dependent on gel setting conditions • Spirulina gels are less structured than gels without microalgae, but this is overcome at lower heating/cooling rates.	Batista et al. (2012)
Gel properties	Chlorella sorokiniana	9.9% protein solution	• Temperature of 61°C • pH of 5.6 resulted in highest gel strength • weak-gel with tan $\delta > 0.1$ • Changing the pH or NaCl negatively affected the gel formation	Grossmann et al. (2019)
Gel properties	Arthrospira platensis (Spirulina)	5%	• Gelation at pH 6.5 • At pH 4 and 5 SPC didn't form gel due to insufficient protein solubility • Phycobiliproteins – the main proteins in SPC, aren't responsible for gel formation. • High-pressure homogenization improved gelatinization (50 MPa)	Shkolnikov Lozober et al. (2021)

(Continued)

TABLE 2.5 (*Continued*)

Technological/functional properties of microalgae biomass and proteins in different applications

Applications	Microalgae species	Microalgae incorporated (%)	Findings/observations	References
Emulsification properties	Chlorella vulgaris	0.1%–2%	• Emulsion stability > 75% and emulsification capacity – 3090 mL oil/g protein • Comparable and higher than soy and casein • Best emulsification capacity in Ph 7	Ursu et al. (2014)
Emulsifying and foaming capacity	Nannochloropsis gaditana	2% protein extract (PE) and proteinhydrolysate (HE)	• The emulsifying capacity for PE > 50% and the emulsion was stable after 24 hours • PH increasing its solubility rate compared to the native protein • PH did not form emulsions	Medina et al. (2015)
Emulsification properties	Chlorella sorokiniana Phaeodactylum tricornutum	1–3.7% protein extract	• *Chlorella* emulsions stabilized with 1% protein • *Phaeodactylum* emulsions stabilized with 3.7% protein. • Salt stabilities up to 500 mM of NaCl levels • *Chlorella* emulsions stability (pH ≥ 5) > *Phaeodactylum* (highest stability at pH 7)	Ebert et al. (2019)
Emulsification properties	*Arthrospira platensis* (Spirulina)	2–5% % protein extract	• Optimal solution 4% protein, O/W weight ratio of 30/70, and pH 7 (zeta potential of –43.83 mV) • Stable emulsions for 30-days • No antimicrobial contamination, microalgae protein may also act as an antimicrobial agent	Silva et al. (2022)
Foaming properties	Tetraselmis sp.	0.01%–0.1% protein solution	• Foam stability was superior to those of whey protein isolate and egg white albumin • Foams stable at pH range 5–7 • Microalgae protein contribute to its superior foam stability	Schwenzfeier et al. (2013)

(*Continued*)

TABLE 2.5 (Continued)

Technological/functional properties of microalgae biomass and proteins in different applications

Applications	Microalgae species	Microalgae incorporated (%)	Findings/observations	References
Foaming properties	*Chlorella pyrenoidosa* Arthospira platensis Nannochloropsis oceanica	71.56–93.63% protein extract	• *Arthospira* > *Chlorella* > *Nannochloropsis* higher solubility, oil absorption capacity, foaming capacity and foam stability • Functional properties of microalgae protein are dependent on their physicochemical characteristics • Optimal conditions near the isoelectric point	Chen et al. (2019)
Foaming, emulsifying activities and stabilities	*Isochrysis galbana* Arthrospira platensis (Spirulina)	71.9–85.5% protein extract	• Better foam stability • Microalgae protein extracts inhibited angiotensin-converting enzyme-I (ACE-I) and renin • *Spiruilina* excellent oil-holding capacity, more than whey and flaxseed proteins • Solubility at pH 10–12	Bleakley and Hayes (2021)
Foaming, emulsifying and films properties	*Arthrospira platensis* (Spirulina)	0.5–20%	• Functional properties were significantly affected by the medium pH. • pH 10 was the most suitable condition to produce emulsions, foams, and films • pH 3 was minimum • The minimal gelation concentration was 12% protein	Benelhadj et al. (2016)
Foaming, emulsifying and gelling properties	*Arthrospira platensis* (Spirulina)	84.17% protein extract	• Techno-functional properties were pH dependent • Improved functionality was at pH far from the isoelectric pH (3.9) • Highest solubility in pH = 10 • Foaming (182.3%) capacities were higher, at pH 2.0 • Emulsifying (80.6%) capacities were higher, at pH 6.0 • Lowest concentration needed to form a gel was 60 g/L at pH = 6.0	Taragini et al. (2022)

hand, in terms of emulsion stability, the *Spirulina* protein extract showed excellent values (85.91%). In addition, the authors also proved the solubility properties and other functional characteristics, including water and oil holding capacity, water activities and foaming for these protein extracts. Positive emulsion-forming effects were also found for water-soluble proteins extracted from the microalgae *Chlorella sorokiniana* and *Phaeodactylum tricornutum* (Ebert et al., 2019). In this study, the concentration of 1.0% by weight of *C. sorokiniana* protein was sufficient to manufacture an emulsion with a monomodal distribution, high stability in relation to pH changes and remained stable for 7 days. In contrast, 3.7% of *P. tricornutum* proteins were required to achieve emulsion stability and stability at pH 7.

Overall, it can be seen that all the technological functional properties of microalgae proteins, exceptionally the emulsifying and foaming capabilities are positively correlated with the protein solubility. In addition, pH is a determining factor for the success of all parameters determined so far. Further clarifications can be found in the bibliography mentioned in Table 2.5.

Other relevant functions and properties are being frequently associated with these essential metabolites. Literature review amply demonstrates that microalgae proteins have properties that go beyond their nutritional value as macronutrients. Some of these protein compounds, as well as peptides, and amino acids have nutraceutical value and biological functional properties prominent to human health, which drive the application of these biopolymers as functional commodities (Jacob-Lopes et al., 2019). A summary of the biological functional properties of bioactive peptides from microalgae is presented in Table 2.6.

Microalgae peptides, units that constitute the microalgae proteins, stand out with their biological functional properties. These molecules are formed by an inherent sequence of amino acids, ranging from 2 to 20 amino acids present in their composition. It is considered that the composition of the peptides, i.e., the number of amino acids present in their chain and the sequence of these compounds, are the determining factors for the success of their biological activity (Kumar et al., 2022). As seen in the literature and presented in Table 2.6, efforts have been made to isolate these biocompounds and evaluate their possible bioactive functions. The research carried out so far shows that therapeutic microalgae peptides are promising to prevent or treat various diseases, including cancer. Many biological activities have already been demonstrated for these important biomolecules, from antioxidant activity that can modulate various biological activities to anti-inflammatory, antihypertensive, immune-modulatory, cardio-protective properties, anti-diabetic activity, among other important physiological functions and therapeutic properties that are being investigated.

Polypeptides from *Arthrospira platensis* (*Spirulina*) demonstrated antiproliferative activities on five cancer cells (HepG-2, A549, MCF-7, SGC-7901, and HT-29), with the IC50 values between <31.25 and 336.57 µg/mL, in addition to low toxicity or stimulatory activity in normal cells. Specifically, YGFVMPRSGLWFR peptide, in addition to demonstrating inhibitory activities in cancer cells, its best activity was observed in A549 cancer cells (IC50 values 104.05 µg/mL) (Wang & Zhang, 2016). In a previous study, purified peptide MGRY derived from *Pavlova lutheri* was shown to attenuate oxidative stress and melanogenesis in B16F10 melanoma cells, reducing the

TABLE 2.6
Bioactive peptides from microalgae and their functional properties

Microalgae species	Peptide production	Peptide sequence	Biological activity	References
Arthrospira platensis (Spirulina)	Trypsin, α-chymotrypsin and Pepsin	LDAVNR MMLDF	Anti-inflammatory, anti-atherosclerotic and adhesion inhibition	Vo and Kim (2013)
Chlorella Pyrenoidosa (CP)	Papain, Trypsin, and Alcalase	Polypeptides of CP anti-tumor polypeptide	Anticancerous and anti-inflammatory	Wang and Zhang (2013)
Nannochloropsis oculata	Pepsin, trypsin, α-chymotrypsin, papain, alcalase, and neutrase	P1:GMNNLTP P2:LEQ	Anti-hypertensive activity	Samarakoon et al. (2013)
Chlorella vulgaris	Pepsin	VECYGPNRPQF	Antioxidant and Anti-cancerous	Fan et al. (2014)
Pavlova lutheri	Fermentation by the yeast Hansenula polymorpha	MGRY	Antioxidant, skin lightning by inhibition of melanogenesis	Oh et al. (2015)
Chlamydomonas reinhardtii	Pepsin, trypsin and chymotrypsin	VLPVP	Antihypertensive and bioactivity	Ochoa-Méndez et al. (2016)
Arthrospira platensis (Spirulina)	Trypsin, alcalase and papain	YGFVMPRSGLWFR	Anticancer activity	Wang and Zhang (2016)
Arthrospira platensis (Spirulina)	Trypsin and chymotrypsin	Not identified	Antibacterial and anticancer activity	Sadeghi et al. (2018)
Scenedesmus obliquus	Pepsin, trypsin or papain	Not identified	Antioxidant and anti-viral activity	Afify et al. (2018)
Tetradesmus obliquus	Cell disruption by bead milling, enzymatic digestion with Alcalase	P1:WPRGYFL P2:GPDRPKFLGPF P3:WYGPDRPKFL P4:SDWDRF	Antihypertensive and antioxidant activity	Montone et al. (2018)
Arthrospira platensis (Spirulina)	Ultrasound-assisted subcritical water extraction	P1. GVPMPNK P2. RNPFVFATKKTVAAR P3. KRSEKAAWSR	Anti-diabetic activity: inhibition of α-glucosidase, α-amylase, and dipeptidyl peptidase-4	Hu et al. (2019)

(Continued)

TABLE 2.6 (Continued)
Bioactive peptides from microalgae and their functional properties

Microalgae species	Peptide production	Peptide sequence	Biological activity	References
Bellerochea, malleus	Trypsin, papain, pepsin, and flavourzyme	Not identified	Antioxidant and antihypertensive activity	Barkia et al. (2019b)
Arthrospira maxima	Pepsin and subtilisin A	Not identified	Antioxidant, anti-inflammatory, anti-bacterial, anti-aging activities, and anti-collagenase	Montalvo et al. (2019)
Dunaliella salina	Pepsin and trypsin	P1. ILTAAIGGL P2. IITPGGL P3. AAPSTVL P4. TVAPPGA	Antioxidant capacity.	Xia et al. (2019)
Dunaliella salina	Pepsin and trypsin	ALVFQAQH (P32)	Antiosteopenic activity	Chen et al. (2021)
Chlorella sorokiniana	α-chymotrypsin, pepsin, trypsin, and thermolysin	P1. LSSATSAPS P2. AGLYGHPQTQEE	Antioxidant activities	Safitri et al. (2022)
Arthrospira platensis (Spirulina)	Tirosinase, arginine, leucine, phenylalanine, valine	P1. AFGRFR P2. MAACLR P3. RCLNGRL P4. RYVTYAVF P5. SPSWY P6. GRF P7. AADQRGKDKCARDIGY	Melanogenesis regulation	Kose and Oncel (2022)
Schizochytrium limacinum	Pepsin, α-chymotrypsin and trypsin	Not identified	Antioxidant activities	Moaveni et al. (2022)

generation of intracellular ROS. The authors also found that the peptide decreased melanogenesis-related proteins; transcription factor associated with TYR protein expressions, activation of extracellular regulated signal kinase, and Microphthalmia-associated transcription factor (Oh et al., 2015).

Recently, Chen et al. (2021), in a study with ovariectomized (OVX) Sprague-Dawley (SD) rats, found that *Dunaliella salina* derived peptide, with ALVFQAQH sequence, protects from bone loss, thus demonstrating antiosteopenic activity. It was also possible to observe that these biomolecules can significantly improve the marker of bone formation, bone microarchitecture, and bone mineral density. The authors concluded that these effects could be attributed to the high proportion of antioxidant-related amino acids in their sequence and their physicochemical properties, including low molecular weight and high hydrophobicity. Furthermore, the ability to stimulate the expression of antioxidant-related genes (Cu/Zn SOD and CAT) and osteoblast-specific gene mRNA (Runx2α, ALP, OC) may have contributed significantly to these potential effects.

Hu et al. (2019) conducted an in vitro study using the insulin-resistant HepG2 cell model to evaluate the inhibitory effect of *Arthrospira platensis (Spirulina)* peptides on three enzymes (α-amylase, α-glucosidase, and dipeptidyl peptidase-4 (DPP-IV). According to the results of this study, peptides were validated with the anti-diabetic activities in vitro; however, the compound with sequence LRSELAAWSR displayed the best activities on α-glucosidase (IC50 = 134.2 µg/mL) and DPP-IV (IC50 = 167.3 µg/mL), but medium activity on α-amylase (IC50 = 313.6 µg/mL).

Among the multivariate biological properties of microalgae peptides, these biomolecules have been shown to exhibit strong bioactivities as antioxidant compounds. A study conducted by Moaveni et al. (2022) evaluated the antioxidant properties of peptides extracted from *Schizochytrium limacinum* by different mechanisms. The methods of 2,2-azino-bis (3-ethylbenzthiazoline)-6-sulfonic acid radical cation (ABTS+) scavenging activity and reducing power activity, and direct free radical scavenging activity such as 2,2-diphenyl-1-picrylhydrazyl Radical (DPPH•) scavenging, as well as the protective effect of purified peptide against oxidation of low-density lipoprotein (LDL) were employed. The synthesized peptides (protein hydrolysates) showed radical scavenging capacity, significantly inhibited LDL oxidation, and showed greater reducing power. In particular, the peptide fraction F_3 (molecular weight cutoff = 5–10 kDa) peptide fraction showed a potent antioxidant capacity for all evaluated methods. Other relevant research exploring the antioxidant capacity of peptides from the microalgae *Chlorella vulgaris, Scenedesmus obliquus, Tetradesmus obliquus, Bellerochea, malleus, Dunaliella salina,* and *Chlorella sorokiniana* also showed potential results (see Table 2.6).

The vital capacity for synthesizing enzymes by microalgae has also been reported. Protease, galactosidases, amylases, phytases, lipase, laccases, cellulases, carbonic anhydrase, and antioxidant enzymes are examples used in the food industry (Brasil et al., 2017). Furthermore, the fraction of antifreeze proteins (AFPs) is found in some microalgae species (Jung et al., 2014). Despite their unknown bioactive capacity, these compounds are considered necessary in the food industry due to their ability for frozen food preservation. Examples of its application are in meat products where they were able to reduce the damage caused by freezing, loss of drip, and loss of proteins and improve the juiciness of the meat after it thawed (Xiang et al., 2020). The application of some microalgae proteins in formulations of anti-obesity functional foods is also explored since they stimulate the production of the cholecystokinin hormone, responsible for regulating appetite suppression (Patias et al., 2018).

Although microalgae proteins may have several applications, ranging from nutritional, technological-functional, and functional components with biological properties, most microalgae proteins are currently explored as food supplements. In addition, they have different forms of consumption and can be found as part of the biomass, whole biomass, or in other forms such as pastes, extracted powders, tablets, capsules, or flakes designed for daily use, and also as food additives, and functional food products (Lafarga, 2019; Kratzer & Murkovic, 2021). As with any emerging technology, much more research is needed to fully understand all potential applications of microalgae proteins, both nutritionally, functionally, and technologically, as well as their benefits and possible regulatory and safety concerns.

2.5 CHALLENGES AND FUTURE TRENDS

Although microalgae are considered perfect and sustainable protein sources, some challenges cannot be ignored, including obstacles in large-scale production to meet nutrient demand, biomass digestibility, and meeting sensory requirements demanded by the consumer.

In terms of the scalability of biomass production, specific approaches need configuration adjustments to achieve excellence (productivity/sustainability/low cost). For example, cultivating *A. platensis* and *C. vulgaris* to obtain protein powder with autotrophic and heterotrophic approaches can potentially have more impacts than classical animal protein production due to high energy consumption and required resources such as lots of light or glucose for maintenance of the microalgae metabolism (Smetana & Heinz, 2019). Many processes developed to produce microalgae biomass still face low efficiency in growing on a larger scale. When this productivity blockage is overcome, microalgae can effectively support the global demand for sustainable food (Tang et al., 2020).

Another issue when growing microalgae is the yield of compounds, including proteins and other essential elements. If cultivation environment conditions are not manipulated, microalgae might not be as efficient as required to generate enough ingredients for the growing food industry demand (Mulders et al., 2014; González et al., 2015). Variations in light intensity, oxygen, salinity, temperature, and radiation can modulate the nutrient profile of microalgae biomass (Mata et al., 2010; Maltsev et al., 2021). Additionally, different taxonomic groups of microalgae can present diverse optimal growth conditions, which are related to biomass composition goals (Maltsev et al., 2021). These variations and flexibility could be beneficial when cultivating microalgae for specific food applications requiring a rigid biomass composition pattern.

Among future trends, meat analogues are a promising application of microalgae biomass. *Spirulina* powder, widely available on the market, can contain more than 50% of dry weight protein. However, strategies that improve absorption by the human body need to be studied. According to a study by Niccolai et al. (2019), different microalgae species can present diverse digestibility. Among other species, marine microalgae tend to be harder to digest. Also, *T. suecica*, *P. tricornutum*, and *P. purpureum* were the species with the lowest digestibility scores. This fact might be explained by their exopolysaccharide cellular wall, which blocks direct contact with digestive enzymes.

On the other hand, *C. vulgaris, C. sorokiniana* IAM-C212 and *A. platensis* tended to be more susceptible to enzyme action, being more digestible. Also, soy, a common element in vegan products, is one of the closest ingredients in terms of protein bio-availability when compared to microalgae (Fu et al., 2021). In this manner, although several applications have been ended tested (see Section 4), there are still challenges in understanding the behavior of microalgae biomass digestibility and improving it.

In addition to digestibility, three enormous challenges in consuming microalgae biomass are exotics odor, taste, and color. Many microalgae species own distinct smells and flavor profiles, presenting fishy and earthy notes that are undesirable in most product formulations (Pérez-Lloréns, 2020). Additionally, these unique sensory characteristics are probably connected to the formation of chemical compounds during carotenoid and lipid oxidation, among others. The fishy taste and aroma are related to the presence of sulfuric compounds. Among them are dimethyl disulfide, dimethyl trisulfide, methional, diketones, α-ionone, and β-ionone. Green aromas might also be present in microalgae, which occurs due to the content of typical aldehydes, such as 2,4-alkadienals and 2,4,6-alkatrienals (Van Durme et al., 2013).

Some techniques can be explored to overcome the unusual odor and taste, consequently improving the acceptability of microalgae-based products. Ethanol pretreatment (Qazi et al., 2021), altering the harvesting period, and providing a source of nitrogen to microalgae are alternatives to decrease the strength of the undesirable notes in the biomass (Hosoglu, 2018).

Color is also a challenging characteristic for some kinds of microalgae-based products. Microalgae varieties are photosynthetic organisms, which means they own pigments, including chlorophyll. Chlorophyll is responsible for the green color that challenges the food industry when developing algae-based products. Because of this color tone, customers that are used to a traditional look in food could reject algae-based items (Fiorentini et al., 2020).

It is important to mention that some species can go from red to brown tones, keeping their photosynthetic properties (Fiorentini et al., 2020). Moreover, heterotrophically cultivated *Auxenochlorella protothecoides* is not green. Instead, this specie has a light tone of yellow, which is much more attractive for many food applications, such as meat analogs (Caporgno et al., 2020). Thus, species with more neutral colors could work as simple alternatives.

Genetic engineering arises as another answer to this issue. Techniques that influence enzymatic conversion can direct the conversion of carotenoids and chlorophylls, turning them into targeted pigments (Anila et al., 2016). However, it is valid to consider changes in protein compositions. Then, there is a need for research and resources when utilizing microalgae in formulations, and it could lead to more expensive algae-based products.

Overcoming these sensory challenges is critical as consumer behavior is a potential blockage to the use of microalgae. For instance, when producing a new meat analog based on microalgae biomass, specific attributes need to be present, as consumers expect. Measuring most of the microalgae-based products' sensory characteristics is impossible by simple tests. It requires extensive sensory analysis and volunteers (Bejaei et al., 2021).

As shown in Section 4, adding microalgae to foods incorporates bioactive compounds and increases food products' nutritional and techno-functional characteristics (Tang et al., 2020). Thus, in addition to the protein appeal of microalgae biomass, the food industry can also take advantage of other qualities applicable in the manufacture of natural products, such as example, the antioxidant capacity, the ability to improve the shelf life of products, pigmentation properties, aggregation, emulsification, and gelation (Souza et al., 2019; Grossmann et al., 2020).

This versatility of microalgae biomass combined with the growing demand for plant-based vegan and sustainable proteins is reasons algae-based products are a promising and possible reality for large-scale supply chains. Thus, as a versatile powder added to many formulations, microalgae biomass is expected to be a first-rate product in this market (Fu et al., 2021).

From a future perspective, the maximum use of all the compounds and properties of the biomass, in addition to protein enrichment, can be an alibi for expanding industrial use and effective use of this valuable, sustainable food resource. Thus, a line of research is designed that deserves to be carefully explored.

2.6 FINAL CONSIDERATIONS

Finally, microalgae proteins are considered promising and sustainable sources of dietary protein. This is partly a consequence of existing production possibilities (especially multi-product biorefineries approaches) that cause less impact than conventional protein sources. In addition, microalgae biomass is an excellent source of essential amino acids and techno-functional properties that allow application in various industrial sectors. Although it is still necessary to overcome obstacles that challenge the effective application of this bioproduct (scalability, improved digestibility, and sensory acceptance), future trends and research efforts in the area outline a promising scenario.

REFERENCES

Abd El-Razik, M. M., & Mohamed, A. G. (2013). Utilization of acid casein curd enriched with *Chlorella vulgaris* biomass as substitute of egg in mayonnaise production. *World Applied Sciences Journal*, 26(7), 917–925. https://doi.org/10.5829/idosi.wasj.2013.26.07.13523

Adhikari, S., Schop, M., de Boer, I. J., & Huppertz, T. (2022). Protein quality in perspective: A review of protein quality metrics and their applications. *Nutrients*, 14(5), 947. https://doi.org/10.3390/nu14050947

Afify, A. E. M. M. R., el Baroty, G. S., el Baz, F. K., Abd El Baky, H. H., & Murad, S. A. (2018). *Scenedesmus obliquus*: Antioxidant and antiviral activity of proteins hydrolyzed by three enzymes. *Journal of Genetic Engineering and Biotechnology*, 16(2), 399–408. https://doi.org/10.1016/j.jgeb.2018.01.002

Aiking, H., & de Boer, J. (2020). The next protein transition. *Trends in Food Science & Technology*, 105, 515–522. https://doi.org/10.1016/j.tifs.2018.07.008

Amorim, M. L., Soares, J., Coimbra, J. S. dos R., Leite, M. de O., Albino, L. F. T., & Martins, M. A. (2021a). Microalgae proteins: Production, separation, isolation, quantification, and application in food and feed. *Critical Reviews in Food Science and Nutrition*, 61(12), 1976–2002. https://doi.org/10.1080/10408398.2020.1768046

Amorim, M. L., Soares, J., Vieira, B. B., de Oliveira Leite, M., Rocha, D. N., Aleixo, P. E., ... & Martins, M. A. (2021b). Pilot-scale biorefining of Scenedesmus obliquus for the production of lipids and proteins. *Separation and Purification Technology*, 270, 118775. https://doi.org/10.1016/j.seppur.2021.118775

Anila, N., Simon, D. P., Chandrashekar, A., Ravishankar, G. A., & Sarada, R. (2016). Metabolic engineering of *Dunaliella salina* for production of ketocarotenoids. *Photosynthesis Research*, 127(3), 321–333.

Banerjee, S., & Bhattacharya, S. (2012). Food gels: Gelling process and new applications. *Critical Reviews in Food Science and Nutrition*, 52(4), 334–346. https://doi.org/10.1080/10408398.2010.500234

Banu, J. R., Preethi, Kavitha S., Gunasekaran, M., Kumar, G. (2020). Microalgae based biorefinery promoting circular bioeconomy-techno economic and life-cycle analysis. *Bioresource Technology*, 302, 2020. https://doi.org/10.1016/j.biortech.2020.122822

Barkia, I., Saari, N., & Manning, S. R. (2019a). Microalgae for high-value products towards human health and nutrition. *Marine Drugs*, 17(5), 304. https://doi.org/10.3390/md17050304

Barkia, I., Al-Haj, L., Abdul Hamid, A., Zakaria, M., Saari, N., & Zadjali, F. (2019b). Indigenous marine diatoms as novel sources of bioactive peptides with antihypertensive and antioxidant properties. *International Journal of Food Science and Technology*, 54(5), 1514–1522. https://doi.org/10.1111/ijfs.14006

Batista, A. P., Gouveia, L., Nunes, M. C., Franco, J. M., & Raymundo, A. (2011). Microalgae biomass as a novel functional ingredient in mixed gel systems. In *Gums and Stabilisers for the Food Industry*, Vol. 14 (pp. 487–494). https://doi.org/10.1016/j.foodhyd.2010.09.018

Batista, A. P., Niccolai, A., Fradinho, P., Fragoso, S., Bursic, I., Rodolfi, L., Biondi, N., Tredici, M. R., Sousa, I., & Raymundo, A. (2017). Microalgae biomass as an alternative ingredient in cookies: Sensory, physical and chemical properties, antioxidant activity and in vitro digestibility. *Algal Research*, 26(June), 161–171. https://doi.org/10.1016/j.algal.2017.07.017

Batista, A. P., Nunes, M. C., Fradinho, P., Gouveia, L., Sousa, I., Raymundo, A., & Franco, J. M. (2012). Novel foods with microalgal ingredients – Effect of gel setting conditions on the linear viscoelasticity of *Spirulina* and *Haematococcus* gels. *Journal of Food Engineering*, 110(2), 182–189. https://doi.org/10.1016/j.jfoodeng.2011.05.044

Becker, E. W. (2007). Micro-algae as a source of protein. *Biotechnology Advances*, 25(2), 207–210. https://doi.org/10.1016/j.biotechadv.2006.11.002

Beheshtipour, H., Mortazavian, A. M., Haratian, P., & Khosravi-Darani, K. (2012). Effects of *Chlorella vulgaris* and *Arthrospira platensis* addition on viability of probiotic bacteria in yogurt and its biochemical properties. *European Food Research and Technology*, 235(4), 719–728. https://doi.org/10.1007/s00217-012-1798-4

Bejaei, M., Stanich, K., & Cliff, M. A. (2021). Modelling and classification of apple textural attributes using sensory, instrumental and compositional analyses. *Foods*, 10(2), 384. https://doi.org/10.3390/foods10020384

Benelhadj, S., Gharsallaoui, A., Degraeve, P., Attia, H., & Ghorbel, D. (2016). Effect of pH on the functional properties of *Arthrospira* (*Spirulina*) *platensis* protein isolate. *Food Chemistry*, 194, 1056–1063. https://doi.org/10.1016/j.foodchem.2015.08.133

Bleakley, S., & Hayes, M. (2021). Functional and bioactive properties of protein extracts generated from *Spirulina platensis* and *isochrysis galbana* T-Iso. *Applied Sciences* 11(9). https://doi.org/10.3390/app11093964

Brasil, B. dos S. A. F., de Siqueira, F. G., Salum, T. F. C., Zanette, C. M., & Spier, M. R. (2017). Microalgae and cyanobacteria as enzyme biofactories. *Algal Research*, 25(February), 76–89. https://doi.org/10.1016/j.algal.2017.04.035

Brown, M. R. (1991). The amino-acid and sugar composition of 16 species of microalgae used in mariculture. *Journal of Experimental Marine Biology and Ecology*, 145(1), 79–99. https://doi.org/10.1016/0022-0981(91)90007-J.

Caporgno, M. P., & Mathys, A. (2018). Trends in microalgae incorporation into innovative food products with potential health benefits. *Frontiers in Nutrition*, 5(July), 1–10. https://doi. org/10.3389/fnut.2018.00058

Caporgno, M. P., Böcker, L., Müssner, C., Stirnemann, E., Haberkorn, I., Adelmann, H., Handschin, S., Windhab, E. J., & Mathys, A. (2020). Extruded meat analogues based on yellow, heterotrophically cultivated *Auxenochlorella protothecoides* microalgae. *Innovative Food Science and Emerging Technologies*, 59, 102275. https://doi.org/10.1016/ j.ifset.2019.102275

Çelekli, A., Alslibi, Z. A., & Bozkurt, H. üseyin. (2019). Influence of incorporated *Spirulina platensis* on the growth of microflora and physicochemical properties of ayran as a functional food. *Algal Research*, 44, 101710. https://doi.org/10.1016/j.algal.2019.101710

Chen, C., & Chaudhary, A. (2019). Dietary change scenarios and implications for environmental, nutrition, human health and economic dimensions of food sustainability. *Nutrients*, 11(4), 856. https://doi.org/10.3390/nu11040856

Chen, Y., Chen, J., Chang, C., Chen, J., Cao, F., Zhao, J., Zheng, Y., & Zhu, J. (2019). Physicochemical and functional properties of proteins extracted from three microalgal species. *Food Hydrocolloids*, 96(886), 510–517. https://doi.org/10.1016/j.foodhyd. 2019.05.025

Chen, Y., Chen, J., Zheng, Y., Yu, H., Zhao, J., Chen, J., & Zhu, J. (2021). *Dunaliella salina*-derived peptide protects from bone loss: Isolation, purification and identification. *LWT*, 137(886), 110437. https://doi.org/10.1016/j.lwt.2020.110437

Chisti, Y. (2007). Biodiesel from microalgae. *Biotechnology Advances*, 25(3), 294–306. https:// doi.org/10.1016/j.biotechadv.2007.02

Cho, E. J., Nam, E. S., & Park, S. I. (2004). Keeping quality and sensory properties of drinkable yoghurt with added *Chlorella* extract. *Korean Journal of Food Nutrition*, 17(2), 128–132. http://www.koreascience.or.kr/article/ArticleFullRecord.jsp?cn=HGSPB1_ 2004_v17n2_128

Chronakis, I. S. (2001). Gelation of edible blue-green algae protein isolate (*Spirulina platensis* strain Pacifica): Thermal transitions, rheological properties, and molecular forces involved. *Journal of Agricultural and Food Chemistry*, 49(2), 888–898. https://doi. org/10.1021/jf0005059

Dineshbabu, G., Goswami, G., Kumar, R., Sinha, A., & Das, D. (2019). Microalgae–nutritious, sustainable aqua- and animal feed source. *Journal of Functional Foods*, 62, 103545. https://doi.org/10.1016/j.jff.2019.103545

Ebert, S., Grossmann, L., Hinrichs, J., & Weiss, J. (2019). Emulsifying properties of water-soluble proteins extracted from the microalgae: *Chlorella sorokiniana* and *Phaeodactylum tricornutum*. *Food and Function*, 10(2), 754–764. https://doi.org/10.1039/c8fo02197j

El-Baz, F. K., Abdo, S. M., & Hussein, A. M. S. (2017). Microalgae *Dunaliella salina* for use as Food Supplement to improve Pasta Quality. *International Journal of Pharmaceutical Sciences Review*, 46(2), 45–51. http://globalresearchonline.net/journalcontents/v46-2/10.pdf

Fan, X., Bai, L., Zhu, L., Yang, L., & Zhang, X. (2014). Marine algae-derived bioactive peptides for human nutrition and health. *Journal of Agricultural and Food Chemistry*, 62(38), 9211–9222. https://doi.org/10.1021/jf502420h

Fiorentini, M., Kinchla, A. J., & Nolden, A. A. (2020). Role of sensory evaluation in consumer acceptance of plant-based meat analogs and meat extenders: A scoping review. *Foods*, 9(9), 1334. https://doi.org/10.3390/foods9091334

Food and Agriculture Organization of the United Nations. (1991). Protein quality evaluation: Report of the Joint FAO/WHO Expert Consultation-Bethesda, MD, 4–8 December 1989. *Food and Agriculture Organization of the United Nations.* https://citeseerx.ist.psu.edu/viewdoc/download?doi=10.1.1.1021.9259&rep=rep1&type=pdf

Fradique, Monica, Batista, A. P., Nunes, M. C., Gouveia, L., Bandarra, N. M., & Raymundo, A. (2010). Incorporation of *Chlorella vulgaris* and *Spirulina maxima* biomass in pasta products. Part 1: Preparation and evaluation. *Journal of the Science of Food and Agriculture,* 90(10), 1656–1664. https://doi.org/10.1002/jsfa.3999

Fu, Y., Chen, T., Chen, S. H. Y., Liu, B., Sun, P., Sun, H., & Chen, F. (2021). The potentials and challenges of using microalgae as an ingredient to produce meat analogues. *Trends in Food Science & Technology,* 112, 188–200. https://doi.org/10.1016/j.tifs.2021.03.050

Gohara-Beirigo, A. K., Matsudo, M. C., Cezare-Gomes, E. A., Carvalho, J. C. M., Danesi, E. D. G. (2022). Microalgae trends toward functional staple food incorporation: Sustainable alternative for human health improvement. *Trends in Food Science & Technology,* 185–199. https://doi.org/10.1016/j.tifs.2022.04.030

González, L. E., Díaz, G. C., Aranda, D. A. G., Cruz, Y. R., & Fortes, M. M. (2015). Biodiesel production based in microalgae: A biorefinery approach. *Natural Science,* 7(07), 358. https://doi.org/10.4236/ns.2015.77039

Gouveia, L., Batista, A. P., Miranda, A., Empis, J., & Raymundo, A. (2007). *Chlorella vulgaris* biomass used as colouring source in traditional butter cookies. *Innovative Food Science & Emerging Technologies,* 8(3), 433–436. https://doi.org/10.1016/j.ifset.2007.03.026

Grossmann, L., Ebert, S., Hinrichs, J., & Weiss, J. (2018). Effect of precipitation, lyophilization, and organic solvent extraction on preparation of protein-rich powders from the microalgae *Chlorella protothecoides. Algal research,* 29, 266–276. https://doi.org/10.1016/j.algal.2017.11.019

Grossmann, L., Hinrichs, J., Goff, H. D., & Weiss, J. (2019). Heat-induced gel formation of a protein-rich extract from the microalga *Chlorella sorokiniana. Innovative Food Science and Emerging Technologies,* 56, 102176. https://doi.org/10.1016/j.ifset.2019.06.001

Grossmann, L., Hinrichs, J., & Weiss, J. (2020). Cultivation and downstream processing of microalgae and cyanobacteria to generate protein-based technofunctional food ingredients. *Critical Reviews in Food Science and Nutrition,* 60(17), 2961–2989. https://doi.org/10.1080/10408398.2019.1672137

Hernández-López, I., Benavente Valdés, J. R., Castellari, M., Aguiló-Aguayo, I., Morillas-España, A., Sánchez-Zurano, A., Acién-Fernández, F. G., & Lafarga, T. (2021). Utilisation of the marine microalgae *Nannochloropsis* sp. and *Tetraselmis* sp. as innovative ingredients in the formulation of wheat tortillas. *Algal Research,* 58. https://doi.org/10.1016/j.algal.2021.102361

Hosoglu, M. I. (2018). Aroma characterization of five microalgae species using solid-phase microextraction and gas chromatography–mass spectrometry/olfactometry. *Food Chemistry,* 240, 1210–1218. https://doi.org/10.1016/j.foodchem.2017.08.052

Hu, J., Nagarajan, D., Zhang, Q., Chang, J. S., Lee, D. J. (2018). Heterotrophic cultivation of microalgae for pigment production: A review. *Biotechnology Advances,* 36(1), 54–67. https://doi.org/10.1016/j.biotechadv.2017.09.009

Hu, S., Fan, X., Qi, P., & Zhang, X. (2019). Identification of anti-diabetes peptides from *Spirulina* platensis. *Journal of Functional Foods,* 56, 333–341. https://doi.org/10.1016/j.jff.2019.03.024

Jacob-Lopes, E., Maroneze, M. M., Deprá, M. C., Sartori, R. B., Dias, R. R., & Zepka, L. Q. (2019). Bioactive food compounds from microalgae: An innovative framework on industrial biorefineries. *Current Opinion in Food Science,* 25, 1–7. https://doi.org/10.1016/j.cofs.2018.12.003

Jacob-Lopes, E., Santos, A. B., Severo, I. A., Deprá, M. C., Maroneze, M. M., Zepka, L. Q. (2020). Dual production of bioenergy in heterotrophic cultures of cyanobacteria: Process performance, carbon balance, biofuel quality and sustainability metrics. *Biomass and Bioenergy*, 142. https://doi.org/10.1016/j.biombioe.2020.105756

Jeon, J. K. (2006). Effect of Chlorella addition on the quality of processed cheese. *Journal of the Korean Society of Food Science and Nutrition*, 35(3), 373–377. https://doi.org/10.3746/jkfn.2006.35.3.373

Jung, W., Gwak, Y., Davies, P. L., Kim, H. J., & Jin, E. S. (2014). Isolation and characterization of antifreeze proteins from the antarctic marine microalga *Pyramimonas gelidicola*. *Marine Biotechnology*, 16(5), 502–512. https://doi.org/10.1007/s10126-014-9567-y

Kell, S. (2022). Editorial foreword for "environment, development and sustainability" journal. *Environment, Development and Sustainability*, 24(3), 2983–2985. https://doi.org/10.1007/s10668-021-02070-z

Khemiri, S., Khelifi, N., Nunes, M. C., Ferreira, A., Gouveia, L., Smaali, I., & Raymundo, A. (2020). Microalgae biomass as an additional ingredient of gluten-free bread: Dough rheology, texture quality and nutritional properties. *Algal Research*, 50. https://doi.org/10.1016/j.algal.2020.101998

Klamczynska, B., & Mooney, W. D. (2017). Heterotrophic microalgae: A scalable and sustainable protein source. In: Nadathur, S. R., Wanasundara, J. P. D., Scanlin, L. (eds), *Sustainable Protein Sources* (pp. 327–339). Academic Press, Cambridge.

Kose, A., & Oncel, S. S. (2022). Design of melanogenesis regulatory peptides derived from phycocyanin of the microalgae *Spirulina platensis*. *Peptides*, 152, 170783. https://doi.org/10.1016/j.peptides.2022.170783

Kratzer, R., & Murkovic, M. (2021). Food ingredients and nutraceuticals from microalgae: Main product classes and biotechnological production. *Foods*, 10(7). https://doi.org/10.3390/foods10071626

Kumar, B. R., Mathimani, T., Sudhakar, M. P., Rajendran, K., Nizami, A. S., Brindhadevi, K., & Pugazhendhi, A. (2021). A state-of-the-art review on the cultivation of algae for energy and other valuable products: Application, challenges, and opportunities. *Renewable and Sustainable Energy Reviews*, 138, 110649. https://doi.org/10.1016/j.rser.2020.110649

Kumar, R., Hegde, A. S., Sharma, K., Parmar, P., & Srivatsan, V. (2022). Microalgae as a sustainable source of edible proteins and bioactive peptides – current trends and future prospects. *Food Research International*, 157, 111338. https://doi.org/10.1016/j.foodres.2022.111338

Lafarga, T. (2019). Effect of microalgal biomass incorporation into foods: Nutritional and sensorial attributes of the end products. *Algal Research*, 41, 101566. https://doi.org/10.1016/j.algal.2019.101566

Lafarga, T., Sánchez-Zurano, A., Villaró, S., Morillas-España, A., & Acién, G. (2021). Industrial production of *Spirulina* as a protein source for bioactive peptide generation. *Trends in Food Science & Technology*, 116, 176–185. https://doi.org/10.1016/j.tifs.2021.07.018

Lam, T. P., Lee, T. M., Chen, C. Y., & Chang, J. S. (2018). Strategies to control biological contaminants during microalgal cultivation in open ponds. *Bioresource Technology*, 252, 180–187. https://doi.org/10.1016/j.biortech.2017.12.088

Lopez-Ridaura, S., Werf, H. V. D., Paillat, J. M., & Le Bris, B. (2009). Environmental evaluation of transfer and treatment of excess pig slurry by life cycle assessment. *Journal of Environmental Management*, 90(2), 1296–1304. https://doi.org/10.1016/j.jenvman.2008.07.008

Lucas, B. F., de Morais, M. G., Santos, T. D., & Costa, J. A. V. (2017). Effect of *Spirulina* addition on the physicochemical and structural properties of extruded snacks. *Food Science and Technology* (Brazil), 37(Special Issue), 16–23. https://doi.org/10.1590/1678-457X.06217

Lum, K. K., Kim, J., & Lei, X. G. (2013). Dual potential of microalgae as a sustainable biofuel feedstock and animal feed. *Journal of Animal Science and Biotechnology*, 4(1), 1–7. https://doi.org/10.1186/2049-1891-4-53

Maltsev, Y., Maltseva, K., Kulikovskiy, M., & Maltseva, S. (2021). Influence of light conditions on microalgae growth and content of lipids, carotenoids, and fatty acid composition. *Biology*, 10(10), 1060. https://doi.org/10.3390/biology10101060

Markou, G., Chentir, I., & Tzovenis, I. (2021). Microalgae and cyanobacteria as food: Legislative and safety aspects. In: *Cultured Microalgae for the Food Industry* (pp. 249–264). Academic Press. https://doi.org/10.1016/B978-0-12-821080-2.00003-4

Martelli, F., Alinovi, M., Bernini, V., Gatti, M., & Bancalari, E. (2020). *Arthrospira platensis* as natural fermentation booster for milk and soy fermented beverages. *Foods*, 9(3). https://doi.org/10.3390/foods9030350

Marti-Quijal, F. J., Zamuz, S., Tomašević, I., Gómez, B., Rocchetti, G., Lucini, L., Remize, F., Barba, F. J., & Lorenzo, J. M. (2019). Influence of different sources of vegetable, whey and microalgae proteins on the physicochemical properties and amino acid profile of fresh pork sausages. *LWT*, 110(October 2018), 316–323. https://doi.org/10.1016/j.lwt.2019.04.097

Mata, T. M., Martins, A. A., & Caetano, N. S. (2010). Microalgae for biodiesel production and other applications: A review. *Renewable and Sustainable Energy Reviews*, 14(1), 217–232. https://doi.org/10.1016/j.rser.2009.07.020

Mazinani, S., Fadaei, V., & Khosravi-Darani, K. (2016). Impact of *Spirulina platensis* on physicochemical properties and viability of *Lactobacillus acidophilus* of probiotic UF Feta Cheese. *Journal of Food Processing and Preservation*, 40(6), 1318–1324. https://doi.org/10.1111/jfpp.12717

Medina, C., Rubilar, M., Shene, C., Torres, S., & Verdugo, M. (2015). Protein fractions with techno-functional and antioxidant properties from *Nannochloropsis gaditana* microalgal biomass. *Journal of Biobased Materials and Bioenergy*, 9(4), 417–425. https://doi.org/10.1166/jbmb.2015.1534

Mehariya, S., Goswami, R. K., Karthikeysan, O. P., & Verma, P. (2021). Microalgae for high-value products: A way towards green nutraceutical and pharmaceutical compounds. *Chemosphere*, 280, 130553. https://doi.org/10.1016/j.chemosphere.2021.130553

Mishra, S., Roy, M., Mohanty, K. (2020). Microalgal bioenergy production under zero-waste biorefinery approach: Recent advances and future perspectives. *Bioresource Technology*, 292. https://doi.org/10.1016/j.biortech.2019.122008

Moaveni, S., Salami, M., Khodadadi, M., McDougall, M., & Emam-Djomeh, Z. (2022). Investigation of *S.limacinum* microalgae digestibility and production of antioxidant bioactive peptides. *LWT*, 154(September 2021), 112468. https://doi.org/10.1016/j.lwt.2021.112468

Mohamed, A. G., Abo-El-Khair, B. E., & Shalaby, S. M. (2013). Quality of novel healthy processed cheese analogue enhanced with marine microalgae *Chlorella vulgaris* biomass. *World Applied Sciences Journal*, 23(7), 914–925. https://doi.org/10.5829/idosi.wasj.2013.23.07.13122

Mok, W. K., Tan, Y. X., & Chen, W. N. (2020). Technology innovations for food security in Singapore: A case study of future food systems for an increasingly natural resource-scarce world. *Trends in Food Science & Technology*, 102, 155–168. https://doi.org/10.1016/j.tifs.2020.06.013

Montalvo, G. E. B., Thomaz-Soccol, V., Vandenberghe, L. P. S., Carvalho, J. C., Faulds, C. B., Bertrand, E., Prado, M. R. M., Bonatto, S. J. R., & Soccol, C. R. (2019). *Arthrospira maxima* OF15 biomass cultivation at laboratory and pilot scale from sugarcane vinasse for potential biological new peptides production. *Bioresource Technology*, 273(October 2018), 103–113. https://doi.org/10.1016/j.biortech.2018.10.081

Montone, C. M., Capriotti, A. L., Cavaliere, C., la Barbera, G., Piovesana, S., Zenezini Chiozzi, R., & Laganà, A. (2018). Peptidomic strategy for purification and identification of potential ACE-inhibitory and antioxidant peptides in *Tetradesmus obliquus* microalgae. *Analytical and Bioanalytical Chemistry*, 410(15), 3573–3586. https://doi.org/10.1007/s00216-018-0925-x

Mulders, K. J., Lamers, P. P., Martens, D. E., & Wijffels, R. H. (2014). Phototrophic pigment production with microalgae: Biological constraints and opportunities. *Journal of Phycology*, 50(2), 229–242. https://doi.org/10.1111/jpy.12173

Nagarajan, D., Varjani, S., Lee, D. J., & Chang, J. S. (2021). Sustainable aquaculture and animal feed from microalgae – nutritive value and techno-functional components. *Renewable and Sustainable Energy Reviews*, 150, 111549. https://doi.org/10.1016/j.rser.2021.111549

Nascimento, T. C., Cazarin, C. B., Marostica Jr, M. R., Risso, É. M., Amaya-Farfan, J., Grimaldi, R., … & Zepka, L. Q. (2019). Microalgae biomass intake positively modulates serum lipid profile and antioxidant status. *Journal of Functional Foods*, 58, 11–20. https://doi.org/10.1016/j.jff.2019.04.047

Nass, P. P., Nascimento, T. C., Fernandes, A. S., Caetano, P. A., de Rosso, V. V., Jacob-Lopes, E., & Zepka, L. Q. (2022). Guidance for formulating ingredients/products from *Chlorella vulgaris* and *Arthrospira platensis* considering carotenoid and chlorophyll bioaccessibility and cellular uptake. *Food Research International*, 111469. https://doi.org/10.1016/j.foodres.2022.111469

Niccolai, A., Zittelli, G. C., Rodolfi, L., Biondi, N., & Tredici, M. R. (2019). Microalgae of interest as food source: Biochemical composition and digestibility. *Algal Research*, 42, 101617. https://doi.org/10.1016/j.algal.2019.101617

Nunes, M. C., Graça, C., Vlaisavljević, S., Tenreiro, A., Sousa, I., & Raymundo, A. (2020). Microalgal cell disruption: Effect on the bioactivity and rheology of wheat bread. *Algal Research*, 45. https://doi.org/10.1016/j.algal.2019.101749

Ochoa-Méndez, C. E., Lara-Hernández, I., González, L. M., Aguirre-Bañuelos, P., Ibarra-Barajas, M., Castro-Moreno, P., González-Ortega, O., & Soria-Guerra, R. E. (2016). Bioactivity of an antihypertensive peptide expressed in *Chlamydomonas reinhardtii*. *Journal of Biotechnology*, 240, 76–84. https://doi.org/10.1016/j.jbiotec.2016.11.001

Oh, G. W., Ko, S. C., Heo, S. Y., Nguyen, V. T., Kim, G., Jang, C. H., Park, W. S., Choi, I. W., Qian, Z. J., & Jung, W. K. (2015). A novel peptide purified from the fermented microalga *Pavlova lutheri* attenuates oxidative stress and melanogenesis in B16F10 melanoma cells. *Process Biochemistry*, 50(8), 1318–1326. https://doi.org/10.1016/j.procbio.2015.05.007

Oliveira, A. P. F., & Bragotto, A. P. A. (2022). Microalgae-based products: Food and public health. *Future Foods*, 6, 100157. https://doi.org/10.1016/j.fufo.2022.100157

Palanisamy, M., Töpfl, S., Berger, R. G., & Hertel, C. (2019). Physico-chemical and nutritional properties of meat analogues based on *Spirulina*/lupin protein mixtures. *European Food Research and Technology*, 245(9), 1889–1898. https://doi.org/10.1007/s00217-019-03298-w

Parniakov, O., Toepfl, S., Barba, F. J., Granato, D., Zamuz, S., Galvez, F., & Lorenzo, J. M. (2018). Impact of the soy protein replacement by legumes and algae based proteins on the quality of chicken rotti. *Journal of Food Science and Technology*, 55(7), 2552–2559. https://doi.org/10.1007/s13197-018-3175-1

Patias, L. D., Maroneze, M. M., Siqueira, S. F., de Menezes, C. R., Zepka, L. Q., & Jacob-Lopes, E. (2018). Single-cell protein as a source of biologically active ingredients for the formulation of antiobesity foods. In: *Alternative and Replacement Foods* (Vol. 17). Elsevier Inc. https://doi.org/10.1016/B978-0-12-811446-9.00011-3

Pereira, H., Silva, J., Santos, T., Gangadhar, K. N., Raposo, A., Nunes, C., … & Varela, J. (2019). Nutritional potential and toxicological evaluation of *Tetraselmis* sp. CTP4 microalgal biomass produced in industrial photobioreactors. *Molecules*, 24(17), 3192. https://doi.org/10.3390/molecules24173192

Perez-Garcia, O., Escalante, F. M. E., de-Bashan, L. E., & Bashan, Y. (2011). Heterotrophic cultures of microalgae: Metabolism and potential products. *Water Research*, 45(1), 11–36. https://doi.org/10.1016/j.watres.2010.08.037

Pérez-Lloréns, J. L. (2020). Microalgae: From staple foodstuff to avant-garde cuisine. *International Journal of Gastronomy and Food Science*, 21, 100221. https://doi. org/10.1016/j.ijgfs.2020.100221

Qazi, W. M., Ballance, S., Uhlen, A. K., Kousoulaki, K., Haugen, J. E., & Rieder, A. (2021). Protein enrichment of wheat bread with the marine green microalgae *Tetraselmis chuii*– Impact on dough rheology and bread quality. *LWT*, 143, 111115

Saadaoui, I., Rasheed, R., Aguilar, A., Cherif, M., al Jabri, H., Sayadi, S., & Manning, S. R. (2021). Microalgal-based feed: Promising alternative feedstocks for livestock and poultry production. *Journal of Animal Science and Biotechnology*, 12(1), 1–15. https://doi. org/10.1186/s40104-021-00593-z

Sadeghi, S., Jalili, H., Siadat, S. O. R., & Sedighi, M. (2018). Anticancer and antibacterial properties in peptide fractions from hydrolyzed *Spirulina* protein. *Journal of Agricultural Science and Technology*, 20, 673–683.

Safafar, H., Uldall Nørregaard, P., Ljubic, A., Møller, P., Løvstad Holdt, S., & Jacobsen, C. (2016). Enhancement of protein and pigment content in two *Chlorella* species cultivated on industrial process water. *Journal of Marine Science and Engineering*, 4(4), 84. https://doi.org/10.3390/jmse4040084

Safitri, N. M., Hsu, J. L., & Violando, W. A. (2022). Antioxidant activity from the Enzymatic Hydrolysates of *Chlorella sorokiniana* and its potential peptides identification in combination with molecular docking analysis. *Turkish Journal of Fisheries and Aquatic Sciences*, 22(4). https://doi.org/10.4194/TRJFAS20316

Samarakoon, K. W., O-Nam, K., Ko, J. Y., Lee, J. H., Kang, M. C., Kim, D., Lee, J. B., Lee, J. S., & Jeon, Y. J. (2013). Purification and identification of novel angiotensin-I converting enzyme (ACE) inhibitory peptides from cultured marine microalgae (*Nannochloropsis oculata*) protein hydrolysate. *Journal of Applied Phycology*, 25(5), 1595–1606. https:// doi.org/10.1007/s10811-013-9994-6

Schwenzfeier, A., Lech, F., Wierenga, P. A., Eppink, M. H. M., & Gruppen, H. (2013). Foam properties of algae soluble protein isolate: Effect of pH and ionic strength. *Food Hydrocolloids*, 33(1), 111–117. https://doi.org/10.1016/j.foodhyd.2013.03.002

Severo, I. A., Siqueira, S. F., Deprá, M. C., Maroneze, M. M., Zepka, L. Q., & Jacob-Lopes, E. (2019). Biodiesel facilities: What can we address to make biorefineries commercially competitive? *Renewable and Sustainable Energy Reviews*, 112. https://doi.org/10.1016/j. rser.2019.06.020

Shkolnikov Lozober, H., Okun, Z., & Shpigelman, A. (2021). The impact of high-pressure homogenization on thermal gelation of Arthrospira platensis (*Spirulina*) protein concentrate. *Innovative Food Science and Emerging Technologies*, 74(October), 102857. https://doi.org/10.1016/j.ifset.2021.102857

Silva, S. C., Almeida, T., Colucci, G., Santamaria-Echart, A., Manrique, Y. A., Dias, M. M., Barros, L., Fernandes, Â., Colla, E., & Barreiro, M. F. (2022). Spirulina (*Arthrospira platensis*) protein-rich extract as a natural emulsifier for oil-in-water emulsions: Optimization through a sequential experimental design strategy. *Colloids and Surfaces A: Physicochemical and Engineering Aspects*, 648(May). https://doi.org/10.1016/j.colsurfa.2022.129264

Smetana, S., & Heinz, V. (2019). Sustainability of meat substitutes: Plant analogues, microalgae, insects and cultured meat. In ICoMST 2019 65th international congress of meat science and technology. Book of Abstracts, Tuesday, 06 August, 2019, Processed Meat & Meat Analogues, Page 323.

Souza, M. P., Hoeltz, M., Gressler, P. D., Benitez, L. B., & Schneider, R. (2019). Potential of microalgal bioproducts: General perspectives and main challenges. *Waste and Biomass Valorization*, 10(8), 2139–2156.

Tang, D. Y. Y., Khoo, K. S., Chew, K. W., Tao, Y., Ho, S.-H., & Show, P. L. (2020). Potential utilization of bioproducts from microalgae for the quality enhancement of natural products. *Bioresource Technology*, 304, 122997. https://doi.org/10.1016/j.biortech.2020.122997

Tańska, M., Konopka, I., & Ruszkowska, M. (2017). Sensory, physico-chemical and water sorption properties of corn extrudates enriched with *Spirulina*. *Plant Foods for Human Nutrition*, 72(3), 250–257. https://doi.org/10.1007/s11130-017-0628-z

Taragjini, E., Ciardi, M., Musari, E., Villaró, S., Morillas-España, A., Alarcón, F. J., & Lafarga, T. (2022). Pilot-scale production of *A. platensis*: Protein isolation following an ultrasound-assisted strategy and assessment of techno-functional properties. *Food and Bioprocess Technology*, 15(6), 1299–1310. https://doi.org/10.1007/s11947-022-02789-1

Tertychnaya, T. N., Manzhesov, V. I., Andrianov, E. A., & Yakovleva, S. F. (2020). New aspects of application of microalgae Dunaliella Salina in the formula of enriched bread. *IOP Conference Series: Earth and Environmental Science*, 422(1). https://doi.org/10.1088/1755-1315/422/1/012021

Thirumdas, R., Brnčić, M., Brnčić, S. R., Barba, F. J., Gálvez, F., Zamuz, S., Lacomba, R., & Lorenzo, J. M. (2018). Evaluating the impact of vegetal and microalgae protein sources on proximate composition, amino acid profile, and physicochemical properties of fermented Spanish "chorizo" sausages. *Journal of Food Processing and Preservation*, 42(11), 1–8. https://doi.org/10.1111/jfpp.13817

Tibbetts, S. M., Milley, J. E., & Lall, S. P. (2015). Chemical composition and nutritional properties of freshwater and marine microalgal biomass cultured in photobioreactors. *Journal of Applied Phycology*, 27(3), 1109–1119. https://doi.org/10.1007/s10811-014-0428-x

Tulli, F., Chini Zittelli, G., Giorgi, G., Poli, B. M., Tibaldi, E., & Tredici, M. R. (2012). Effect of the inclusion of dried *Tetraselmis suecica* on growth, feed utilization, and fillet composition of European sea bass juveniles fed organic diets. *Journal of Aquatic Food Product Technology*, 21(3), 188–197. https://doi.org/10.1080/10498850.2012.664803

Ursu, A. V., Marcati, A., Sayd, T., Sante-Lhoutellier, V., Djelveh, G., & Michaud, P. (2014). Extraction, fractionation and functional properties of proteins from the microalgae *Chlorella vulgaris*. *Bioresource Technology*, 157, 134–139. https://doi.org/10.1016/j.biortech.2014.01.071

USDA. U.S. Department of Agriculture (https://fdc.nal.usda.gov/).

Van Durme, J., Goiris, K., De Winne, A., De Cooman, L., & Muylaert, K. (2013). Evaluation of the volatile composition and sensory properties of five species of microalgae. *Journal of Agricultural and Food Chemistry*, 61(46), 10881–10890. https://doi.org/10.1021/jf403112k

Varga, L., Szigeti, J., Kovács, R., Földes, T., & Buti, S. (2002). Influence of a *Spirulina platensis* biomass on the microflora of fermented ABT milks during storage (R1). *Journal of Dairy Science*, 85(5), 1031–1038. https://doi.org/10.3168/jds.S0022-0302(02)74163-5

Veerabadhran, M., Natesan, S., MubarakAli, D., Xu, S., & Yang, F. (2021). Using different cultivation strategies and methods for the production of microalgal biomass as a raw material for the generation of bioproducts. *Chemosphere*, 285, 131436. https://doi.org/10.1016/j.chemosphere.2021.131436

Vo, T. S., & Kim, S. K. (2013). Down-regulation of histamine-induced endothelial cell activation as potential anti-atherosclerotic activity of peptides from *Spirulina maxima*. *European Journal of Pharmaceutical Sciences*, 50(2), 198–207. https://doi.org/10.1016/j.ejps.2013.07.001

Wang, X., & Zhang, X. (2013). Separation, antitumor activities, and encapsulation of polypeptide from *Chlorella pyrenoidosa*. *Biotechnology Progress*, 29(3), 681–687. https://doi.org/10.1002/btpr.1725

Wang, Y., Tibbetts, S. M., & McGinn, P. J. (2021). Microalgae as sources of high-quality protein for human food and protein supplements. *Foods*, 10(12), 1–18. https://doi.org/10.3390/foods10123002

Wang, Z., & Zhang, X. (2016). Inhibitory effects of small molecular peptides from *Spirulina (Arthrospira) platensis* on cancer cell growth. *Food and Function*, 7(2), 781–788. https://doi.org/10.1039/c5fo01186h

Xia, E., Zhai, L., Huang, Z., Liang, H., Yang, H., Song, G., Li, W., & Tang, H. (2019). Optimization and identification of antioxidant peptide from underutilized *Dunaliella salina* protein: Extraction, in vitro gastrointestinal digestion, and fractionation. *BioMed Research International*. https://doi.org/10.1155/2019/6424651

Xia, S., Xue, Y., Xue, C., Jiang, X., & Li, J. (2022). Structural and rheological properties of meat analogues from *Haematococcus pluvialis* residue-pea protein by high moisture extrusion. *LWT*, 154(5), 112756. https://doi.org/10.1016/j.lwt.2021.112756

Xiang, H., Yang, X., Ke, L., & Hu, Y. (2020). The properties, biotechnologies, and applications of antifreeze proteins. *International Journal of Biological Macromolecules*, 153, 661–675. https://doi.org/10.1016/j.ijbiomac.2020.03.040

Yaakob, Z., Ali, E., Zainal, A., Mohamad, M., & Takriff, M. S. (2015). Yaakob_2013_*biomolecules from microalgae for animal feed and aquaculture*.pdf. 1–10.

Žugčić, T., Abdelkebir, R., Barba, F. J., Rezek-Jambrak, A., Gálvez, F., Zamuz, S., Granato, D., & Lorenzo, J. M. (2018). Effects of pulses and microalgal proteins on quality traits of beef patties. *Journal of Food Science and Technology*, 55(11), 4544–4553. https://doi.org/10.1007/s13197-018-33

3 Microbial production of acetic acid

Bellamkonda Ramesh, Srinivasan Kameswaran,
Gopi Krishna Pitchika, Gujjala Sudhakara,
B. Swapna, and M. Ramakrishna

CONTENTS

3.1 INTRODUCTION

Acetic acid is a common platform chemical that has been employed as a food preservative in the past. It's a corrosive carboxylic acid with a sour taste and a pungent odour that's clear, colourless, and corrosive (Wang et al. 2013a,b). Acetic acid can be made synthetically or through bacterial fermentation. Methanol, acetaldehyde,

DOI: 10.1201/9781003394600-4

butane, and ethylene are the most common petroleum-derived stocks used in synthetic manufacturing. Biological routes account for only about 10% of global production now (Ragsdale and Pierce 2008). It is still significant for vinegar manufacturing since many food purity rules throughout the world require that vinegar used in meals be of biological origin. Vinegar, which is mostly a diluted acetic acid solution of 4–6%, is used as a taste ingredient and also as a food preservative (Ho et al. 2017).

The global acetic acid market was 13 million tonnes in 2015, and it is expected to grow to around 16 million tonnes by 2020. According to industry data from 2015, the market price of acetic acid in various countries ranges between US$1200 and US$1600 per tonne (Shah and Consultant 2014; Christodoulou and Velasquez-Orta 2016; Pal and Nayak 2016). The creation of vinyl acetate, acetic anhydride, acetate esters, monochloroacetic acid, and as a solvent in the manufacturing of dimethyl terephthalate and terephthalic acid are the most common end uses of acetic acid. Latex emulsion resins are made from vinyl acetate and are utilised in paints, adhesives, paper coatings, and textile treatment. Cellulose acetate textile fibres, cigarette filter tow, and cellulose plastics are all made with acetic anhydride (Christodoulou and Velasquez-Orta 2016).

3.2 MICROORGANISMS THAT PRODUCE ACETIC ACID

Acetogenesis, or the excretion of acetate into the environment, is caused by the requirement to replenish the NAD^+ depleted by glycolysis and to recycle the coenzyme A (CoASH) required in the conversion of pyruvate to acetyl-CoA. Acetogenesis occurs when the TCA cycle is disrupted or when the carbon input into cells exceeds the capacity of the cells. As a result, during mixed-acid fermentation, acetate is excreted anaerobically. Under aerobic conditions, the Crabtree effect occurs when growth on extra glucose (or other highly digestible carbon sources) suppresses respiration. Acetic acid bacteria use the aerobic pathway to produce acetic acid, whereas acetogens use the anaerobic pathway to produce acetic acid (Wolfe 2005).

3.2.1 AEROBICS

Acetic acid bacteria (AAB) are well-known for their capacity to oxidise ethanol as a substrate into acetic acid in both neutral and acidic environments under aerobic conditions, making them valuable to the vinegar business (Sengun and Karabiyikli 2011; Li et al. 2015). AAB are polymorphic, with Gram-negative cells that are ellipsoidal to rod-shaped, straight or slightly curved, and 0.6–0.8 m long, and can be found individually, in pairs, or in chains. With polar or peritrichous flagella, there exist nonmotile and motile forms. They are obligate aerobic organisms that create pigments and cellulose, respectively (Raspor and Goranovic 2008; Sakurai et al. 2012). They are also involved in the creation of commercially essential fine chemicals, as well as the production of fermented foods, either beneficially (chocolate goods, coffee, vinegar, and specialty brews) or detrimentally (spoilage of beers, wines, and ciders). The primary genera in AAB for aerobic acetic acid fermentations are *Acetobacter, Gluconacetobacter,* and *Gluconobacter* (Li et al. 2015). *Acetobacter* members were once distinguished from those of the genus *Gluconobacter* by their predilection for ethanol and the ability to overoxidize acetate to CO_2 when ethanol was scarce (Xu et al. 2011).

In recent years, the AAB taxonomy has changed significantly in response to the development and implementation of new technologies. Several AAB taxa and species have been discovered for the first time. AAB are divided into ten genera and 45 species, as follows: *Acetobacter* (16 species), *Gluconobacter* (five species), *Acidomonas* (one species), *Gluconacetobacter* and *Komagatabacter* (newly called as *Komagateibacter* (Aidan and Parte 2017)) (15 species), *Asaia* (three species), *Saccharibacter* (one species), *Kozakia* (one species), *Neoasaia* (one species), *Swaminathania* (one species), and *Granulibacter* (one species), in the family Acetobacteraceae (Cleenwerck and De Vos 2008; Raspor and Goranovic 2008; Yamada et al. 2012; Mamlouk and Gullo 2013). The first phylogenetic investigations of AAB based on 16S rDNA sequences were published, revealing that these species belonged to the Proteobacteria's α-subclass. Species were distinguished based on their appearance in fluid medium, iodine reaction, and a variety of molecular features, including DNA–DNA hybridizations and PCR-based genomic finger printings (Xu et al. 2011).

3.2.2 ANAEROBIC

Under anaerobic conditions, acetogenic bacteria (acetogens) have been discovered to convert hexose into three molecules of acetic acid (Ljungdahl 1986). Acetogenic bacteria are not the same as AAB bacteria. These prokaryotes were first investigated for their new CO_2-fixing abilities. Over 100 acetogenic species from 23 genera have been isolated from various habitats so far (Schuchmann and Müller 2014). *Acetobacterium* and *Clostridium* are the two genera with the most known acetogenic species among the 23 genera. Acetogens include *Spirochaetes*, δ-proteobacteria like *Desulfotignum phosphitoxidans*, and acidobacteria like *Holophaga foetida*, all of which belong to the phylum *Firmicutes*. They all play a vital part in the biology of soil, lakes, and seas, respectively. Acetogens have been isolated from a variety of sources, including animal and termite gastrointestinal tracts, rice paddy soils, hypersaline fluids, surface soils, and deep subterranean sediments (Xu et al. 2011). *Moorella thermoacetica* was utilised to clarify the mechanism of homofermentation of acetic acid, which turns 1 mol of glucose into 3 mol of acetic acid in anaerobic synthesis of acetic acid. *Moorella thermoacetica* is the only homoacetogen that has been extensively researched for anaerobic acetic acid fermentation out of all the homoacetogens identified to date.

Acetogenic bacteria are a type of bacteria that may use the Wood–Ljungdahl pathway to create acetate from two molecules of carbon dioxide (CO_2) (WLP). Acetogens are facultative autotrophs that can grow by oxidising a wide range of organic substrates, such as hexoses, pentoses, alcohols, methyl groups, and formic acid, or inorganic substrates, such as hydrogen (H_2) or carbon monoxide (CO), which is generally associated with CO_2 reduction (Wang et al. 2013a,b). Acetogens are sometimes known as "homoacetogens" (because they solely create acetate as a byproduct of fermentation) or "CO_2 reducing acetogens." They have the ability to convert glucose to three moles of acetic acid practically stochiometrically (Ragsdale and Pierce 2008).

3.3 PRODUCTION OF ACETIC ACID

3.3.1 Two stages from ethanol

Using membrane-bound quinoproteins (ethanol dehydrogenase and acetaldehyde dehydrogenase), acetic acid bacteria may convert ethanol to acetate in two stages (Sakurai et al. 2012; Gullo et al. 2014). Alcohol dehydrogenase converts ethanol to acetaldehyde, which is then converted to acetic acid by aldehyde dehydrogenase. *Acetobacter* prefers the hydrogen acceptor pyrroloquinoline for transferring electrons generated by these processes. Initially, electrons are transferred to ubiquinone, which is then reoxidized by a membrane-associated oxidase. Finally, oxygen is the last electron acceptor, resulting in H_2O and a proton motive force required for energy production via a membrane-bound adenosine triphosphatase (ATPase). As a result, AAB are classified as obligatory aerobes since they have an absolute necessity for oxygen (Sengun and Karabiyikli 2011).

Except for the genus *Asaia*, all AAB undergo ethanol oxidation. In addition to the membrane-bound dehydrogenases that catalyse oxidations, the cytoplasm contains a second group of dehydrogenases that use NAD(P) as a cofactor (Saichana et al. 2015). While acetic acid is cytotoxic, it has been discovered that the *Acetobacter aceti* proteins aconitase and putative ATP-binding cassette (ABC) transporter are involved in acetic acid resistance. AAB kills competing organisms by secreting acetic acid, a membrane-permeable organic acid that poisons and disrupts proton gradients by acidifying the cytoplasm of vulnerable microbes. The cytoplasm of *A. aceti* becomes acidic during this process, but the cells continue to proliferate and oxidise ethanol even when the cytoplasmic pH dips to as low as 3.7 (Nakano et al. 2006; Nakano and Fukaya 2008).

3.3.2 Wood–Ljungdahl trail

In acetogens, hexose fermentation is processed through glycolysis to produce pyruvate, which is subsequently oxidised to produce acetyl-CoA and CO_2. Phosphotransacetylase converts acetyl-CoA to acetyl phosphate, which is ultimately transformed to acetate by acetate kinase. Instead of the two ATP molecules ordinarily created in glycolysis, this metabolic alternative allows for the creation of four.

$$C_6H_{12}O_6 + 4ADP + 4Pi \rightarrow 2CH_3COOH + 2CO_2 + 4ATP + 2H_2O + 8[H]$$

The reducing equivalents and 2 moles of CO_2 obtained in this reaction are transferred to the Wood–Ljungdahl trial, where they generate the third acetate molecule:

$$2CO_2 + 8[H] + nADP + nPi \rightarrow CH_3COOH + nATP + (2+n)H_2O$$

Glucose is oxidised to 3 mole of acetate in total:

$$C_6H_{12}O_6 + (4+n)ADP + (4+n)Pi \rightarrow 3CH_3COOH + (4+n)ATP + (4+n)H_2O$$

The reducing equivalents are produced by the oxidation of sugars, but they can also be produced by the oxidation of H_2, making acetogens facultative autotrophs capable of converting both CO_2 and H_2 (Ragsdale and Pierce 2008; Bengelsdorf et al. 2013; Schuchmann and Müller 2014).

$$2CO_2 + 4H_2 + nADP + nPi \rightarrow CH_3COOH + nATP + (2+n)H_2O$$

3.3.2.1   The Wood–Ljungdahl pathway is described as follows

A methyl and a carbonyl branch are characterised as essential reactions in the Wood–Ljungdahl route of CO_2 fixation. In the methyl branch, one molecule of CO_2 is reduced by six electrons to a methyl group, while the carbonyl branch involves the reduction of the other CO_2 molecule to carbon monoxide. The attached methyl group is then condensed with CO and coenzyme A (CoA) to produce acetyl-CoA. Acetyl-CoA is subsequently absorbed into cell carbon or transformed to acetyl phosphate, whose phosphoryl group is transferred to ADP to generate ATP and acetate, aceto-genic bacteria's principal growth product.

One molecule of CO_2 is converted to CO in the carbonyl branch of the WLP by the carbon monoxide (CO) dehydrogenase/acetyl-CoA synthase (CODH/ACS) enzyme (EC 2.3.1.169).

The conversion of CO_2 to formate by a formate dehydrogenase is the first step in the methyl branch (EC 1.2.1.43). A formyl-THF synthetase (EC 6.3.4.3) binds the formyl group to tetrahydrofolate (THF), generating formyl-THF in a process that involves the hydrolysis of ATP (Ragsdale and Pierce 2008).

Formyl-THF cyclohydrolase (EC 3.5.4.9) and methylene-THF dehydrogenase catalyse the next two steps in the Ljungdahl–Wood pathway (EC 1.5.1.15). Thus, formyl-THF is transformed to methenyl-THF by formyl-THF cyclohydrolase, and then methenyl-THF is reduced to methylene-THF by methylene-THF dehydrogenase in a NAD(P)H-dependent reaction. The methylene-THF reductase (EC 1.1.99.15) uses NAD(P)H as an electron donor to convert methylene-THF to methyl-THF in the following step (Bengelsdorf et al. 2013).

Finally, via a corrinoid iron-sulphur protein (CoFeSP), a methyltransferase (EC 2.1.1.245) transfers the methyl group from methyl-THF to the CODH/ACS. This bifunctional enzyme lowers CO_2 to CO and fuses it with the methyl group from the methyl branch and CoA to form acetyl-CoA in the carbonyl branch. A phosphotrans-acetylase (EC 2.3.1.8) converts it to acetyl phosphate, which is then converted to acetate by an acetate kinase (EC 2.7.2.1) (Bar-Even et al. 2012).

3.3.3   THE GLYCINE SYNTHASE ROUTE IS ONE WAY TO GET GLYCINE

Clostridium acidiurici, *Clostridium cylindrosporum*, *Clostridium purinolyticum*, and *Eubacterium angustum* exhibit a route for synthesis of acetate from CO_2 and one-carbon molecules that is almost entirely reliant on tetrahydrofolate (Ljungdahl 1986).

The CO_2 fixation via the glycine synthase pathway is merely an electron sink for recycling reduced electron carriers created during purine and amino acid fermentation. By reducing two CO_2 molecules to glycine, which is then transformed to acetate and excreted from the cell, the route recycles the electron carriers. The glycine cleavage system, which lies at the heart of the glycine synthase pathway, is a multi-protein complex that catalyses the reversible synthesis of glycine (Ljungdahl 1986; Bar-Even et al. 2012).

The metabolic mechanism for glycine dismutation to CO_2 and acetate has been extensively studied in the past, and it is likely to be used in the reductive direction as well, as several non-homoacetogenic clostridia have suggested: CO_2 is reduced through formate, ATP-dependent tetrahydrofolate linkage, and subsequent reduction to methylene tetrahydrofolate. The methylene derivative is reduced to acetate, which releases ammonia, after being reductively carboxylated and aminated. This phase also releases one ATP in a substrate-level phosphorylation process, allowing the ATP previously invested to be recovered (Ljungdahl 1986; Bar-Even et al. 2012).

The general reaction sequences of the reductive acetyl-CoA pathway and the glycine synthase pathway are very similar. Despite their apparent similarities, the reductive acetyl-CoA pathway is far more adaptable than the glycine synthase pathway in terms of application. For autotrophic development, energy conservation, and as an electron sink, the reductive acetyl-CoA pathway is utilised. The glycine synthase route, on the other hand, only serves as an electron sink. The glycine synthase route is unlikely to allow for energy saving (Schneeberger et al. 1999; Bar-Even et al. 2012).

3.4   PROCESSES OF FERMENTATION

Since ancient times, acetic acid has been made from ethanol as vinegar by sour wine and beer (Hailu and Jha 2012). Vinegar is made using a two-stage fermentation process that includes alcoholic and acetous fermentation. Most sugars are normally depleted within the first three weeks of alcoholic fermentation. The action of yeasts, usually *Saccharomyces cerevisiae* strains, converts fermentable carbohydrates into ethanol. The AAB are mostly members of the genus *Acetobacter* in acetous fermentation and can further oxidise ethanol into acetic acid. Alcoholic fermentation takes place in anaerobic settings, while acetic fermentation takes place in aerobic conditions (Ho et al. 2017).

There are two well-defined methods for vinegar production from a technological standpoint: slow processes (such as the Orleans method and the generator method) and rapid ones (such as submerged method and method using immobilised cells). The first is surface culture fermentation, in which the AAB are in contact with ambient air at the air–liquid interface (oxygen). Because the bacteria's presence is limited to the surface of the acidifying liquid, it is often referred to as a static technique. This method is now used to make conventional and selected vinegars, and it takes a long time to get a high acetic degree (acetic acid concentration %) in vinegar using this approach. As a result, the time and cost of production have increased. These technologies allow acetification and ageing to take place at the same time (Tesfaye et al. 2002; Hidalgo et al. 2013; Ho et al. 2017). Because of the low cost of the product, traditional vinegar manufacturing often uses indigenous starter culture rather than chosen (monophyletic) starter culture (SSC). However, a recent study found that using SSC to establish a systematic approach can confirm the feasibility of using a specific strain (Gullo et al. 2016).

3.4.1 THE METHOD OF ORLEANS

The Orleans process is one of the oldest and most well-known ways to make vinegar. It's a time-consuming, continual procedure that started in France. As a beginning culture, high-quality vinegar is utilised, and wine is added at weekly intervals. The vinegar is fermented in barrels with a volume of 200 litres. In the barrel, 65–70 litres of high-quality vinegar are added, along with 15 litres of wine. After one week, another 10–15 litres of wine is poured, and the process is repeated every week. As new wine is poured to replace the vinegar, the vinegar can be taken from the barrel after about four weeks (10–15 litres per week). Acetous fermentation is a sluggish process that occurs only at the liquid's surface, where there is enough dissolved oxygen to ensure that the alcohol is converted to acetic acid. One of the difficulties with this procedure is determining how much liquid to add to the barrel without disrupting the floating culture. A glass tube that reaches the bottom of the barrel can be used to solve this problem. The bacteria are not disturbed because more liquid is put in through the tube. This fermentation can continue anywhere from 8 to 14 weeks, depending on the fermentation temperature, the initial composition of the alcoholic solution, the type of the microbes, and the oxygen supply (Raspor and Goranovic 2008; Xu et al. 2011; Hailu and Jha 2012; Ho et al. 2017). The "Orleans-method" is being improved by focusing on a thin microbial coating and a high surface area for effective oxygen absorption (Ebner et al. 2008; Rogers et al. 2013).

3.4.2 THE GENERATOR METHOD IS USED TO PRODUCE ACETIC ACID

Generator procedures are sometimes known as "trickling" or "German" processes because they were developed in Germany and have been in use for about 200 years. This approach increases acetification surface contact by using wood shavings as a bacterial supportive material to promote faster vinegar production rates. In this method, the generator is utilised, which is an upright tank filled with beech wood shavings or grape stalks and equipped with devices that allow the alcoholic solution to trickle down through the shavings. The microbial population is immobilised on wood shavings in this process, which is mostly a surface process. The cylindrical tank features a perforated false bottom that supports beech-wood shavings or other similar material to aid in air passage from the bottom to the top exit (Raspor and Goranovic 2008; Hidalgo et al. 2013; Ho et al. 2017).

The liquor is made out of an adjusted alcohol solution acidified with acetic acid, as well as specific nutrients for AAB development. The liquid is poured into a trough at the chamber's top and let to drip down over the shavings. The liquid is collected at the bottom of the generator and re-circulated over the shavings, causing more alcohol to be oxidised until the appropriate concentration of vinegar is achieved. Bacteria may oxidise alcohol, resulting in a rise in temperature high enough to kill them. Cooling coils are required to keep the temperature between 25 and 30°C (Raspor and Goranovic 2008; Xu et al. 2011; Hailu and Jha 2012).

The accumulation of dead AAB over the wood shavings, the growth of cellulose producing bacteria (*Acetobacter xylinum*) on the shaving woods, the infection of the vinegar with anguillulas (vinegar eels), and the difficulty of controlling temperature, oxygen supply, and the evaporation of the substrate (ethanol) are all disadvantages of this method (Tesfaye et al. 2002).

3.4.3 METHOD OF SUBMERSION

The AAB are suspended in the acetifying liquid with high aeration to ensure that the oxygen demand is met in the submerged culture system. For the manufacturing of vinegar, this technology was invented around 1952. This system consists of stainless steel fermentation tanks with capacities ranging from 10,000 to 40,000 litres, as well as an air supply system, cooling system, foam control system, and loading and unloading valves. The batch process has three steps: loading the raw material and inoculating it into the fermentation medium, fermentation, and then complete unloading of the fermented medium. A semi-continuous process is identical to a continuous process, except that part of the completed product is unloaded and the rest is kept in the vessel to inoculate the following cycle. A continuous process comprises continually discharging a small aliquot of the bio-transformed product with a constant composition over time, as well as continuous substrate delivery to maintain the fermentation medium volume in the bioreactor constant. The ability to keep the bacterial culture in the exponential growth phase is crucial. It is vital to offer nutrition and oxygen to the bacteria at this stage in order for them to survive (Tesfaye et al. 2002; Hailu and Jha 2012; Ho et al. 2017).

Within 24–48 hours, this method provides for a high acetification rate (an increase in acetic degree with time) of roughly 8–9 acetic degrees. The Frings Acetator, Cavitator, Bubble column fermenter, and fermenters with various aeration systems, such as the Jet or Effigas Turbine, are only a few of the methods and bioreactors for submerged acetification that have been described and patented (vinegator). The oxygen mass transfer coefficients in each system are highly diverse. In general, these fermenters ensure the maximum oxygen transfer coefficients and do not include an agitation mechanism. In the vinegar-making industry, the Bubble Column Fermenter and Frings Acetator are commonly utilised.

The bubble column fermenter is made up of a 1:5 diameter-to-height ratio column. The upper half of the column has a larger diameter, allowing for a lower foam height at the surface and easier sedimentation of the bacterial culture, resulting in less removal of these along the completed product. Aeration and mechanical agitation are achieved by passing air through a diffuser (perforated plate) at the foot of the column, which is provided by a compressor.

The Frings Acetator is a fully automated system that ensures speedy acetification and a homogenous result. A stainless steel aerator, charging pump, alkalograph, cooling water valve, thermostat cooling, rotameter, cooling coil, air line, air exhaust line, and defoamer are all included (Tesfaye et al. 2002; Raspor and Goranovic 2008; Gullo et al. 2014). The aerator is the most specialised portion of the Acetator, because this distinctive impeller is self-aspirating and provides the largest oxygen transfer through excellent dispersion, as discussed in detail above (Hailu and Jha 2012).

3.4.4 FERMENTATION OF IMMOBILISED CELLS

Fibrous-bed bioreactors were commonly used to produce acetic acid from immobilised cells. The immobilised cell bioreactor consisted of a glass column filled
with spiral-wound terry cloth or cotton towel, with a working volume of 0.4–1 litres
that could be adjusted depending on the strains. A 5 litre recirculation reactor with
around 4 litres of the basal medium was attached to the bioreactor. The pH control
was the major purpose of the recirculation reactor. The increased reactor productivity that results from the high cell density is one of the key advantages of utilising
immobilised cells. The reactor was revealed to have cell densities more than 30 g/L–1
and to be self-renewing, finally settling into a dynamic steady state due to a balance
between new cell growth and dead cell removal (Huang and Yang 1998; Schwitzgue
and Pe 2000; Huang et al. 2002). Several diverse vinegar and balsamic vinegar
manufacturing processes emerge from the above fermentations, as documented in a
recent study (Giudici et al. 2009). Ageing plays a crucial part in these processes in
order to get the correct quality and composition.

3.5 PURIFICATION AND PRODUCT RECOVERY

Product separation is the most energy-intensive and expensive phase in the fermentation of bulk chemicals and liquid fuels, according to energy budget studies and
economic calculations. Most acidogenic fermentations work best when the pH is
between 6.0 and 7.5, which implies the acids are ionised and hence entirely nonvola-
tile. Before substantial concentrations of free unionised acid may be formed, the pH
must be reduced to about the pK_a (4.76) values of acetic acid. In addition to acetic
acid, which is commonly present in concentrations around 100 g/L^{-1}, fermentation
broths contain substrates, microbial biomass, inorganic salts, colloids, dissolved
ammonia and carbon dioxide, and other organic nonelectrolytes. In an ideal world, a
process would isolate the acids and recycle the other materials to earlier phases of the
process. For many years, various technologies such as fractional distillation, azeotropic dehydration distillation (Schniepp 1948), solvent extraction (Shin et al. 2009),
a combination of the above methods, extractive distillation (Lei et al. 2003; Bhatt
et al. 2012), and adsorption have been developed to separate acetic acid from water.

3.5.1 EXTRACTION OF LIQUID-LIQUID METHOD

Three primary aspects play a role in the design of an extraction process. To begin,
the pH should be adjusted to keep the acid in its undissociated state, as only undissociated acid is extracted to the solvent phase. Second, the extraction solvent should
have a high carboxylic acid partition coefficient, indicating a strong preference for
the organic phase over the aqueous phase, as well as a high selectivity for carboxylic
acid extraction over water, to avoid water co-extraction. Finally, the extraction system should be reversible, allowing the solvent to be recovered easily (Jipa et al. 2009;
Kersten et al. 2014). The three types of extraction solvents are (I) carbon-bonded oxygen carrying extractants like alcohols and ethers, (II) organophosphorous extractants
like trioctylphosphine oxide (TOPO) and tributylphosphate, and (III) aliphatic amine

extractants like trin-octylamine. The types of extractants (II) and (III) display strong complexation and are commonly utilised with one or more diluents, which primarily serve as a solvent for the complexes that these extractants form with the carboxylic acid and to alter the organic phase's viscosity and density. For the first category (I), the recovery of acids from dilute aqueous acid solutions prevalent in most fermentation streams has poor distributions (Usman et al. 2011).

Ionic liquids, which are viewed as possible green solvents, are a novel form of solvent that has recently been proposed. Ionic liquids are appealing because of their temperature stability and near-zero vapour pressure (Lateef et al. 2012; Sun et al. 2012). Because low-molecular-weight solvents (esters, ethers, and ketones) have relatively high distribution coefficients for acids at low concentrations, liquid–liquid extraction is used for intermediate concentrations (10–50%) and is usually followed by azeotropic distillation. Helsel (1977) has developed a cost-effective method for recovering acids from a dilute aqueous stream by extracting them with a hydrocarbon and then distilling them.

The following are the inherent benefits of this technology: (1) a high distribution coefficient of acetic acid in very dilute aqueous solution allows for a small solvent usage; (2) good phase separation reduces the size of the extraction equipment; (3) the stability and high boiling point of the solvent allow for a small volume of acid to be recovered from a much larger solvent flow; and (4) the low solubility of the solvent in water allows for higher selectivity and eliminates the n When compared to other recovery procedures, these advantages imply significant energy and capital savings, especially when the acetic acid content is less than 5% (Kumar and Babu 2008).

3.5.2 ADSORPTION

Adsorption on solid adsorbents such as activated carbon (Munson et al. 1987; Berhe et al. 2015), Anatase TiO_2 (Grinter et al. 2012), and cross-linked polymer adsorbent of pyridine skeleton structure can all be used to recover acetic acid. Even in the presence of inorganic salts, the adsorbents must exhibit strong selectivity and high adsorption capacity for acetic acid (Munson et al. 1987). Aliphatic alcohol, aliphatic ketones, and carboxylic esters were chosen as elutants (Patil and Kulkarni 2014). Because it is not employed in the industry, acetic acid recovery via adsorption is ineffective and also uneconomical.

3.5.3 PRECIPITATION

Precipitation is a traditional process for recovering organic acids from broth that has been used in industry to isolate lactic acid and citric acid since the nineteenth century. Precipitation can efficiently recover organic acids from a large volume of fermentation broths, making it more competitive, particularly in early purification.

Organic acids are frequently separated by four processes, for example, in calcium precipitation. To obtain the mother liquor and remove impurities, the fermentation liquid is filtered first, and then $Ca(OH)_2$ or $CaCO_3$ is added to the mother liquor while agitating. Second, the organic acid's calcium salt is filtered out of the fermentation. To liberate the required acid, the calcium salt is treated with a high concentration of sulphuric acid. Purification techniques are then used to produce the pure acid (Li et al. 2016).

Precipitation is now a well-established technology after years of research. Precipitation's primary advantages are its excellent selectivity, lack of phase change, and high product purity. Meanwhile, the most important aspect of this procedure is locating appropriate precipitants for the products. When a one molar amount of organic acid is converted, an equivalent amount of $Ca(OH)_2/CaCO_3$ and H_2SO_4 are used, and low value calcium sulphate is generated in the industrial-scale calcium precipitation process (Cheol and Chang 2011). Despite the widespread usage of calcium-based precipitation techniques, different precipitants should be used for acetic acid purification since the solubility of Ca-acetate is rather high, resulting in essentially no precipitation at the acetic acid concentration attained during fermentation.

3.5.4 DISTILLATION

Water has a lower boiling point than acetic acid and has a low relative volatility. Although acetic acid and water do not form an azeotrope, obtaining glacial grade acid by simple distillation requires a huge number of equilibrium stages and a very high reflux ratio (Bhatt et al. 2012). Azeotropic dehydration with the addition of another liquid can be used as an alternative to fractionation to save energy. The water is carried overhead in the distillation column by the entrainer, with the mixture being phase separated after condensation and the entrainer being returned to the column. It's only useful when there's a lot of acid in the environment (Kumar and Babu 2008; Patil and Kulkarni 2014). The entrainer, a water-insoluble "withdrawing" liquid, is introduced in Othmer's (1957) approach, which lowers the effective boiling point of water compared to that of acetic acid by forming a low boiling point azeotrope. Entrainers are typically low molecular weight esters like butyl acetate (Xu et al. 2011).

In extractive distillation, a descending stream of a high boiling point liquid, which is selectively solvent for one of the components, is used to wash mixed vapours in a distillation column. Suida (1929) in Austria was the first to discover this process for removing acetic acid from pyroligneous acid, which requires more expensive equipment and uses more steam than other methods.

3.5.5 REACTIVE DISTILLATION

Traditional physical separation processes like as distillation and extraction have been surpassed by reactive distillation (RD). Distillation is linked to the high costs of vaporising highly volatile water that exists in large quantities and has a high latent heat of vaporisation. Because of the dispersion of components in the responding system, extraction is limited (Lei et al. 2003).

Reactive distillation decreases capital and operational expenses while also allowing for a wider range of operating circumstances. RD is gaining popularity, and it has a lot of potential for recovering acetic acid. RD, which combines a chemical reaction with distillative product separation in a single piece of equipment, has several advantages over traditional processes in which the reaction and product separation are carried out in parallel, particularly for reactions that are constrained by equilibrium constraints. To name a few advantages, RD provides enhanced selectivity, higher conversion, better heat control, efficient use of reaction heat, flexibility for difficult separations, and avoidance of azeotrope. The ability to combine reaction and distillation separation in one unit enables for continuous production and shorter processing times. This result in consistent high product quality while also making maintenance and process management easier, which is especially beneficial for greater production capacity. The breakdown of fatty acids and fatty acid esters is minimised throughout the plant by well-defined and short residence duration under mild circumstances. There is no need for catalyst neutralisation, separation, or recycling. There is no need to empty and clean the equipment, which reduces waste streams to a bare minimum. The RD technique uses half the amount of energy as a traditional batch procedure. The plant's size could also be dramatically lowered (Saha et al. 2000; Sharma and Mahajani 2003).

3.5.6 MEMBRANE PROCESSES METHOD

Because of their versatility and selectivity, membrane technologies have been applied in the recovery of organic acids. Membrane separation is gaining popularity as recovery procedures and material technology advance, particularly in the field of in situ product removal. A membrane is a thin artificial or natural barrier that allows selective mass movement of solutes or solvents over it for the purposes of physical separation and enrichment. Membrane separation for acetic acid can produce high purity and yield. Microfiltration, ultrafiltration, nanofiltration (NF) (Baruah and Hazarika 2014), reverse osmosis, pervaporation, and electrodialysis are the most common membrane filtrations used to separate organic acids (Gorri et al. 2005).

3.5.7 IN SITU METHOD OF PRODUCT REMOVAL

ISPR stands for in situ product removal, which is the rapid elimination of organic acid without interfering with cellular or medium components. To achieve a continuous process, it combines fermentation with separation, which includes extraction, resin, and membrane. Furthermore, removing products can reduce metabolite toxicity to bacteria.

Ion exchange, solvent extraction, and membrane separation are all recommended extraction procedures. TOPO solvent extraction may be particularly useful in eliminating acids at concentrations ranging from 1–3% to 20–25%. It was claimed that a novel extractive fermentation technique was used to achieve high efficiency propionic acid synthesis and purification, with benefits such as better pH control and a purer product (Xu et al. 2011).

Membrane fouling is an issue in these operations, which necessitates frequent dialyzer cleaning. It allows for a greater degree of acid separation, but at the cost of increased power and energy consumption. The disadvantages include hampered implementation, namely complexity of operation and swelling in liquid surfactant membranes, as well as membrane instability in supported liquid membranes (Patil and Kulkarni 2014).

The acid is distilled in the presence of $KMnO_4$, $K_2Cr_2O_7$, or other oxidants for final purification. Although $KMnO_4$ is more expensive, it is capable of oxidising a larger spectrum of contaminants. At the summit, acetic acid is obtained. The bottom product is processed through a solvent extraction system containing toluene or butyl acetate, which removes the oxidised organic contaminants with the solvent layer while leaving the MnO_4^- in the water layer to be reused (Xu et al. 2011).

3.6 CONCLUSIONS

Fermentation has been discovered to be an effective approach for producing acetic acids from renewable biomass. However, properly separating organic acids from a mixture of numerous diluted components while minimising the impurity of other organic acids with similar properties remains a challenge. All of the current recovery technologies have limits, and improvements are notably needed in terms of yield, purity, and energy usage. As a result, a procedure must be developed that is simple to carry out and allows the purification of acetic acid straight from fermentation broths. Aside from that, the development of new materials and technologies would increase recovery processes, making biological procedures more competitive than chemical approaches and encouraging the development of green chemistry (Li et al. 2016). Due to the depletion of natural gas and petroleum supplies, as well as an increasing demand for these materials worldwide, a significant reduction in the capacity to synthesis industrial acetic acid from methanol and CO or by other chemical processes may occur by the end of the century. Bacterial-based methods have the potential to become big participants in the glacial acetic acid industry, given the price rises that have accompanied this situation (Rogers et al. 2013) (Table 3.1).

TABLE 3.1
Microorganisms engaged for acetic acid synthesis

S. no.	Name of the species	References
1	*Acetobacter aceti*	Lisdiyanti et al. (2000)
2	*Acetobacter ascendens*	Kim et al. (2018)
3	*Acetobacter cerevisiae*	Iino et al. (2012)
4	*Acetobacter cibinongensisc*	Lisdiyanti et al. (2001)
5	*Acetobacter estunensis*	Lisdiyanti et al. (2000)
6	*Acetobacter fabarum*	Iino et al. (2012)
7	*Acetobacter farinalis*	Iino et al. (2012)
8	*Acetobacter ghanensis*	Iino et al. (2012)
9	*Acetobacter indonesiensis*	Lisdiyanti et al. (2000)
10	*Acetobacter lambici*	Spitaels et al. (2013)
11	*Acetobacter lovaniensis*	Iino et al. (2012)
12	*Acetobacter malorum*	Iino et al. (2012)
13	*Acetobacter musti*	Ferrer et al. (2016)
14	*Acetobacter nitrogenifigens*	Dutta and Gachhui (2006)
15	*Acetobacter oeni*	Silva et al. (2006)

(Continued)

TABLE 3.1 (*Continued*)

Microorganisms engaged for acetic acid synthesis

S. no.	Name of the species	References
16	*Acetobacter okinawensis*	Iino et al. (2012)
17	*Acetobacter orientalis*	Lisdiyanti et al. (2001)
18	*Acetobacter orleanensis*	Lisdiyanti et al. (2000)
19	*Acetobacter oryzoeni*	Baek et al. (2020)
20	*Acetobacter oryzifermentans*	Kim et al. (2018)
21	*Acetobacter pasteurianus*	Lisdiyanti et al. (2000)
22	*Acetobacter papayae*	Iino et al. (2012)
23	*Acetobacter peroxydans*	Iino et al. (2012)
24	*Acetobacter persicus*	Iino et al. (2012)
25	*Acetobacter pomorum*	Sokollek et al. (1998)
26	*Acetobacter senegalensis*	Ndoye et al. (2007)
27	*Acetobacter sicerae*	Li et al. (2014)
28	*Acetobacter syzygii*	Iino et al. (2012)
29	*Acetobacter tropicalis*	Lisdiyanti et al. (2000)
30	*Acidomonas methanolica*	Ramírez-Bahena et al. (2013)
31	*Ameyamaea chiangmaiensis*	Yukphan et al. (2009)
32	*Asaia astilbes*	Suzuki et al. (2010)
33	*Asaia bogorensis*	Yamada et al. (2000)
34	*Asaia krungthepensis*	Yukphan et al. (2004)
35	*Asaia lannaensis*	Malimas et al. (2008)
36	*Asaia platycodi*	Suzuki et al. (2010)
37	*Asaia prunellae*	Suzuki et al. (2010)
38	*Asaia siamensis*	Katsura et al. (2001)
39	*Asaia spathodeae*	Kommanee et al. (2010)
40	*Bombella apis*	Yun et al. (2017)
41	*Commensalibacter intestini*	Kim et al. (2012)
42	*Commensalibacter papalotli*	Servin-Garciduenas et al. (2014)
43	*Endobacter medicaginis*	Ramírez-Bahena et al. (2013)
44	*Gluconacetobacter aggeris*	Nishijima et al. (2013)
45	*Gluconacetobacter asukensis*	Nishijima et al. (2013)
46	*Gluconacetobacter azotocaptans*	Nishijima et al. (2013)
47	*Gluconacetobacter diazotrophicus*	Yamada et al. (1997)
48	*Gluconacetobacter entanii*	Lisdiyanti et al. (2006)
49	*Gluconacetobacter johannae*	Nishijima et al. (2013)
50	*Gluconacetobacter liquefaciens*	Yamada et al. (1997)
51	*Gluconacetobacter sacchari*	Franke et al. (1999)
52	*Gluconacetobacter takamatsuzukensis*	Nishijima et al. (2013)
53	*Gluconacetobacter tumulicola*	Nishijima et al. (2013)
54	*Gluconacetobacter tumulisoli*	Nishijima et al. (2013)
55	*Gluconobacter albidusf*	Malimas et al. (2007)
56	*Gluconobacter cerinus*	Malimas et al. (2007)
57	*Gluconobacter frateurii*	Malimas et al. (2007)

(Continued)

TABLE 3.1 *(Continued)*
Microorganisms engaged for acetic acid synthesis

S. no.	Name of the species	References
58	*Gluconobacter japonicus*	Malimas et al. (2009)
59	*Gluconobacter kanchanaburiensis*	Tanasupawat et al. (2011)
60	*Gluconobacter kondonii*	Malimas et al. (2007)
61	*Gluconobacter nephelii*	Kommanee et al. (2010)
62	*Gluconobacter oxydans*	Malimas et al. (2007)
63	*Gluconobacter roseus*	Tanasupawat et al. (2011)
64	*Gluconobacter sphaericus*	Tanasupawat et al. (2011)
65	*Gluconobacter thailandicus*	Malimas et al. (2007)
66	*Gluconobacter wancherniae*	Tanasupawat et al. (2011)
67	*Gluconobacter uchimurae*	Tanasupawat et al. (2011)
68	*Granulibacter bethensis*	Ramírez-Bahena et al. (2013)
69	*Komagataeibacter europaeus*	Yamada et al. (1997)
70	*Komagataeibacter hansenii*	Yamada et al. (1997)
71	*Komagataeibacter intermedius*	Yamada et al. (2000)
72	*Komagataeibacter kakiaceti*	Škraban et al. (2018)
73	*Komagataeibacter kombuchae*	Škraban et al. (2018)
74	*Komagataeibacter maltaceti*	Slapšak et al. (2013)
75	*Komagataeibacter medellinensis*	Castro et al. (2013)
76	*Komagataeibacter nataicola*	Lisdiyanti et al. (2006)
77	*Komagataeibacter oboediens*	Yamada et al. (2000)
78	*Komagataeibacter pomaceti*	Škraban et al. (2018)
79	*Komagataeibacter rhaeticus*	Dellaglio et al. (2005)
80	*Komagataeibacter swingsii*	Dellaglio et al. (2005)
81	*Komagataeibacter sucrofermentans*	Škraban et al. (2018)
82	*Komagataeibacter saccharivorans*	Lisdiyanti et al. (2006)
83	*Komagataeibacter xylinus*	Yamada et al. (1997)
84	*Kozakia baliensis*	Ramírez-Bahena et al. (2013)
85	*Neoasaia chiangmaiensisg*	Ramírez-Bahena et al. (2013)
86	*Neokomagataea tanensis*	Yukphan et al. (2011)
87	*Neokomagataea thailandica*	Yukphan et al. (2011)
88	*Nguyenibacter vanlangensis*	Vu et al. (2013)
89	*Saccharibacter floricola*	Jojima et al. (2004)
90	*Swaminathania salitolerans*	Ramírez-Bahena et al. (2013)
91	*Swingsia samuiensis*	Malimas et al. (2013)
92	*Tanticharoenia sakaeratensis*	Yukphan et al. (2008)

REFERENCES

Aidan C. Parte. "Genus Komagataeibacter." LPSN bacterio.net. 2017. [Online]. Available from: http://www.bacterio.net/komagataeibacter. html [Accessed: 22nd August 2017].

Baek, J. H., Kim, K. H., Moon, J. Y., Yeo, S. H., Jeon, C. O. "Acetobacter oryzoeni sp. nov., isolated from Korean rice wine vinegar." *International Journal of Systematic and Evolutionary Microbiology.* 70(3), pp. 2026–2033. 2020.

Bar-Even, A., Noor, E., Milo, R. "A survey of carbon fixation pathways through a quantitative lens." *Journal of Experimental Botany.* 63(6), pp. 2325–2342. 2012.

Baruah, K., Hazarika, S. "Separation of acetic acid from dilute aqueous solution by nanofiltration membrane." *Journal of Applied Polymer Science.* 131(15), pp. 40537–40546. 2014.

Bengelsdorf, F. R., Straub, M., Dürre, P. "Bacterial synthesis gas (syngas) fermentation." *Environmental Technology.* 34(13–14), pp. 1639– 1651. 2013.

Berhe, A., Jeevan, A., Lijalem, T. "Removal of acetic acid from aqueous solution by using activated carbon." *International Journal of Innovation and Scientific Research.* 17(2), pp. 443–450. 2015.

Bhatt, R. P., Thakore, P. S. B., Engg, L. D. C. "Extractive distillation of acetic acid from its dilute solution using lithium bromide." *International Journal of Scientific Engineering and Technology.* 1(2), pp. 46–50. 2012.

Castro, C., Cleenwerck, I., Trček, J., Zuluaga, R., De Vos, P., Caro, G., Aguirre, R., Putaux, J. L., Gañán, P. "*Gluconacetobacter medellinensis* sp. nov., cellulose- and non-cellulose-producing acetic acid bacteria isolated from vinegar." *International Journal of Systematic and Evolutionary Microbiology.* 63(3), pp. 1119–1125. 2013.

Cheol, S., Chang, H. "Batch and continuous separation of acetic acid from succinic acid in a feed solution with high concentrations of carboxylic acids by emulsion liquid membranes." *Journal of Membrane Science.* 367, pp. 190–196. 2011.

Christodoulou, X., Velasquez-Orta, S. B. "Microbial electrosynthesis and anaerobic fermentation: An economic evaluation for acetic acid production from CO_2 and CO." *Environmental Science & Technology.* 50, pp. 11234–11242. 2016.

Cleenwerck, I., De Vos, P. "Polyphasic taxonomy of acetic acid bacteria: An overview of the currently applied methodology." *International Journal of Food Microbiology.* 125(1), pp. 2–14. 2008.

Dellaglio, F., Cleenwerck, I., Felis, G. E., Engelbeen, K., Janssens, D., Marzotto, M. "Description of *Gluconacetobacter swingsii* sp. nov. and *Gluconacetobacter rhaeticus* sp. nov., isolated from Italian apple fruit." *International Journal of Systematic and Evolutionary Microbiology.* 55(6), pp. 2365–2370. 2005.

Dutta, D., Gachhui, R. Novel nitrogen-fixing *Acetobacter nitrogenifigens* sp. nov., isolated from Kombucha tea. *International Journal of Systematic and Evolutionary Microbiology.* 56(8), pp. 1899–1903. 2006.

Ebner, H., Sellmer, S., Follmann, H. "Acetic acid." In: *Biotechnology* (pp. 381–401). Wiley-VCH Verlag GmbH, Federal Republic of Germany, Weinheim. 2008.

Ferrer, S., Mañes-Lázaro, R., Benavent-Gil, Y., Yépez, A., Pardo, I. "Acetobacter musti sp. nov., isolated from Bobal grape must." *International Journal of Systematic and Evolutionary Microbiology.* 66(2), pp. 957–961. 2016.

Franke, I. H., Fegan, M., Hayward, C., Leonard, G., Stackebrandt, E., Sly, L. I. "Description of *Gluconacetobacter sacchari* sp. nov. a new species of acetic acid bacterium isolated from the leaf sheath of sugar cane and from the pink sugar-cane mealy bug." *International Journal of Systematic and Bacteriology.* 49(4), pp. 1681–1693. 1999.

Giudici, P., Gullo, M., Solieri, L., Falcone, P. M. "Technological and microbiological aspects of traditional balsamic vinegar and their influence on quality and sensorial properties." In: *Traditional Balsamic Vinegar and Related Products.* 1st edition, Vol. 58 (pp. 137–182). Academic Press, Elsevier, USA. 2009.

Gorri, D., Urtiaga, A., Ortiz, I. "Pervaporative recovery of acetic acid from an acetylation industrial effluent using commercial membranes." *Industrial and Engineering Chemistry Research.* 44, pp. 977–985. 2005.

Grinter, D. C., Nicotra, M., Thornton, G. "Acetic acid adsorption on anatase TiO2 (101)." *The Journal of Physical Chemistry.* 116(21), pp. 11643–11651. 2012.

Gullo, M., Verzelloni, E., Canonico, M. "Aerobic submerged fermentation by acetic acid bacteria for vinegar production: Process and biotechnological aspects." *Process Biochemistry.* 49(10), pp. 1571–1579. 2014.

Gullo, M., Zanichelli, G., Verzelloni, E., Lemmetti, F., Giudici, P. "Feasible acetic acid fermentations of alcoholic and sugary substrates in combined operation mode." *Process Biochemistry.* 51(9), pp. 1129–1139. 2016.

Hailu, S., Jha, S. A. Y. K. "Vinegar production technology–an overview." *Beverage and Food World.* 3(2), pp. 139–155. 2012.

Helsel, R. W. "Waste recovery: Removing carboxylic acids from aqueous wastes." *Chemical Engineering Progress.* 73(5), pp. 55–59. 1977.

Hidalgo, C., García, D., Romero, J., Mas, A., Torija, M. J., Mateo, E. "Acetobacter strains isolated during the acetification of blueberry (Vaccinium corymbosum L.) wine." *Letters in Applied Microbiology.* 57(3), pp. 227–232. 2013.

Ho, C. W., Lazim, A. M., Fazry, S., Kalsum, U., Zaki, H. H., Lim, S. J. "Varieties, production, composition and health benefits of vinegars: A review." *Food Chemistry.* 221, pp. 1621–1630. 2017.

Huang, Y., Yang, S. "Acetate production from whey lactose using co-immobilized cells of homolactic and homoacetic bacteria in fibrous-bed bioreactor." *Biotechnology and Bioengineering.* 60(4), pp. 498–507. 1998.

Huang, Y. L., Wu, Z., Zhang, L., Cheung, C. M., Yang, S. "Production of carboxylic acids from hydrolyzed corn meal by immobilized cell fermentation in a fibrous-bed bioreactor." *Bioresource Technology.* 82, pp. 51–59. 2002.

Iino, T., Suzuki, R., Kosako, Y., Ohkuma, M., Komagata, K., Uchimura, T. "*Acetobacter okinawensis* sp. nov., *Acetobacter papayae* sp. nov., and *Acetobacter persicus* sp. nov.; novel acetic acid bacteria isolated from stems of sugarcane, fruits, and a flower in Japan." *Journal of General and Applied Microbiology.* 58(3), pp. 235–243. 2012.

Jipa, I., Dobre, T., Stroescu, M., Stoica, A. "Acetic acid extraction from fermentation broth experimental and modelling studies." *Revista de Chimie.* 60(10), pp. 1084–1090. 2009.

Jojima, Y., Mihara, Y., Suzuki, S., Yokozeki, K., Yamanaka, S., Fudou, R. "Saccharibacter floricola gen nov, sp. nov., a novel osmophilic acetic acid bacterium isolated from pollen." *International Journal of Systematic and Evolutionary Microbiology.* 54(6), pp. 2263–2267. 2004.

Katsura, K., Kawasaki, H., Potacharoen, W., Saono, S., Seki, T., Yamada, Y., Uchimura, T., Komagata, K. "*Asaia siamensis* sp. nov., an acetic acid bacterium in the α-*Proteobacteria*." *International Journal of Systematic and Evolutionary Microbiology.* 51(2), pp. 559–563. 2001.

Kersten, S. R. A., Ham, A. G. J. Van Der, Schuur, B., Ijmker, H. M., Gramblic, M. "Acetic acid extraction from aqueous solutions using fatty acids." *Separation and Purification Technology.* 125, pp. 256–263. 2014.

Kim, E. K., Kim, S. H., Nam, H. J., Choi, M. K., Lee, K. A., Choi, S. H., Seo, Y. Y., You, H., Kim, B., Lee, W. J. "Draft genome sequence of *Commensalibacter intestini* A911T, a symbiotic bacterium isolated from drosophila melanogaster Intestine." *Journal of Bacteriology.* 194(5), p. 1246. 2012.

Kim, K. H., Cho, G. Y., Chun, B. H., Weckx, S., Moon, J. Y., Yeo, S. H., Jeon, C. O. "*Acetobacter oryzifermentans* sp. nov., isolated from Korean traditional vinegar and reclassification of the type strains of *Acetobacter pasteurianus* subsp *ascendens* (Henneberg1898) and *Acetobacter pasteurianus* subsp. *paradoxus* (Frateur1950) as *Acetobacter ascendens* sp. nov., comb nov." *Systematic and Applied Microbiology.* 41(4), pp. 324–332. 2018.

Kommanee, J., Tanasupawat, S., Yukphan, P., Malimas, T., Muramatsu, Y., Nakagawa, Y., Yamada, Y. "*Asaia spathodeae* sp. nov., an acetic acid bacterium in the α-*Proteobacteria*." *Journal of General and Applied Microbiology.*56(1), pp. 81–87. 2010.

Kumar, S., Babu, B.V. "Separation of carboxylic acids from waste water via reactive extraction." In: *International Proceedings of Convention on Water Resources Development and Management (ICWRDM).* 23–26 October 2008 (pp. 1–9). Birla Institute of Technology and Science (BITS), Pilani-333031, Rajasthan, India. 2008.

Lateef, H., Gooding, A., Grimes, S. "Use of 1-hexyl-3-methylimidazolium bromide ionic liquid in the recovery of lactic acid from wine." *Journal of Chemical Technology and Biotechnology.* 87(8), pp. 1–8. 2012.

Lei, Z., Li, C., Chen, B. "Extractive distillation: A review." *Separation & Purification Reviews.* 32(2), pp. 121–213. 2003.

Li, L., Wieme, A., Spitaels, F., Balzarini, T., Nunes, O. C., Manaia, C. M., Van Landschoot, A., De Vuyst, L., Cleenwerck, I., Vandamme, P. "*Acetobacter sicerae* sp. nov., isolated from cider and kefir, and identification of species of the genus *Acetobacter* by dnaK, groEL and rpoB sequence analysis." *International Journal of Systematic and Evolutionary Microbiology.* 64(7), pp. 2407–2415. 2014.

Li, Q.-Z., Jiang, X.-L., Feng, X.-J., Wang, J.-M., Sun, C., Zhang, H.-B., Xian, M., Liu, H.-Z. "Recovery processes of organic acids from fermentation broths in the biomass-based industry." *Journal of Microbiology and Biotechnology.* 26(1), pp. 1–8. 2016.

Li, S., Li, P., Feng, F., Luo, L. X. "Microbial diversity and their roles in the vinegar fermentation process." *Applied Microbiology and Biotechnology.* 99(12), pp. 4997–5024. 2015.

Lisdiyanti, P., Kawasaki, H., Seki, T., Yamada, Y., Uchimura, T., Komagata, K. "Systematic study of the genus Acetobacter with descriptions of Acetobacter indonesiensis sp. nov.. Acetobacter tropicalis sp. nov.. Acetobacter orleanensis (Henneberg (1906) comb nov Acetobacter lovaniensis (Frateur 1950) comb nov and Acetobacter estunensis (Carr 1958) comb nov." *Journal of General and Applied Microbiology.* 46(3), pp. 147–165. 2000.

Lisdiyanti, P., Kawasaki, H., Seki, T., Yamada, Y., Uchimura, T., Komagata, K. "Identification of *Acetobacter* strains isolated from Indonesian sources, and proposals of *Acetobacter syzygii* sp. nov.., Acetobacter cibinongensis sp. nov., and Acetobacter orientalis sp. nov." *Journal of General and Applied Microbiology.* 47(3), pp. 119–131. 2001.

Lisdiyanti, P., Navarro, R. R., Uchimura, T., Komagata, K. "Reclassification of *Gluconacetobacter hansenii* strains and proposals of *Gluconacetobacter saccharivorans* sp. nov. and *Gluconacetobacter nataicola* sp. nov." *International Journal of Systematic and Evolutionary Microbiology.* 56(9), pp. 2101–2111. 2006.

Ljungdahl, L. G. "The autotrophic pathway of acetate synthesis in acetogenic bacteria." *Annual Review of Microbiology.* 40, pp. 415–450. 1986.

Malimas, T., Chaipitakchonlatarn, W., Vu, H. T. L., Yukphan, P., Muramatsu, Y., Tanasupawat, S., Potacharoen, W., Nakagawa, Y., Tanticharoen, M., Yamada, Y. "Swingsia samuiensis gen nov, sp. nov., an osmotolerant acetic acid bacterium in the α-Proteobacteria." *Journal of General and Applied Microbiology.* 59(5), pp. 375–384. 2013

Malimas, T., Yukphan, P., Lundaa, T., Muramatsu, Y., Takahashi, M., Kaneyasu, M., Potacharoen, W., Tanasupawat, S., Nakagawa, Y., Suzuki, K. I., Tanticharoen, M., Yamada, Y. "*Gluconobacter kanchanaburiensis* sp. nov., a brown pigment-producing acetic acid bacterium for Thai isolates in the *Alphaproteobacteria.*" *Journal of General and Applied Microbiology.* 55(3), pp. 247–254. 2009.

Malimas, T., Yukphan, P., Takahashi, M., Kaneyasu, M., Potacharoen, W., Tanasupawat, S., Nakagawa, Y., Tanticharoen, M., Yamada, Y. "*Gluconobacter kondonii* sp. nov.., an acetic acid bacterium in the *alpha-Proteobacteria.*" *Journal of General and Applied Microbiology.* 53, pp. 301–307. 2007.

and Biochemistry. 72(3), pp. 666–671. 2008.

Mamlouk, D., Gullo, M. "Acetic acid bacteria: Physiology and carbon sources oxidation." *Indian Journal of Microbiology.* 53(4), pp. 377– 384. 2013.

Munson, C. L., Garcia, A. A., Kuo, Y. "Use of adsorbents for recovery of acetic acid from aqueous solutions part II—factors governing selectivity." *Separation & Purification Reviews.* 16(1), pp. 65–89. 1987.

Nakano, S., Fukaya, M. "Analysis of proteins responsive to acetic acid in Acetobacter: Molecular mechanisms conferring acetic acid resistance in acetic acid bacteria." *International Journal of Food Microbiology*. 125(1), pp. 54–59. 2008.

Nakano, S., Fukaya, M., Horinouchi, S. "Putative ABC transporter responsible for acetic acid resistance in Acetobacter aceti." *Applied and Environmental Microbiology*. 72(1), pp. 497–505. 2006.

Ndoye, B., Cleenwerck, I., Engelbeen, K., Dubois-Dauphin, R., Guiro, A. T., Van Trappen, S., Willems, A., Thonart, P. "Acetobacter senegalensis sp. nov., a thermotolerant acetic acid bacterium isolated in Senegal (sub-Saharan Africa) from mango fruit (Mangifera indica L)." *International Journal of Systematic and Evolutionary Microbiology*. 57(7), pp. 1576–1581. 2007.

Nishijima, M., Tazato, N., Handa, Y., Tomita, J., Kigawa, R., Sano, C., Sugiyama, J. "Gluconacetobacter tumulisoli sp. nov., Gluconacetobacter takamatsuzukensis sp. nov. and Gluconacetobacter aggeris sp. nov., isolated from Takamatsuzuka Tumulus samples before and during the dismantling work in 2007." *International Journal of Systematic and Evolutionary Microbiology*. 63(11), pp. 3981–3988.

Othmer, D. F. "Process for recovering acetic acid." *United States Patent Office*, (635863). 1957. https://patents.google.com/patent/WO2019002240A1/en

Pal, P., Nayak, J. "Acetic acid production and purification: Critical review towards process intensification." *Separation & Purification Reviews*. 21(11), pp. 44–61. 2016.

Patil, K. D., Kulkarni, B. D. "Review of recovery methods for acetic acid from industrial waste streams by reactive distillation." *Journal of Water Pollution & Purification Research*. 1(2), pp. 13–18. 2014.

Ragsdale, S. W., Pierce, E. "Acetogenesis and the Wood-Ljungdahl pathway of CO2 fixation." *Biochimica et Biophysica Acta – Proteins and Proteomics*. 1784(12), pp. 1873–1898. 2008.

Ramírez-Bahena, M. H., Tejedor, C., Martín, I., Velázquez, E., Peix, A. "Endobacter medicaginis gen nov, sp. nov., isolated from alfalfa nodules in an acidic soil." *International Journal of Systematic and Evolutionary Microbiology*. 63(5), pp. 1760–1765. 2013.

Raspor, P., Goranovic, D. "Biotechnological applications of acetic acid bacteria." *Critical Reviews in Biotechnology*. 28, pp. 101–124. 2008.

Rogers, P., Jiann-Shin, C., Mary, J. Z. "Organic acid and solvent production part I: Acetic, lactic, gluconic, succinic and polyhydroxyalkanoic acids." In: Rosenberg, E., DeLong, E., Lory, S., Stackebrandt, E., Thompson, F. B., Heidelberg (eds.). *The Prokaryotes* (pp. 3–75). Springer, New York. 2013.

Saha, B., Chopade, S. P., Mahajani, S. M. "Recovery of dilute acetic acid through esterification in a reactive distillation column." *Catalysis Today*. 60, pp. 147–157. 2000.

Saichana, N., Matsushita, K., Adachi, O., Frébort, I., Frébortová, J. "Acetic acid bacteria: A group of bacteria with versatile biotechnological applications." *Biotechnology Advances*. 33(6), pp. 1260–1271. 2015.

Sakurai, K., Arai, H., Ishii, M., Igarashi, Y. "Changes in the gene expression profile of Acetobacter aceti during growth on ethanol." *Journal of Bioscience and Bioengineering*. 113(3), pp. 343–348. 2012.

Schneeberger, A., Frings, J., Schink, B. "Net synthesis of acetate from CO2 by Eubacterium acidaminophilum through the glycine reductase pathway." *FEMS Microbiology Letters*. 177(1), pp. 1–6. 1999.

Schniepp, L. E. "Process for the purification of acetic acid by azeotropic distillation." *United States Patent Office*, 2438300. 1948.

Schuchmann, K., Müller, V. "Autotrophy at the thermodynamic limit of life: A model for energy conservation in acetogenic bacteria." *Nature Reviews Microbiology*. 12(12), pp. 809–821. 2014.

Schwitzgue, J., Pe, P. "Acetic acid production from lactose by an anaerobic Thermophilic Coculture immobilized in a fibrous-bed bioreactor." *Biotechnology Progress*. 16, pp. 1008–1017. 2000.

Sengun, I. Y., Karabiyikli, S. "Importance of acetic acid bacteria in food industry." *Food Control.* 22(5), pp. 647–656. 2011. https://doi.org/10.1016/j.foodcont.2010.11.008

Servin-Garciduenas, L. E., Sanchez-Quinto, A., Martinez-Romero, E. "Draft genome sequence of *Commensalibacter papalotli* MX01, a symbiont identified from the guts of overwintering monarch butterflies." *Genome Announcements.* 2(2), pp. e00128–e1114. 2014.

Shah, K., Consultant, S. "Acetic acid: Overview & market outlook." In: *Indian Petrochem Conference*, pp. 1–23. Elite Conferences Pvt. Ltd., Mumbai, India. 2014.

Sharma, M. M., Mahajani, S. M. "Reactive distillation process development in the chemical process industries." In: Sundmacher, K., Kienle, A. (eds.). *Reactive Distillation Status and Future Directions* (pp. 30–47). Wiley-VCH, Weinheim, Germany. 2003.

Shin, C.-H., Kim, J., Kim, J., Kim, H., Lee, H., Mohapatra, D., Ahn, J.-W., Ahn, J.-G., Bae, W. "A solvent extraction approach to recover acetic acid from mixed waste acids produced during semiconductor wafer process." *Journal of Hazardous Materials.* 162, pp. 1278–1284. 2009.

Silva, L. R., Cleenwerck, I., Rivas, R., Swings, J., Trujillo, M. E., Willems, A., Velazquez, E. "*Acetobacter oeni* sp. nov., isolated from spoiled red wine." *International Journal of Systematic and Evolutionary Microbiology.* 56(1), pp. 21–24. 2006.

Škraban, J., Cleenwerck, I., Vandamme, P., Fanedl, L., Trček, J. "Genome sequences and description of novel exopolysaccharides producing species *Komagataeibacter pomaceti* sp. nov. and reclassification of *Komagataeibacter kombuchae* (Dutta and Gachhui 2007) Yamada et al. 2013 as a later heterotypic synonym of *Komagataeibacter hansenii* (Gosselé et al. 1983) Yamada et al. 2013." *Systematic and Applied Microbiology.* 41(6), pp. 581–592. 2018.

Slapšak, N., Cleenwerck, I., De Vos, P., Trček, J. "*Gluconacetobacter maltaceti* sp. nov., a novel vinegar producing acetic acid bacterium." *Systematic and Applied Microbiology.* 36(1), pp. 17–21. 2013.

Sokollek, S. J, Hertel, C., Hammes, W. P. "Description of *Acetobacter oboediens* sp. nov. and *Acetobacter pomorum* sp. nov. two new species isolated from industrial vinegar fermentations." *International Journal of Systematic and Evolutionary Microbiology.* 48(3), pp. 935–940. 1998.

Spitaels, F., Li, L., Wieme, A., Balzarini, T., Cleenwerck, I., Van Landschoot, A., De Vuyst, L., Vandamme, P. "*Acetobacter lambici* sp. nov., isolated from fermenting lambic beer." *International Journal of Systematic and Evolutionary Microbiology.* 64(4), pp. 1083–1089. 2013.

Suida, H. "Process for the recovery of concentrated acetic acid." *United States Patent Office*, 175832. 1929.

Sun, X., Luo, H., Dai, S. "Ionic liquids-based extraction: A promising strategy for the advanced nuclear fuel cycle." *Chemical Reviews.* 112(4), pp. 2100–2128. 2012.

Suzuki, R., Zhang, Y., Iino, T., Kosako, Y., Komagata, K., Uchimura, T. "*Asaia astilbes* sp. nov., *Asaia platycodi* sp. nov., and *Asaia prunellae* sp. nov., novel acetic acid bacteria isolated from flowers in Japan." *Journal of General and Applied Microbiology.* 56(4), pp. 339–346. 2010.

Tanasupawat, S., Kommanee, J., Yukphan, P., Moonmangmee, D., Muramatsu, Y., Nakagawa, Y., Yamada, Y. "*Gluconobacter uchimurae* sp. nov., an acetic acid bacterium in the α-*Proteobacteria*." *Journal of General and Applied Microbiology.* 57(5), pp. 293–301. 2011.

Tesfaye, W., Morales, M. L., García-Parrilla, M. C., Troncoso, A. M. "Wine vinegar: Technology, authenticity and quality evaluation." *Trends in Food Science and Technology.* 13(1), pp. 12–21. 2002.

Usman, M. R., Hussain, S. N., Asghar, H. M. A., Sattar, H., Ijaz, A. "Liquid-Liquid extraction of acetic acid from an aqueous solution using a laboratory scale sonicator." *Journal of Quality and Technology Management.* 7(2), pp. 115–121. 2011.

Vu, H. T. L., Yukphan, P., Chaipitakchonlatarn, W., Malimas, T., Muramatsu, Y., Bui, U. T. T., Tanasupawat, S., Duong, K. C., Nakagawa, Y., Pham, H. T., Yamada, Y. "Nguyenibacter vanlangensis gen nov sp. nov. an unusual acetic acid bacterium in the α-Proteobacteria." *Journal of General and Applied Microbiology.* 59(2), pp. 153–166. 2013.

Wang, Z., Sun, J., Zhang, A., Yang, S.-T. "Propionic acid fermentation." In: Yang, S.-T., El-Enshasy, H. A., Nuttha, T. (eds.). *Bioprocessing Technologies in Biorefinery for Sustainable Production of Fuels, Chemicals, and Polymers* (pp. 331–350). Wiley, AIChe, New York, USA. 2013a.

Wang, Z., Yan, M., Chen, X., Li, D., Qin, L., Li, Z., Yao, J., Liang, X. "Mixed culture of Saccharomyces cerevisiae and Acetobacter pasteurianus for acetic acid production." *Biochemical Engineering Journal.* 79, pp. 41–45. 2013b.

Wolfe, A. J. "The acetate switch." *Microbiology and Molecular Biology Reviews.* 69(1), pp. 12–50. 2005.

Xu, Z., Shi, Z., Jiang, L. "Acetic and propionic acids." In: Moo-Young, M. (ed.). *Comprehensive Biotechnology.* 2nd Edition (pp. 189–199). Academic Press, Burlington. 2011.

Yamada, Y., Hoshino, K. I., Ishikawa, T. "The phylogeny of acetic acid bacteria based on the partial sequences of 16S ribosomal RNA: The elevation of the subgenus *Gluconoacetobacter* to the generic level." *Bioscience, Biotechnology, and Biochemistry.* 61(8), pp. 1244–1251. 1997.

Yamada, Y., Katsura, K., Kawasaki, H., Widyastuti, Y., Saono, S., Seki, T., Uchimura, T., Komagata, K. "Asaia bogorensis gen nov, sp. nov., an unusual acetic acid bacterium in the alpha-Proteobacteria." *International Journal of Systematic and Evolutionary Microbiology.* 50(2), pp. 823–829. 2000.

Yamada, Y., Yukphan, P., Vu, H. T. L., Muramatsu, Y., Ochaikul, D., Nakagawa, Y. "Subdivision of the genus Gluconacetobacter Yamada, Hoshino and Ishikawa 1998: The proposal of Komagatabacter gen. nov., for strains accommodated to the Gluconacetobacter xylinus group in the α-Proteobacteria." *Annals of Microbiology.* 62(2), pp. 849–859. 2012.

Yukphan, P., Malimas, T., Muramatsu, Y., Potacharoen, W., Tanasupawat, S., Nakagawa, Y., Tanticharoen, M., Yamada, Y. "Neokomagataea gen nov, with descriptions of Neokomagataea thailandica sp. nov. and Neokomagataea tanensis sp. nov., osmotolerant acetic acid bacteria of the α-Proteobacteria." *Bioscience, Biotechnology, and Biochemistry.* 75(3), pp. 419–426. 2011.

Yukphan, P., Malimas, T., Muramatsu, Y., Takahashi, M., Kaneyasu, M., Potacharoen, W., Tanasupawat, S., Nakagawa, Y., Hamana, K., Tahara, Y., Suzuki, K. I., Tanticharoen, M., Yamada, Y. "Ameyamaea chiangmaiensisgen nov, sp. nov., an Acetic Acid Bacterium in the α-Proteobacteria." *Bioscience, Biotechnology, and Biochemistry.* 73(10), pp. 2156–2162. 2009.

Yukphan, P., Malimas, T., Muramatsu, Y., Takahashi, M., Kaneyasu, M., Tanasupawat, S., Nakagawa, Y., Suzuki, K.-I., Potacharoen, W., Yamada, Y. "Tanticharoenia sakaeratensisgen nov, sp. nov., a new osmotolerant acetic acid bacterium in the α-Proteobacteria." *Bioscience, Biotechnology, and Biochemistry.* 72(3), pp. 672–676. 2008.

Yukphan, P., Potacharoen, W., Tanasupawat, S., Tanticharoen, M., Yamada, Y. "*Asaia krungthepensis* sp. nov., an acetic acid bacterium in the α-*Proteobacteria.*" *International Journal of Systematic and Evolutionary Microbiology.* 54(2), pp. 313–316. 2004.

Yun, J. H., Lee, J. Y., Hyun, D. W., Jung, M. J., Bae, J. W. "*Bombella apis* sp. nov., an acetic acid bacterium isolated from the midgut of a honey bee." *International Journal of Systematic and Evolutionary Microbiology.* 67(7), pp. 2184–2188. 2017.

4 Conventional and green pharmaceutical products – a review

Srinivasan Kameswaran, Bellamkonda Ramesh, and P. Sudhakar Reddy

CONTENTS

4.1 INTRODUCTION

The success story of pharmaceutical sciences throughout history is astounding. Everywhere in daily life, pharmaceutical industry products can be found. Chemicals used in pharmaceuticals have more or less specialised biological activities. They support the current way of life while also enhancing our wellbeing and high standard of living. Chemical and pharmaceutical manufacture, as well as their use and application, have long been linked to significant environmental contamination and negative health effects. Significant progress was made in the second half of the 20th century in preventing environmental contamination and minimising its negative effects on human health. In affluent nations today, proper, efficient treatment and prevention of emissions into the air, water, and land are in place and spreading globally. In the 1990s, it was discovered that 50–100 kg of waste were produced for every kg of an active pharmaceutical molecule synthesised (Sheldon 2007). This realisation sparked initiatives within the pharmaceutical industries to limit waste generation through a variety of means, including the use of new, better (i.e., "greener") solvents and the creation of alternative synthesis methods to prevent intensive waste. However, since the turn of the century, it has come to light that the pharmaceutical businesses' own products, or the pharmaceuticals themselves, constitute a brand-new

DOI: 10.1201/9781003394600-5

kind of environmental pollution and potential health hazards for the user. Because of increasing living standards, longer life spans, and increased drug use as individuals age, it is anticipated that pharmaceutical consumption would rise in the future. Pharmaceuticals have been discovered in the environment at low concentrations (ng L^{-1} to g L^{-1}) due to the ever-increasing sensitivity of analytical tools, along with other micro-pollutants like disinfectants, pesticides, flame retardants, de-icing fluids, and others (Heberer 2002; Fatta-Kassinos et al. 2009; Kümmerer 2010). Not because of incorrect use, but rather because of appropriate use, these compounds frequently wind up in the environment. Despite concerns about overuse, medications are occasionally disposed of improperly by flushing them down the toilet, pouring them down the drain, or throwing them in the trash. It is commonly acknowledged that drugs are present in the environment. It is difficult to determine the relative rates of pharmaceutical production, release, and usage in the ensuing 10–50 years, although pharmaceutical loading into the environment is anticipated to rise for a number of reasons. First, there will be a rise in the intensive use of drugs (such as using many drugs at once) as the population of elderly people grows. Additionally, when living standards rise and drugs become more accessible, their use will rise everywhere, but especially in nations with fast-growing economies. When evaluating the risks of drugs in the environment, these modifications must be taken into account. It is undoubtedly incorrect to assume that pharmaceutical output and use will remain roughly constant. The possibility exists that medications, their metabolites, and transformation products could penetrate the aquatic environment and eventually make their way into drinking water supplies if they are not removed during sewage treatment or sorbed in soil. Human drug use leaves behind traces has been a hot topic for a while. There are worries regarding these substances' existence in drinking water and the aquatic environment due to their biological activity. Do these substances have a negative impact on people or other living things in the environment? Research initially concentrated on analysing and locating these micro-pollutants. Later, studies into the fate and (eco)toxic repercussions took centre stage. The importance of risk management and risk assessment is now growing. Despite the fact that the subject of medicines in the environment is still relatively new, a substantial body of literature has already been written on the subject. In this article, some significant topics are discussed and a very quick summary of the state of the art is provided. The reader is recommended to check the many books and more specialist reviews that have already been published for more extensive information and findings (Heberer 2002; Daughton and Ternes 1999; Kümmerer 2001, 2004/2008, 2009a,b,c; Thiele-Bruhn 2003; Williams 2005; Ternes and Joss 2006).

4.2 WHAT ARE ACTIVE PHARMACOLOGICAL INGREDIENTS?

A pharmaceutical often consists of one or more active pharmaceutical ingredients (APIs), excipients, and additives, as well as inorganic salts or other organic molecules, such sugars, fragrances, pigments, and colours. They frequently have negligible environmental impact. However, some medications may have chemical excipients and additives that disturb the endocrine system. Excipients and additives are only known to be important if they are compounds that are also utilised for other things, which is unfortunate. The APIs are the main topic of this study and ongoing research. Chemically speaking, APIs include a broad spectrum of so-called small molecules with various physicochemical and biological properties (their molecular weights typically fall between 200 and 500 Da). Even minor modifications to an API's chemical structure can have a big impact on how that API interacts with the environment (Cunningham 2008). Therefore, even when considering various environmental factors like pH, individual analyses for each type of API may occasionally be required. Some medications contain biopharmaceutical molecules, which are pharmaceuticals made utilising biotechnology methods other than direct extraction from an unmodified (i.e., native) biological source. Examples include recombinant human insulin, nucleic acids, and proteins (including antibodies). They are not yet the subject of environmental research or risk management, and their environmental relevance is unclear. According to research, some are digested by humans or can degrade during sewage treatment (Straub 2010). Environmentally relevant substances, like plasmids, have been discovered. Additionally, it is understood that sewage treatment maintains the protein structures of prions, which is particularly stable1 (Hinckley et al. 2008). By sorption onto sewage sludge and subsequent entry into sludge treatment and disposal procedures, prions are removed from the water. There is now much insufficient information on the role that biopharmaceuticals play in the environment to draw any firm judgments. Biopharmaceuticals do not provide a solution because they are frequently used in conjunction with traditional small molecules, such as antineoplastic drugs. X-ray contrast media, contrast materials used in magnetic resonance imaging (MRI), and disinfectants are examples of other classes of compounds that are of interest and employed in medical contexts in diagnostic applications.

4.3 TRANSFORMATION PRODUCTS, METABOLITES, AND PARENT COMPOUNDS

Drugs can be categorised based on their biological action and intended usage (e.g., antibiotics are used topically to treat bacterial infections, analgesics are used to lessen pain, and antineoplastics are used in anticancer therapy). The APIs within subgroups of medications, such as the group, or subgroups, of antibiotics, such as-lactams, cephalosporins, penicillins, and quinolones, are primarily classified according to chemical structure. Other classifications are based on the mode of action (MOA), for example, alkylating agents or antimetabolites that fall within the category of cytotoxics and/or antineoplastics. When it comes to MOA classification, molecules within the same group may have quite distinct chemical structures, which can lead to various environmental outcomes. All chemicals are interesting in terms of their biological action, generally speaking. Despite their decreased usage, some of them are of exceptional interest since they are active in very low quantities. These medications have endocrine effects, making them hormones like ethinylestradiol, the primary ingredient in most birth control pills (Christiansen et al. 2009; Williams et al. 2009; Sumpter et al. 2006). Anticancer medications are also very important because some of them can actually lead to cancer (Johnson et al. 2008; Rowney et al. 2009; Kümmerer and Al-Ahmad 2010). Since antibiotics may encourage resistance, they are of particular importance. Before being excreted, many medications undergo a structural change in the bodies of both humans and animals, producing metabolites. Only very seldom is this transformation complete, meaning that often some amount of the original substance (the API) is eliminated along with the metabolites. For instance, certain antibiotics are digested up to 95% while others are only metabolised at 5%. According to a research of API levels utilised and excretion rates, antibiotics used in Germany account for 75% of excretion rates as still active APIs (Kümmerer and Henninger 2003). Furthermore, bacterial activity in sewage treatment plants (STPs) and the environment might release compounds like glucoronides and others once again. Both parent chemicals and metabolites can undergo structural alterations through a range of biotic and abiotic mechanisms after their excretion and introduction into the environment. Pharmaceuticals can undergo incomplete transformation by environmental organisms like bacteria and fungus, as well as by light and other abiotic chemical reactions (Haiß and Kümmerer 2006; Groning et al. 2007; Trautwein et al. 2008). Structures may also change as a result of technical processes, such as the oxidation, hydrolysis, and photolysis of effluent (Ravina et al. 2002; Zuhlke et al. 2004; Lee et al. 2007; Perez-Estrada et al. 2007; Méndez-Arriaga et al. 2008). Transformation products are the chemicals that result (Langin et al. 2008). New chemical entities with new properties are created as a result of such structural modifications.

4.4 RESOURCES FOR ENVIRONMENTALLY ACTIVE PHARMACEUTICAL INGREDIENTS

Excipients, additives, and APIs used in pharmaceuticals may reach the environment through a variety of nonpoint sources, including production facilities, STP effluents, household trash, and landfill effluent. Animals excrete pharmaceuticals used in veterinary care, such as growth boosters and other objectives, and these pharmaceuticals end up in the soil. They might get into the groundwater if they are not anchored to soil components. Some may also be carried to surface water from runoff during strong rain occurrences. Typically, it is believed that North America and Europe have minimal pharmaceutical manufacturing and production emissions. However, pharmaceutical manufacturing facilities have just recently been discovered to be a significant source of pharmaceuticals entering the environment in the United States (http://toxics.usgs.gov/highlights/PMFs.html). For a particular antibiotic, Norway's local manufacturer contributed significantly more than hospitals and the general public did (Thomas and Langford 2010). Other recent study indicated that industrial facility effluents in Asian nations can include amounts of API up to several mg L1 (Larsson et al. 2007; Li et al. 2008a,b). The need for further data is evident from these findings. To my knowledge, there are no data on emissions during storage and transportation.

Pharmaceuticals are present in hospital effluent as predicted (Kümmerer 2001; Steger-Hartmann et al. 1996; Hartmann et al. 1999; Gomez et al. 2006; Brown et al. 2006; Seifrtová et al. 2008; Schuster et al. 2008). Hospital wastewater has a higher concentration of drugs than municipal sewage. However, the overall load is significantly smaller than that associated with municipal wastewater because hospitals in affluent nations utilise medications at a much lower proportion than the general public (Thomas and Langford 2010; Schuster et al. 2008). Municipal wastewater significantly dilutes hospital wastewater, by a factor of at least 100 (Kümmerer and Helmers 2000). In other words, hospitals are a source of modest relevance; thus, it may not be necessary to treat this waste separately, but there are still ways to reduce the amount of APIs in hospital effluent, such as by using them properly. Since 1994, it has been legal under EU law to dispose of unwanted medications in household waste. These are perhaps the most efficient and environmentally responsible solutions if the rubbish is burnt. If the garbage is landfilled, the APIs will eventually surface in the landfill's wastewater (Eckel et al. 1993; Holm et al. 1995; Ahel and Jelicic 2001; Metzger 2004). If the wastewater is not collected, it could contaminate groundwater or surface waters if there is no collection system in place. According to research, people flush leftover and expired tablets and liquid medications down the toilet to dispose of them (Götz and Keil 2007; Bound and Voulvoulis 2005; Seehusen and Edwards 2006; Abahussain et al. 2006; Ruhoy and Daughton 2007). These data imply that patient education regarding the correct disposal of unused and expired drugs is necessary in all nations. There are take-back programmes in existence in various nations (Niquille and Bugnon 2008; Vollmer 2010). According to Ruhoy and Daughton (2007) alone, orphaned pharmaceuticals from the deceased population may account for 19.7 tonnes of APIs dumped into sewage systems in the United States each year. One simple strategy to reduce pharmaceutical pollution is to get people to stop flushing away their unused medication. The U.S. Office of National Drug Control Policy (ONDCP) recommends removing unused, unnecessary, or expired prescription medications from their original containers and throwing them in the garbage. The ONDCP advises mixing the medications with an unwanted item, such as coffee grounds or kitty litter, to prevent accidental poisonings or potential drug misuse. Before being disposed of, the mixture should be put into imperme-able, unassuming containers like discarded cans or sealable plastic bags. However, the U.S. Food and Drug Administration advise against flushing away several regu-lated medications, including the analgesics OxyContin® and Percocet®. Individual municipalities are implementing pharmaceutical take-back programmes that collect unused prescriptions and then burn them in order to ensure that drugs are disposed of in the most ecologically responsible way possible.

4.5 FATE AND OCCURRENCE IN THE ENVIRONMENT

In the meanwhile, there is proof that the aquatic environment contains about 160 different APIs. APIs have been discovered in surface water, seawater, groundwater, and drinking water as well as in the effluent from medical care facilities, landfills, municipal sewage, and STPs (Heberer 2002; Daughton and Ternes 1999; Kümmerer 2001, 2004/2008, 2009a,b,c; Benotti et al. 2009). Seasonal changes in sewage, reclaimed waste, and finished water (the water that is produced following technologi-cal treatment) have all been investigated (Alexy et al. 2006; Loraine and Pettigrove 2006). Additionally, they are found in the Arctic climate (Kallenborn et al. 2008). It has been determined that pharmaceutical concentrations in surface water and effluent from STPs range from ng L^{-1} to g L^{-1}. Only recently have illicit and psy-choactive substances been found in surface water and wastewater, including amphet-amine, cocaine and its metabolite benzoylecgonine, morphine, 6-acetylmorphine,

11-nor-9-carboxy-9-tetrahydrocannabinol, and methadone and its primary metabolite 2-ethylidene-1,5-dimethyl3,3-diphenylpyrrolidine (Zuccato et al. 2005; Boleda et al. 2007; Huerta-Fontela et al. 2008; Castiglioni et al. 2008). Despite the fact that understanding the fate of pharmaceuticals in sewage sludge and biosolids is crucial for risk assessment, there is a paucity of literature on the subject (Jones-Lepp and Stevens 2007). The occurrence, fate, and activity of metabolites are poorly understood. It is crucial to determine whether the glucoronides, methylates, glycinates, acetylates, and sulphates are still active as well as whether they can be broken down again by bacteria in the environment and during sewage treatment. The active component would then be released once more as a result of this. Additionally, other kinds of metabolites are expelled and are visible in wastewater (Miao et al. 2005). They might have less of an impact on environmental organisms than their parent chemicals did. However, as has been demonstrated with norfluoxetin, the situation may be different in the case of prodrugs, as it may be for the metabolites of a number of other medications (Nałecz-Jawecki 2007). Suppression (important, for example, for tetracyclines and quinolones) (Golet et al. 2002) and (bio)degradation are the main processes for the removal of pharmaceuticals from the various environmental compartments. Additionally important in surface water and technical treatment methods are photo-degradation and hydrolysis. The spread (particle-bound transit) and (bio)availability of medications in the environment may be impacted by sorption. A drop in activity may accompany a loss in detectability due to complex formation or binding to particles. Sorbing APIs may diffuse into sludge flocks, stones in rivers and lakes, or bio-films found in sewage pipelines. Results from Maskaoui and Zhou (2009) show that sorption to colloids provides a major sink for pharmaceuticals in the aquatic environment, indicating that colloids appear to be an essential sink for pharmaceuticals. As a result of these potent pharmaceutical/colloid interactions, drugs may be stored for a long time, increasing their persistence while lowering their bioavailability in the environment. Since the concentration in these so-called "reservoirs" may be significantly higher than when they are free in water, sorption generally leads to biassed risk estimation. It is anticipated that aquatic colloids will have an effect on the bio-availability and bioaccumulation of medications due to their abundance, universality, and power as sorbents (Maskaoui and Zhou 2009). As a result, nothing is known about the effects and behaviour of antibiotics in biosolids with such a high bacterial density and unique environmental conditions. In this regard, even less is known about conjugates, additional metabolites, and transformation products. The two types of organisms most able to break down organic molecules are bacteria and fungi. While fungi are particularly significant in soils, they often have little impact on the aquatic environment. Therefore, it is considered that bacteria are in charge of the majority of biodegradation activities in STPs as well as in surface, ground, and ocean. The kind and quantity of microorganisms present as well as the API itself determine the rate and extent of API biodegradation in sewage treatment and the aquatic environment. The presence of pharmaceuticals in the aquatic environment indicates sluggish removal processes in the aquatic environment as well as, at the very least, partial degradation and elimination in sewage treatment. It is frequently believed that the metabolism and modification of APIs, such as through improved effluent treatment, reduces their toxicity. However, in other circumstances, metabolism results in more active substances (e.g., in the case of prodrugs). For phototransformation and other oxidising processes, the same has been discovered. However, frequently a single test was used to monitor only one end point. The reaction products of photolysis and (photo)oxidation processes have been shown to have mutagenic and other hazardous qualities (Lee et al. 2007; Méndez-Arriaga et al. 2008; Isidori et al. 2005, 2007; DellaGreca et al. 2007; Wei-Hsiang and Young 2008; Calza et al. 2008; Rizzo et al. 2009; Naddeo et al. 2009).

4.6 EFFECTS

Environmental hygiene appears to be more important to human health concerns from drugs in the environment than toxicology and pharmacology (Jones et al. 2004). There are a few exceptions, though: Human sexual development may be affected by endocrine-active substances and hormones since they are potent substances that interact with hormone systems (Hannah et al. 2009; Lange et al. 2001; Kidd et al. 2007; Caldwell et al. 2008). Additionally, antibiotics may contribute to the selection of bacteria that are resistant to antibiotics and some anticancer medications may actually cause cancer directly, even at very low dosages (Kümmerer 2009a,b,c). These problems are not well understood. The greatest amount of contaminated water that a person can consume in their lifetime (2 litres of drinking water per day during a lifespan of 70 years) is much less than the dosages used in general therapy. This assertion, however, is predicated on a few suppositions, including the following: (a) that the quality and quantity of the effects and side effects experienced during therapeutic use (short-term, high dosage) are equivalent to those experienced during a lifelong ingestion (long-term ingestion, low dosage); (b) that the effects are the same for foetuses, babies, children, healthy adults, and elderly people; and (c) that the risk presented by a single compound. In toxicology and ecotoxicology, the question of how to extrapolate data from a high-dose short-term ingestion during therapy to a low-dose long-term ingestion, i.e., drug taken by drinking water, is still open. Elderly adults, young children, and expectant women may be at risk (Kümmerer and Al-Ahmad 2010; Collier 2007), although risk evaluations frequently do not particularly address these groups. Although it is growing, there is still insufficient knowledge regarding how active compounds affect species in both aquatic and terrestrial habitats (Fent et al. 2006). In long-term experiments, effects on fish, daphnia, algae, and bacteria have been shown at low doses (Holten-Lützhøft et al. 1999; Yamashita et al. 2006). Diclofenac's effective concentration for chronic fish toxicity was in the wastewater concentration range (Schwaiger et al. 2004; Triebskorn et al. 2004, 2005; Hoeger et al. 2007), whereas propranolol and fluoxetine's were close to the highest recorded STP effluent values for zooplankton and benthic species (Fent et al. 2006). Concentrations and environmental dangers are lower in surface water. But there are very few focused ecotoxicological studies, and they are required to concentrate on minor environmental effects (Fent et al. 2006). Chronic effects frequently don't produce blatantly obvious effects. Instead, they could result in more subtle alterations over a longer period of time. As a result, these effects are frequently disregarded and an ecosystem-level cause-and-effect relationship cannot be formed. Although single compounds constitute the basis for all risk evaluations (Jones et al. 2004), it has been discovered that combinations may have effects that are different from those of single compounds (Christiansen et al. 2009; Silva et al. 2002; Pomati et al. 2007; Pohl et al. 2009). A standardised test has also been demonstrated to underestimate the impact of (Kümmerer et al. 2004). Additionally, it has been discovered that the transfer of chemicals within the food web may have negative impacts. Diclofenac, for instance, has unquestionably had a very severe effect on vulture populations in Southeast Asia and maybe abroad. In veterinary medicine, this non-steroidal anti-inflammatory medication is applied to sick animals in Asia. Old world vultures eat dead livestock and consume any diclofenac that is still present in the carcase. Sadly, vultures are extremely sensitive to this medication, and a dose of just 1 mg results in rapid kidney failure and death in a matter of days (Oaks et al. 2004). The oriental white-backed vulture experienced substantial yearly adult and subadult mortality (5–86%) between 2000 and 2003, and the population reductions that followed (34–95%) were linked to renal failure and visceral gout. By giving oriental white-backed vultures direct oral exposure to diclofenac residues and treated livestock remains, researchers were able to replicate diclofenac residues and kidney illness in vultures in an experimental setting (Oaks et al. 2004). Studies on one of these species have shown substantial evidence that death brought on by ingesting leftovers of the veterinary non-steroidal anti-inflammatory medication diclofenac. Other research indicates that the dramatic decreases in vulture populations across the subcontinent are most likely the result of veterinary usage of diclofenac (Swan et al. 2006; Taggart et al. 2007). Tens of millions of vultures in Asia have died as a result of this one drug, and three species are now in danger of going extinct. Ketoprofen, a non-steroidal anti-inflammatory medicine, was just recently discovered to have the same effects on these animals. The condor and other new world vultures do not appear to be adversely impacted by diclofenac. Other structurally comparable non-steroidal anti-inflammatory medications have little effect in old world vultures. Therefore, it is unlikely that testing could have predicted this ecological disaster. Other surprises cannot be ruled out in the future because we will never be able to test all pharmaceuticals against all creatures due to a lack of knowledge, resources, and time. This illustration makes it very evident that there is a need to lessen the input and presence of drugs in the environment. Hahn and Schulz (2007) presented a nuanced example of antibiotics' unintended side effects. Results of food choice trials with *Gammarus pulex* showed distinct preferences for leaves that had been conditioned without two antibiotics, oxytetracycline and sulfadiazine, as opposed to those that had been conditioned with them. One of the most significant classes of APIs used in medicine, antibiotics are now found in aquatic environments (Kümmerer 2009a,b,c). Antimicrobials like antibiotics have long been used to prevent the negative consequences of bacteria proliferation. In general, the development of antimicrobial resistance is a highly complex process that, even in a medical setting, is still poorly understood in terms of the importance of the interaction between bacterial populations and antibiotics. Antibiotic-resistant bacteria have been discovered in the aquatic environment (Kümmerer 2009a,b,c; Watkinson et al. 2007; Kim and Aga 2007; Schluter et al. 2007; Caplin et al. 2008). Is the introduction of antibiotics into the environment a significant element in the development of resistant bacteria, or is the spread of resistance from existing resistant bacteria as a result of incorrect use of antibiotics much more significant? At concentrations as low as those observed for antimicrobials in the environment, the relationships between the presence of antimicrobials and the promotion of resistant bacteria as well as the transfer of resistance have not yet been established. There is limited and conflicting information regarding subinhibitory concentrations of antibiotics and how they affect environmental bacteria. In any case, using antibiotics responsibly is essential to prevent resistance. There are extremely few documented instances where a pharmaceutical has been demonstrably proved to have a negative impact on an ecosystem, despite the fact that there are many different human medications present in the environment. Sex steroids, and in particular estrogens like 17-ethinylestradiol (EE2), have been discovered in the aquatic environment by the 1990s. They could have an impact on fish reproduction. The tablet is the only product that contains EE2It is abundant in the aquatic environment, albeit most rivers only have sub-ng L^{-1} concentrations (Hannah et al. 2009). However, fish are so sensitive to this substance that even very low ng L^{-1} concentrations in their water can cause population collapses by shifting the male-to-female sex ratio in favour of females due to males' lack of sexual differentiation (Lange et al. 2001; Kidd et al. 2007). Fortunately, although the margin of safety is not very big, real-world EE2 concentrations are lower than those that significantly impact fish (Hannah et al. 2009; Caldwell et al. 2008). One of these instances of the effects of pharmaceuticals on wildlife, EE2, may have and probably should have been anticipated (Sumpter et al. 2006), but the other—diclofenac killing off old world vultures—came as a complete shock.

4.7 RISKS AND HAZARDS

There is currently very little information available about potential risks posed by pharmaceuticals present in water for animals and plants. However, the examples of EE2 and diclofenac, along with the problem of antibiotic resistance, show that we are currently unable to forecast the environmental effects of widespread environmental contamination by a large number of human medications. The instances of EE2 and diclofenac also show that, despite pharmaceutical concentrations being relatively low in almost all sites, they can nonetheless have negative impacts on wildlife under specific conditions. Choosing which medications to be concerned about and which animals are particularly vulnerable are the main uncertainties. There are just a few compounds for which comprehensive ecotoxicological research are available. The rationale is due to the enormous quantity of compounds as well as the fact that only extensive research can yield relevant conclusions. But higher time and financial expenditures are associated with the discovery of chronic impacts. Only at concentrations much above the present observations in surface water do the majority of the APIs that have been examined so far display acute toxicity for aquatic species. For metabolites and transformation products, data necessary for a reliable risk evaluation are mostly absent. Furthermore, risk evaluations for mixes have not yet been done; just for single compounds. Some APIs are classified as CMR substances because they have potential to cause cancer, mutation, or reproductive harm. How these substances should be handled during the risk assessment procedure is unknown (Kümmerer et al. 2008). The drugs used to treat cancer are a subgroup of this particular set of chemicals. There is a dearth of knowledge regarding the course, impacts, and importance to humans and the ecosystem. Risk analyses are provided, for example, for cyclophosphamide and ifosfamide. The outcome varies depending on the databases and methods used. While some claim there is no risk, others give facts and extra information that leads to the conclusion that "risk cannot be ruled out" (Kümmerer and Al-Ahmad 2010). Longer exposures may lead to multiple contaminations of the environment, and persistent organic pollutants increase the possibility for long-term, diverse consequences. With the current test systems available, these issues cannot be identified beforehand (Cairns and Mount 1990).

4.8 ASSESSING RISK

Combinations of management techniques, such as pharmaceutical return programmes, stakeholder consultation, enhanced effluent treatment, and financial incentives for the creation of "green" drugs, will probably be most successful in reducing the dangers posed by pharmaceuticals (Daughton 2003). To effectively reduce the input of pharmaceuticals into the environment, they are all required. Advanced effluent treatment is the one tactic that has received the most attention recently. In order to find practical solutions for the third strategy, environmental protection must consider the stockholders, stakeholders, and users of the compounds, such as patients, nurses, physicians, and pharmacists. The fourth tactic is being developed in the area of green chemistry. It looks to offer the best long-term prospects for sustainability. The advanced treatment of effluents has been studied utilising filtration (Drewes et al. 2002; Heberer and Feldmann 2005), application of powdered charcoal (Metzger et al. 2005; Nowotny et al. 2007), (photochemical) oxidation processes (Isidori et al. 2007; Watkinson et al. 2007; Strässle 2007), and artificial wetlands (Drewes et al. 2002; Heberer and Feldmann 2005; Matamoros and Bayona 2006). Reviews of the benefits and drawbacks of various technologies are available (Ternes and Joss 2006; Schulte-Oehlmann et al. 2007; Jones et al. 2007; Wenzel et al. 2008). Each kind of advanced effluent treatment has been discovered to have certain significant downsides, some of which are distinctive (Jones et al. 2007; Wenzel et al. 2008; Kümmerer 2007). For instance, the reaction products of (photo)oxidation processes have been reported to have hazardous and mutagenic characteristics (Lee et al. 2007; Méndez-Arriaga et al. 2008; Isidori et al. 2005, 2007; Wei-Hsiang and Young 2008; Calza et al. 2008). As a result, the use of advanced effluent treatment should not be viewed as a generic solution. Instead, it should only be taken into account as a piece of the solution when necessary. For the majority of compounds, hospital contributions to the overall load of pharmaceuticals in municipal wastewater are below 10% and frequently even below 3% (Kümmerer and Henninger 2003; Thomas and Langford 2010; Schuster et al. 2008; Heberer and Feldmann 2005; Bayerisches Landesamt für Umwelt 2005; Heinzmann et al. 2006; Thomas et al. 2007). Therefore, it is debatable whether separate treatment of hospital effluent is a viable environmental objective from an economic perspective. What proportion of drug residues that are present in the ambient environment are caused by discarded leftover drugs is a significant unanswered question with regard to the use of medicines as pollutants (Götz and Keil 2007; Ruhoy and Daughton 2007). Therefore, giving doctors, pharmacists, and patients the right advice can help reduce the amount of APIs that are introduced into the aquatic environment (Wennmalm and Gunnarson 2010). Initial findings indicate that this strategy may help to limit the input of medications with undesirable environmental qualities. Data on the types, amounts, and frequency of drug accumulation in homes, hospitals, centres for the elderly, and rehabilitation hospitals are required (Ruhoy and Daughton 2007). Mid- to long-term changes should be made to increase environmental sensitivity in the prescription, therapy, and consultation processes of doctors and pharmacists as well as in the usage and disposal of medications by patients. In this technique, the relationship between doctors and patients is crucial: Physicians are more vigilant during patient consultations when they are aware of the environmental significance and issues with medications. The environmental consequences of pharmaceuticals should be covered in advanced training for health education and policy teachers as well as medical education to make it easier for doctors to incorporate the issues into their daily practice. By altering the funding of or reimbursement for medications and therapies, sources of health funds might encourage the demand for better ecological alternatives. The pharmaceutical sector can furnish a viable product line with the help of this rising demand. The variety of products is decreased, leading to savings, if a hospital's internal commission publishes a list of suggested medications that forms the foundation for hospital purchasing activities. Reducing the amount of expired medications and, consequently, the environmental burden, in the wards' drug storage area. The internal system ought to enable the wards to return unopened, damaged, and unopened goods to the pharmacy. These leftovers can be properly managed by the pharmacist. A doctor with experience treating infectious infections should be present to offer guidance on how antibiotics should be used. The appropriate amount of hygiene—not too much, nor too little—at the right time and place can also help prevent infections and lessen the need for antibiotics and disinfectants. Currently, drug design places a lot of emphasis on synthesis enhancement while undervaluing the molecules' environmental characteristics. The functionality of a chemical should, in accordance with the principles of green chemistry, contain both easy and quick degradability after usage as well as the features of a chemical necessary for its use. A distinct comprehension of the functionality required for a pharmaceutical results from taking into account the whole life cycle of medications (Kümmerer 2007). By "benign by design," it is meant that consideration is given to facile degradability after usage or application even before a pharmaceutical is created. Such a strategy is not entirely novel. For instance, it happens frequently when developing drugs in terms of undesirable side effects. Long-term, this may also lead to economic benefits (Daughton 2003; Kümmerer 2007). These guidelines and the understanding of green chemistry should be used by researchers and pharmaceutical corporations to medications.

4.9 SUSTAINABLE AND GREEN PHARMACY

The design of pharmaceutical products and procedures that remove or greatly minimise the use and creation of potentially dangerous compounds, as well as the avoidance and/or mitigation of adverse effects on the environment's safety and health, are all part of "green pharmacy" (Clark 2006). "Green" isolates the environmental factors and concentrates only on the medication itself. These are crucial details. There are extra things to think about. For instance, a new drug may be green in terms of the type and volume of waste produced during its synthesis, and it may have utilised renewable feedstock, but the drug may linger in the environment after excretion and cause environmental or health issues. It may not be sustainable if the reproduction of the renewable feedstock requires a lot of water and fertiliser, competes with food production, or depends on an endangered species. The same is true if a compound is environmentally friendly or even sustainable, but its manufacturing, distribution, and use involve large-scale (quantity- or quality-wise) or nonrenewable resource-based material flows. Another concern is that when produced and used in large quantities, greener products or chemicals may start to look less green or even harmful. For instance, if a green packaging material includes an antimicrobial to prevent its contents, such as food, from spoiling, this antimicrobial can contribute to resistance by passing through the food or by being released into the environment when the packing is thrown away. A package may be environmentally friendly but not sustainable. A much broader perspective on sustainability (Kümmerer and Hempel 2010). Additionally, it would inquire as to the food's origin and the length of time we want versus need to preserve it. Instead of raising the possibility of antibiotic resistance through antibiotic packaging, educating people on safe food handling and storage techniques would be a much more sustainable way to prevent food spoilage. These instances highlight the importance to keep the entire life cycle in mind. This also covers additional elements like corporate social responsibility and ethical dilemmas. One can claim, for instance, that we cannot afford sustainable medications since fewer active molecules would reach the market. Or, given that each drug has a benefit, is it all sustainable? A second perspective, though, is to think about how many people (Triggle 2010), for instance, due to financial constraints, do not have any access to appropriate treatment. Or why are there ailments for which there are no novel medications? What are the costs in relation to the advantages? What are the negative effects? In 2002, the European Parliament and the European Commission came to an agreement that within a generation, no-impact chemicals should be manufactured and used (EU Parliam./EU Comm. 2002). This need to apply to medications as well.

REFERENCES

Abahussain, E.A., Ball, D.E. and Matowe, W.C. 2006. Practice and opinion towards disposal of unused medication in Kuwait. *Medical Principles and Practice* 15:352–357.

Ahel, M. and Jelicic, I. 2001. Phenazone analgesics in soil and groundwater below a municipal solid waste landfill. In Daughton, C.G. and Jones-Lepp, T.L. (Eds.), *Pharmaceuticals and Personal Care Products in the Environment*. Scientific and Regulatory Issues. Washington, DC: Am. Chem. Soc. Publ., 6:100–115.

Alexy, R., Sommer, A., Lange, F.T. and Kummerer, K. 2006. Local use of antibiotics and their input and fate in a small sewage treatment plant—significance of balancing and analysis on a local scale vs. nationwide scale. *Acta hydrochimica et hydrobiologica* 34:587–592.

Bayerisches Landesamt für Umwelt. 2005. Arzneistoffe in der Umwelt. http://www.lfu.bayern. de/lfu/umweltberat/data/chem/stoff/arznei_2005.pdf

Benotti, M.J., Trenholm, R.A., Vanderford, B.J., Holady, J.C., Stanford, B.D. and Snyder, S.A. 2009. Pharmaceuticals and endocrine disrupting compounds in U.S. drinking water. *Environmental Science & Technology* 43:597–603.

Boleda, M.R., Galceran, M.T. and Ventura, F.J. 2007. Trace determination of cannabinoids and opiates in wastewater and surface waters by ultra-performance liquid chromatography-tandem mass spectrometry. *Journal of Chromatography A* 1175:38–48.

Bound, J.P. and Voulvoulis, N. 2005. Household disposal of pharmaceuticals as a pathway for aquatic contamination in the United Kingdom. *Environmental Health Perspectives* 113:1705–1711.

Brown, K.D., Kulis, J., Thomson, B., Chapman, T.H. and Mawhinney, D.B. 2006. Occurrence of antibiotics in hospital, residential, and dairy effluent, municipal wastewater, and the Rio Grande in New Mexico. *Science of the Total Environment* 366:772–783.

Cairns, J.J. and Mount, D.I., 1990. Aquatic toxicology. Part 2. *Environmental Science & Technology*, 24(2):154–161.

Caldwell, D.J., Mastrocco, F., Hutchinson, T.H., Lange, R., Heijerick, D., et al. 2008. Derivation of an aquatic predicted no-effect concentration for the synthetic hormone, 17alpha-ethinyl estradiol. *Environmental Science & Technology* 42:7046–7054.

Calza, P., Massolino, C., Monaco, G., Medana, C. and Baiocchi, C. 2008. Study of the photolytic and photocatalytic transformation of amiloride in water. *Journal of Pharmaceutical and Biomedical Analysis* 48:315–320.

Caplin, J.L., Hanlon, G.W. and Taylor, H.D. 2008. Presence of vancomycin and ampicillin-resistant *Enterococcus faecium* of epidemic clonal complex-17 in wastewaters from the south coast of England. *Environmental Microbiology* 10:885–892.

Castiglioni, S., Zuccato, E., Chiabrando, C., Fanelli, R. and Bagnati, R. 2008. Mass spectrometric analysis of illicit drugs in wastewater and surface water. *Mass Spectrometry Reviews* 27:378–384.

Christiansen, S., Scholze, M., Dalgaard, M., et al. 2009. Synergistic disruption of external male sex organ development by a mixture of four antiandrogens. *Environmental Health Perspectives* 117:1839–1846.

Clark, J.H. 2006. Green chemistry: today (and tomorrow). *Green Chemistry* 8:17–21.

Collier, A.C. 2007. Pharmaceutical contaminants in potable water: potential concerns for pregnant women and children. *EcoHealth* 4:164–171.

Cunningham, V. 2008. Special characteristics of pharmaceuticals related to environmental fate. In Kummerer, K. (Ed.), *Pharmaceuticals in the Environment. Sources, Fate, Effects and Risk*, third ed. Berlin/Heidelberg: Springer, 23–34.

Daughton, C.G. 2003. Cradle-to-cradle stewardship of drugs for minimizing their environmental disposition while promoting human health. I. Rationale for and avenues toward a green pharmacy. *Environmental Health Perspectives* 111:757–774.

Daughton, C.G. and Ternes, T.A. 1999. Pharmaceuticals and personal care products in the environment: agents of subtle change? *Environmental Health Perspectives* 107:907–938.

DellaGreca, M., Lesce, M.R., Isidori, M., Montanaro, S., Previtera, L. and Rubino, M. 2007. Phototransformation of amlodipine in aqueous solution: toxicity of the drug and its photoproduct on aquatic organisms. *International Journal of Photoenergy* 2007:63459.

Drewes, J.E, Heberer, T. and Reddersen, K. 2002. Fate of pharmaceuticals during indirect potable use. *Water Science & Technology* 46:73–80.

Eckel, W.P., Ross, B. and Isensee, R. 1993. Pentobarbital found in ground water. *Ground Water* 31:801–804.

EU Parliam./EU Comm. 2002. Decision No. 1600/2002/EC of the European Parliament and of the Council of 22 July 2002 laying down the Sixth Community Environment Action Programme. 10.9.2002. Off. J. Eur. Communities L242/1–15, 2002. 24. http://eur-lex.europa. eu/LexUriServ/LexUriServ.do?uri=OJ:L:2002:242:0001:0015:EN:PDF

Fatta-Kassinos, D., Bester, K. and Kummerer K, eds. 2009. *Xenobiotics in the Urban Water Cycle: Mass Flows, Environmental Processes, Mitigation and Treatment Strategies*. Vol. 16: Environmental Pollution. Dordrecht: Springer.

Fent, K., Weston, A.A. and Caminada, D. 2006. Ecotoxicology of human pharmaceuticals. *Aquatic Toxicology* 76:122–159.

Golet, E.M., Alder, A.C. and Giger, W. 2002. Environmental exposure and risk assessment of fluoroquinolone antibacterial agents in wastewater and river water of the Glatt Valley Watershed, Switzerland. *Environmental Science & Technology* 36:3645–3651.

Gomez, M.J., Petrović, M., Fernández-Alba, A.R. and Barceló, D. 2006. Determination of pharmaceuticals of various therapeutic classes by solid-phase extraction and liquid chromatography-tandem mass spectrometry analysis in hospital effluent wastewaters. *Journal of Chromatography A* 1114:224–233.

Götz, K. and Keil, F. 2007. Medikamentenentsorgung in privaten Haushalten: Ein Faktor bei der Gewässerbelastung mit Arzneimittelwirkstoffen?. *UWSF - Z Umweltchem Ökotox* 19:180–188. https://doi.org/10.1065/uwsf2007.07.201

Groning, J., Held, C., Garten, C., Claussnitzer, U., Kaschabek, S.R. and Schlömann, M. 2007. Transformation of diclofenac by the indigenous microflora of river sediments and identification of a major intermediate. *Chemosphere* 69:509–516.

Hahn, T. and Schulz, R. 2007. Indirect effects of antibiotics in the aquatic environment: a laboratory study on detrivore food selection behaviour. *Human and Ecological Risk Assessment* 13:535–542.

Haiß, A. and Kümmerer, K. 2006. Biodegradability of the X-ray contrast compound diatrizoic acid, identification of aerobic degradation products and effects against sewage sludge micro-organisms. *Chemosphere* 62:294–302.

Hannah, R., D'Aco, V.J., Anderson, P.D., et al. 2009. Exposure assessment of 17αethinylestradiol in surface waters of the United States and Europe. *Environmental Toxicology and Chemistry* 28:2725–2732.

Hartmann, A., Golet, E.M., Gartiser, S., Alder, A.C., Koller, T. and Widmer, R.M. 1999. Primary DNA damage but not mutagencity correlates with ciprofloxacin concentrations in German hospital wastewaters. *Archives of Environmental Contamination and Toxicology* 36:115–119.

Heberer, T. 2002. Occurrence, fate, and removal of pharmaceutical residues in the aquatic environment: a review of recent research data. *Toxicology Letters* 131:5–17.

Heberer, T. and Feldmann, D. 2005. Contribution of effluents from hospitals and private households to the total loads of diclofenac and carbamazepine in municipal sewage effluents—modelling versus measurements. *Journal of Hazardous Materials* 122:211–218.Heinzmann, B., Schwarz, R.J. and Pineau, C. 2006. Getrennte Erfassung von jodorganischen Röntgenkontrastmitteln in Berliner Krankenhäusern und deren Transformation. Presented at Getrennte Erfass. von jodorganischen Rontgenkontrastmitteln in Berliner Krankenh. und deren Transform., Worksh., Berlin.

Hinckley, G.T., Johnson, C.J., Jacobson, K.H., et al. 2008. Persistence of pathogenic prion protein during simulated wastewater treatment processes. *Environmental Science & Technology* 42:5254–5259.

Hoeger, B., Kollner, B., Dietrich, D.R. and Hitzfeld, B. 2007. Water-borne diclofenac affects kidney and gill integrity and selected immune parameters in brown trout (*Salmo trutta* f. *fario*). *Analytical and Bioanalytical Chemistry* 387:1405–1416.

Holm, J.V., Rugge, K., Bjerg, P.L. and Christensen, T.H. 1995. Occurrence and distribution of pharmaceutical organic compounds in the groundwater down gradient of a landfill (Grinsted Denmark). *Environmental Science & Technology* 29:1415–1420.

Holten-Lützhøft, H.C., Halling-Sørensen, B. and Jørgensen, S.E. 1999. Algal toxicity of antibacterial agents applied in Danish fish farming. *Archives of Environmental Contamination and Toxicology* 36:1–6.

Huerta-Fontela, M., Galceran, M.T., Martin-Alonso, J. and Ventura, F. 2008. Occurrence of psychoactive stimulatory drugs in wastewaters in north-eastern Spain. *Science of the Total Environment* 397:31–40.

Isidori, M., Lavorqna, M., Nardelli, A., Parella, A., Previtera, L. and Rubino, M. 2005. Ecotoxicity of naproxen and its phototransformation products. *Science of the Total Environment* 348:93–101.

Isidori, M., Nardelli, A., Pascarella, M., Rubino, M. and Pascarella, A. 2007. Toxic and genotoxic impact of fibrates and their protoproducts on non-target organisms. *Environment International* 33:635–641.

Johnson, A.C., Jurgens, M.D., Williams, R.J., Kümmerer, K., Kortenkamp, A. and Sumpter, J.P. 2008. Do cytotoxic chemotherapy drugs discharged into rivers pose a risk to the environment and human health? An overview and UK case study. *Journal of Hydrology* 348:167–175.

Jones, O.A.H., Voulvoulis, N. and Lester, J.N. 2004. Potential ecological and human health risks associated with the presence of pharmaceutically active compounds in the aquatic environment. *Critical Reviews in Toxicology* 34:335–350.

Jones, O.H.A., Green, P.G., Voulvoulis, N. and Lester, J.N. 2007. Questioning the excessive use of advanced treatment to remove organic micro-pollutants from waste water. *Environmental Science & Technology* 41:5085–5089.

Jones-Lepp, T.L. and Stevens, R. 2007. Pharmaceuticals and personal care products in biosolids/sewage sludge: the interface between analytical chemistry and regulation. *Analytical and Bioanalytical Chemistry* 387:1173–1183.

Kallenborn, R., Fick, J., Lindberg, R., Moe, M., Nielsen, K.M., Tysklind M. and Vasskog T. 2008. Pharmaceuticals in the Environment. Berlin/Heidelberg: Springer. Pharmaceutical Residues in Northern European Environments: Consequences and Perspectives; pp. 61–74.

Kidd, K.A., Blanchfield, P.J., Mill, K.H., et al. 2007. Collapse of a fish population after exposure to a synthetic estrogen. *Proceedings of the National Academy of Sciences USA* 10:8897–8901.

Kim, S. and Aga, D.S. 2007. Potential ecological and human health impacts of antibiotics and antibioticresistant bacteria from wastewater treatment plants. *Journal of Toxicology and Environmental Health B* 10:559–573.

Kümmerer, K. 2001. Drugs, diagnostic agents and disinfectants in waste water and water—a review. *Chemosphere* 45:957–969.

Kümmerer, K. 2004/2008. *Pharmaceuticals in the Environment: Sources, Fate, Effects and Risks,* 2nd/3rd ed. Berlin/Heidelberg/New York: Springer.

Kümmerer, K. 2007. Sustainable from the very beginning: rational design of molecules by life cycle engineering as an important approach for green pharmacy and green chemistry. *Green Chemistry* 9:899–907.

Kümmerer, K. 2009a. Antibiotics in the aquatic environment—a review—part I. *Chemosphere* 75:417–434.

Kümmerer, K. 2009b. Antibiotics in the environment—a review—part II. *Chemosphere* 75:435–441.

Kümmerer, K. 2009c. Pharmaceuticals from human use in the environment—present knowledge and future challenges. *Journal of Environmental Management* 90:2354–2366.

Kümmerer, K. 2010. Emerging contaminants. In Wilderer, P. and Frimmel, F. (Eds.), *Treatise on Water Science*. Amsterdam: Elsevier, 69–87.

Kümmerer, K. and Al-Ahmad, A. 2010. Estimation of the cancer, risk to humans from cyclophoshamide and ifosfamide excreta emitted into surface water via hospital effluents. *Environmental Science and Pollution Research* 17:486–496.

Kümmerer, K., Alexy, R. and Hüttig, J. 2004. Standardized tests fail to assess the effects of antibiotics against environmental bacteria because of delayed effects. *Water Research* 38:2111–2116.

Kümmerer, K. and Helmers, E. 2000. Hospital effluents as a source for gadolinium in the aquatic environment. *Environmental Science & Technology* 34:573–577.

Kümmerer, K. and Hempel M, eds. 2010. *Green and Sustainable Pharmacy*. Heidelberg/New York: Springer.

Kümmerer, K. and Henninger, A. 2003. Promoting resistance by the emission of antibiotics from hospitals and households into effluent. *European Journal of Clinical Microbiology & Infectious Diseases* 9:1203–1214.

Kümmerer, K., Schuster, A., Haiss, A., Günther, A. and Jacobs, J. 2008. Umweltrisikobewertung von Zytostatika (Environmental risk assessment for cytotoxic compounds). Studie im Auftrag des Umweltbundesamtes Freiburg. German Federal Environ. Agency Rep., Berlin.

Lange, R., Hutchinson, T.H., Croudace, C.P. and Siegmund, F. 2001. Effects of the synthetic estrogen 17 alphaethinylestradiol on the life-cycle of the fathead minnow (*Pimephales promelas*). *Environmental Toxicology and Chemistry* 20:1216–1227.

Langin, A., Schuster, A. and Kümmerer, K. 2008. Chemicals in the environment—the need for a clear nomenclature: parent compounds, metabolites, transformation products and their elimination. *Clean* 36:349–350.

Larsson, D.G., de Pedro, C. and Paxeus, N. 2007. Effluent from drug manufactures contains extremely high levels of pharmaceuticals. *Journal of Hazardous Materials* 148:751–755.

Lee, C., Lee, Y., Schmidt, C., Yoon, J. and von Gunten, U. 2007. Oxidation of N-nitrosodimethylamine (NDMA) with ozone and chlorine dioxide: kinetics and effect on NDMA formation potential. *Environmental Science & Technology* 41:2056–2063.

Li, D., Yang, M., Hu, J., et al. 2008a. Determination and fate of oxytetracycline and related compounds in oxyteracycline production wastewater and the receiving river. *Environmental Science & Technology* 27:80–86.

Li, D., Yang, M., Hu, J., Zhang, Y., Chang, H. and Jin, F. 2008b. Determination of penicillin G and its degradation products in a penicillin production wastewater treatment plant and the receiving river. *Water Research* 42:307–317.

Loraine, G.A. and Pettigrove, M.E. 2006. Seasonal variations in concentrations of pharmaceuticals and personal care products in drinking water and reclaimed wastewater in southern California. *Environmental Science & Technology* 40:5811–5816.

Maskaoui, K. and Zhou, J.L. 2009. Colloids as a sink for certain pharmaceuticals in the aquatic environment. *Environmental Science and Pollution Research International* 17:898–907.

Matamoros, V. and Bayona, J.M. 2006. Elimination of pharmaceuticals and personal care products in subsurface flow constructed wetlands. *Environmental Science & Technology* 40:5811–5816.

Méndez-Arriaga, F., Esplugas, S. and Giménez, J. 2008. Photocatalytic degradation of non-steroidal anti-inflammatory drugs with TiO2 and simulated solar irradiation. *Water Research* 42:585–594.

Metzger, J.W. 2004. Drugs in municipal landfills and landfill leachates. In Kümmerer, K. (Ed.), *Pharmaceuticals in the Environment: Sources, Fate, Effects and Risks*, 2nd ed. Berlin/Heidelberg/New York: Springer, 133–138.

Metzger, S., Kapp, H., Seitz, W., Weber, W.H., Hiller, G. and Süßmuth, W. 2005. Entfernung von iodierten Rontgenkontrastmitteln bei der kommunalen Abwasserbehandlung durch den Einsatz von Pulverak- tivkohle. *GWF Wasser Abwasser* 9:638–645.

Miao, X.S., Yang, J.J. and Metcalfe, CD. 2005. Carbamazepine and its metabolites in wastewater and in biosolids in a municipal wastewater treatment plant. *Environmental Science & Technology* 39:7469–7475.

Naddeo, V., Meric, S., Kassinos, D., Belgiorno, V. and Guida, M. 2009. Fate of pharmaceuticals in contaminated urban wastewater effluent under ultrasonic irradiation. *Water Research* 43:4019–4027.

Nałecz-Jawecki, G. 2007. Evaluation of the in vitro biotransformation of fluoxetine with HPLC, mass spectrometry and ecotoxicological tests. *Chemosphere* 70:29–35.

Niquille, A. and Bugnon, O. (2008). Pharmaceuticals and environment: role of community pharmacies. In Kümmerer, K. (Eds.), *Pharmaceuticals in the Environment*. Berlin/Heidelberg: Springer. https://doi.org/10.1007/978-3-540-74664-5_29

Nowotny, N., Epp, B., von Sonntag, C. and Fahlenkamp, H. 2007. Quantification and modeling of the elimination behavior of ecologically problematic wastewater micropollutants by adsorption on powdered and granulated activated carbon. *Environmental Science & Technology* 41:2050–2055.

Oaks, J.L., Gilbert, M., Virani, M.Z., et al. 2004. Diclofenac residues as the cause of vulture population decline in Pakistan. *Nature* 427:630–633.

Perez-Estrada, L.A., Malato, S., Agüera, A. and Fernández-Alba, A.R. 2007. Degradation of dipyrone and its main intermediates by solar AOPs: identification of intermediate products and toxicity assessment. *Catalysis Today* 129:207–214.

Pohl, H.R., Mumtaz, M.M., Scinicariello, F. and Hansen, H. 2009. Binary weight-of-evidence evaluations of chemical interactions—15 years of experience. *Regulatory Toxicology and Pharmacology* 54:264–271.

Pomati, F., Orlandi, C., Clerici, M., Luciani, F. and Zuccato, E. 2007. Effects and interactions in an environmentally relevant mixture of pharmaceuticals. *Toxicological Sciences* 102:129–137.

Ravina, M., Campanella, L. and Kiwi, J. 2002. Accelerated mineralization of the drug diclofenac via Fenton reactions in a concentric photo-reactor. *Water Research* 36:3553–3560.

Rizzo, L., Meric, S., Guida, M., Kassinos, D. and Belgiorno, V. 2009. Heterogenous photocatalytic degradation kinetics and detoxification of an urban wastewater treatment plant effluent contaminated with pharmaceuticals. *Water Research* 43:4070–4078.

Rowney, N.C., Johnson, A.C. and Williams, R.J. 2009. Cytotoxic drugs in drinking water: a prediction and risk assessment exercise for the Thames catchment in the United Kingdom. *Environmental Toxicology and Chemistry* 28:2733–2743.

Ruhoy, I.S. and Daughton, C.G. 2007. Types and quantities of leftover drugs entering the environment via disposal to sewage—revealed by coroner records. *Science for the Total Environment* 388:137–148.

Schluter, A., Szczepanowski, R., Kurz, N., Schneiker, S., Krahn, I. and Pühler, A. 2007. Erythromycin resistance-conferring plasmid pRSB105, isolated from a sewage treatment plant, harbors a new macrolide resistance determinant, an integron-containing Tn402-like element, and a large region of unknown function. *Applied and Environmental Microbiology* 73:1952–1960.

Schulte-Oehlmann, U., Oehlmann, J. and Puttman, W. 2007. Arzneimittelwirkstoffe in der Umwelt—Einträge, Vorkommen und der Versuch einer Bestandsaufnahme. *Umweltwissenschaften und Schadstoffforschung* 19:168–179.

Schuster, A., Hadrich, C. and Kümmerer, K. 2008. Flows of active pharmaceutical ingredients originating from health care practices on a local, regional, and nationwide level in Germany—Is hospital effluent treatment an effective approach for risk reduction? *Water Air Soil Pollution* 8:457–471.

Schwaiger, J., Ferling, H., Mallow, U., Wintermayr, H. and Negele, R.D. 2004. Toxic effects of the non-steroidal anti-inflammatory drug diclofenac. Part I: histopathological alterations and bioaccumulation in rainbow trout. *Aquatic Toxicology* 68:141–150.

Seehusen, D.A. and Edwards, J. 2006. Patient practices and beliefs concerning disposal of medications. *Journal of the American Board of Family Medicine* 19:542–547.

Seifrtová, M., Pena, A., Lino, C.M. and Solich, P. 2008. Determination of fluoroquinolone antibiotics in hospital and municipal wastewaters in Coimbra by liquid chromatography with a monolithic column and fluorescence detection. *Analytical and Bioanalytical Chemistry* 391:799–805.

Sheldon, R.A. 2007. The E factor: fifteen years on. *Green Chemistry* 9:1273–1283.

Silva, E., Rajapakse, N. and Kortenkamp, A. 2002. Something from "nothing"—eight weak estrogenic chemicals combined at concentrations below NOECs produce significant mixture effects. *Environmental Science & Technology* 36:1751–1756.

Steger-Hartmann, T., Kümmerer, K. and Schecker, J. 1996. Trace analysis of the antineoplastics ifosfamide and cyclophosphamide in sewage water by two-step solid-phase extraction and GC/MS. *Journal of Chromatography A* 726:179–184.

Strässle, R. 2007. Mikroverunreinigung: Ozon als Lösung? Umw. Perspekt.—Das Fachmag. *für Erfolgreiches Umweltmanag* 4:34–37.

Straub, J.O. 2010. Protein and peptide therapeuticals: an example of 'benign by nature' active pharmaceutical ingredients. In Klaus, K. and Maximilian, H. (Eds.), *Green and Sustainable Pharmacy*. Heidelberg: Springer Berlin, 118, 127–33.

Sumpter, J.P., Johnson, A.C., Williams, R.J., Kortenkamp, A. and Scholze, M. 2006. Modeling effects of mixtures of endocrine disrupting chemicals at the river catchment scale. *Environmental Science & Technology* 40:5478–5489.

Swan, G.E., Cuthbert, R., Quevedo, M., et al. 2006. Toxicity of diclofenac to Gyps vultures. *Biology Letters* 2:279–282.

Taggart, M.A., Cuthbert, R., Das, D., et al. 2007. Diclofenac disposition in Indian cow and goat with reference to Gyps vulture population declines. *Environmental Pollution* 147:60–65.

Ternes, T.A. and Joss, A, eds. 2006. *Human Pharmaceuticals, Hormones and Fragrances*. The Challenge of MicroPollutants in Urban Water Management. London: IWA Publ. https://doi.org/10.2166/9781780402468

Thiele-Bruhn, S. 2003. Pharmaceutical antibiotic compounds in soils—a review. *Journal of Plant Nutrition and Soil Science* 166:145–167.

Thomas, K.V., Dye, C., Schlabach, M. and Langford, K.H. 2007. Source to sink tracking of selected human pharmaceuticals from two Oslo city hospitals and a wastewater treatment works. *Journal of Environmental Monitoring* 9:1410–1418.

Thomas, K.V. and Langford, K.H. 2010. Point sources of human pharmaceuticals into the aquatic environment. Klaus, K. and Maximilian, H. (Eds.), *Green and Sustainable Pharmacy*. Heidelberg: Springer Berlin, 118, pp. 211–23.

Trautwein, C., Metzger, J. and Kummerer, K. 2008. Aerobic biodegradability of the calcium channel antagonist verapamil and identification of a microbial dead-end transformation product studied by LC-MS/MS. *Chemosphere* 72:442–450.

Triebskorn, R., Casper, H., Heyd, A., Eikemper, R., Kohler, H.R. and Schwaiger, J. 2004. Toxic effects of the non-steroidal anti-inflammatory drug diclofenac. Part II: cytological effects in liver, kidney, gills and intestine of rainbow trout (Oncorhynchus mykiss). *Aquatic Toxicology* 68:151–166.

Triebskorn, R., Casper, H., Scheil, V. and Schwaiger, J. 2005. Ultrastructural effects of pharmaceuticals (carbamazepine, clofibric acid, metoprolol, diclofenac) in rainbow trout (*Oncorhynchus mykiss*) and common carp (*Cyprinus carpio*). *Aquatic Toxicology* 75:53–64.

Triggle, D.J. 2010. Pharmaceuticals in society. In Klaus, K. and Maximilian, H. (Eds.), *Green and Sustainable Pharmacy*. Heidelberg: Springer Berlin, 118, 23–35.

Vollmer, G. 2010. Disposal of pharmaceutical waste in households—a European survey. In Klaus, K. and Maximilian, H. (Eds.), *Green and Sustainable Pharmacy*. Heidelberg: Springer Berlin, 118, pp. 165–178.

Watkinson, A.J., Murby, E.J. and Costanzo, S.D. 2007. Removal of antibiotics in conventional and advanced wastewater treatment: implications for environmental discharge and wastewater recycling. *Water Research* 41:4164–4176.

Wei-Hsiang, C. and Young, T.M. 2008. NDMA formation during chlorination and chloramination of aqueous diuron solutions. *Environmental Science & Technology* 42:1072–1077.

Wennmalm, A. and Gunnarson, B. 2010. Experiences with the Swedish environmental classification scheme. In Klaus, K. and Maximilian, H. (Eds.), *Green and Sustainable Pharmacy*. Heidelberg: Springer Berlin, 118, 243–249.

Wenzel, H., Larsen, H.F., Clauson-Kaas, J., Høibye, L. and Jacobsen, B.N. 2008. Weighing environmental advantages and disadvantages of advanced wastewater treatment of micro-pollutants using environmental life cycle assessment. *Water Science & Technology* 57:27–32.

Williams, R. 2005. *Science for Assessing the Impacts of Human Pharmaceuticals on Aquatic Ecosystems*. Pensacola: SETAC Press.

Williams, R.J., Keller, V.D.J., Johnson, A.C., et al. 2009. National risk assessment for intersex in fish arising from steroid estrogens. *Environmental Toxicology and Chemistry* 28:220–230.

Yamashita, N., Yasojima, M., Nakada, N., et al. 2006. Effects of antibacterial agents, levofloxacin and clarithromycin, on aquatic organisms. *Water Science and Technology* 53:65–72.

Zuccato, E., Chiabrando, C., Castiglioni, S., et al. 2005. Cocaine in surface waters: a new evidence-based tool to monitor community drug abuse. *Environmental Health* 5:4–14.

Zuhlke, S., Dünnbier, U. and Heberer, T. 2004. Detection and identification of phenazone-type drugs and their microbial metabolites in ground- and drinking water applying solid-phase extraction and gas chromatography with mass spectrometric detection. *Journal of Chromatography A* 1050:201–209.

5 Green pharmaceutical production and its benefits for sustainability

*Zobaidul Kabir, Sharmin Zaman Emon,
and SMA Karim*

CONTENTS

DOI: 10.1201/9781003394600-6

5.1   INTRODUCTION

Green manufacturing is a production process where sustainability (social, economic, environmental, and cultural) issues are considered by the companies during the operation of their industries to produce products those are usually called green products. Green manufacturing includes a process that is resource efficient, energy efficient, and cost-effective. The concept of green manufacturing is an arrangement where the negative impacts of the manufacturing process or system on the environment and society will be minimum. This is a sustainable process that may ensure the environmental and socioeconomic sustainability by using minimum natural resources (for example, water and energy), adopting green technologies, and the production process is harmless for employees, local communities, and consumers and enable the companies to be more profitable than those who do not have such arrangements for sustainable development (Miettinen and Khan, 2021). It is a shift from the traditional dealing of the 'end of products to the continuous control of products and production process' where environment-friendly products are produced, rational use of resources and minimum generation of waste is ensured (Adam et al., 2020). Green manufacturing include resource efficient, energy efficient, and cost-effective process.

This also includes that the operation of the industry will not affect surrounding environment and society. Importantly, green manufacturing with proper initiatives brings benefits for the society, environment, and economy. Although additional investment is required for green manufacturing and apparently this increases the cost of production, it is well recognized that the companies' financial performance is enhanced in the long run. The objectives of green manufacturing are to produce environment-friendly quality products given the increasing demand of green consumers, to improve corporate image and to achieve competitive advantage in the market (Zheneng, 2010). While traditional manufacturing is harmful to the environment, for example, air, water, and soil, and affects the surrounding community's quality of life, green manufacturing tends to ensure that surrounding environment is protected from pollution and the day-to-day life of residents is not disrupted. Green manufacturing use green materials, renewable energy or energy efficient technologies and minimize the use of resources as much as possible to ensure sustainability (Lin and Hao, 2020). Overall, the green manufacturing can be characterized by innovative manufacturing system, prevention of environmental pollution, conservation of resources, reduction of waste generation and management of waste generated, environment-friendly product, the use of modern technology, reduced consumption of raw materials, water, and energy. Green manufacturing also includes the reduction of the cost of waste treatment, the use of energy, enhancement of the image of the company and increasing the competitive advantage in the market (Acharya et al., 2014; Paul et al., 2014). A green manufacturing companies with characteristics mentioned above may be able to produce a product not only cost-effective but also cost-efficient.

Attention to green production has been increasing over time both in developed and developing countries while the quality of life of people in the society depends on the better environment (Drake and Spinler, 2013; Zhang et al., 2019) and the number of green consumers is increasing. The importance of green production is emphasized by stakeholders, academics, and decision-makers who work for government.

Importantly, conferences and seminars on green production are held worldwide at national and international levels, for example, at Rio +20 Summit and conferences organized by United Nations in different countries (Griggs et al., 2013; Colglazier, 2015; Gong et al., 2018; Zhang et al., 2019). These conferences and seminars underscore the importance of green and sustainable development where the manufacturing companies can play a vital role. Manufacturing companies are now increasingly focusing on green production and taken initiatives with necessary policies and strategies to make their production green. For example, IKEA a Swedish company underlined the importance of sustainable development and ecological conservation and taken necessary strategies and approaches for green product (Gong et al., 2018) and General Motors has incorporated the notion of greening business in its strategic policy and thereby focused on the green production process with green materials, more research and development, green procurement, and even green human resource management (HRM) practice (Zhai et al., 2014).

Manufacturing companies with the concept of sustainability are also adopting new technologies to make the production process green, on the one hand, and to curb the emission of greenhouse gases (GHGs), on the other hand (Dong et al., 2016). Manufacturing industries are generating significant amount of hazardous waste and pollutants. For example, chemical industries generate pollutants and hazardous waste such as washed-out reagents, GHGs, and poisonous solvents and these may pollute environment and cause ecological imbalance if the pollutants are not treated and directly released to environment. Among the manufacturing industries pharmaceutical companies also generate significant amount of hazardous waste and pollutants those may affect not only ecosystem but also public health (Mahender et al., 2019). Therefore, it is imperative to develop appropriate strategies in the line of the country's rules and regulations and identify the production process approaches to produce green product (Arora et al., 2021).

In response to the national environmental laws, green chemistry principles, increasing green consumers and their pressure on the polluters to ensure sustainable development, the manufacturing companies including the pharmaceutical companies have taken various steps. The pharmaceutical companies are now advancing with new technologies, the use of renewable energy, and green materials for green products and thereby to minimize generation of waste and other pollutants. The corporate social responsibility (CSR) framework are now in place made by all most all transnational companies to ensure sustainability. Awareness is growing among manufacturing companies on sustainability (economic, social, and environmental), and therefore, companies are competing to attract green consumers through the resource efficient green production process (Mishra et al., 2021; Silva et al., 2021).

Among the manufacturing industries, pharmaceutical industries are important for any country to ensure public health as they produce medicines. Pharmaceutical industries, although important for medicine production, are considered as environment-sensitive industries globally (Yang and Yulianto, 2022). Therefore, according to Grougiou et al. (2016) there are negative public perceptions about pharmaceutical industries because the operational activities of pharmaceutical industries may pollute the environment including water, soil, and air (to be elaborated in detail in the next section). Industries fails to comply with environmental and social obligations may face higher cost of production in long run as well as financing risks due to less investors support (Breuer et al., 2018). On the other hand, the management of companies complies with the environmental rules and regulations may improve the corporate image and make the companies profitable. There are around 700,000 chemicals are produced with different nature and the global market value of these products is around $500 billion (Asthana 2014).

It has been estimated that manufacturing sector is responsible for nearly one-fifth of the total carbon emission globally. The manufacturing sector also responsible for the consumption of 54% of Words' total supplied energy. While there is no available global data on the emission of GHGs, studies elsewhere, for example, Belkhir and Elmeligi (2019) shows that pharmaceutical industries may emit significant amount of GHGs. This means the pharmaceutical companies are significant contributors to global warming. A study in the USA undertaken by Eckelman and Sherman (2016) finds that manufacturing companies are responsible for 12% of acid rain, 10% of GHG emission and 10% of smog formation in addition to release of air pollutants to the atmosphere (Eckelman and Sherman, 2016). It has been estimated that the pharmaceutical sector emitted around 52 mega tonnes of CO_2 equivalent GHGs in 2015. In the same year, the amount of CO_2 equivalent GHGs was generated by automotive industry. This indicates that pharmaceutical industries release more GHGs than automotive industries although the market value of automotive industry is higher than the pharmaceutical industry. Importantly, the pharmaceutical industries are more pollutant intensive than automotive industry sector (Belkhir and Elmeligi, 2019).

The aim of this chapter is to explore the benefits of sustainability where green pharmaceutical manufacturing is available. This chapter is divided into five sections. The introduction section is followed by section two where section two pharmaceutical production process and discharge of pollutants. The third section has identified and elaborated the benefits of sustainability of green pharmaceuticals production. The fourth section has offered a discussion on the sustainability issues associated with green pharmaceutical industries. This is followed by a conclusion and recommendations.

5.2 PRODUCTION PROCESS AND DISCHARGE OF POLLUTANTS FROM PHARMACEUTICAL PRODUCTION

5.2.1 PRODUCTION PROCESS

The production of pharmaceutical drugs goes through the process of synthesis at an industrial scale. The process of drug manufacturing can be broken down into series of unit operations such as milling, granulation, coating, tablet processing, and others. In industrial production process raw materials are charged or separated into pharmaceutically useful products such as drugs and excipients. These conventions include a wide-ranging selection of procedures. Several processes are:

5.2.1.1 Production of dosage forms

In this process, drugs are converted into medicines (final product) to be used by the patients. Here the chemicals are raw materials, and the chemicals are transformed into medicine appropriate for the use of patients. Diclofenac Sodium, for example, is converted, into medicines or dosage forms such as pills, capsules, suspensions, and injections.

5.2.1.2 Production of bulk drugs

In this process, chemicals are used as raw materials, and these are finally transformed into medicines. For example, to get acetylsalicylic acid or aspirin, salicylic acid is converted to acetylated through the production of bulk drugs. In this area of production, chemicals are also transformed into intermediates, and these are used for the manufacture of drugs for commercial purpose.

5.2.1.3 Production of antibiotics

Manufacturing of drugs (antibiotics) using microbes with the aid of processors is another area of interest to the pharmacist. Fermentation technique is used to produce antibiotics. For example, penicillin G is produced using penicillium chrysogenum with the aid of precursor, phenylacetic acid.

5.2.1.4 Production of biological

Extraction of drugs from animals, plants, and minerals from native raw materials into purified (or semi-purified) products are of interest to the pharmacist. Some examples are vaccines, insulin, streptokinase, and recombinant DNA technology products.

Each process is developed systematically from a laboratory scale to pilot scale and finally to an industrial scale. In general, every process involves a series of steps. Each step is performed individually.

5.2.2 Unit operations

Each chemical process frequently consists of a fewer number of distinct individual steps. Each step is called unit operation. Each unit operation is based on one type of scientific principle. Some examples of unit operations and underlying principles are given below.

5.2.2.1 Drying

It is a unit operation used to remove liquid or moisture from solids by evaporating with the aid of heat. For example, drying process is employed to remove excess moisture (above equilibrium moisture content) from the wet granules in the production of tablets.

5.2.2.2 Size reduction

This is a unit operation in which drugs (plant or chemical origin) are reduced to smaller pieces, coarse particles, or fine powder. This process is extensively used in the manufacture of talcum powders and tooth powders (cosmetic industry).

5.2.2.3 Distillation

It is a unit operation of converting liquid into vapour by heating and reconverting vapour again into liquid by condensing the vapour. This unit operation is used to obtaining essential oils from various parts of the plants.

5.2.2.4   Evaporation

It is a unit operation which involves free escape of vapour from the surface of a liquid below its boiling point. For example, evaporation technique is extensively used for concentrating solution in pharmaceutical industry.

5.2.2.5   Solvent extraction

It is a unit operation where solvent is used to separate one component from others depending on their chemical nature. Powder feeding in continuous manufacturing: in continuous manufacturing, input raw materials and energy are fed into the system at a constant rate, and at the same time a constant extraction of output products is achieved. The process performance is heavily dependent on the stability of material flowrate.

5.2.2.6   Powder blending

In the pharmaceutical industry, a wide range of excipients may be blended with the active pharmaceutical ingredient (API) to create the final blend used to manufacture the solid dosage form.

5.2.2.7   Milling

During the drug manufacturing process, milling is often required in order to reduce the average particle size in a drug powder.

5.2.2.8   Granulation

Granulation can be thought of as the opposite of milling; it is the process by which small particles are bound together to form large particles, called granules. Granulations are of two types: wet granulation and dry granulation. granulation is used for several reasons. Granulation prevents the 'demixing' the components in the mixture, by creating a granule which contains all the components in their required proportions, improves flow characteristics of powders and improves compaction properties for tablet formation.

5.2.2.9   Hot melt extrusion

Hot melt extrusion is utilized in pharmaceutical solid oral dose processing to enable delivery of drugs with poor solubility and bioavailability. Hot melt extrusion has been shown to molecularly disperse poorly soluble drugs in a polymer carrier increasing dissolution rate and bioavailability. Each of the above unit operations is a common technique and employed in diverse chemical and pharmaceutical industries.

5.2.3   RAW MATERIALS

Generally, pharmaceutical raw materials can be categorized into three categories.

5.2.3.1   Active pharmaceutical ingredients

API is one of the main parts of the pharma drug which is pharmaceutically active and responsible for the drug action. Accuracy and precision with the raw materials are two main must aspects for while making API. API is the core ingredient of every finished drug product. In Bangladesh, for example, the APIs account for 30% of total drug costs in case of small molecules and can go up to 55% for generic products. Currently, Bangladesh meets 98% of the demand for finished-form pharmaceutical products locally. Though nearly self-sufficient in finished drugs, more than 99% of the API and raw materials must be imported (Government of Bangladesh, 2020).

5.2.3.2 Inactive ingredients or excipients

Excipients also called as inactive ingredients or drug carriers. Pharmaceutical raw materials used as excipients consist of solvents and other such carriers. Excipients bring bulkiness and stability in the drug formulation, along with facilitating absorption and preventing denaturation of drugs.

5.2.3.3 Packaging raw materials

Packaging in the pharmaceutical industry need to be perfect and precise. Raw materials used for packaging in this sector includes plastics, polymer, glass, aluminium foil, paper, and more. Packaging is made a separate category for the pharma because of the use of diversified raw materials.

5.2.4 DISCHARGE OF POLLUTANTS BY PHARMACEUTICAL INDUSTRIES

There are pharmaceutical industries in almost all countries all over the world. For example, Bangladesh is one of the medicine countries and according to Directorate General of Drug Administration of Bangladesh, there are 257 allopathic companies are operating at present in Bangladesh. The pharmaceutical sector in Bangladesh is mostly developed in the areas of drug formulation and manufacturing of finished products. The pharmaceutical plants generate many pollutants during manufacturing, housekeeping, and maintenance operations everyday which are being directly discharged into the surrounding waterbodies including lakes, rivers and canals, agricultural fields, and surface water. Different classes of drugs have been documented as environmental **pollutants such as analgesics**, antibiotics, antiepileptic, antihypertensive, antiseptics, heart drugs, contraceptives, hormones, and psychotherapeutics (Larsson, 2014). Furthermore, pollutants are discharged from the pharmaceutical industry depending on the methods used to produce a product (medicine). In general, four methods are used in the manufacturing of pharmaceuticals are considered. These include (a) Research and Development, (b) Chemical Synthesis, (c) Natural Product Extraction, and (d) Formulation (Stewart et al., 2016). The most common pollutants generated from research and development department includes halogenated and non-halogenated solvents, photographic chemicals, radionuclide, bases, and oxidizers. During chemical synthesis, manufacturers use many solvents listed as priority pollutants and these are used for product recovery, purification, and reaction media. The manufacture of the pharmaceutical products, for example, organic medicinal chemicals, medicine from animal glands, inorganic medicinal chemicals, antibiotics, biological products, and botanicals may generate hazardous wastes and pollutants.

The production of pharmaceutical products may affect the environment in many ways. In general, there are three key pathways through which pollutants generated from pharmaceutical production may reach the environment. First, the discharges of wastewater at different stages of process during the manufacturing of pharmaceutical products. Second, when generation of solid and liquid wastes is disposed with or without treatment in an improper way, for example, open dumping. Third, the harmful pollutants may enter human and animal bodies through food chain and when they are exposed to the pollutants generated from pharmaceutical industries (Götz et al., 2019; Miettinen and Khan, 2021). The solvents used for pharmaceutical production have residues and may spill over into nearby waterbodies through rainwater or drainage. If the waste generated from pharmaceutical production is disposed in the open field which is common practice in developing countries may pollute soil, air, and ground water. The active APIs may go into the food chain and consequently can be consumed by human, animals, and plants and thereby can be badly affected by the pollutants (Sharma et al., 2022).

It is to be noted that the use of water is essential to produce drugs although the diversity of pharmaceutical industry may be very associated with the production of the quantity of drugs, size of the industry and methods used for a particular product. The wastewater is generated from the production process of a pharmaceutical plant. This wastewater contains effluents that may degraded the quality of environment particularly when the wastewater is not treated at all or treated using conventional technologies (Tiwari et al., 2020). Some substances or waste are difficult to treat and reduce the intensity of the pollutant at safety level for environment even using wastewater plant with modern technologies. For example, antibiotics released from pharmaceutical industries are found available in wastewater and solid waste dumped by pharmaceutical companies (Li et al., 2013). Because of the persistent nature of antibiotics, it not always possible to eliminate entirely. The residue of antibiotics therefore after transformation may affect the environment including both aquatic environment and terrestrial environment and gradually deposit in soil and ground water (Karkman et al., 2017; Götz et al., 2019).

In pharmaceutical industries water is used to produce drugs and water discharged by the pharmaceutical plants after the production process may contain residues of solvents and other pollutants. Furthermore, pharmaceutical industries usually do not recycle the solvents to reuse for further production process and therefore wastewater may contain hazardous chemicals with different degrees of quantity. This wastewater if not treated using, for example, effluent treating plant (ETP) may be released to nearby waterbodies and pollute ground water too through rainwater. Study shows that pollutants generated from pharmaceutical process are found in ocean and river water (Sharma et al., 2022). The wastewater released from pharmaceutical industries can be discharged to nearby waterbodies such as lakes, rivers, and even sea waters (Rogowska et al., 2020). It should be noted that only substances that are commonly found in the wastewater are known or listed by regulatory bodies. However, there are some other pollutants out of list with significant concentration levels and my pose an environmental and social risks. Knowledge about these pollutants present in wastewater their combination and potential impacts on human and environment, however, is limited (Rogowska et al., 2020).

The synthesis of medicinal chemicals may be done in a very small facility producing only one chemical or in a large integrated facility producing many chemicals by various processes. Organic chemicals are used as raw materials and as solvents. Volatile organic compounds, which are the main pollutants, may be emitted from a variety of sources within plants synthesizing pharmaceutical products.

It is well recognized that the exposure to both liquid and solid waste may negatively affect human health as well as animal life. The largest quantities of hazardous waste and pollutants are generated from the production of organic medicinal chemicals and antibiotics (Yaqub et al., 2012). Table 5.1 shows the major pollutants and its impacts on environment and public health.

TABLE 5.1
Impacts of major pollutants produced from pharmaceutical manufacturing

	Pollutants	Impacts
1	Acetone	Acetone is a hazardous pollutant as it is very inflammable. It can dissolve other substances and may easily mix with water and other chemical substances and thereby may pollute water. This organic compound can cause the damage of membrane, may significantly affect the size of plants, and reduces the in development of farming and various agricultural and decorative both in aquatic and terrestrial environment.
		The exposure to acetone may affect public health by causing the damage of skin and case damage of Acetone exposure can also result in damage to the skins and eyes with eyesight. This compound may affect reproductive ability of animals because of exposure for extended time (EPA, 1985).
2	Amyl alcohol	This pollutant may affect human health when someone breathe in and inhaling the substance may affect lung, nose, and throat. Exposure to this compound may worsen eyes and skins. It is a flammable chemical and known as a hazardous element.
3	Benzene	Benzene can create foggy atmosphere by reacting with other substances. This may contaminate the water and soil when it is released without treatment. It causes the damage of aquatic ecosystem in rivers, oceans and lakes and the pollution of water by benzene may cause sickness of fish species and affect fertility capacity of fishes. Exposure to benzene in the soil may affect the growth of plants where the growth can be slowed down and the plants may die due to exposure for long-term. Benzene is harmful for human health. Excessive exposure to benzene may cause reduction in red blood cells and may lead to anaemia. Benzene can be responsible for excess bleeding and the reduction of immune system in human bodies (Masekameni et al., 2021).
4	Carbon tetrachloride	This pollutant can stay in the atmosphere for long time, for example, around 100 years. This can damage the ozone layer in the atmosphere when it breaks down and forms chemicals. This has impact on public health. Highly exposure to this compound, the organs such as liver and kidney of human bodies can be damaged (Unsal et al., 2021).
5	Dimethyl formamide	At workplaces, for long-term exposure to this compound may affect liver and digestion capacity of workers. High exposure to this chemical may damage livers of animals and humans (Hu et al., 2022).

(Continued)

TABLE 5.1 (*Continued*)
Impacts of major pollutants produced from pharmaceutical manufacturing

	Pollutants	Impacts
6	Ethyl acetate	This pollutant is very active when it is broken down in both air and water. The incorrect handing of this chemical can be very hazardous to human health and may cause serious accident as this is highly flammable and toxic. Inhaling or ingesting this element for a long period may cause damage badly to inside organs of human and animals (Yeşilyurt et al., 2022).
7	Isopropanol	This pollutant with high intensity may have significant negative impacts on environment including both terrestrial and aquatic lives. This compound may affect public health causing throat irritation, coughing, headache, dizziness and even death if the person is exposed to isopropanol for a long time (Slaughter et al., 2014).
8	Methanol	Wildlife including animals, birds, and fish species may die to exposure to methanol. Exposure to this element, may slow down the rate of growth of plants. The fertility and behaviour of biota can be affected by methanol when the biota is exposed methanol for a long time. The burning of methanol may emit GHGs such as carbon dioxide which is responsible for sea level rise (Yadav et al., 2020).
9	Methylene chloride	This pollutant is mostly emitted to air. Some of the element is released to water and soil. The spilling, leaking, and evaporating of this element may pollute the environment particularly when the chemical waste is dumped in open space. Because of evaporative character, most of the methylene chloride is emitted to atmosphere and is subsequently converted to carbon dioxide (CO_2), a harmful gas responsible for lobal worming. Animals can be unconscious or even die if they are exposed to this element with high intensity, for example, 8,000–20,000 ppm (EPA, 2016).
10	Toluene	Causes damage of membrane to the leaves of plants. Long-term exposure to high concentration of toluene may result in the health sickness including dizziness, sleepiness, unconsciousness and in some cases death. Also, high level of toluene may affect human health including damage of brain, eyesight, and memory loss (Masekameni et al., 2021).
11	Xylene	Because of easy evaporation, xylene goes into air and broken down under sunlight. Released by pharmaceutical plants, xylene goes to watercourse and spills on land. Exposure to this element may cause irritation of eyes, nose, and skin of human being. Also, this may cause loss of muscle and even death. Given the level of exposure including dose and duration, workers' health in the workplace can be affected by this element (Masekameni et al., 2021).

It is available practice of waste management that include prevention, minimization, reuse, and recycling in addition to energy recovery and disposal (Jaseem et al., 2017). The disposal of pharmaceutical waste is not well organized globally with a few exceptions in developed countries. Usually, the practice of waste disposal in different countries is comingling the waste with household waste. The second most options are to dispose of waste into sewerage system (Götz et al., 2019) and this is the worst kind of practice. Study shows that the key reasons of pollution of water is to enter the residues from pharmaceutical production those enter water when flushed by companies in sewage system (Miettinen and Khan, 2021).

5.3   STRATEGIES FOR GREEN PRODUCTION AND BENEFITS FOR SUSTAINABILITY

5.3.1   ENVIRONMENTAL BENEFITS FROM GREEN PRODUCTION

Pharmaceuticals have an influence on the environment along the whole value chain, i.e., from raw material extraction and processing to API manufacturing, formulation, and packaging, to distribution, usage, and product end-of-life (Ott et al., 2014). Environmental considerations must be incorporated into various chain-related activities, such as product development and innovation, product approval, production development, procurement, distribution, administration, and use, where various value-chain stakeholders play different roles in this work, to manage and control the impacts (Cespi et al., 2015). Therefore, there is a growing demand for trustworthy, relevant, and comparable information concerning the environmental consequences of medicines from a life cycle perspective to allow environmental considerations and choices for improvements by various stakeholders along the pharmaceutical value chain. Recently, some pharmaceutical products industry started green pharmaceutical production and most of the industry gain a double amount of yield than chemical process production and all the industry used less quantity of natural resources, i.e., water, energy, and less time needed.

At the same time, the elimination of the use of metal catalysts, reduced amount of wastage generation in the environment or biodegradable wastes (Valavanidis et al., 2012; Menges, 2017; Mishra et al., 2021). For these reasons, these green pharmaceutical production leads to sustainable in the environment. In most of the countries, governments have created guidelines for green regulations that must be followed by all organizations because the ISO 14001 environmental management system is now required for business operations. To maintain environmental practices and increase their profits, businesses must abide by environmental regulations set forth by the government. Organizations adopted green supply chain management (GSCM) techniques into their operations to lessen environmental issues in the sector because of environmental rules and liability for hazardous items (Khaksar et al., 2016). Governments across the world are becoming more conscious of environmental challenges, and some of them are providing financial support to groups that engage in environmentally friendly activities (Martusa, 2013), which could demonstrate lower potential for global warming, ozone depletion, and smog formation.

The following environmental benefits are attained in recent examples of some pharmaceutical products production:

- **Sertraline hydrochloride**: used in depression, uses green technique in their production process and this process prevented pollution, used reduced amount of water and energy and the product yield is double in amount than the commercial one.
- **Sitagliptin**: used in type-2 diabetes, this is the enzymatic process with double yield, elimination of metal catalysts and reduces the quantity of wastes to the environmental.

- **Doramectin**: used in parasite infection and in this production process, they use biocatalysts. 40% increase in efficiency, reduced generation of the byproducts and reduced generation of wastes in purification (Valavanidis et al., 2012).
- **Gemifloxacin**: used as antibiotics and they used bio catalysis in the manufacturing process. This process uses less energy and less process time and produces less wastes (Valavanidis et al., 2012).
- **Simvastatin**: used in treatment of high cholesterol. Optimization of chemical and enzymatic process with the use of engineered enzyme with low-cost feedstock. This process has signification increase of yield of 97%, use of elimination of hazardous compounds and they used biological process for biodegradable wastes (Mishra et al., 2021).

Although there are limited studies, literature show that traditional pharmaceutical manufacturing may emit more carbon than green pharmaceuticals (Belkhir and Elmeligi, 2019). A study undertaken by Belkhir and Elmeligi (2019) on 15 leading pharmaceutical industries shows that these industries in this sector release significant amount of carbon. The release of the amount of carbon emission in pharmaceutical sector may be more than that of in other, for example, automotive sector (Belkhir and Elmeligi, 2019). This indicates that the green pharmaceutical manufacturing including green material, energy efficient technology will reduce the carbon emission and thereby improve the quality of environment and the impacts of climate change on human can be reduced. Table 5.2 shows the summary of and strategies for green production and environmental sustainability benefits of green production.

5.3.2 Social benefits from green production

5.3.2.1 Public health benefits

The pharmaceutical sector is committed to giving its clients happier, longer lives. However, a foundation for human health and wellbeing is environmental sustainability in general. Thus, for a very long time, in the pharmaceutical sector, cost and quality control were equally vital to green engineering (Ahmed et al., 2020).

The followings are the public health benefit due to green pharmaceutical production:

- Patient have safe medicines made of nontoxic chemical free substrates for their recovering.
- Less dangerous chemical emission into the atmosphere means less lungs' damage.
- Less harmful chemical waste is released into the environment, which results in cleaner drinking and recreational water.
- A reduction in the usage of dangerous materials, a reduction in the need for personal protective equipment, and a reduction in the risk of accidents for chemical industry employees (e.g., fires or explosions).

TABLE 5.2

Strategies for green production and environmental benefits

		Strategies for green production and environmental sustainability benefits	
	Strategies	Benefits	Comments
1	Green policy and guidelines	Conservation of natural resources	The pharmaceutical industries without green approach may produce product at the cost natural resources such as water and forest resources. Therefore, companies with green manufacturing could protect the natural resources from depletion.
2	Use of renewable energy	Reduction of GHGs	Pharmaceutical companies use electricity generated from coal other sources of carbon. Such industries consume huge amount of energy for production process to produce products and other purposes and thereby create GHGs. On the other hand, look for suitable sources for renewable energy and the use of renewable energy will release less amount of GHGs.
3	Use of green materials/chemicals	Less harmful pollutants	Green pharmaceutical companies use green materials or chemicals for pharmaceutical production. The use of green materials has relatively less impacts on environment and public health than raw materials or chemicals. The use of green chemicals for manufacturing process results in green products.
4	Green Technology and Innovation	Generation of less hazardous waste	Green technology or Best Available Technology (BAT) for pharmaceutical production can play an important role in sustainability context. The use of green technologies by pharmaceutical companies is supportive to produce green pharmaceutical products, generate less hazardous waste and make the whole production process energy efficient.
5	Prevention of environmental pollution	Protection of biophysical environment	Green manufacturing by pharmaceutical companies produces relatively a smaller number of pollutants and protect the environment from pollution. The green initiatives taken by pharmaceutical companies may prevent water and soil from pollution. The loss of flora and fauna in the water and the loss of production of food due to loss of soil fertility can be prevented if the companies adopt technologies for green manufacturing process.

- Safer consumer goods of all kinds will be available for purchase; certain goods (like medications) will be produced with less waste; others (like pesticides and cleaning supplies) will replace goods that are not as safe.
- Safer pesticides that are poisonous exclusively to certain pests and disintegrate quickly after application; removal of persistent toxic compounds that can infiltrate the food chain.
- Less exposure to harmful substances like endocrine disruptors (Özkan et al., 2014).

Green pharmaceutical production now a days a social responsibility enforced by government laws and regulations in every country and as because yield is much higher than the chemical process, so more manpower is needed for this type of production process to implement the ISO regulations (Sharabati, 2021). Unemployment problem is also solved by this technique and industries are way too keen the training process to make their staffs more skilful and experienced. Otherwise, making environmental protection a duty of every person, group, and nation, the implementation of such rules and laws is not discretionary but rather mandatory. So, to maintain environmentally friendly practices, including the use of recycled raw materials, clean energy sources, pollution prevention, eco-friendly packaging, and production of waste reduction, more adroit professionals and technologists are needed and at the same time, worker health and safety is also important for the continuation of this sustainable process (Famiyeh et al., 2018). In this regard, good supply chain management training and operational competitive performance (quality, flexibility, cost, and delivery time) is much needed and essential for enhancing environment-friendly performance, which impacts competitive advantages in the market (Ananda et al., 2018).

One of the key aspects of green pharmaceutical manufacturing is human resources management (HRM) and development of a pharmaceutical company where the company underscores the importance of green HRM practices. Literature shows that there is a strong relationship between green human resources development practice and sustainability (Masri and Jaaron, 2017). The pharmaceutical companies with corporate social and environmental responsibility are increasingly focusing on this area (Veleva and Cue, 2017). The training imparted to employees of a company on sustainability issues may improve the knowledge, skills, and attitude about sustainability and show more pro-environmental behaviour (Saeed et al., 2019), through, for example, management of water, energy, and other input materials in a sustainable way while producing pharmaceuticals. In Palestine, for example, the employees of a pharmaceutical company trained on sustainability issues showed better resource management performance than companies without such green HRM practice (Masri and Jaaron, 2017). In addition, green HRM practice may create an environment-friendly culture within the companies, the image or goodwill of the company can be improved and may save cost of production in the long run through the reduction of waste (Muster and Schrader, 2011; Arulrajah and Opatha, 2016). With this practice, the society become beneficial due to the reduction of pollutants and stakeholders of the company also gain confidence about the company's sustainability. The companies may develop a collaboration with NGOs and local community for environmental management and company may avoid conflicts that could be raised due to environmental pollution.

With increasing attention of employees to social issues the participation of employees in volunteer program allows companies to make relationship with local communities who are the ultimate victims of environmental pollution (Mahmud et al., 2021). Companies with green production process not only adopt CSR policies for the socio-economic improvement of local communities and relevant stakeholders but also for the wellbeing of employees. One of the key business strategies is to achieve customer satisfaction and therefore manufacturing company's delivery their quality products in such a way that customers become happy with the products. Since the manufacturing companies discharge a significant number of pollutants, a company needs to show their social responsibility not only by adopting the green manufacturing but also by taking initiatives such as insurance for local community if they are affected by the company's activities. A society can expect social benefits such as better relationship with the green manufacturing company, creation of jobs for the community people, improvement of the living condition of local community, and comply with the environmental regulation together with the relevant stakeholders (Momtaz and Kabir, 2018; Mahmud et al., 2021; Yang and Yulianto, 2022). Through the implementation of social responsibility and protection of environment, a green manufacturing company many achieve 'social licence to operate' its activities. Table 5.3 shows the summary of green production strategies and social sustainability benefits from pharmaceutical production.

5.3.3 GREEN PRODUCTION STRATEGIES AND ECONOMIC BENEFITS

It is apparent that green pharmaceutical production may offer substantial economic value through the green **HRM** practice as well as improved environmental performance. In developing countries, companies may enhance economic diversity by adding value of HRM practice to their businesses (UNIDO, 2013). Importantly, companies may have opportunities to increase their access to global markets through good practice HRM polices and complying with local sustainability standards where the company operates. By doing this good practice, companies may generate more income since green HRM may increases the efficiency of the employees of the company and thereby improve the global image. The green HRM practice enable the employees to be more committed to the company and green activities during, for example, the extraction of materials from waste and the use of renewable energy and energy efficient technology. Also, the green HRM practice enable companies to penetrate in the new markets of other parts of the world.

It is evident that companies adopted a CSR policy may have better financial performance than those companies who do not have a CSR policy in place. A robust CSR policy helps companies to reduce the ecological footprint on the one hand and improve the communities' and consumers confidences on the other hand. The CSR policy may significantly increase the retention of employees, and satisfaction of stakeholders. This may positively affect the financial performance of the company. Green manufacturing under CSR program may offer companies competitive advantage in the market and thereby companies can make more profits (Porter and Kramer, 2006). Importantly, the competitive advantage of a company with green manufacturing can achieve due to the satisfaction of the of green consumers, local communities,

TABLE 5.3

Social sustainability benefits from green pharmaceutical production

	Strategies	Benefits	Comments
			Strategies for green production and social sustainability benefits
1	Green products	Increased number of green consumers	In developed counties consumers strongly believe that green products are the most reliable solutions for environmental sustainability. The value of green products may significantly affect the attitude of customers to the environmental and social sustainability and thereby the company will find more green consumers in the market. The consumer will be happy with the companies' responsible consumption of resources and green production activities which is related to the achievement of SDG goal 12.
2	Green production process	Customer satisfaction	When companies take initiative to green their production process and produce green products, this may make happy the green customers. Green customers are aware of environmental and social impacts of industrial production and therefore company with green manufacturing can achieve customer satisfaction and trust. Customer satisfaction increase the reliability on the company and support to increase the sale of products in the market.
3	Community Engagement	Community awareness	Green manufacturing program aware local community about the environmental pollution and four subsequently mitigation of pollution. The socially responsible companies engage communities in identifying the social and environmental impacts of manufacturing pharmaceutical products and thereby become aware of environmental issues. Importantly, engagement of community by companies encourage community to participate in the mitigation of environmental pollution willingly.
4	Programs for local community	Quality of life of local community	By addressing the environmental impacts, the quality of life of local community is prevented from disruption. The company may provide jobs local community members and thereby improve their quality of life. Companies may compensate the people of local community to be affected by environmental pollution and thereby improve their livelihoods specially the marginal people.
5	Communication with stakeholders	Relationship with stakeholders	When companies take initiative to green their production process, the stakeholders including local communities, environmental activist and others become happy. By taking green initiatives company may avoid the potential conflicts those might arise in the absence of such initiatives. Over time, stakeholders trust the company and their activities and a healthy relationship between the companies and stakeholders developed.
6	Compliance with local culture and rules	Corporate image or reputation	Pharmaceutical companies with green manufacturing activities have opportunity to enhance their image or reputation in the market. This will allow companies to create confidence and trust among customers where the companies are caring for the sustainable development. Customers will purchase the green products from the companies with green activities. The sale of products of the companies will therefore increase.

and other stakeholders. Companies without CSR policy for environmental mitigation can make profit for the short term but loose profit substantially in the long run due to lack of competitive advantage. The consumers may boycott the products of the company's polluting environment and those do not show responsibility to the society. Companies with CSR policies many secure long-term economic performances by replacing old traditional practice of manufacturing harmful to environmental and society. Overall, to comply with the local regulations and environmental standard company with green manufacturing may avoid the pollution tax and thereby can save money (Galbreath, 2001).

Another aspect of economic sustainability for a green manufacturing company with CSR policy is the avoidance of both social and financial cost of conflicts. Companies without greening business may have potential conflict with stakeholders. The polluting companies may affect the land of local communities and the fertility of the land can be lost and thereby communities may lose their sources of livelihoods. The environmentally affected communities may go for litigation against the polluting companies. In the long run, the polluter companies may need to compensate more money that could be avoided if the companies could take care of the affected communities. The compensation often may surpass the profit that the companies made at the cost of social and environmental degradation. Due to litigation, the company may lose its corporate image and branding in the global market.

One of the key aspects of economic sustainability benefits is avoidance of risks. There is a potential risk of business for a company when there is a conflict between the companies and other groups in the society. Therefore, the management of risks is an important issue for a company. The potential social and financial risks can be avoided by engaging relevant groups from the local community and negotiating with the relevant groups. A risk management plan prepared by the company and endorsed by the local communities may encourage the community to participate in the management of risks voluntarily and the company may operate its business without any barriers. Companies with low risks may increase share price and thereby shareholders can be benefitted. Also, the companies have opportunity to increase market share with the support of relevant stakeholders.

Similarly, green companies can reduce the amount of waste by using green technology and use modern technologies to manage the wastes in a sustainable way. It is well recognized that the waste generated from manufacturing may considerably affect the local environment and society. Therefore, the reduction of waste generation to avoid the environmental pollution is one of the key objectives of green manufacturing. Studies elsewhere show that investments in clean technologies typically have relatively short payback periods and lead to lower annual costs (Owens, 2016). This means that, after an initial cost, these investments help enterprises save money through decreased resource use or generate more money through improved productivity. Importantly, the waste generated from pharmaceutical manufacturing can be used for renewable energy generation and there are various waste-to-energy technologies (Khan and Kabir, 2020) to generate renewable energy from waste. This is one of the smart ways to improve the quality of green manufacturing. The treatment of generated waste not only protects the environment from pollution but also generates employment for local community. It was found that the green manufacturing companies by reducing waste generation can significantly cut the costs on the consumption of materials and treatment of hazardous waste (Dornfeld, 2013) and subsequently can increase the profit margin. Studies in the US shows that where industries adopted waste management program such as treatment of water and use of solid waste for energy generation rather than putting in the landfills could success-fully reduce their overall manufacturing costs (Heal, 2005; Kabir and Khan, 2020). The byproducts generated from the treatment of solid waste, for example, fly ash and bottom ash are useful to road construction (Khan and Kabir, 2020). The untreated waste affects the local environment by polluting water and soil thereby affect the quality of life of surrounding community.

Another economic sustainability aspect can be ensured by green manufacturing companies through the achievement of **competitive advantage** in the market. In fact, the achievement of competitive advantage is possible when there is a support from consumers to the green industries. With the support of consumers and stakeholders a company with green manufacturing may take the lead in the market (Delmas, 2001). A study on 4,000 manufacturing facilities shows that companies with green manu-facturing practice tend to make profit than other companies without green manufac-turing activities (Johnstone, 2007). The companies with green production may have variety in products and may have capacity to compete with other companies (Yang and Liu, 2021). It has been evidenced that companies with green production process and better environmental performance would benefit from customers with enhanced awareness on environmental issues. On the other hand, companies with traditional production process and poor environmental protection initiatives would profit only when competition with the counterpart companies is low (Yang and Liu, 2021). This indicates that there are economics incentives for green manufacturing industries and for green manufacturing industries and therefore green manufacturing industries are coming up with innovation and research to achieve competitive advances and they want to see them in the market different from others in relation to environmental performance (Seidel et al., 2007).

There is an intractably link between green chemistry approach and environmental sustainability. There are significant environmental benefits of green chemistry (Gao et al., 2019; Adam et al., 2020; Jimenez-Gonzalez and Lund, 2022). Green chemistry is a potential support to the pharmaceutical industry. Green chemistry may consid-erably benefit to achieve environmental improvement. It also helps companies to achieve their business sustainability and enable the companies to produce and deliver quality products at lower cost (Koenig and Dillon, 2017). Similarly, the successful implementation of a green chemistry initiative enables a company to gain external recognition, for example, the United Nations Sustainability Development Goals. The implementation of green chemistry program using green chemicals or materials may decrease the consumption of natural resources. Furthermore, the use of green chem-istry may reduce the amount of GHGs emission and thereby decrease the impacts of global warming or sea level rise. Overall, the adoption of green chemistry approach in pharmaceutical companies provides a remarkable benefit including the release of reduced toxic waste and public health safety (Koenig and Dillon, 2017; Zhang et al., 2019). Table 5.4 shows the summary of economic sustainability benefits of green manufacturing by pharmaceuticals companies.

TABLE 5.4

Green production strategies and economic sustainability benefits from pharmaceutical industries

	Strategies	Benefits	Comments
			Strategies for green production and economic sustainability benefits
1	Environmental management system	Decreased overhead expenses	The overhead cost of operation and production can be reduced through green manufacturing in pharmaceutical sector. Going green initiatives including environment-friendly packaging, the use of renewable energy, green materials, green chemicals, green technologies, building relationship with local community to avoid conflict, retention of employees, generation of minimum waste and management of waste for energy production will save costs. These initiatives will increase the overhead cost of companies for the short-term but in the long run the overhead cost will decrease. Consequently, the companies will be able to take more initiatives to ensure sustainability.
2	Policy for employees' welfare	Increased employee moral/Well motivated employees	Initiatives for green manufacturing provide the employees of a company with additional motivation. Employees feel more secure with their jobs because they understand that their company is not unmanful to the environment, and therefore, there is no potential conflict with government or local communities. With increased motivation employees work with more productivity because they understand that they work for a company that has green credentials in the society and market. Recruitments of employees becomes easy, and the new employees also care about environment.
3	Expansion of company's green activities	Job creation	Green manufacturing creates job in many ways. The management of waste using modern technologies may create job for local community and people from elsewhere. The initiatives for greening production process and reduction of environmental pollution may requires human resources. The increase of sale of green products in the long run will require more workers or staff.
4	Continuous relationship with communities	Avoidance costs of conflicts with community	The minimization of potential environmental and social risks may turn the business green, and companies may sale their product with a reputation. There is a huge cost that might be borne by the company that is polluting environment and affecting local community. The community may sue against the company, protests to ban the product to sale and the company may face loss instead of benefits.
6	Compliance with environmental laws	Tax benefits	Following the polluter pay principle, companies may follow the government rules and regulations to reduce the environmental impacts. In return, government may offer incentives for tax benefits and the company who comply with environmental laws may enjoy these benefits. This is a win-win situation. The lower tax paid to government may lower the cost of product as well. There are financial incentives in different forms available in many countries to offer the pharmaceutical industries those comply with local environmental laws using green technologies and methods.

5.4 CONCLUSION AND RECOMMENDATIONS

The aim of this chapter was to identify the strategies for green pharmaceutical manufacturing and sustainability benefits. Industrial production is one of the key reasons of environmental pollutions, conflict with community and regulators, and global worming by emitting GHGs. Among the industries, pharmaceutical companies are also one of the key polluters. However, in response to pressure from green consumers, international and local laws, community trust, and competition, the pharmaceutical companies are paying attention to green their green manufacturing. The green manufacturing strategies may include CSR program, use of green energy, treatment of waste management, and use of rational number of resources. By greening the production process, pharmaceutical companies may offer several sustainability benefits including social, environmental, and economic. The social benefits may include increased number of green consumers, customer satisfaction, increased community awareness, improved quality of life of local community, improved relationship with stakeholders, and improved corporate image or reputation. The environmental benefits may include conservation of natural resources, prevention of environmental pollution, and improved quality of environment.

The economic benefits may include lower material costs, decreased overhead expenses of companies, increased employee morale and motivated employees, creation of more jobs, better financial performance, avoidance of cost of conflicts with communities and regulators, tax benefits, and enhanced access to market. However, there are areas of improvement for greening the business of pharmaceutical companies and recommendations cam be made as follows.

5.4.1 RECOMMENDATIONS

It is to be noted that while developed or industrialist countries already have taken several initiatives and for green economy through green production, the developing countries although have manufacturing industries have just started to conceptualize green production where the enforcement of environmental laws is one of the key challenges (Kabir et al., 2010; Climent and Soriano, 2011; Chirambo, 2018). With is in view, the following recommendations can be made for developing countries and elsewhere:

Guidelines: In many developing countries there are no guidelines for pharmaceuticals industries to comply with the environmental standards. In Bangladesh, for example, there is a general guideline for industries but no specific guidelines for pharmaceutical companies (Kabir et al., 2010). It is to be noted that there are 257 pharmaceutical companies and there may be potential significant impacts of these industries on environment and local communities. However, there is a lack of Guidelines to assess the impacts of production process these companies. A comprehensive assessment at strategic level in this sector is necessary to understand the impacts on sustainability (Kabir and Morgan, 2020; Kabir et al., 2020). In the USA, for example, there are guidelines for different sectors limiting the discharge of pollutants including pharmaceutical sector and criteria to assess the sustainability of companies and environmental standards where there the limit of pollutants to be released by a pharmaceutical industry is mentioned in detail.

Renewable energy use: For green manufacturing, the use of renewable energy is one of the key aspects of environmental sustainability. While developed countries have been adopting technologies for renewable energy production, the developing countries are still looking for suitable sources to produce renewable energy (Ashwath and Kabir, 2019; Kabir et al., 2022a,b). Given the increased and available renewable energy in a country companies will be able to use the renewable energy and thereby enhance the environmental sustainability.

Environmental governance: Enforcement of law for the protection of environmental pollution is a challenge in developing countries. For the enforcement of laws effectively, it is necessary to have adequate rules and regulations in place, adequate manpower, and incentives for the pharmaceutical companies with green manufacturing. The close monitoring by the competent agencies is important so that companies comply with the local environmental standards. Importantly an Environmental Impact Assessment need to be undertaken to understand the potential impacts of the pharmaceutical industry (Kabir et al., 2010) to address the impacts through environmental management plan.

Research and innovation: Undertaking research for innovative design for production process is important for green production. Overall, the key research issues of green manufacturing may include the design of product, development of sustainability evaluation of the company where it operates, green technologies, waste management technologies, effective community engagement, CSR, green chemistry to identify environment-friendly solvent and green materials.

Waste management: In addition to treatment of wastewater using ETP, the disposal of solid waste management by pharmaceutical companies needs to be improved. In developing counties, the disposal of pharmaceutical waste is unsustainable, and pollutants are released to environment from open dumping of waste management practice. The companies need to find a cost-effective solution and adopt good practice for the reduction of waste and avoid the risks from hazardous waste (Kabir and Kabir, 2021; Kabir et al., 2022a,b).

REFERENCES

Acharya, PSG., Vadher, JA., and Acharya, GD. 2014. A Review on Evaluating Green Manufacturing for Sustainable Development in Foundry Industries, *International Journal of Emerging Technology*, Vol. 4 (1): 232–237.

Adam, DH., Supriadi, YN., Ende, and Siregar, ZME. 2020. Green Manufacturing, Green Chemistry and Environmental Sustainability: A Review, *International Journal of Scientific & Technology Research*, Vol. 9 (4): 2209–2211.

Ahmed, W., Asim, M., and Manzoor, S. 2020. Importance and Challenges of Green Supply Chain Management in Healthcare, *European Journal of Business and Management Research*, Vol. 5 (2): 1–20.

Ananda, ARW., Astuty, P., and Nugroho, YC. 2018. Role of Green Supply Chain Management in Embolden Competitiveness and Performance: Evidence from Indonesian Organizations, *International Journal of Supply Chain Management*, 7 (5): 437–442.

Arora, G., Shrivastava, R., Kumar, P., Krishnamurthy, D., Sharma, RK., Matharu, AS., and Rizwan, M. 2021. Recent Advances Made in the Synthesis of Small Drug Molecules for Clinical Applications: An Insight, *Current Research in Green and Sustainable Chemistry*, Vol. 4: 2666–2865.

Arulrajah, A., and Opatha, HHDNP. 2016. Analytical and Theoretical Perspectives on Green Human Resource Management: A Simplified Underpinning, *International Business Research*, Vol. 9 (12): 153–164.

Ashwath, N., and Kabir, Z. 2019. Environmental, Economic, and Social Impacts of Biofuel Production from Sugarcane in Australia, in Khan, MT., and Khan, IA. (eds), *Sugarcane Biofuels*, Springer, Cham, pp. 267–284.

Asthana, AN. 2014. Thirty Years after the Cataclysm: Toxic Risk Management in the Chemical Industry, American-Eurasian, *The Journal of Toxicological Sciences*, Vol. 1: 1401–1408.

Belkhir, L., and Elmeligi, A. 2019. Carbon Footprint of the Global Pharmaceutical Industry and Relative Impact of Its Major Players, *Journal of Cleaner Production,* Vol. 214: 185–194.

Breuer, W., Müller, T., Rosenbach, D., and Salzmann, A. 2018. Corporate Social Responsibility, Investor Protection, and Cost of Equity: A Cross-Country Comparison, *Journal of Banking and Finance*, Vol. 96: 34–55.

Cespi, D., Beach, ES., Swarr, TE., Passarini, F., Vassura, I., Dunn, P. J., and Anastas, PT. 2015. Life Cycle Inventory Improvement in the Pharmaceutical Sector: Assessment of the Sustainability Combining PMI and LCA Tools, *Green Chemistry*, Vol. 17 (6): 3390–3400.

Chirambo, D. 2018. Towards the Achievement of SDG 7 in Sub-Saharan Africa: Creating Synergies between Power Africa, Sustainable Energy for All and Climate Finance in-Order to Achieve Universal Energy Access before 2030, *Renewable and Sustainable Energy Reviews*, Vol. 94: 600–608.

Climent, F., and Soriano, P. 2011. Green and Good? The Investment Performance of US Environmental Mutual Funds, *Journal of Business Ethics*, Vol. 103: 275–287.

Colglazier, W. 2015. Sustainable Development Agenda: 2030, *Science*, Vol. 349: 1048–1050.

–Delmas, M. 2001. Stakeholders and Competitive Advantage: The Case of ISO 14001, *Production and Operations Management*, Vol. 10 (3): 343–358.

Dong, C., Shen, B., Chow, PS., Yang, L., and Ng, CT. 2016. Sustainability Investment under Cap-and-Trade Regulation. *Annals of Operations Research*, Vol. 240: 1–23.

Dornfeld, D., Yuan, C., Diaz, Zhang, NT., and Vijayaraghavan, A. 2013. Introduction to Green Manufacturing, in Dornfeld, DA. (ed.), *Green Manufacturing: Fundamentals and Applications*, Chapter 1, Springer, New York, pp. 1–25.

Drake, DF., and Spinler, S. 2013. Sustainable Operations Management: An Enduring Stream or a Passing Fancy? *Manufacturing & Service Operations Management,* Vol. 15: 689–700.

Eckelman, MJ., and Sherman, J. 2016. Environmental Impacts of the U.S. Health Care System and Effects on Public Health, *PLoS One*, Vol. 11 (6): e0157014. https://doi.org/10.1371/journal.pone.0157014

EPA, 1985. *Health and Environmental Effects Profile for Acetone Cyanohydrin*. U.S. Environmental Protection Agency, Washington, DC.

Famiyeh, S., Kwarteng, A., Asante-Darko, D., and Dadzie, SA. 2018. Green Supply Chain Management Initiatives and Operational Competitive Performance. *Benchmarking: An International Journal*, Vol. 25(2): 607–631.

Galbreath, DJ. 2001. *The Benefits of Corporate Social Responsibility: An Empirical Study, Graduate School of Business.* Curtin University of Technology, Perth. https://www.anzam.org/wp-content/uploads/pdf-manager/1279_GALBREATH_JEREMY-13.PDF

Gao, Z., Geng, Y., Wu, R., Chen W., Wu F. and Tian, X. 2019. Analysis of Energy-Related CO2 Emissions in China's Pharmaceutical Industry and Its Driving Forces, *Journal of Cleaner Production*, Vol. 223: 94–108.

Gong, Y., Jia, F., Brown, S., and Koh, L. 2018. Supply Chain Learning of Sustainability in Multi-Tier Supply Chains: A Resource Orchestration Perspective, *International Journal of Operations & Production Management*, Vol. 38: 1061–1090.

Götz, K., Courtier, A., Stein, M., Strelau, L., Sunderer, G., Vidaurre, R., Winker, and M., Roig, B. 2019. Risk Perception of Pharmaceutical Residues in the Aquatic Environment and Precautionary Measures, in Roig, B., Weiss, K., and Thireau, V. (eds), *Management of Emerging Public Health Issues and Risks: Multidisciplinary Approaches to the Changing Environment*, Chapter 8, Academic Press, Cambridge, MA, pp. 189–224.

Government of Bangladesh (GOB). 2020. Bangladesh Pharma Market & Regulatory Report, available at https://www.lightcastlebd.com, accessed on 23/8/2022.

Griggs, D., Stafford-Smith, M., and Gaffney, O. 2013. Policy: Sustainable Development Goals for People and Planet, *Nature*, Vol. 495: 305–307.

Grougiou, V., Dedoulis, E., and Leventis, S. 2016. Corporate Social Responsibility Reporting and Organizational Stigma: The Case of "Sin" Industries, *Journal of Business Research*, Vol. 69: 905–914.

Heal, G., 2005. Corporate Social Responsibility: An Economic and Financial Framework, *The Geneva Papers*, Vol. 30: 387–409.

Hu, ZY., Chang, J., Guo, FF., Deng, HY., Pan, GT., Li, BY., and Zhang, ZL. 2022. The Effects of Dimethylformamide Exposure on Liver and Kidney Function in the Elderly Population: A Cross-Sectional Study, *Medicine (Baltimore)*, Vol. 99 (27): 20749.

Jaseem, M., Pramod Kumar, P., and John, RM. 2017. An overview of waste management in pharmaceutical industry, *The Pharma Innovation Journal*, 6(3): 158–161.

Jimenez-Gonzalez, C., and Lund, C. 2022. Green Metrics in Pharmaceutical Development, *Current Opinion in Green and Sustainable Chemistry*, Vol. 33: 100564.

Johnstone, N. 2007. *Environmental Management, Performance and Innovation: Evidence from OECD Manufacturing Facilities*. OECD Workshop on Sustainable Manufacturing and Competitiveness, Copenhagen, Denmark.

Kabir, SMZ., Momtaz, S., and Gladstone, W. 2010. The Quality of Environmental Impact Statement (EIS) in Bangladesh, *Proceedings of the 30th Annual Conference of the International Association for Impact Assessment*, pp. 1–5, 6–11 April, Geneva.

Kabir, Z., and Kabir, M. 2021. Solid Waste Management in Developing Countries: Towards a Circular Economy, in Baskar, C., Ramakrishna, S., Baskar, S., Sharma, R., Chinnappan, A., Sehrawat, R. (eds), *Handbook of Solid Waste Management: Sustainability through Circular Economy*, Springer, Gateway East, pp. 1–34.

Kabir, Z., Kabir, M., Rahman, MA., and Rahman, M. 2022a. Operational Tools and Techniques for Municipal Solid Waste Management, in Gupta, RK., and Nguyen, TA. (eds), *Energy from Waste*, CRC Press, Boca Raton, FL, pp. 1–14.

Kabir, Z., and Khan, I. 2020. Environmental Impact Assessment of Waste to Energy Projects in Developing Countries: A Guideline in the Context of Bangladesh, *Sustainable Energy Technologies and Assessments*, Vol. 37: 1–13.

Kabir, Z., and Morgan, R. 2020. Strategic Environmental Assessment of Urban Plans in New Zealand: Current Practice and Future Directions, *Journal of Environmental Planning and Management*, Vol. 64 (1): 1–24.

Kabir, Z., Salim, M., and Morgan, R. 2020. Strategic Environmental Assessment of Urban Plans in Australia: The Case Study of Melbourne Urban Extension Plan, *Impact Assessment and Project Appraisal*, Vol. 38 (5): 368–381.

Kabir, Z., Sultana, N., and Khan, I. 2022b. Environmental, Social, and Economic Impacts of Renewable Energy Sources, in Khan, I. (ed.), *Renewable Energy and Sustainability*, Chapter 3, Elsevier, Amsterdam, pp. 57–85.

Karkman, A., Do, T., Walsh, F., Fiona and Virta, M., 2017. Antibiotic-Resistance Genes in Wastewater, *Trends in Microbiology*, Vol. 26 (3): 220–228, https://doi.org/10.1016/j.tim.2017.09.005

Khaksar, E., Abbasnejad, T., Esmaeili, A., and Tamošaitienė, J. 2016. The Effect of Green Supply Chain Management Practices on Environmental Performance and Competitive Advantage: A Case Study of the Cement Industry, *Technological and Economic Development of Economy*, Vol. 22 (2): 293–308.

Khan, I., and Kabir, Z., 2020, Sustainability Assessment of Waste-to-Energy (Electricity) Generation Technologies and Its Prospect in the Developing World: A Case of Bangladesh, *Renewable Energy*, Vol. 150: 320–333.

Koenig, SG., and Dillon, B. 2017. Driving toward Greener Chemistry in the Pharmaceutical Industry, *Current Opinion in Green and Sustainable Chemistry*, Vol. 7: 56–59

Larsson, DGJ. 2014. Pollution from Drug Manufacturing: Review and Perspectives, *Philosophical Transactions of the Royal Society B*, Vol. 369: 20130571.

Li, YX., Zhang, Xl., Li, W., Lu, X., Liu, B., and Wang, J. 2013. The Residues and Environmental Risks of Multiple Veterinary Antibiotics in Animal Faeces, *Environ Monitoring Assess*, Vol. 185: 2211–2220.

Lin, G., and Hao, B., 2020. Research on Green Manufacturing Technology, *Journal of Physics: Conference Series*, Vol. 1601: 042046, https://doi.org/10.1088/1742-6596/1601/4/042046

Mahender, M., Yakambaram, B., Pandey, J., Chandrashekar, ERR., Amarnath, L., Jayashree, RA., and Bandichhor, R. 2019. Stereoselective Synthesis for Potential Isomers of ticagrelor Key Starting Material, *Journal of Heterocyclic Chemistry*, Vol. 56: 2866–2872.

Mahmud, A., Ding, D., and Hasan, MM., 2021. *Corporate Social Responsibility: Business Responses to Coronavirus (COVID-19) Pandemic*. SAGE Open, https://doi.org/10.1177/2158244020988710 journals.sagepub.com/home/sgo

Martusa, MR. 2013. Green Supply Chain Management: Strategy to Gain Competitive Advantage, *Journal of Energy Technologies and Policy*, Vol. 3 (11): 334–341. https://doi.org/10.1177/2158244020988710 journals.sagepub.com/home/sgo

Masekameni, MD., Moolla, R., Gulumian, M., and Brouwer, D. 2021. Risk Assessment of Benzene, Toluene, Ethyl Benzene, and Xylene Concentrations from the Combustion of Coal in a Controlled Laboratory Environment, *International Journal of Environmental Research and Public Health*, Vol. 16 (95), https://doi.org/10.3390/ijerph16010095

Masri, HA., and Jaaron, AAM., 2017. Assessing Green Human Resources Management Practices in Palestinian Manufacturing Context: An Empirical Study, *Journal of Cleaner Production*, Vol. 143: 474–489. https://doi.org/10.1016/j.jclepro.2016.12.087

Menges, N. 2017. The Role of Green Solvents and Catalysts at the Future of Drug Design and of Synthesis, *Green Chemistry*. https://doi.org/10.5772/intechopen.71018

Miettinen, M., and Khan, SA. 2021. Pharmaceutical Pollution: A Weakly Regulated Global Environmental Risk, *Review of European, Comparative International Environmental Law*, Vol. 31 (1): 75–88. https://doi.org/10.1111/reel.12422

Mishra, M., Sharma, M., Dubey, R., Kumari, P., Ranjan, V., and Pandey, J. 2021. Green Synthesis Interventions of Pharmaceutical Industries for Sustainable Development, *Current Research in Green and Sustainable Chemistry*, Vol. 4: 100174.

Momtaz, S., and Kabir, Z. 2018. Evaluating Community Participation in Environmental Impact Assessment, in Momtaz, S., and Kabir, Z. (eds), *Evaluating Environmental and Social Impact Assessment in Developing Countries*, 2nd ed., Elsevier, Amsterdam.

Muster, V., and Schrader, U. 2011. Green Work-Life Balance: A New Perspective for Green HRM, *German Journal of Research in Human Resource Management*, Vol. 25 (2): 140–156.

Ott, D., Kralisch, D., Denčić, I., Hessel, V., Laribi, Y., Perrichon, P.D., Berguerand, C., Kiwi-Minsker, L., and Loeb, P. 2014. Life Cycle Analysis within Pharmaceutical Process Optimization and Intensification: Case Study of Active Pharmaceutical Ingredient Production, *ChemSusChem*, Vol. 7 (12): 3521–3533.

Owens, G., 2016. *Best Practices Guide: Economic & Financial Evaluation of Renewable Energy Projects*, USAID, Washington DC.

Özkan, O., Akyürek, ÇE., and Toygar, SA. 2014. Green Supply Chain Method in Healthcare Institutions, in *Chaos, Complexity and Leadership* (Ed.) Sefika Sule Erçetin, Springer International Publishing AG, Cham, pp. 285–293.

Paul, ID., Bohle, GP., and Chaudhari, JR. 2014. A Review on Green Manufacturing: It's Important, Methodology and Its Application, *Procedia Materials Science*, Vol. 6: 1644–1649.

Porter, ME., and Kramer, MR. 2006. Strategy and Society: The Link Between Competitive Advantage and Corporate Social Responsibility, *Harvard Business Review*, available at https://hazrevista.org/wp-content/uploads/strategy-society.pdf, accessed on 16/8/2022.

Rogowska, J., Cieszynska-Semenowicz, M., Ratajczyk, W., et al. 2020. Micropollutants in Treated Wastewater, *Ambio*, Vol. 49: 487–503. https://doi.org/10.1007/s13280-019-01219

Saeed, BB., Afsar, B., Hafeez, S., Khan, I., Tahir, M., and Afridi, MA. 2019. Promoting Employee's Pro-Environmental Behaviour through Green Human Resource Management Practices, *Corporate Social Responsibility and Environmental Management*, Vol. 26: 424–438.

Seidel, RHA., Shahbazpour, M., and Seidel, MC. 2007. Establishing Sustainable Manufacturing Practices in SMEs. *Proceedings of the Second International Conference on Sustainability Engineering and Science*, Auckland.

Sharabati, AAA. 2021. Green Supply Chain Management and Competitive Advantage of Jordanian Pharmaceutical Industry, *Sustainability*, Vol. 13 (23): 13315.

Sharma, J., Joshi, M., Amit Bhatnagar, A., Chaurasia, AK., and Nigam, S. 2022. Pharmaceutical Residues: One of the Significant Problems in Achieving '*Clean Water for All*' and Its Solution, *Environmental Research*, Vol. 215 (1): 114219.

Silva, E., Pires, FCS., Ferreira, MCR., I.Q. da Silva, IQ., G.C.M. Aires, GCM., and Ribeiro, TM. 2021. Case Studies of Green Solvents in the Pharmaceutical Industry, in Inamuddin, Boddula, R., Ahamed, MI., and Asiri, AM. (eds), *Green Sustainable Process for Chemical and Environmental Engineering and Science*, Chapter 7, Elsevier, Amsterdam, pp. 51–159.

Slaughter, RJ., Mason, RW, Beasley, DMG., Vale, JA., and Schep, LJ. 2014. Isopropanol Poisoning, *Clinical Toxicology*, Vol. 52 (5): 470–478.

Stewart, KD., Johnston, JA., Matza, LS., Curtis, SE., Havel, HA., Sweetana, SA., and Gelhorn, HL. 2016. Preference for Pharmaceutical Formulation and Treatment Process Attributes, *Patient Prefer Adherence*, Vol. 10: 1385–1399.

Tiwari, B., Drogui, P., and Tyagi, RD. 2020. Removal of Emerging Micro-Pollutants from Pharmaceutical Industry Wastewater, in Varjani. S., Pandey, A., Tyagi, RD., Ngo, HH., Larroche, C (eds), *Current Developments in Biotechnology and Bioengineering: Emerging Organic Micro-Pollutants*, Chapter 18, Elsevier, Amsterdam, pp. 457–480.

UNIDO, 2013. *Sustaining Employment Growth: The Role of Manufacturing and Structural Change*, Industrial Development Report 2013, Vienna.

Unsal, V., Cicek, M., and Sabancilar, I. 2021. Toxicity of Carbon Tetrachloride, Free Radicals, and Role of Antioxidants, *Reviews on Environmental Health*. https://doi.org/https://doi.org/10.1515/reveh-2020-0048

Valavanidis, A., and Vlachogianni, T. 2012. Pharmaceutical Industry and Green Chemistry: New Developments in the Application of Green Principles and Sustainability, *Pharmakeftiki*, Vol. 24 (3): 44–56.

Veleva, V., and Cue, BW. 2017. Benchmarking Green Chemistry Adoption by "Bigpharma" and Generics Manufacturers, *Benchmark*, Vol. 24: 1414–1436. https://doi.org/10.1108/BIJ-01-2016-0003

Yadav, P., Athanassiadis, D., Yacout, DMMY., Tysklind. M., Venkata K.K., and Upadhyayul, VKK. 2020. Environmental Impact and Environmental Cost Assessment of Methanol Production from wood biomass, *Environmental Pollution*, Vol. 265 (Part A): 114990.

Yang, AS., and Yulianto, FA. 2022. Cost of Equity and Corporate Social Responsibility for Environmental Sensitive Industries: Evidence from International Pharmaceutical and Chemical Firms, *Finance Research Letters*, Vol. 47: 102532.

Yang, G., and Liu, B. 2021. Research on the Impact of Managers' Green Environmental Awareness and Strategic Intelligence on Corporate Green Product Innovation Strategic Performance, *Annals of Operations Research.* https://doi.org/org/10.1007/s10479-021-04243-5

Yaqub, G., Hamid, A., and Iqbal, S. 2012. Pollutants Generated from Pharmaceutical Processes and Microwave Assisted Synthesis as Possible Solution for Their Reduction-A Mini Review, *Nature Environment and Pollution Technology*, Vol. 11 (1): 29–36.

Yeşilyurt, MK., Erol, D., Yaman, H. et al. 2022. Effects of Using ethyl acetate as a Surprising Additive in SI Engine Pertaining to an Environmental Perspective, *International Journal of Environmental Science and Technology.* https://doi.org/10.1007/s13762-021-03706-3

Zhai, Q., Cao, H., Zhao, X., and Yuan, C. 2014. Assessing Application Potential of Clean Energy Supply for Greenhouse Gas Emission Mitigation: A Case Study on General Motors Global manufacturing, *Journal of Cleaner Production*, Vol. 75: 11–19.

Zhang, Z., Wang, Y., Meng, Q., and Luan, X. 2019. Impacts of Green Production Decision on Social Welfare, *Sustainability*, Vol. 11: 453. https://doi.org/10.3390/su11020453

Zheneng, W. 2010. Development Direction of Modern Manufacturing- Green Manufacturing, *Equipment Manufacturing Technology*, Vol. 3: 35–38.

6 Current trends in microbial production of citric acid, applications, and perspectives

*Srinivasan Kameswaran, Bellamkonda Ramesh,
Gopi Krishna Pitchika, Gujjala Sudhakara,
B. Swapna, and M. Ramakrishna*

CONTENTS

6.1 INTRODUCTION

Citric acid (CA) is an organic acid present in a wide range of fruits, including limes, lemons, oranges, pineapples, and grapefruits. It's a natural component that helps with cleansing, energy maintenance, and good digestion and renal function. It's used to balance the sweetness in soft drinks, juices, and other beverages, and it has a somewhat

DOI: 10.1201/9781003394600-7

tangy and refreshing flavour. Because of its antioxidant characteristics, CA is used in the food and beverage business to preserve food or as an acidifier to enhance the flavours and fragrances of fruit juices, ice cream, and marmalades. It's employed as an antioxidant to preserve vitamins, an effervescent, a pH corrector, a blood preservative, and iron citrate tablets as a supply of iron for the body, ointments and cosmetic preparations, and so on in the pharmaceutical business (Max et al. 2010). It is used as a foaming agent in the chemical industry for softening and treating fabrics. Certain metals are used in metallurgy in the form of citrate. CA is utilized as a phosphate alternative in the detergent industry since it has a lower eutrophic effect (Max et al. 2010). CA is also commonly used in facial packs and masks since it brightens and lightens the skin tone naturally, reduces acne and oiliness, and regenerates dead skin cells. Currently, the global CA market is estimated to reach US$3.2 billion by 2023, with a Compound Annual Growth Rate (CAGR) of 5.1% throughout that time period (Market report world.com 2020). The global production of CA is projected to be around 736,000 tonnes per year, with fermentation handling the majority of the process. Almost all of Brazil's CA need is fulfilled through imports. The volume of CA produced by fermentation is gradually expanding at a high yearly rate of 5% (Finogenova et al. 2005; Francielo et al. 2008) due to its diverse applications, and demand/consumption is also steadily increasing. It has been authorized by the Joint Food and Agriculture Organization/World Health Organization Expert Committee on Food Additives as a generally regarded as safe (Carlos et al. 2006; Rohr et al. 1996). The rising use in numerous sectors is the main element behind the worldwide CA market's rise. Biotechnology, which provides proper knowledge of fermentation techniques and product recovery; biochemistry, which provides knowledge of different factors that affect synthesis and blockage of CA production; molecular regulatory mechanisms; and strategies that enhance CA production have all contributed to the rapid growth of CA production over the last century. In the last 60 years, thousands of reports and extensive reviews of the literature have been published in relation to CA production (Max et al. 2010; Papagianni 2007; Prescott and Dunn 1959; Show et al. 2015; Tong et al. 2019; Vandenberghe et al. 1999b). However, with current advancements in the previous few years, the enhancement of CA output has not been updated.

As a result, this study provides a brief overview of current breakthroughs in CA production, including a description of microorganisms, production procedures, and substrates, among other things.

6.1.1 Background of citric acid

Wehmer (1893) discovered CA fermentation as a fungal product in a culture of *Penicillium glaucum* on sugar medium. He discovered two new fungal strains with the ability to accumulate CA after a few years, which he named Citromyces (*Penicillium*). Industrial testing, on the other hand, failed due to contamination issues and a protracted fermentation period. Currie's work paved the door for successful CA manufacturing in the industrial sector. He discovered that a variety of *Aspergillus niger* strains produced substantial amounts of CA in 1916. The most important conclusion was that *A. niger* thrived at pH levels of 2.5–3.5, and that high sugar concentrations favour CA synthesis (Table 6.1).

Surface cultures were used to carry out the first CA fermentations. For commercial production, certain units were seeded in England, the Soviet Union, and Germany in the 1930s. In general, CA is produced commercially by submerged microbial molasses fermentation; nonetheless, *Aspergillus niger* fermentation remains the primary source of CA worldwide. Although chemical methods for producing CA have been well researched, microbial fermentations have shown to be more successful, and this technology has now been the method of choice for commercial production over chemical synthesis (Mattey 1992).

Despite this, submerged fermentation posed a number of challenges, including selecting productive strains with minimal sensitivity to trace elements. It was important to pay much more attention to the raw materials. Several studies have been conducted to improve the circumstances for the use of low-cost materials such as sugar cane molasses, beet molasses, starch, and hydrolysate starch (Sarangbin and Watanapokasin 1999). Various methods for treating and purifying molasses, particularly for the removal of trace metals, have been developed. Furthermore, a minor excess of copper ions was discovered to be useful in achieving large CA yields. Despite this, submerged fermentation posed a number of challenges, including selecting productive strains with minimal sensitivity to trace elements. It was important to pay much more attention to the raw materials. Several studies have been conducted to improve the circumstances for the use of low-cost materials such as sugar cane molasses, beet molasses, starch, and hydrolysate starch (Sarangbin and Watanapokasin 1999). Various methods for treating and purifying molasses, particularly for the removal of trace metals, have been developed. Furthermore, a minor excess of copper ions was discovered to be useful in achieving large CA yields.

6.2 CITRIC ACID-PRODUCING MICROORGANISMS

6.2.1 MICROORGANISMS

The *Aspergillus* species like *A. wenti*, *A. aculeatus*, *A. foetidus*, *A. fonsecaeus*, *A. awamori*, *A. carbonaries*, and *A. phoenicis*, as well as *Trichoderma viride* and *Mucor pyriformis*, have been discovered to produce substantial levels of CA (Berovic and Legisa 2007).

Candida tropiclaus (Legisa and Mattey 1986), *Candida oleophilis* (Käppeli et al. 1978), *Candida guilermondi* (Angumeenal et al. 2003), *Yarrowia lipolytica* (Angumeenal and Venkappayya 2013), *Hansenula, Torulopsis, Torula, Debaromyces, Pichia* (Table 6.1) (Weyda et al. 2014). Because yeast produces a lot of isocitric acid, which is an undesired by-product, mutant strains with lower aconitase activity are required. Furthermore, the rising cost of oil makes it less economically viable, as oils are now used as the primary carbon source, similar to how alkanes were previously used (Mazinanian et al. 2015). Because it has benefits over other bacterial microbes such as *Arthrobacter paraffinens, Bacillus licheniformis, Bacillus subtilis, Brevibacterium flavum, Corynebacterium* spp., and *Penicillium janthinellum, Aspergillus niger* has kept its place in CA production thus far (Ikram-ul et al. 2004). It's simple to use, can ferment a wide range of low-cost ingredients, and produces excellent yields (Themelis and Tzanavaras 2001). Mutagenesis has been employed in recent years to develop citric-acid-producing strains so that they can be used in industrial applications. The use of mutagens to generate mutations in the parental strains is one of the most popular ways. Gamma radiation, ultraviolet (UV) radiation, and chemical mutagens are among the mutagens used for enhancement. A hybrid approach combining ultraviolet and chemical mutagens is used for super production strains (Ratledge and Kristiansen 2001). The passage and single spore approaches are used for selection. The passage approach is favoured because the single-spore method has the disadvantage of simulating the presence of CA with organic acids (oxalic and gluconic acids) and mineral acids (Soccol et al. 2006).

The yield of CA is also influenced by the fermentation method. For example, a strain might provide an excellent yield in submerged fermentation but a poor yield in solid-state fermentation. To determine the optimal fermentation method, the producer strains must be evaluated in each of the fermentation methods as well as the industrial substrates (Chen et al. 2014a,b). Citric microorganisms must be infected with spores, which are then delivered to the fermentation medium. Air is one of the several transfer media, and it can be in the form of a suspension that is then injected into bottles containing the substrate. *A. niger* requires a 7-day incubation period in order to produce excellent yields. However, after 7 days of incubation, the ability to germinate decreases with time (Vergano et al. 1996).

6.3 IMPROVEMENTS TO CITRIC ACID-PRODUCING STRAINS

For the enhancement of industrially relevant microorganisms, many approaches such as mutations, protoplast fusion, recombinant DNA technology, and gene cloning are used (Table 6.1) (Parekh et al. 2000). The simplest and most widely utilized approaches are random mutagenesis and protoplast fusion. Physical, chemical, and site-directed mutagenesis is among the mutagenic techniques used to enhance strains. The commercial fermentation process has taken into account the overproduction of industrial products caused by strain improvement (Parekh et al. 2000; Vu et al. 2010). Strain improvement can also be accomplished by altering the microorganism's metabolism by introducing mutations in them via physical or chemical mutagens (Swain et al. 2012). Gamma and UV radiations are the most commonly employed physical mutagens (Pelechova et al. 1990). Diethyl sulphonate, *N*-methyl-*N*-nitroso-guanidine, ethidium bromide aziridine, *N*-nitroso-*N*-methylurea, and ethyl methane-sulphonate are among the most prevalent chemical mutagens (Musilkova et al. 1983). Pontecorvo et al. were the first to describe strain improvement through the parasexual cycle (1953). CA productions were higher in diploids than in their parent haploids, according to Das and Roy (1978). Kirimura et al. described genetic modification of *A. niger* with regard to CA synthesis by protoplast fusion (1988a). The single-spore technique and the passage method are two more well-known alternate strategies for selecting better strains (Soccol et al. 2006).

TABLE 6.1
List of citric acid synthesis microorganisms

Micro-organisms	References
Bacteria	
Acetobacter xylinum	Lu et al. (2016)
Arthrobacter paraffinens	Kroya Fermentation Industry (1970)
Bacillus licheniformis	Sardinas (1972), Kapoor et al. (1983)
Corynebacterium sp.	Fukuda et al. (1970), Kapoor et al. (1983)
Fungi	
Aspergillus aculeatus	El Dein and Emaish (1979)
A. awamori	Grewal and Kalra (1995)
A. carbonarius	El Dein and Emaish (1979)
A. foetidus	Chen (1994), Tran et al. (1998)
Aspergillus niger	Hang and Woodams (1984, 1985, 1987), Roukas (1991), Garg and Hang (1995), Lu et al. (1997), Pintado et al. (1998), Vandenberghe et al. (1999a), Wang et al. (2017)
Aspergillus niger ATCC12846	Yu et al. (2018)
A. niger H915-1	Yin et al. (2017)
A. niger MTCC 282	Ganne et al. (2008)
A. wentii	Karow and Waksman (1947)
Penicillium janthinelum	Grewal and Kalra (1995)
Penicillium oxalicum	Li et al. (2016)
Yeasts	
Candida tropicalis	Kapelli et al. (1978)
C. oleophila	Ishi et al. (1972)
C. guilliermondii	Miall and Parker (1975), Gutierrez et al. (1993)
C. parapsilosis	Omar and Rehm (1980)
C. citroformans	Uchio et al. (1975)
C. liplytica	Crolla and Kennedy (2001)
Hansenula anamola	Oh et al. (1973)
Saccharomicopsis lipolytica	Ikeno et al. (1975), Maddox et al. (1985), Kautola et al. (1992), Wojtatowicz et al. (1993), Rane and Sims(1993), Good et al. (1985)
Yarrowia lipolytica Wratislavia 1.31	Rywińska et al. (2010)
Yarrowia lipolytica 1.31	Rymowicz et al. (2005)
Yarrowia lipolytica NG40/UV5	Morgunov et al. (2018)

6.4 PRETREATMENT AND SUBSTRATES

In fermentation, the substrate plays a key role in lowering costs and achieving optimal yield. As a result, substrate plays a larger role in productivity and fermentation yield (Lesniak 1999). Increased yield and shorter fermentation times are directly proportional to the quality of the substrate employed (Lesniak 1999). For CA production, *Aspergillus niger* ferments molasses, sucrose, beet or cane sugar syrups, hydrol produced as a by-product during crystalline glucose production, and palm oil (Gutcho 1973; Kutermankiewicz et al. 1980; Lesniak 1999; Lesniak et al. 1986). Pretreatment of the substrate used in CA fermentation is required to remove trace metals (Kristiansen et al. 1999). Calcium, magnesium, manganese, iron, and zinc have been discovered in cane molasses used in fermentation, all of which slow down CA production. Potassium ferrocyanide, which can effectively precipitate zinc and iron, is commonly used for chemical pretreatment of substrate prior to fermentation. Enzymatic hydrolysis was also used to remove heavy metal concentrations below essential levels in starch materials such as corn, wheat, and potato prior to the synthesis of CA (Pietkiewicz et al. 1996).

Agricultural waste and by-products are used in CA production to minimize production costs. Rice bran, carrot waste, coffee husk, cassava bagasse, wheat bran, banana peel, tapioca, vegetable wastes, cheese whey, sugarcane bagasse, rice straw, brewery wastes, coconut husk, decaying fruits, orange peel, corn cob, kiwifruit peel, grape pomaces, pineapple peel, and apple pomaces are some of the most commonly used agricultural residues (Dutta et al. 2019; Sawant et al. 2018; Soccol et al. 2006).

6.5 CITRIC ACID PRODUCTION FROM A BIOCHEMICAL PERSPECTIVE

During glycolysis, the pyruvate generated is oxidized and mixed with coenzyme A to produce CO_2, acetyl coenzyme A (acetyl-CoA), and nicotinamide adenine dinucleotide (NAD) with hydrogen (H) (NADH). Following that, the produced acetyl-coA is mixed with the oxaloacetate to generate citrate. Pyruvate carboxylase can also carboxylate pyruvate generated during glycolysis to make oxaloacetate. CA is created by combining acetyl-coA and oxaloacetate, which is subsequently converted into two molecules of CO_2 and four-carbon oxaloacetate. At each cycle turn, one molecule of acetic acid enters, two molecules of ATP and CO_2 are produced, and a molecule of oxaloacetate is used to produce citrate (Prescott and Dunn 1959).

Enzymes play a critical part in the production of CA. *A. niger* uses the pentose phosphate and glycolytic pathways to metabolize glucose and accumulate CA (Cleland and Johnson 1954; Martin and Wilson 1951). Citrate synthase, according to Mischak et al. (1985), is an enzyme that catalyses reversible catalysis between acetyl-CoA and oxaloacetate, preferring citrate synthesis. CA buildup occurred only when enzymes such as aconitase (ACO), isocitrate dehydrogenase, and succinic dehydrogenase were significantly inhibited during the tricarboxylic acid (TCA) cycle, according to Ramakrishnan et al. (1955). However, Kubicek and Rohr (1977) found that these enzymes were present in very little levels throughout the CA fermentation process. As a result, rather than being a result of slowed degradation, CA buildup could be the result of increased biosynthesis (Max et al. 2010). Outside the cell, a membrane-bound enzyme called invertase transforms sucrose to glucose and fructose, which are then delivered into the cell via glucose transporters (Rubio and Maldonado 1995). Hexokinase converts transporter glucose into glucose-6-phosphate inside the cell, kicking off the glycolysis process. Because some enzymes are inhibited during glycolysis, there is a high flux through glycolysis, resulting in CA accumulation by *A. niger* (Rohr et al. 1992). For example, Manganese deficits, as well as phosphate and nitrogen deprivation, restrict fungal anabolism, resulting in protein breakdown and a rise in intracellular NH_4^+ concentration (Habison et al. 1983; Rohr and Kubicek 1981). The enzyme phosphofructokinase (PFK), which requires magnesium as a cofactor and converts fructose 6-phosphate to fructose 1,6-bisphosphate in glycolysis, is inhibited by the high concentration of intracellular NH_4^+. When PFK is inhibited, a flow through glycolysis occurs, resulting in the buildup of CA. In contrast, Papagianni et al. (2005) found that *A. niger* accumulates CA as a result of an internal ammonium pool that inhibits the enzyme PFK. Scientists discovered that instead of ammonium ions accumulating or depositing inside the cell to form an ammonium pool, ammonium ions entered the cell and interacted with glucose to form glucosamine, the phosphofructokinase inhibition is caused by a protein that is secreted outside the cell or into the fermentation broth. As a result, the precise link between glucose, ammonium ion concentrations, TCA cycle enzymes, and CA buildup is unknown and warrants additional exploration.

Isocitrate dehydrogenase is a $NADP^+$ dependent enzyme that is activated in the presence of Mg^{2+} and Mn^{2+} in both mitochondria and cell cytoplasm. CA buildup is favoured when this enzyme is inhibited by α-ketoglutarate and citrate (Ratledge and Kristiansen 2001). High levels of glucose and ammonium in the pool slowed the synthesis of α-ketoglutarate dehydrogenase, a key enzyme for regulating the TCA cycle, and hence slowed the movement of the CA cycle, resulting in the buildup of CA (Rohr and Kubicek 1981).

6.6 PRODUCTION OF CITRIC ACID

The type of microbe utilized and the raw material used determine which fermentation method is best for CA synthesis. Rohr et al. (1983) divided raw materials utilized in the synthesis of CA into two categories: (i) raw materials with a high ash content and high amounts of other nonsugar substances from which the cations could be removed using standard procedures (e.g., cane or beet sugar, dextrose syrups, and crystallized dextrose) and (ii) raw materials with a high ash content and high amounts of other nonsugar substances from which the cations could be removed using standard procedures (e.g., cane and beet molasses, crude unfiltered starch hydro-lysates). Depending on the kind of fermentation, a wide range of substrates could be used to produce CA efficiently and economically. Industrial CA fermentation can be done on a variety of substrates in three main ways: surface, submerged, and solid-state fermentation, each with its own set of benefits and drawbacks. Each of these fermentation procedures, however, involves media preparation, inoculation, fermentation, and CA recovery (Figure 6.1).

6.6.1 SURFACE FERMENTATION

Surface fermentation, also known as the start of the CA fermentation process, is a stationary batch fermentation procedure that takes 8–12 days to complete. It's done in stainless steel fermentation chambers with a number of trays stacked in stainless

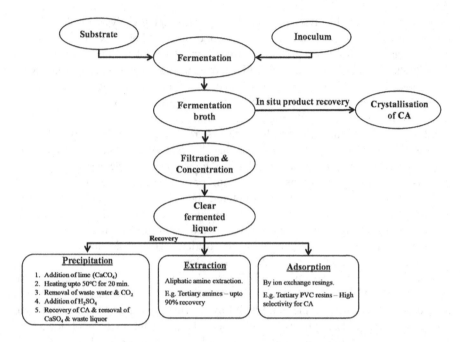

FIGURE 6.1 Current technologies in citric acid bio-production and recovery.

steel shelves (Bauweleers et al. 2014). On those trays, the fungal mycelium grows on the surface of the media. After sterilizing the fermentation medium in trays (Manzoni 2006), it is infected with spore suspension and incubated for 24 hours at 28–30°C (Marzona 1996). The germination of spores began with a decline in the pH of the medium from 6.0–6.5 to 1.5–2.0, which was visible with the naked eye due to continuous mycelium on the surface. During fermentation, a large quantity of heat is generated, which can be regulated with sufficient aeration. In quantities more than 10%, CO_2 produced during the fermentation process would hinder the development of CA. The fermentation chambers include an efficient air circulation system that circulates over the medium surface through a bacteriological filter, allowing humidity and temperature to be adjusted by evaporative cooling (Max et al. 2010; Soccol et al. 2006). Surface fermentation, on the other hand, has significant drawbacks, such as low yield and labour intensiveness, as well as greater maintenance costs, when compared to submerged fermentation (Drysdale and McKay 1995). It's also susceptible to changes in the media's makeup (Benghazi et al. 2014). The benefits of surface fermentation include lower energy usage and the absence of froth (Moyer 1953).

6.6.2 SUBMERGED FERMENTATION

The broth medium inside the nutrient substratum is liquid in the submerged fermentation process, and the organism can proliferate throughout the broth medium (Reddy 2002). The fermentation takes place in bioreactors and takes 5–12 days to finish (Socool et al. 2006). After 1–2 days of inoculation, the organism grows into pellets about 0.5 cm in diameter that float freely in the medium. As a result, an organism with a large contact surface area is able to absorb nutrients and oxygen. To raise oxygen levels, a high speed air flow is delivered into the tank, and agitation equipment mixes and breaks the air bubbles. Microorganisms present in the media breakdown the carbon source anaerobically or partially anaerobically (Swain et al. 2012). Submerged fermentation has several advantages, including better control of the fermentation process and the ability to use a wide variety of substrates (Max et al. 2010). It produces higher yields and is more ideal for CA production because to lesser capital, maintenance, labour, and contamination hazards (Rohr et al. 1983). Submerged fermentation, on the other hand, has the problem of foam generation, which can be avoided by employing antifoam agents such animal or vegetable fats and chambers with a volume of up to one-third of the entire fermenter volume.

6.6.3 SOLID-STATE FERMENTATION

Microorganisms are cultivated in a low-water environment containing insoluble material that serves as both a physical support and a source of nutrients in solid-state fermentation (Pandey 1992; Vandenberghe et al. 2000). Solid-state fermentation requires a substrate that is solid and hydrated to about 70% moisture, a pH between 4.5 and 6.0, and a temperature between 28 and 30°C. The procedure takes 4–5 days to complete (Drysdale and McKay 1995). When compared to submerged fermentation, solid-state fermentation has the following advantages: low energy needs, higher yield, less danger of contamination, less effort in downstream processing, less effluent generation, simple operation, operable under less water, and lower operating expenses.

Furthermore, because the system is less sensitive to the presence of trace elements than submerged fermentation, cheap and readily available agro-industrial substrates can be used without any pre-treatment in solid-state fermentation (Berovic and Legisa 2007). This approach, on the other hand, has some drawbacks, including difficulty scaling up, limited amenability of the process to standardization, challenging control of process parameters, and heat build-up issues. Due of insufficient heat and oxygen transmission in the substrate, it is unable to fully utilize available nutrients (Kapilan 2015).

6.7 CITRIC ACID RECOVERY

At the end of the fermentation, CA contains a variety of undesired by-products such as mycelium, other organic acids, mineral salts, proteins, and other contaminants. As a result, only pure CA must be obtained (Grewal and Kalra 1995). Precipitation, extraction, and purification are the three main techniques for recovering CA from fermented broth. During the fermentation of CA, oxalic acid is produced, which can be eliminated by raising the pH to 3.0 using calcium hydroxide at 72–75°C (Sawant et al. 2018). Calcium oxalate is generated as a result, which can be precipitated and removed via centrifugation or filtration. The CA that remained in the original solution in the form of calcium salt (calcium citrate) can be recovered through precipitation by adding calcium oxide at 90°C and a pH close to 7.0 (Soccol et al. 2006). The tetrahydrate of tri-calcium citrate is generated, which is subsequently treated with 70% sulphuric acid to produce CA and insoluble calcium sulphate (gypsum). After filtering off the gypsum, a solution containing 25%–30% CA is obtained. The filtrate can be processed in ion exchange columns or treated with activated carbon to eliminate remaining contaminants. It can also be concentrated by evaporation in a vacuum at 40°C, and crystals of CA monohydrate or anhydrous CA can be produced in a vacuum crystallizer at 20–25°C or at crystallization temperatures over 36.5°C (Grewal and Kalra 1995; Kubicek 1986).

Alternative solvent extraction methods for purification and crystallization of CA from fermentation broth include n-octyl alcohol, tridodecylamine, and isoalkane (Soccol et al. 2006), alanine 336 in heptane or xylene (Sirman et al. 1990), mixture of butylacetate and N,N-disubstituted alkylamide (Yi et al. 1987). The solvent extraction method has the benefit of avoiding the usage of lime and sulphuric acid (H_2SO_4) (Prochazka et al. 1994), and therefore avoiding the formation of gypsum (Grewal and Kalra 1995). CA is extracted from the aqueous solution using this method, which involves washing the extract with water, crystallization, and concentration. The compressed CO_2 is passed through a concentrated CA solution dissolved in acetone. CO_2's anti-solvent properties eliminate any remaining contaminants, resulting in food-grade CA after simple decolorization and crystallization (Shishikura et al. 1992).

6.8 FACTORS AFFECTING THE PRODUCTION OF CITRIC ACID

Carbon source, nitrogen and phosphate limitation, pH, aeration, trace element content, and the shape of the producing microbe are the key parameters impacting citric fermentation (Max et al. 2010).

During CA fermentation, the pH of the medium is especially critical during the initial stage of fermentation and towards the conclusion, before the product is recovered (Papagianni 2007). The germinating fungal spores in the fermentation medium after inoculation require a pH greater than five to germinate in the germination stage (Papagianni 2007). The germinating spores take ammonia and release protons, increasing the medium's acidity and promoting CA synthesis (Papagianni 2007). The initial pH of the medium, on the other hand, is determined by the substrate used. For chemically defined media, a pH value of 2.5–4.0 was shown to be optimal (Jernejc et al. 1982), whereas for molasses medium, an initial pH value of 6.0–7.5 was found to be optimal (Berry et al. 1977). The pH during the CA recovery stage, on the other hand, should be less than 2. The developments of undesirable compounds like oxalic and gluconic acid is hindered at this low pH, as is the potential of contamination by other microbes, making CA recovery easier (Max et al. 2010).

Glycolysis and the TCA cycle enzyme are required for CA formation. Temperature affects the activity of these enzymes. These enzyme systems will not work at higher or lower temperatures, affecting CA synthesis. Temperatures between 25 and 30°C were shown to be more favourable for high yields and rapid rates of CA production (Prescott and Dunn 1959). According to Kapoor et al. (1982), the CA yield decreases beyond 30°C due to an increase in oxalic acid synthesis, whereas temperatures below 25°C reduce the organism's growth and fermentation rates.

Because *A. niger's* extracellular mycelium-bound invertase can rapidly hydrolyse sucrose at low pH, Angumeenal and Venkappayya (2013) found that sucrose is the most suited carbon source for CA synthesis above glucose, fructose, and lactose (Kubicek-Pranz et al. 1990). The carbon source's concentration is also crucial to the effectiveness of CA generation. According to Xu et al. (1989), *A. niger* strains require an initial sugar concentration of 10%–14%, but no CA was formed at a sugar concentration of less than 2.5%, and maximal CA production was recorded at a sugar concentration of 14%–22%. Because of the high sugar concentration, α-ketoglutarate dehydrogenase is suppressed, resulting in maximum CA buildup (Hossain et al. 1983). The size of the mycelium is diminished and its form is also changed at low sugar concentrations (Papagianni et al. 1999). It was also discovered that immobilized *A. niger* cells required lower sugar concentrations than loose cells for optimal CA formation (Honecker et al. 1989).

Oxalic acid builds in the fermentation medium at nitrogen concentrations larger than 0.25%, lowering the CA output (Gupta et al. 1975). A high nitrogen concentration has also been shown to promote sugar consumption and fungal development while lowering the quantity of CA generated (Hang et al. 1977). In order to make CA, ammonium salts such ammonium nitrate and sulphate, urea, peptone, malt extract, and so on are used (Grewal and Kalra 1995). Acid ammonium compounds are favoured because they generate a drop in pH, which is necessary for citric fermentation. The recommended salt is ammonium sulphate, which does not produce the undesired oxalic acid while lowering the pH of the medium as it is digested. Molasses is often nitrogen rich, and it is employed in industrial fermentation without the addition of any ammonium salts. Low quantities of phosphate have been discovered to be ideal for CA generation, however high levels can lead to the development of sugar acids, a reduction in CO_2 fixation, and growth stimulation (Grewal and Kalra 1995). At a concentration of 0.5–5 g/L, potassium dihydrogenphosphate (KH_2PO_4) has been found to be the best suited for CA synthesis (Shu and Johnson 1948).

According to Moyer (1953), adding ethanol doubles citrate synthetase activity while decreasing ACO activity, resulting in greater CA buildup. Because of its inhibitory action on metal ions, alcohols like methanol have a favourable influence on CA formation. The amount of methanol/ethanol required is determined by the medium's composition and the microorganism strain utilized (Moyer 1953). Alcohols enhance CA synthesis through altering cell proliferation and sporulation due to changes in the lipid composition of the cell membrane, according to Ingram and Buttke (1984). Lower alcohols introduced to pure material restrict CA production, but when added to crude carbohydrates, these alcohols boost CA production, according to Kubicek and Rohr (1986). A concentration of ethanol, methanol, isopropanol, *n*-propanol, or methylacetate between 1% and 5% neutralizes the detrimental effect of metal ions in CA formation and promotes maximum CA formation (Kubicek and Rohr 1986).

Copper ion (Cu^{2+}), zinc ion (Zn^{2+}), ferrous ion (Fe^{2+}), and manganese (Mn^{2+}) are the metal ions that are most vulnerable to CA formation and hence must be limited (Dronawat et al. 1995). Clark et al. (1966) found that concentrations of Mn^{2+} less than 3 g/L dramatically lowered the yield of CA generation, however Mattey and Bowes (1978) found that adding 10 mg Mn^{2+} per litre reduced CA accumulation by 50% compared to control culture. Manganese shortage impairs cellular anabolism in *A. niger*, resulting in excessive intracellular ammonium and protein degradation. The enzyme PFK (which is required for the conversion of glucose and fructose to pyruvate) is inhibited by the high intracellular ammonium concentration, resulting in a flux via the glycolysis pathway and the creation of CA.

The presence of various copper concentrations in the pellet formation medium is critical for enhancing a suitable structure for CA generation that is related to cellular function (Benuzzia and Segovia 1996). A high quantity of Cu^{2+} can counteract the harmful effects of Fe^{2+} (Rohr et al. 1983). The optimum zinc concentrations for CA formation, according to Tomlinson et al. (1950), are 0.3 ppm. When zinc was combined with KH_2PO_4, it encouraged the synthesis of CA (Vandenberghe et al. 1999b). On the other hand, its excessive presence could promote fungal development even if no CA is accumulated (Grewal and Kalra 1995). The addition of iron reduced CA buildup while simultaneously having an influence on mycelia development. Iron values of 1.3ppm are ideal for CA formation (Tomlinson et al. 1950). Magnesium is essential for both growth and the generation of CA. The optimal magnesium sulphate concentration for CA formation was found to be between 0.02 and 0.025% (Kapoor et al. 1982). Other trace metals that have been linked to CA accumulation in *A. niger* include nickel, molybdenum, and cobalt (Habison et al. 1983).

According to Vandenberghe et al. (1999b), higher oxygen concentrations are required for improved CA synthesis. The critical dissolved oxygen tension for the growth phase is 9–12% of air saturation, while the critical dissolved oxygen tension for the production phase is 12–13% of air saturation (Grewal and Kalra 1995). Even if the dry mass has not increased by more than 5%, the dissolved oxygen tension soon falls to less than 50% of its prior value when the organism begins to form filaments. As a result, during CA formation, small compact pellets are favoured mycelial forms of *A. niger*. Low oxygen environments cause growth limitation, which is important for CA formation, whereas heavily aerated cultures (0.3m³/kg dry CB/h) enhance sporulation, reducing CA buildup (Vandenberghe 2000). Carbon dioxide is a vital substrate for PC because it replenishes the oxaloacetate supply for citrate synthase. The reaction mediated by pyruvate decarboxylase produces enough CO_2, although excessive aeration causes some losses. Increased CO_2 levels, on the other hand, harm the ultimate biomass and citrate concentrations (McIntyre and McNeil 1997). It has been discovered that a high partial pressure of CO_2 slows filamentous fungus spore liberation and aids CA buildup. As a result, elevated CO_2 concentrations in the environment have a favourable effect on CA synthesis (Vandenberghe 2000).

The creation of compact aggregates or pellets by the fungus during CA fermentation has been found to favour higher CA production than the filamentous form. The morphology of *A. niger* in submerged fermentation is influenced by agitation rate, pH of the medium, medium composition, and inoculum concentration (Papagianni 2007). At pH values of 2.0±0.2, small aggregates of short filaments form, which are linked to an increase in CA formation. The aggregated short filamentous form of the fungus transformed to bulbous hyphae at lower pH (pH 1.6), resulting in very little CA generation, whereas aggregates with longer perimeters and oxalic acid formation were detected at higher pH (pH > 3.0) (Papagianni 2007).

Lipids as groundnut oil (Souza et al. 2014) and sodium monofluoro acetate also affect CA production (Meixner-Monori et al. 1984). Lipid can increase CA yield while having little influence on mycelium dry weight (Millis et al. 1963). The effects of calcium fluoride, sodium fluoride, and potassium fluoride on the industrial production of CA were studied by Kareem et al. (2010).

6.9   CITRIC ACID PRODUCTION THROUGH METABOLIC ENGINEERING

To boost CA synthesis, several metabolic engineering strategies have been tried, including altering genes and metabolic pathways (Ruijter et al. 1999; Yin et al. 2015). Systems metabolic engineering is a key method for developing and introducing a new biochemical pathway in *A. niger* to boost CA output (Tong et al. 2019). Meijer et al. (2009) found that deleting the gene *acl1*, which is responsible for ATP-citrate lyase synthesis and potentially boost CA production in *A. niger*, increased CA production. Chen et al. (2014a,b) discovered that knocking down two cytosolic ATP citrate lyase (ACL) subunits (*Acl1* and *Acl2*) reduces not only CA synthesis but also vegetative growth, pigmentation, conidial germination, and asexual development in *A. niger*. According to de Jongh and Nielsen (2008) created the cytoplasmic reductive TCA reverse tricarboxylic acid cycle (*rTCA*) by combining heterogeneous malate dehydrogenase (*mdh2*), fumarase (*FumR*), and fumarate reductase (*Frds1*). It has been discovered that the *mdh2* over expressing strain can speed up the first stages of CA synthesis. Over expression of cytosolic *FumR* converted fumarate to malate, supplying more substrate to the mitochondrial malate-citrate antiporter and thus improving CA secretion and productivity, whereas over expression of *Frds1* converted fumarate to succinate, a potential substrate for the mitochondrial CA antiporter. More CA synthesis was reported when both the *FumR* and *Frds1* genes were co-expressed in the same strain (de Jongh and Nielsen 2008).

The important enzyme PFK converts fructose 6-phosphate to fructose 1,6-bisphosphate with magnesium as a cofactor, and is a critical regulating step for glycolysis metabolic flow via allosteric inhibition or activation. Legisa and Mattey (2007) discovered that the shorter PFK1 segment is not only resistant to citrate inhibition but also more responsive to positive effectors including adenosine monophosphate, ammonium ions, and fructose 2,6-bisphosphate, all of which reduce ATP inhibition. It was discovered that an *A. niger* strain with an active shorter PFK1 fragment mtpfkA10 with T89D single-site mutation (to avoid phosphorylation requirement) produced 70% more CA than the control strain (Capuder et al. 2009). According to Ruijter et al. (1999), an *A. niger* mutant strain lacking both glucose oxidase (goxC) and oxaloacetate acetylhydrolase (prtF) could easily create CA instead of oxalic acid in the medium at pH 5, and production was entirely unaffected by Mn^{2++}. It has been found that part of the residual sugar, such as iso-maltose, remains unutilized at the end of the fermentation. This residual sugar causes a significant loss in the day-to-day large-scale production process and has a negative impact on production profit. This leftover sugar (iso-maltose) is produced by the enzyme α-glucosidase during CA fermentation, and deletion of the α-glucosidase producing gene *agdA* might effectively reduce the iso-maltose concentration when corn starch as is used as the raw carbon source. It was also discovered that over expression of glucoamylase *glaA* with deletion of the α-glucosidases producing gene *agdA* could reduce 88.2% of the residual sugar, resulting in a 16.9% rise in CA synthesis.

Hexokinase transforms glucose to glucose-6-phosphate and fructose-6-phosphate in the same way. Increased sugar intake leads to higher concentrations of glucose 6-phosphate and, as a result, trehalose 6-phosphate, both of which can block hexokinase and CA buildup (Arisan-Atac et al. 1996). However, deletion of the *GgsA* gene, which encodes trehelose-6-phosphate synthase, resulted in a reduction in trehelose-6-phosphate levels and CA buildup (Arisan-Atac et al. 1996).

Hou et al. (2018) altered *A. niger* strain CGMCC10142 by deleting and over expressing the mitochondrial AOX gene *aox1*. They discovered that over expression of the *aox1* gene improves CA synthesis, and CRR respiration was identified in *A. niger* mycelia under CA production conditions. As a result of the presence of AOX, the suppression of PFK induced by excess ATP is alleviated, as is the damage produced by ROS (Campos et al. 2015). When CA builds up, cytochrome-dependent respiration is replaced by the CRR system, which allows NADH oxidation without ATP generation (Papagianni 2007). Mycelial morphology of *A. niger* has a crucial impact in CA synthesis during submerged fermentation. Sun et al. (2018) looked into RNA interference through the silence of the chitin synthase gene (*chsC*). They found that after suppressing the *chsC* gene, the mutant strain *chsC*-3 of *A. niger* could demonstrate a dramatic change in mycelial shape, resulting in 42.6% greater CA production than the normal strain (Sun et al. 2018).

Similarly, the *Brsa-25* gene is implicated in the regulation of *A. niger* morphology development in response to Mn^{2+}. Down regulation of the *Brsa-25* gene expression using antisense RNA resulted in a 10% increase in CA synthesis, according to Dai et al. (2004). In the presence of Mn^{2+}, this downregulation pathway favoured pelleted development and increased CA generation. The orotidine-5′-decarboxylase gene (*pyrG*) is a crucial enzyme in the production of uridine (van Hartingsveldt et al. 1987). During submerged fermentation, gene disruption and down-regulation of *pyrG* has been shown to greatly improve industrial CA production (Zhang et al. 2020). Zhang et al. (2020) used a highly efficient clustered regularly interspaced short palindromic repeats (CRISPR)/CRISPR associated protein 9 (*Cas9*) system based on ribosomal 5s ribonucleic acid (5S rRNA) as a promoter to disrupt both *pyrG* and ku70 homologous gene (*kusA*) in CA producing isolates wild type (WT-D) and D353. They discovered that disrupting *pyrG* disrupted intracellular central metabolism and increased intracellular levels of CA and its precursors including acetyl-CoA and oxaloacetate, possibly leading to extracellular CA buildup. In another study, Zhang et al. (2019) used the modified CRISPR/Cas9 system to disrupt and replace the *agdF* gene with the glucoamylase gene, resulting in a 25.9% increase in glucoamylase synthesis and a 25% increase in CA production over the wild strain (Zhang et al. 2019).

6.10 CITRIC ACID'S NEW APPLICATIONS

CA is widely utilized as a safe acidulant in the food, sugar, confectionery, and beverage industries due to its pleasant flavour, high water solubility, chelating, and buffering capabilities. It is used to give fruit and berry flavours to carbonated beverages (Fukui and Tanaka 1980); as a flowing agent in confectionery (Buchard and Merrit 1979); to prevent turbidity in wines (Soccol et al. 2006); in candies to provide dark colour and tartness; as an antioxidant synergism in fats, oils, and fat-containing foods (Buchard and Merrit 1979); as an antioxidant synergis; as a flavouring agent in sherbets (Buchard and Merrit 1979); and as an emulsifying agent in ice cream (Buchard and Merrit 1979). CA esters, such as triethyl, tributyl, and acetyltributyl, are also used as nontoxic plasticizers in food-safe plastic films (Buchard and Merrit 1979).

CA's chelating and pH-adjusting properties are used in the food sector to improve the durability of frozen food products by keeping the colour and flavour of frozen fruit from deteriorating. It also aids in the preservation of frozen seafood and shellfish. CA and its salts are widely used in the pharmaceutical industry as oral pharmaceutical liquids, elixirs, and suspensions to buffer and maintain the stability of active ingredients in pharmaceutical products due to their sequestering action, such as ascorbic acid stabilization and good buffering capacity (Moledina et al. 1977). Trisodium citrate is a common blood preservative that works by complexing calcium to avoid clotting (Ciriminna et al. 2017; Vandenberghe et al. 1999b).

Acetic acid is used to prevent kidney stones when combined with sodium citrate (Gul and Monga 2014). It can be found in ultrafine protein fibres for biomedical applications (Reddy et al. 2015), polyols for making biodegradable films like bio-plastic suitable for eco-friendly packaging (Seligra et al. 2016), and hydroxyapatite for making bioceramic composites for orthopaedic tissue engineering (Sun et al. 2014).

CA is used to soften water, making it suitable for use in home detergents, dishwashing cleansers, and soap (Ciriminna et al. 2017). It helps to build foam and operate better without the requirement for water softening by chelating metal ions like Ca^{2+} and Mg^{2+} ions in hard water.

Because of its high metal chelating capabilities, CA can be employed as a cleaning agent. It binds to metals and dissolves them. As a result, CA solutions are utilized in the cleaning of power plant boilers and other similar installations to remove and prevent limescale buildup (Verhoff and Hugo 2015). CA was initially employed to extract pectin from apple pomace by Brazilian researchers in 2005 (Canteri-Schemin et al. 2005). The greatest average figure was for pectin extraction yield using CA (13.75%).

CA has high metal chelating characteristics, hence it's commonly employed to clean radionuclide-contaminated nuclear sites and bioremediate heavy metal-contaminated soils (Ates et al. 2002; Kantar and Honeyman 2006). CA in conjunction with rhamnolipid biosurfactants has been shown to provide outstanding outcomes in bio-based chemical soil remediation. This mixture is not only environmentally friendly, but it also aids in the restoration of soil ecology following remediation (Wan et al. 2015).

A mixture of 70% ethanol, 1% urea, and 1.5% CA showed a considerable inactivation impact on poliovirus, but was insufficient for adenovirus and polyomavirus. Ionidis et al. (2016) discovered that a mixture of 70% ethanol, 2% urea, and 2% CA was effective in deactivating all enveloped viruses on surfaces, including polyomavirus, norovirus, and adenovirus vaccinia virus. As a component of various bleaches, fixers, and stabilizers in printing plate emulsions. CA and its esters are utilized in the following industries: photography (France Patient 1937), oil well treatment and cements (Buchard and Merrit 1979), textiles (US Patent 1949), paper and tobacco industries (Hushedeck 1965), and cosmetics (Wells et al. 1972).

6.11 CITRIC ACID'S ECONOMIC BENEFITS

Schierholt (1977) conducted an economic evaluation of surface and submerged fermentation processes for CA generation. The capacity of surface and submerged fermentation processes were examined in nine days of fermentation time at 72 and 12 tonnes per day productivity. He discovered that the expenses of construction associated with submerged fermentation are 2.5 times lower than those associated with surface fermentation.

However, the costs of equipment for submerged fermentation are significantly greater, with more than 60% of those expenditures consisting of technical components such as bioreactors and more advanced instrumental control, both of which are prone to relatively high wear. For larger capacities, overall investment costs for the submerged process are around 25% lower than for surface fermentation, whereas for lesser capacities, total investment costs are about 15% lower. The submerged process consumes around 30% more electrical energy than surface fermentation, whereas labour costs in highly developed countries are significantly greater for surface fermentation. In nations where the cooling water temperature is above 20°C, additional costs for chilling the bioreactors are required for the submerged process. Submerged fermentation, on the other hand, is sensitive to short delays or breakdowns in aeration, resulting in production losses as well as the whole breakdown of the batch.

6.12   PERSPECTIVES FOR THE FUTURE

A significant goal is to boost CA production while lowering costs. New technology and innovations are required not only for bioreactor designs that may solve scale-up issues in fermentation processes but also for online monitoring and control of many parameters. To lower costs and environmental issues, less expensive substrates such as agro-industry by-products and residues might be employed for CA manufacturing. High CA producing strains must be investigated through strain improvement, which can be accomplished using a variety of high-throughput technologies. End-product recovery and purification are key difficulties that have prompted academics to work tirelessly to discover answers. The combined fermentation and product recovery approach, which can be used in large-scale CA production, should be used. The above-mentioned creative solutions can help conserve the environment by reducing the cost of production and recovery while also reducing the usage of toxic chemicals in traditional CA manufacturing.

6.13   CONCLUSION

CA has a wide range of applications in several industrial sectors, which contributes to its broad characteristics. To meet the growing global demand for CA, a cost-effective industrial production process is required. This process is highly complex, sensitive, and dependent on a number of variables, including the microorganism used, the raw materials used, the types of fermentation technique used, the design of appropriate bioreactors with precise control over process parameters, biochemical pathways, and factors affecting CA production, quantification techniques, recovery techniques, and strain selection. Using agricultural wastes for CA production not only solves the waste disposal problem, but it also saves significant foreign exchange by reducing the amount of CA imported from other countries. Adoption of sophisticated technology and metabolic engineering can also aid in the reduction of several important parameters raised during fermentation, although more research is required.

REFERENCES

Angumeenal AR, Kamalakannan P, Prabhu HJ, Venkappayya D (2003) Effect of transition metal cations on the production of citric acid using mixed cultures of *Aspergillus niger* and *Candida gulliermondii*. *Journal of the Indian Chemical Society* 80:903–906.

Angumeenal AR, Venkappayya D (2013) An overview of citric acid production. *LWT-Food Science Technology* 50:367–370.

Arisan-Atac I, Wolschek MF, Kubicek CP (1996) Trehalose-6-phosphate synthase A affects citrate accumulation by *Aspergillus niger* under conditions of high glycolytic flux. *FEMS Microbiology Letters* 140(1):77–83.

Ates S, Dingil N, Bayraktar E, Mehmetoglu U (2002) Enhancement of citric acid production by immobilized and freely suspended *Aspergillusniger* using silicone oil. *Process Biochemistry* 38:433–436.

Bauweleers HMK, Groeseneken DR, Van Peij NNME (2014) Genes useful for the industrial production of citric acid. Google Patents. *United States Patent 8637280.* https://www. sciencedirect.com/science/article/pii/S1387265607130118?via%3Dihub.

Benghazi L, Record E, Suarez A, Gomez-Vidal JA, Martinez J, de la Rubia T (2014) Production of the Phanerochaete flavido-alba laccase in *Aspergillus niger* for synthetic dyes decolorization and biotransformation. *World Journal of Microbiology and Biotechnology* 30:201–211.

Benuzzia DA, Segovia RF (1996) Effect of copper concentration on citric acid productivity by an *Aspergillus niger* strain. *Applied Biochemistry and Biotechnology* 61:393–397.

Berovic M, Legisa M (2007) Citric acid production. *Biotechnology Annual Review* 13:303–343.

Berry DR, Chemiel A, Alobaidi Z (1977) Citric acid production by *Aspergillus niger*. In J. E. Smith & J. A. Pateman (Eds.), *Genetics and physiology of Aspergillus niger*, pp. 405–426. Academic Press, Minneapolis, MN.

Buchard EF, Merrit EG (1979) Citric acid. In Kirk-Othmer (Ed.), *Kirk-Othmers encyclopedia of chemical technology* (3rd edition, volume 6), pp. 1–150. Wiley.

Campos C, Cardoso H, Nogales A, Svensson J, Lopez-Raez JA, Pozo MJ, Nobre T, Schneider C, Arnholdt-Schmitt B (2015) Intra and inter-spore variability in *Rhizophagus irregularis AOX* gene. *PLoS One* 10:e0142339

Canteri-Schemin MH, Ramos FHC, Waszczynskyj N, Wosiacki G (2005) Extraction of pectin from apple pomace. *Brazilian Archives of Biology and Technology* 48:259–266.

Capuder M, Solar T, Bencina M, Legisa M (2009) Highly active, citrate inhibition resistant form of *Aspergillus niger* 6-phosphofructo-1-kinase encoded by amodified *pfkA* gene. *Journal of Biotechnology* 144:51–57.

Carlos RS, Vandenberghe LPS, Rodrigues C, Pandey A (2006) New perspectives for citric acid production and application. *Food Technology and Biotechnology* 44(2):141–149.

Chen H, He X, Geng H, Liu H (2014a) Physiological characterization of ATP-citrate lyase in *Aspergillus niger*. *Journal of Industrial Microbiology and Biotechnology* 41:1–11.

Chen H, He XH, Geng HR, Liu H (2014b) Physiological characterization of ATP-citrate lyase in *Aspergillus niger*. *Journal of Industrial Microbiology and Biotechnology* 41:721–731.

Chen HC (1994) Response-surface methodology for optimizing citric acid fermentation by *Aspergillus foetidus*. *Process Biochemistry* 29:399–405.

Ciriminna R, Meneguzzo F, Delisi R, Pagliaro M (2017) Citric acid: Emerging applications of key biotechnology industrial product. *Chemistry Central Journal* 11:1–9.

Clark DS, Ito K, Horitsu H (1966) Effect of manganese and other heavy metals on submerged citric acid fermentation of molasses. *Biotechnology and Bioengineering* 8:465–471.

Cleland WW, Johnson MJ (1954) Tracer experiments on the mechanism of citric acid formation by *Aspergillus niger*. *Journal of Biological Chemistry* 208:679–689.

Crolla A, Kennedy KJ (2001) Optimization of citric acid production from *Candida lipolytica* Y-1095 using n-paraffin. *Journal of Biotechnology* 89(1):27–40.

Dai Z, Mao X, Magnuson JK, Lasure LL (2004) Identification of genes associated with morphology in *Aspergillus niger* by using suppression subtractive hybridization. *Applied and Environmental Microbiology* 70:2474–2485.

Das A, Roy P (1978) Improved production of citric acid by diploid strain of *Aspergillus niger*. *Canadian Journal of Microbiology* 24:622–625.

de Jongh WA, Nielsen J (2008) Enhanced citrate production through gene insertion in *Aspergillus niger*. *Metabolic Engineering* 10:87–96.

Dronawat SN, Svihla CK, Hanley TR (1995) The effects of agitation and aeration on the production of gluconic acid by *Aspergillus niger*. *Applied Biochemistry and Biotechnology* 51–52:347–354.

Drysdale C, McKay A (1995) Citric acid production by *Aspergillus niger* in surface culture on inulin. *Letters in Applied Microbiology* 20:252–254.

Dutta A, Sahoo S, Mishra RR, Pradhan B, Das A, Behera BC (2019) A comparative study of citric acid production from different agro-industrial wastes by *Aspergillus niger* isolated from mangrove forest soil. *Environmental Experimental Biology* 17:115–122.

El Dein SMN, Emaish GMI (1979) Effect of various conditions on production of citric acid from molasses in presence of potassium ferrocyanide by *A. aculeatus* and *A. carbonarius*, *Indian Journal of Experimental Biology* 17:105–106.

Finogenova TV, Morgunov IG, Kamzolova SV, Chernyavskaya OG (2005) Organic acid production by the yeast *Yarrowia lipolytica*: A review of prospects. *Applied Biochemistry and Microbiology* 41:418–425.

France Patient (1937) French Patent No. 813,548. https://www.sciencedirect.com/science/article/pii/S1387265607130118?via%3Dihub.

FrancieloV, Patricia M, Fernanda SA (2008) Apple pomace: A versatile substrate for biotechnological applications. *Critical Reviews in Biotechnology* 28:1–12.

Fukuda H, Susuki T, Sumino Y, Akiyama S (1970) Microbial preparation of citric acid. Ger Pat 2,003,221.

Fukui S, Tanaka A (1980) Production of useful compound from alkanes media in Japan. *Advances in Biochemical Engineering* 17:1–35.

Ganne KK, Dasari VRR, Garapati HR (2008) Production of citric acid by *Aspergillus niger* MTCC 282 in submerged fermentation using *Colocassia antiquorum*. *Research Journal of Microbiology* 3(3):150–156.

Garg N, Hang YD (1995) Microbial production of organic acids from carrot processing waste. *Journal of Food Science and Technology* 32:119–121.

Good DW, Droniuk R, Lawford, RG, Fein JE (1985) Isolation and characterization of a Saccharomycopsis lipolytica mutant showing increased production of citric acid from canola oil. *Canadian Journal of Microbiology* 31:436–440.

Grewal HS, Kalra KL (1995) Fungal production of citric acid. *Biotechnology Advances* 13:209–234.

Gul Z, Monga M (2014) Medical and dietary therapy for kidney stone prevention. *Korean Journal of Urology* 55:775–779.

Gupta J, Heding L, Jorgensen O (1975) Effect of sugars, hydrogen ion concentration and ammonium nitrate on the formation of citric acid by *Aspergillus niger*. *Acta Microbiologica et Immunologica Hungarica* 23:63–67.

Gutcho SJ (1973) *Chemicals by fermentation*. Park Ridge, NY: Noyes Data Corporation.

Gutierrez NA, Mckay IA, French CE, Brooks J, Maddox IS (1993) Repression of galactose utilization by glucose in the citrate-producing yeast *Candida guilliermondii*. *Journal of Industrial Microbiology* 11:143–146.

Habison A, Kubicek CP, Rohr M (1983) Partial purification and regulatory properties of phosphofructokinase from *Aspergillus niger*. *Biochemical Journal* 209:669–676.

Hang YD, Splittstoesser D, Woodams E, Sherman R (1977) Citric acid fermentation of brewerywaste. *Journal of Food Science* 42:383–384.

Hang YD, Woodams EE (1984) Apple pomace: a potential substrate for citric acid production by *Aspergillus niger*. *Biotechnology Letters* 6:763–764.

Hang YD, Woodams EE (1985) Grape pomace: A novel substrate for microbial production of citric acid. *Biotechnology Letters* 7:253–254.

Hang YD, Woodams EE (1987) Microbial production of citric acid by solid state fermentation of kiwifruit peel. *Journal of Food Science* 52:226–227.

Honecker S, Bisping B, Yang Z, Rehm HJ (1989) Influence of sucrose concentration and phosphate limitation on citric acid production by immobilized cells of *Aspergillus niger*. *Applied Microbiology and Biotechnology* 31:17–24.

Hossain M, Brooks JD, Maddox IS (1983) Production of citric acid from whey permeate by fermentation using *Aspergillus niger*. *New Zealand Journal of Dairy Science and Technology* 18:161–168.

Hou L, Liu L, Zhang H, Zhang L, Zhang L, Zhang J, Gao Q, Wang D (2018) Functional analysis of the mitochondrial alternative oxidase gene (aox1) from *Aspergillus niger* CGMCC 10142 and its effects on citric acid production. *Applied Microbiology and Biotechnology* 102(18):7981–7995.

Hushedeck HR (1965) U.S. Patent No. 3,212,928. https://www.sciencedirect.com/science/article/pii/S1387265607130118?via%3Dihub.

Ikeno Y, Masuda M, Tanno K, Oomori I, Takahashi N (1975) Citric acid production from various raw materials by yeasts. *Journal of Fermentation Technology* 53:752–756.

Ikram-ul H, Ali S, Qadeer M, Iqbal J (2004) Citric acid production by selected mutants of *Aspergillus niger* from cane molasses. *Bioresource Technology* 93:125–130.

Ingram LO, Buttke TM (1984) Effects of alcohols on microorganisms. *Advances in Microbial Physiology* 25:253–300.

Ionidis G, Hubscher J, Jack T, Becker B, Bischoff B, Todt D, Hodasa V, Brill FHH, Steinmann E, Steinmann J (2016) Development and virucidal activity of a novel alcohol-based hand disinfectant supplemented with urea and citric acid. *BMC Infectious Disease* 16:77.

Ishi K, Nakajima Y, Iwakura T (1972) Citric acid by fermenetation of waste glucose. *German Patent* 2:157–847.

Jernejc K, Cimerman A, Perdih A (1982) Citric acid production in chemically defined media by *Aspergillus niger*. *Europian Journal of Microbiology and Biotechnology* 14:29–33.

Kantar C, Honeyman BD (2006) Citric acid enhanced remediation of soils contaminated with uranium by soil flushing and soil washing. *Journal of Environmental Engineering* 132:247–255.

Kapelli O, Muller M, Fiechter A (1978) Chemical and structural alterations at cell surface of *Candida tropicalis*, induced by hydrocarbon substrate. *Journal of Bacteriology* 133:952–958.

Kapilan R (2015) Solid state fermentation for microbial products: A review. *Archives of Applied Science Research* 7:21–25.

Kapoor KK, Chaudhery K, Tauro P (1982) Citric acid. In G. Reed (Ed.), *Prescott and Dunn's industrial microbiology* (4th ed., pp. 709–747). A VI Publishing Company Inc.

Kapoor KK, Chaudhary K, Tauro P (1983) Citric acid. In G. Reed (Ed.), *Prescott and Dunn's industrial microbiology*, pp. 709–747. AVI Publishing Company, Westport, CT.

Käppeli O, Müller M, Fiechter A (1978) Chemical and structural alterations at the cell surface of *Candida tropicalis*, induced by hydrocarbon substrate. *Journal of Bacteriology* 133:952–958.

Kareem S, Akpan I, Alebiowu O (2010) Production of citric acid by *Aspergillus niger* using pineapple waste. *Malaysian Journal of Microbiology* 6:161–165.

Karow EO, Waksman SA (1947) Production of citric acid in submerged culture. *Industrial & Engineering Chemistry Research* 39:821–825.

Kautola H, Rymowicz W, Linko YY, Linko P (1992) The utilization of beet molasses in citric acid production with yeast. *Science des Aliments* 12:383–392.

Kirimura K, Lee SP, Kawajima I, Kawabe S, Usami S (1988a) Improvement in citric acid production by haploidization of *Aspergillus niger* diploid strains. *Journal of Fermentati on Technology* 66:375–382.

Kristiansen B, Mattey M, Linden J (1999) *Citric acid biotechnology*, pp. 7–9. Taylor & Frances Ltd.

Kroya Ferment Ind (1970) Citric acid prepared by fermentation. *Br Pat* 1:187,610.

Kubicek CP (1986) The role of the citric acid cycle in fungal organic acid fermentations. *Proceedings of Biochemical Society Symposium* 54:113–126.

Kubicek CP, Rohr M (1977) Influence of manganese on enzyme synthesis and citric acid accumulation in *Aspergillus niger*. *Europian Journal of Applied Microbiology and Biotechnology* 4:167–175.

Kubicek CP, Rohr M (1986) Citric acid. *CRC Critical Review in Biotechnology* 3:331–373.

Kubicek-Pranz EM, Mozelt M, Rohr M, Kubicek CP (1990) Changes in the concentration of fructose 2,6-bisphosphate in *Aspergillus niger* during stimulation of acidogenesis by elevated sucrose concentration. *Biochimica et Biophysica Acta* 1033:250–255.

Kutermankiewicz M, Lesniak W, Bolach E (1980) Wykorzystanie cukierniczego soku gęstego do fermentacji wgłębnej kwasu cytrynowego. *Przem Ferm Owoc-Warzyw* 6:27–31.

Legisa M, Mattey M (1986) Glycerol as an initiator of citric acid accumulation in *Aspergillus niger*. *Enzyme and Microbial Technology* 8:258–259.

Legisa M, Mattey M (2007) Changes in primary metabolism leading to citric acid overflow in *Aspergillus niger*. *Biotechnology Letter* 29:181–190.

Lesniak W (1999) Fermentation substrates. In B. Kristiansen, M. Mattey, & J. Linden (Eds.), *Citric acid biotechnology*, pp. 149–159. Taylor Francis Ltd.

Lesniak W, Podgorski W, Pietkiewicz J (1986) Możliwości zastosowania hydrolu glukozowego do produkcji kwasu cytrynowego. *Przem Fern Owoc-Warzyw* 6:22–25.

Li Q, Jiang X, Feng X, Wang J, Sun C, Zhang H, Xian M, Liu H (2016) Recovery processes of organic acids from fermentation broths in the biomass-based industry. *Journal of Microbiology and Biotechnology* 26(1):1–8.

Lu M, Brooks JD, Maddox IS (1997) Citric acid production by solid-state fermentation in a packed-bed reactor using *Aspergillus niger*. Enzyme and Microbial Technology 21:392–397.

Maddox IS, Spencer K, Greenwood JM, Dawson MW, Brooks JD (1985) Production of citric acid from sugars present in wood hemicellulose using *Aspergillus niger* and *Saccharomycopsis lipolytica*. *Biotechnology Letters* 7:815–818.

Manzoni M (2006) Microbiologia industriale (11th edition). Milan: Casa editrice ambrosiana.

Market Report World (2020) Global anhydrous citric acid market research report 2020. Retrieved from https://www.marketreportsworld.com/global-anhydrous-citric-acid-market-14538496

Martin SM, Wilson PW (1951) Uptake of C14O2 by *Aspergillus niger* in the formation of citric acid. *Archives of Biochemistry and Biophysics* 32:150–157.

Marzona M (1996) *Chimica delle fermentazioni & microbiologia industrial* (2nd edition). Padua: Piccin.

Mattey, M (1992) The production of organic acids. *Critical Reviews in Biotechnology* 12:87–132.

Mattey M, Bowes I (1978) Citrate regulation of NADP+-specific isocitrate dehydrogenase of *Aspergillus niger*. *Biochemical Society Transactions* 6:1224–1226.

Max B, Salgado JM, Rodriguez N, Cortes S, Converti A, Dominguez JM (2010) Biotechnological production of citric acid. *Brazilian Journal of Microbiology* 41:862–875.

Mazinanian N, Odnevall Wallinder I, Hedberg Y (2015) Comparison of the influence of citric acid and acetic acid as simulant for acidic food on the release of alloy constituents from stainless steel AISI 201. *Journal of Food Engineering* 145: 51–63.

McIntyre M, McNeil B (1997) Dissolved carbon dioxide effects on morphology, growth and citrate production in *Aspergillus niger* A60. *Enzyme and Microbial Technology* 20:135–142.

Meijer S, Nielsen ML, Olsson L, Nielsen J (2009) Gene deletion of cytosolic ATP: Citrate lyase leads to altered organic acid production in *Aspergillus niger*. *Journal of Industrial Microbiology and Biotechnology* 36:1275–1280.

Meixner-Monori B, Kubicek CP, Rohr M (1984) Pyruvatekinase from *Aspergillus niger*: A regulatory enzyme in glycolysis? *Canadian Journal of Microbiology* 30:16–22.

Miall M, Parker GF (1975) Continuous preparation of citric acid by *Candida lipolytica*. *Ger Pat* 2(429):224.

Millis NF, Trumpy BH, Palmer BM (1963) The effect of lipids on citric acid production by an *Aspergillus niger* mutant. *Journal of General Microbiology* 30:365–379.

Mischak H, Kubicek CP, Rohr M (1985) Formation and location of glucose oxidase in citric acid producing mycelia of *Aspergillus niger*. *Applied Microbiology and Biotechnology* 21:27–31.

Moledina KH, Regenstein JM, Baker RC, Steinkraus KH (1977) Effects of antioxidants and chelators on the stability of frozen stored mechanically deboned flounder meat from racks after filleting. *Journal of Food Science* 42:759–764.

Morgunov IG, Kamzolova SV, Lunina JN (2018) Citric acid production by *Yarrowia lipolytica* yeast on different renewable raw materials. *Fermentation* 4(2):36.

Moyer AJ (1953) Effect of alcohols on the mycological production of citric acid in surface and submerged culture: I. Nature of the alcohol effect. *Applied Microbiology* 1:1–7.

Musilkova M, Ujcova E, Seichert L, Fencl Z (1983) Effect of changed cultivation conditions of the morphology of *Aspergillus niger* and citric acid biosynthesis in laboratory cultivation. *Folia Microbiologia* 27:382–332.

Oh MJ, Park YJ, Lee SK (1973) Citric acid production by *Hansenula anamola* var. *anamola*. *Hanguk Sikpum Kawahakhoe Chi* 5:215–223.

Omar SH, Rehm HJ (1980) Physiology and metabolism of two alkane oxidizing and citric acid producing strains of *Candida parapsilosis*. *Europian Journal of Applied Microbiology and Biotechnology* 11:42–49.

Pandey A (1992) Recent process developments in solid state fermentation. *Process Biochemistry* 27:109–117.

Papagianni M (2007) Advances in citric acid fermentation by *Aspergillus niger*: Biochemical aspects, membrane transport and modeling. *Biotechnology Advances* 25:244–263.

Papagianni M, Mattey M, Kristiansen B (1999) Hyphal vacuolation and fragmentation in batch and fed-batch culture of *Aspergillus niger* and its relation to citric acid production. *Process Biochemistry* 35:359–366.

Papagianni M, Wayman FM, Mattey M (2005) Fate and role of ammonium ions during fermentation of citric acid by *Aspergillus niger*. *Applied and Environmental Microbiology* 71:7178–7186.

Parekh S, Vinci VA, Strobel RJ (2000) Improvement of microbial strains and fermentation processes. *Applied Microbiology and Biotechnology* 54:287–301.

Pelechova J, Petrova L, Ujcova E, Martinkova L (1990) Selection of a hyperproducing strain of *Aspergillus niger* for biosynthesis of citric acid on unusual carbon substrates. *Folia Microbiologia* 35:138–142.

Pietkiewicz J, Podgorski W, Lesniak W (1996) Proceedings of the International Conference on Advances in Citric Acid Technology Bratislava, Slovak Republic, 9.

Pintado J, Lonsane BK, Gaime-Perraud I, Roussos S (1998) On-line monitoring of citric acid production in solid-state culture by respirometry. *Process Biochemistry* 33:513–518.

Pontecorvo G, Roper JA, Forbes E (1953) Genetic recombination without sexual reproduction in *Aspergillus niger*. *Journal of General Microbiology* 8:198–210.

Prescott SC, Dunn EG (1959) The citric acid fermentation in industrial microbiology (3rd edition), pp. 533–577. McGraw Hill Book Company, Inc., New York.

Prochazka J, Heyberger A, Bizek V, Kousova M, Volaufova E (1994) Amine extraction of hydroxy-carboxylic acids. 2. Comparison of equilibria for lactic, malic and citric acids. *Industrial & Engineering Chemistry Research* 33:1565–1573.

Ramakrishnan CV, Steel R, Lentz CP (1955) Mechanism of citric acid formation and accumulation in *Aspergillus niger*. *Archives of Microbiology* 55:270–273.

Rane KD, Sims KA (1993) Production of citric acid by *Candida lipolytica* Y-1095: Effect of glucose concentration on yield and productivity. *Enzyme and Microbial Technology* 15:646–651.

Ratledge C, Kristiansen B (2001) *Basic biotechnology*. Cambridge University Press.

Reddy M (2002) Coconut cake: A novel substrate for citric acid production under solid substrate fermentation. *Indian Journal of Microbiology* 42:347–349.

Reddy N, Reddy R, Jiang Q (2015) Crosslinking biopolymers for biomedical applications. *Trends in Biotechnology* 33:362–369.

Rohr M, Kubicek CP (1981) Regulatory aspects of citric acid fermentation by *Aspergillus niger*. *Process Biochemistry* 16:34–37.

Rohr M, Kubicek CP, Kominek J (1983) Citric acid. In G. Reed & H. J. Rehm (Eds.), *Biotechnology* (volume 3), pp. 419–454. Wiley-VCH Verlag, Weinheim, Germany.

Rohr M, Kubicek CP, Kominek J (1992) Industrial acids and other small molecules. In J.W. Bennett & M. A. Klich (Eds.), *Aspergillus: Biology and industrial applications*, pp. 93–131. Butterworth-Heinemann, Oxford, United Kingdom.

Rohr M, Kubicek CP, Kominek J (1996) Citric acid. In H.-J. Rehm & G. Reed (Eds.), *Biotechnology* (2nd edition, volume 6), pp. 307–345. Wiley-VCH Verlag, Weinheim, Germany.

Roukas T (1991) Production of citric acid from beet molasses by immobilized cells of *Aspergillus niger*. *Journal of Food Science* 56:878–880.

Rubio MC, Maldonado MC (1995) Purification and characterization of invertase from *Aspergillus niger*. *Current Microbiology* 31:80–83.

Ruijter GJ, van de Vondervoort PJ, Visser J (1999) Oxalic acid production by *Aspergillus niger*: An oxalate-non-producing mutant produces citric acid at pH 5 and in the presence of manganese. *Microbiology* 145:2569–2576.

Rymowicz W, Juszczyk P, Rywinska A, Żarowska B, Musaial I (2005) Citric acid production from raw glycerol by *Yarrowia lipolytica* yeast (in polish). Biotechnologia Monografie 2(2):46–54.

Rywińska A, Rymowicz W, Marcinkiewicz M (2010) Valorization of raw glycerol for citric acid production by *Yarrowia lipolytica* yeast. *Electronic Journal of Biotechnology* 13(4):0717–3458.

Sarangbin S, Watanapokasin Y (1999) Yam bean starch: A novel substrate for citric acid production by the protease-negative mutant strain of *Aspergillus niger*. *Carbohydrate Polymers* 38:219–224.

Sardinas JL (1972) Fermentative production of citric acid. *Fr Pat* 2:113,668.

Sawant O, Mahale S, Ramchandran V, Nagaraj G, Bankar A (2018) Fungal citric acid production using waste materials: A mini-review. *Journal of Microbiology Biotechnology and Food Science* 8:821–828.

Schierholt J (1977) Fermentation processes for the production of citric acid. *Process Biochemistry* 20:34–42.

Seligra PG, Medina JC, Fama L, Goyanes S (2016) Biodegradable and non-retrogradable ecofilms based on starch–glycerol with citric acid as cross linking agent. *Carbohydrate Polymers* 138:66–74.

Shishikura A, Takuhashi H, Hirohama S, Arai K (1992) Citric acid purification process using compressed carbon dioxide. *Journal of Supercritical Fluids* 5:303–312.

Show PL, Oladele KO, Siew QY, Zakry FAA, Lan JCW, Ling TC (2015) Overview of citric acid production from *Aspergillus niger*. *Frontier in Life Science* 8:271–283.

Shu P, Johnson MJ (1948) The interdependence of medium constituents in citric acid production by submerged fermentation. *Journal of Bacteriology* 56:577–585.

Sirman T, Pyle DL, Grandison AS (1990) Extraction of citric acid using a supported liquid membrane. *Separations for Biotechnology* 22:245–254.

Soccol CR, Vandenberghe LPS, Rodrigues C, Pandey A (2006) A new perspective for citric acid production and application. *Food Technology and Biotechnology* 44:141–149.

Souza KST, Schwan RF, Dias DR (2014) Lipid and citric acid production by wild yeasts grown in glycerol. *Journal of Microbiology and Biotechnology* 24:497–506.

Sun D, Chen Y, Tran RT, Xu S, Xie D, Jia C, Wang Y, Guo Y, Zhang Z, Guo J, Yang J, Jin D, Bai X (2014) Citric acid-based hydroxyapatite composite scaffolds enhance calvarial regeneration. *Scientific Report* 4:6912.

Sun X, Wu H, Zhao G, Li Z, Wu X, Liu H, Zheng Z (2018) Morphological regulation of *Aspergillus niger* to improve citric acid production by *chsC* gene silencing. *Bioprocess and Biosystems Engineering* 41:1029–1038.

Swain MR, Ray RC, Patra JK (2012) *Citric acid: Microbial production and applications in food and pharmaceutical industries*, pp. 1–22. Nova Science Publishers, Inc., Hauppauge, NY.

Themelis DG, Tzanavaras PD (2001) Reagent-injection spectrophotometric determination of citric acid in beverages and pharmaceutical formulations based on its inhibitory effect on the iron (III) catalytic oxidation of 2,4-diaminophenol by hydrogen peroxide. *Analytica Chimica Acta* 428:23–30.

Tomlinson N, Campbell JJR, Trussell PC (1950) The influence of zinc, iron, copper, andmanganese on the production of citric acid by *Aspergillus niger*. *Journal of Bacteriology* 59:217–227.

Tong Z, Zheng X, Tong Y, Shi YC, Sun J (2019) Systems metabolic engineering for citric acid production by *Aspergillus niger* in the postgenomic era. *Microbial Cell Factory* 18:1–15.

Tran CT, Sly LI, Mitchell DA (1998) Selection of a strain of *Aspergillus* for the production of citric acid from pineapple waste in solid state fermentation. *World Journal of Microbiology and Biotechnology* 14:399–404.

Uchio R, Maeyashiki I, Kikuchi K, Hirose Y (1975) Citric acid production by yeast. *Jp Kokai* 76,63,059.

US Patient (1949). US Patent No. 2,474,092. https://www.sciencedirect.com/science/article/pii/S1387265607130118?via%3Dihub

van Hartingsveldt W, Mattern IE, van Zeijl CM, Pouwels PH, van den Hondel CA (1987) Development of a homologous transformation system for *Aspergillus niger* based on the *pyrG* gene. *Molecular and General Genetics* 206:71–75.

Vandenberghe LPS (2000) Development of process for citric acid production by solid-state fermentation using cassava agro–industrial residues (PhD thesis), Universite de Technologie de Compiegne, Compiegne, France.

Vandenberghe LPS, Pandey A, Soccol CR, Lebeault JM (1999a) Citric acid production by *Aspergillus niger* in solid-state fermentation on coffee husk. In *III Internatl Sem biotechnol coffee agroindustry May*, pp. 26–29, Londrina, Brazil.

Vandenberghe LPS, Soccol CR, Pandey A, Lebeault JM (1999b) Microbial production of citric acid. *Brazilian Archives of Biology and Technology* 42:263–276.

Vandenberghe LPS, Soccol CR, Pandey A, Lebeault JM (2000) Solidstate fermentation for the synthesis of citric acid by *Aspergillus niger*. *Bioresource Technology* 74:175–178.

Vergano MF, Soria M, Kerber N (1996) Short communication: Influence of inoculum preparation on citric acid production by *Aspergillus niger*. *World Journal of Microbiology and Biotechnology* 12:655–656.

Verhoff FH, Hugo B (2015) Citric acid. In Claudia Ley and Barbara Elvers (Eds.), *Ullmann's encyclopedia of industrial chemistry* (7th edition), pp. 1–9. Wiley-VCH. https://doi.org/10.1002/14356007.a07_103.pub3.

Vu VH, Pham TM, Kim K (2010) Improvement of a fungal strain by repeated and sequential mutagenesis and optimization of solid state fermentation for the hyper-production of raw starch digesting enzyme. *Journal of Mirobiology and Biotechnology* 20:718–726.

Wan J, Meng D, Long T, Ying R, Ye M, Zhang S, Li Q, Zhou Y, Lin Y (2015) Simultaneous removal of lindane, lead and cadmium from soils by rhamnolipids combined with citric acid. *PLoS One* 10:e0129978.

Wang B, Li H, Zhu L, Tan F, Li Y, Zhang L, Shi G (2017) High-efficient production of citric acid by *Aspergillus niger* from high concentration of substrate based on the staged-addition glucoamylase strategy. *Bioprocess and Biosystems Engineering* 40(6):891–899.

Wehmer C (1893) Note sur la fermentation citrique. *Bulletin de la Société Chimique de France* 9:728–732.

Wells CE, Martin DC, Tichenor DA (1972) Effect of citric acid on frozen whole strawberries. *Journal of the American Dietetic Association* 6:665–668.

Weyda I, Lübeck M, Ahring BK, Lübeck PS (2014) Point mutation of the xylose reductase (XR) gene reduces xylitol accumulation and increases citric acid production in *Aspergillus carbonarius*. *Journal of Industrial Microbiology and Biotechnology* 41:1–7.

Wojtatowicz M, Marchin GL, Erickson LE (1993) Attempts to improve strain A-10 of *Yarrow lipolytica* for citric acid production from n-paraffins. *Process Biochemistry* 28:453- 460.

Xu D-B, Madrid CP, Röhr M, Kubicek CP 1989 The influence of type and concentration of the carbon source on production of citric acid by *Aspergillus niger*. *Applied Microbiology and Biotechnology* 30:553–558.

Yi M, Pen Q, Chen D, Pen L, Zhang M, Wen R, Wang W (1987) Extraction of citric acid by *N,N*-disubstituted alkyl amides from fermentation aqueous solution. *Beijing Dax Xue* 4:30–37.

Yin X, Hyun-dong Shin H, Li J, Du G, Liu L, Chem J (2017) Comparative genomics and transcriptome analysis of *Aspergillus niger* and metabolic engineering for citrate production. *Scientific Reports* 7:41040.

Yin X, Li J, Shin HD, Du G, Liu L, Chen J (2015) Metabolic engineering in the biotechnological production of organic acids in the tricarboxylic acid cycle of microorganisms: Advances and prospects. *Biotechnology Advances* 33:830–841.

Yu B, Zhang X, Sun W, Xi X, Zhao N, Huang Z, Ying H (2018) Continuous citric acid production in repeated-fed batch fermentation by *Aspergillus niger* immobilized on a new porous foam. *Journal of BiotechnologyJournal of Biotechnology* 276–277:1–9.

Zhang L, Zheng X, Cairns TC, Zhang Z, Wang D, Zheng P, Sun J (2020) Disruption or reduced expression of the orotidine-5′-decarboxylase gene pyrG increases citric acid production: A new discovery during recyclable genome editing in *Aspergillus niger*. *Microbial Cell Factory* 19:76.

Zhang Y, Ouyang L, Nan Y, Chu J (2019) Efficient gene deletion and replacement in *Aspergillus niger* by modified in vivo CRISPR/Cas9 systems. *Bioresources and Bioprocessing* 6:4.

7 Anaerobic microbial communities for bioenergy production

Daniela Peña-Carrillo, Rebeca Díez-Antolínez,
Rubén González, and Xiomar Gómez

CONTENTS

7.1   INTRODUCTION

The search for new energy sources is becoming an acute topic of research, given the serious concern of society regarding climate change and the need for attaining a higher degree of energetic independence. Although several technologies are already available for the production of renewable energy, some of them still present a high cost and their dependence on climatic conditions limits the reliability of the energetic system. Biological processes capable of producing fuels or electricity have been developed; some have high technological maturity, whereas others are still subject to extensive research finding great difficulties in scaling up. Nevertheless, many of these processes still present several challenges regarding profitability and energetic efficiency, as it is the well-known anaerobic digestion technology. There is a spreading interest in demonstrating the feasibility of integrating different biological technologies into the economy for revalorising waste streams, increasing the value of side streams, and thus approaching to a circular economy model.

Well-known processes such as ethanol fermentation and anaerobic digestion have been widely studied and currently find extensive application. Anaerobic systems present the advantage of having low energetic requirements given that these communities lack oxygen demand. However, the profitability of this process still confronts several challenges regarding the scale of the process and biogas upgrading

DOI: 10.1201/9781003394600-8

costs. Other biological processes, still subject to intensive research, are the ancient acetone–butanol–ethanol (ABE) fermentation, fermentative hydrogen production and bioelectrochemical systems (BES).

Increasing yields, lowering operating costs and energy demand are imperative for these processes to reach once again commercial scale in the case of the ABE process and find real industrial application when considering biological hydrogen production and electricity generation. BES are still in their infancy despite the great efforts in understanding biofilm performance, overcoming mass transfer limitations and ohmic resistances. Small pilot-scale systems have been tested for the direct production of electricity by microbial fuel cells (MFC) or hydrogen when operating microbial electrolysis cells (MEC). Dark fermentation as it is also known the process of hydrogen production by fermentative consortia, is a promising technology. This fermentation shares similarities with conventional digestion and therefore can take advantage of using already developed industrial equipment. However, several issues are still pending a solution, as it may be the stability of the fermentation and increasing reactor productivities.

This chapter describes microbial communities present in anaerobic systems dedicated to fuel and energy production. Different processes that are currently under intensive research are reviewed along with a description of main organisms responsible for a successful outcome. Different operating experiences are also described along with main parameters influencing performance and improving efficiency.

7.2 ANAEROBIC DIGESTION

Digestion technology finds extensive application worldwide thanks to its capacity to degrade different organic compounds under high loadings and produce biogas as a final valuable fuel. This gas can be easily valorised for thermal and electric energy production by installing a boiler or combined heat and power units. Biogas can also be upgraded to produce biomethane. The excellent capacity for treating a wide variety of wastes makes this technology capable of reintroducing low-quality materials into the production chain, attaining their transformation into energy, organic amendments or any other type of goods. Therefore, wastes can be used to generate new products and should be considered as "renewable resources" (Akturk and Demirer 2020).

Biogas contains methane (CH_4) and carbon dioxide (CO_2) as majoritarian constituents. Digestate is a by-product stream of this process, containing anaerobic biomass, partially degraded organics and residual components which are recalcitrant to the degradation route. Digestate is rich in nutrients, humic and fulvic substances making it a suitable raw material to produce organic fertilisers (Pecorini et al. 2020). Digestate is recently considered an organic soil improver, growing medium or organic non-microbial plant biostimulant (Stürmer et al. 2020). The performance of the anaerobic reactor is influenced by microbial activity, but it is also evident that there exists a close relationship between reactor operating conditions, the type of substrate and the availability of trace elements and supplements which can modify the outcome of these communities (González et al. 2018).

A schematic representation of anaerobic digestion is shown in Figure 7.1. In this figure, different substrates are schematised along with digestion stages and the use of gas and digestate. Hydrolysis of complex molecules is the first stage of this process. When particulate substrates or rich cellulosic materials are used as feeding stream, the hydrolysis step is then recognised as the limiting one. Hydrolysis is characterised by surface and transport phenomena which are closely dependent on biomass activity and concentration (González et al. 2018). This stage is carried out by extracellular enzymes responsible for transforming complex organics into soluble compounds such as fermented sugars and amino acids (Wainaina et al. 2019). Therefore, the solubilisation of organics is frequently studied as an enzymatic phenomenon rather than the result of microbial interaction (Cirne et al. 2007).

The analysis of microbial biomass during this initial stage is usually associated with the enhancement of the biological degradation of cellulosic and lignocellulosic biomass, where this process may limit global performance. Within the eubacteria, Actinomycetales and Clostridiales have been recognised as microorganisms with cellulolytic capabilities (Lynd et al. 2002). Recently, Bao et al. (2019) confirmed that *Firmicutes*, *Actinobacteria*, *Verrucomicrobia* and *Fibrobacteres* were the dominant bacterial phyla-degrading cellulose, corroborating that anaerobic degradation requires the presence of a diverse community. However, it is essential to keep in mind that cellulose is a material with a diverse structure; that is, cellulose can be found forming a crystalline or amorphous structure, with the latter being more easily degraded by enzyme complexes (Park et al. 2010) and therefore requiring different types of enzymes to complete the full degradation.

The way anaerobes perform cellulose degradation can be exemplified by the degradative capabilities of *Clostridium thermocellum* having enzymes distributed in the liquid phase and on the cell surface highly specialised for growth on cellulose and cellodextrins (Schwarz 2001). This organism contains protuberances called cellulosomes which attach to the substrate, limiting diffusion losses during cellulose hydrolysis (Lamed et al. 1987; Mayer et al. 1987). The effectiveness in cell attachment during degradation is demonstrated by the better performance of recirculating

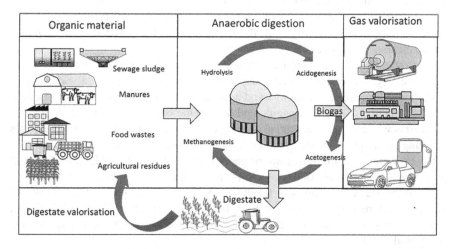

FIGURE 7.1   Schematic representation of substrates, stages of the anaerobic digestion process, gas and digestate valorisation.

leachate reactors. These reactors are often used to digest lignocellulosic material and high cellulosic-containing wastes.

Proteins are also degraded under anaerobic conditions. In this case, amino acids are usually transformed by two main routes, one is the Stickland pathway which is based on coupled oxidation/reduction reactions, but if the partial pressure of hydrogen is sufficiently low, then the pathway proceeds through an uncoupled oxidation route. The oxidation of amino acids leads to the release of ammonia and sulphide (Schnürer 2016). The class *clostridia* play a relevant role in the different stages of digestion, with some members participating in the conversion of proteins in addition to its role in the hydrolysis of cellulose, already described. Some species such as *Acetoanaerobium sticklandii* (Fonknechten et al. 2010) and *Butyrivibrio proteoclasticus* (Attwood et al. 1996) have been isolated and reported as amino acid-fermenting organisms. Table 7.1 shows a list of some of the main organisms reported in the literature as dominating in different types of anaerobic reactors.

After the hydrolysis stage and the formation of soluble compounds, volatile fatty acids (VFAs) and hydrogen are produced, serving as precursors in methane forming reactions. This subsequent stage denoted as acidogenesis, may be considered the quickest step, thus being crucial a proper balance between the different reaction rates (Vavilin et al. 2008) to avoid the accumulation of intermediaries, which may adversely affect the continuity of the process. The rapid formation of short-chain fatty acids leads to a decrease in pH when acid accumulation surpasses the buffer capacity of the system. The pH is an important parameter for keeping hydrolysis and acidogenesis at a good pace. Kim et al. (2003) and Zhang et al. (2005) reported pH values around 6.5–7.0 as optimum when studying thermophilic and mesophilic digestion.

Some clostridia are capable of carrying out acetogenesis and fatty acid conversion reactions, in addition the phylum *Proteobacteria* contains many syntrophs capable of living in association with *Archaea* (Hassa et al. 2018). The conversion of fatty acids with a carbon number greater than two into acetic acid and the final transformation into methane are the two final steps in biogas production, but only some methanogenic *Archaea* have the ability to synthesise methane as the end product of the previous fermentation sequence (Hassa et al. 2018).

The reduction of carbon dioxide characterises acetogenesis. These reactions are performed by a diverse bacterial group called acetogens. They use a wide variety of carbon sources, electron donors and acceptors and grow as autotrophs or heterotrophs (Schnürer 2016). Acidogenic and acetogenic bacteria are a diverse group of organisms which can be facultative or obligate anaerobes. The first ones are capable of living in aerobic and anaerobic environments whereas the second ones as indicated by its name cannot tolerate oxygen (Bajpai 2017). Some industrial digestion processes introduce an initial aerobic stage to accelerate the solubilisation of particulate material due to the ability of facultative organisms to survive in the presence of air and therefore accelerate the first degradation stage (Jang et al. 2014). *Escherichia*, *Acinetobacter* and *Clostridium* are the predominant genera of the acidogenesis stage (Jung et al. 2015).

Acetogens also produce acetate from organic acids and waste electrons as hydrogen. The genus *Pelotomaculum, Smithllela* and *Syntrophobacter* are involved in

TABLE 7.1
Shows predominant organisms based on substrate and reactor configuration

Reactor	Substrate	Bacteria	Methanogens	Reference
Leachate bed – solid state	Energy crop (*Pennisetum* hybrid) co-digestion	Bacteroidetes, Firmicutes, Proteobacteria	*Methanothrix, Methanobacterium Methanosarcina*	Xing et al. (2020)
Batch reactor		Bacteroidetes, Firmicutes,	*Methanosarcina, Methanomassiliicoccus*	Xing et al. (2020)
Hydrolytic/ acidogenic first phase (leached bed system)	Maize silage	*Lactobacillus,* Clostridiales	–	Sträuber et al. (2016)
Methanogenic phase (Hybrid reactor)	Maize silage from hydrolytic phase	–	Aceticlastic *Methanosaeta,* hydrogenotrophic *Methanoculleus* and *Methanobacteriaceae*	Sträuber et al. (2016)
CSTR reactor fatty acid degradation	Manure and sodium oleate	*Syntrophomonas* species Clostridiales, Rykenellaceae, *Halothermothrix and Anaerobaculum*	*Methanoculleus* genus and *Methanosarcina* sp.	Treu et al. (2016)
Industrial biogas plants mesophilic and thermophilic conditions	Plant 1: Maize silage, sugar beet, poultry manure Plant 2: Maize silage, grass, manure Plant 3: Maize silage, pig manure Plant 4: Maize silage, grass, pig manure	*Firmicutes* and *Bacteroidetes.* *Thermotogae* in thermophilic system and *Spriochaetes* in Plants 2 and 3	*Methanoculleus* genus, *Methanothermobacter* genus in thermophilic reactor	Stolze et al. (2016)
Two-stage: Aerobic thermophilic and Mesophilic	Sewage sludge	Firmicutes, Proteobacteria, Actinobacteria	*Methanosarcinales, Methanomicrobiales, Methanobacteriales*	Jang et al. (2014)
Continuous CSTR: mesophilic and thermophilic conditions	Food waste: Evaluate the increase in OLR	*Firmicutes* was dominant for both reactors *Thermotogae* in thermophilic reactor *Bacteroidetes* in mesophilic reactor	*Methanosarcina* was dominant for both reactors *Methanothermobacter* and *Methanoculleus* in thermophilic reactor *Methanosaeta* in mesophilic reactor	Guo et al. (2014)

(Continued)

TABLE 7.1 (*Continued*)
Shows predominant organisms based on substrate and reactor configuration

Reactor	Substrate	Bacteria	Methanogens	Reference
Continuous CSTR: mesophilic conditions	Food waste at increasing total solid (TS) content	*Chloroflexi, Bacteroidetes* and *Firmicutes.* *Bacteroidetes* enriched with TS increase *Proteobacteria, Spirochaetes* and *Tenericutes* increased with TS content	*Methanosarcina* was dominant due to high VFA concentration in higher TS working systems	Yi et al. (2014)

CSTR: Completely stirred tank reactor, OLR: organic loading rate.

the oxidation of propionate, whereas *Syntrophus* and *Syntrophomonas* participate in the oxidation of butyrate (Liu et al. 1999; Imachi et al. 2007; Sousa et al. 2007; Venkiteshwaran et al. 2015). Hydrogen released is subsequently assimilated by hydrogenotrophic methanogens, whereas acetoclastic methanogens use acetic acid to produce methane (Angelidaki and Batstone 2010). There is a syntrophic relationship between acetogens and methanogens. This syntrophy is necessary to keep hydrogen partial pressure at low levels so that the thermodynamics for volatile acids and alcohol conversion into acetate becomes favourable, allowing the process to proceed until the final methane production (Bajpai 2017).

Methanogens, capable of assimilating acetate, must confront competition with acetate oxidising bacteria, which transform acetate back to hydrogen and carbon dioxide, or they can also use the reverse reaction to produce acetate. Therefore, when high hydrogen levels are kept in the system (e.g., greater than 500 Pa) acetogenesis is favoured, but on the contrary, when low hydrogen levels are found (lower than 40 Pa), then the oxidation of acetate is the preferential transformation (Demirel and Scherer 2008).

Some fermenting bacteria capable of producing fatty acids along with H2 and CO2 are *Clostridium bornimense* sp. and *Proteiniborus indolifex* sp. (Hahnke et al. 2014, 2018). They participate in the acidogenesis phase with some species from the genus *Methanobacterium*. The great majority of rod-shaped methanogens belong to the order of Methanobacteriales, which in addition to *Methanobacterium* also include the genus of *Methanobrevibacter* and *Methanosphaera*, all of these growing under mesophilic conditions whereas under thermophilic and hyperthermophilic conditions are the genus of *Methanothermobacter* and *Methanothermus* (Ma et al. 2005). Some examples include *Methanobacterium aggregans* sp. and *Methanobacterium congolense* sp. which use H_2 and CO_2 as sole substrate, whereas *Methanobacterium beijingense* sp. can also use formate (Cuzin et al. 2001; Ma et al. 2005; Kern et al. 2015).

The genus *Methanosarcina* is known by the formation of cell clusters during growth (Kern et al. 2015). These are a very versatile member of methanogens capable of using several C1 substrates in addition to acetate, H_2 and CO_2, capable of acting as acetoclastic and hydrogenotrophic methanogens (Venkiteshwaran et al. 2015). This flexibility explains their capacity for outcompeting other members. *Methanosaeta* is a strict acetoclastic methanogen that usually dominates in mesophilic systems. About two-thirds of methane is derived from the degradation of acetate unless inhibitory conditions such as high ammonia levels and high temperatures are present (Lim et al. 2020).

Finally, Li et al. (2017) has documented the presence of direct interspecies electron transfer (DIET) as a mechanism for favouring syntrophic oxidation of short-chain acids, thus explaining the enhancement in biogas production obtained by different authors when supplementing digestion systems with carbon conductive materials or metals. This enhancement has been associated with the increase in hydrogenotrophic methanogens (Zhou et al. 2021). The introduction of these materials into digestion systems submitted to extreme inhibitory conditions also allowed methane production to be successfully attained (Cuetos et al. 2017). Milán et al. (2010) studied the use of adsorbents and the effects of metals (nickel and cobalt). Based on the analyses of anaerobic biofilms, these authors reported that the presence of nickel and cobalt favoured *Methanosaeta*, while the same dose of magnesic zeolite stimulated the presence of *Methanosarcina* and sulphate-reducing bacteria. These two genera were also reported to be enriched in a system used to evaluate DIET in the presence of pyrene (a toxic compound), along with the enrichment of extracellular electron-transfer bacteria (e.g. *Pseudomonas*, *Cloastridia* and *Synergistetes*) when bio-nano-FeS or magnetic carbon was added as supplement (L. Li et al. 2021).

7.3   FERMENTATIVE HYDROGEN PRODUCTION

Fermentative hydrogen production is one of the available processes to derive hydrogen from organics using specific microbial biomass. This process has a great possibility of becoming a mature technology in the short term thanks to technical similarities with the anaerobic digestion process (Moreno and Gómez 2012) and the capacity of shearing already developed equipment. However, these systems are also complex, just as their homologous anaerobic digestion. The fermentation is affected by multiple factors, including operational conditions (e.g. temperature, pH, retention time), substrate type, reactor design etc. These factors greatly influence microbial predominance, activity and productivity and are also responsible for the adverse effects associated with the accumulation of end products in the fermenting liqueur (Wang and Yin 2019).

The production of hydrogen is mediated by the activity of hydrogenase enzymes ([NiFe]-hydrogenase and [FeFe]-hydrogenase) catalysing the reduction of protons with the use of electrons derived from the oxidation of organics. The theoretical yield of 12 mol of H_2 derived from the reaction of glucose and water is never reached since microbial biomass needs energy for growth and maintenance. The main production pathways involve the generation of acetic and butyric acid yielding lower values (4 moles of H_2 in the case of the acetic pathway and 2 for the latter) (Hallenbeck 2005; Osman et al. 2020). The outcome is a fermenting liqueur containing a mixture of acid products needing additional treatment to attain the total conversion of the substrate. The production of these acids also translate in the requirement of alkalinity to keep pH at optimum conditions for continuous hydrogen evolution. The effluent stream may also be submitted to extractive distillation units for acid recovery, but this latter option may be unfeasible due to the high energy demand and difficulties found in separating the distinct acid components of the mixture. A reasonable approach would be the coupling with other biological processes such as, digestion, photo-fermentation or BES for metabolising acid by-products (Martínez et al. 2019a).

The dark fermentation process has been studied by several authors evaluating the performance of different cultures using synthetic substrates (Rachman et al. 1997; Logan et al. 2002; Liu et al. 2006; Moreno et al. 2015). However, its applicability at a real scale needs to guarantee fermentation stability when using wastes as carbon sources. It is also necessary to avoid sterilisation and the application of extreme pretreatments either for the influent or for generating the inoculum. The co-culture of different organisms offers various advantages against the use of a single strain, as it is the reduction in lag phase and resistance to environmental fluctuations providing stability to the process (Pachapur et al. 2015). When started up with mixed microflora, the hydrogen-producing fermentation can be seen as a clever approach to increase the efficiency of already well-developed treatment technologies, making it easier to integrate this process into an existing waste valorisation chain, thus producing continuous H_2 as a valuable by-product.

The direct conversion of carbohydrates into H_2 is performed by microorganisms with the ability to transform pyruvate obtained from glycolysis into formic acid and an acetyl group, in the case of enteric bacteria, and the subsequent degradation of formic acid into CO_2 and H_2 (Hallenbeck 2005). *Clostridia* produce hydrogen with higher yields thanks to the re-oxidation of NADH generated during glycolysis (Hallenbeck 2009). The fermentative production of hydrogen can be categorised into the kind of process where strict anaerobiosis is necessary, then being clostridium the predominant species, or a partial tolerance to the presence of oxygen, then being enterobacter predominant. However, this is just a gross simplification since the participation of a variety of organisms has been demonstrated by Etchebehere et al. (2016). These authors studied microbial communities of hydrogen fermenting bioreactors operating in different countries. Their results indicated that *Firmicutes* was the predominating phylum. High yield hydrogen-producing genera detected were *Clostridium, Kosmotoga, Enterobacter*, whereas Veillonelaceae was detected as the most common low hydrogen yield producer.

Other authors studied the effect of micro-aeration, reporting that air addition in a fermentative hydrogen system originally inoculated with coal mining water, resulted in the shift of microbial populations, from *Clostridium* under the butyric acid-type fermentation to a propionic type fermentation where *Propionibacterium* was dominant (Guo et al. 2022). Small temperature changes may also cause a shift in the microbial population, as shown by Lin et al. (2008), who observed a change from *Clostridium intestinale* and *Klebsiella pneumoniae* to *Clostridium sulfatireducens* and *Bacillus sp.*, when increasing the temperature from 35 to 40 °C.

The hydrogen-producing inoculum is usually obtained by applying pretreatments to mixed anaerobic cultures. Acid and alkaline shock, chloroform addition, heat pretreatment, gamma irradiation, ultrasounds or induced acidification have been actively tested (Kim et al. 2003; Gómez et al. 2009; Yin et al. 2014; Mañunga et al. 2019). However, the posterior performance will be determined by the organisms capable of surviving and subsequently by their capacity to adapt to operating conditions, the type of substrate, hydraulic retention time and pH, among others. Figure 7.2 presents a schematisation of the process and possible alternatives for treating final acid products. The application of acidic conditions for obtaining the inoculum leads to the preferential elimination of cellulolytic and methanogenic organisms. In contrast, heat shock favours the predominance of endospore-producers such as *Clostridia* genus (Viana et al. 2019) as well as it does the application of extreme acid conditions (Kim et al. 2011). In addition, heat shock reduces species diversity, Nissilä et al. (2011) reported that this type of pretreatment provoked an enrichment in *Thermoanaerobacterium thermosaccharolyticum* and thermophilic and cellulolytic microorganisms related to *Clostridium caenicola*.

Carillo et al. (2012) studied the use of buffalo dung as inoculum for hydrogen-producing systems. They also indicated that heat pretreatment allowed the accumulation in *Clostridium* species. The sequences analysis of the pretreated material revealed the abundance of the phyla *Bacteroidetes* and *Firmicutes*, with dominance of a very active hydrogen-producing bacteria belonging to *Clostridium cellulosi* species. The evolution of microbial communities during the start-up under batch conditions was evaluated by Yang and Wang (2019) reporting an increasing predominance of *Clostridium sensu stricto* 1 after 12 hours of fermentation starting from a heat pretreated inoculum. At the beginning of the fermentation, the community at phylum level was dominated by *Firmicutes* and *Bacteroidetes* with other five phyla detected at a lower proportion, but the former became dominant just after 6 hours.

El-Bery et al. (2013) studied an anaerobic baffled reactor under continuous conditions using alkali hydrolysed rice straw as substrate. They indicated that the reactor inoculated with a heat pretreated inoculum presented better performance, reporting a hydrogen yield of 1.19 mol H_2/mol glucose. However, the value reported for the no-pretreated system was still interesting (0.97 mol H_2/mol glucose). The presence of *Clostridium*, *Prevotella*, *Paludibacter*, *Ensifer* and *Petrimonas* was found

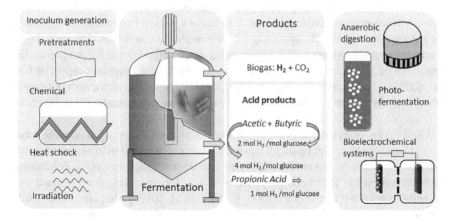

FIGURE 7.2 Scheme representing fermentative hydrogen-producing systems. Common pretreatments applied for generating inoculum from mix microflora. Valorisation alternatives for treating acid by-products.

in both reactors, and acetic and butyric were the main acid products. These results are particularly interesting because the elimination of pretreatment significantly reduces operating costs and energy demand. Rosa et al. (2020) reported the presence of the genus *Ruminococcus* when optimising H_2 fermentation conditions using as substrate palm oil mill effluent (POME). These authors compared the fermentation performance of mixed anaerobic consortia and a pure strain of *Clostridium beijirinckii (ATCC 8260), indicating that the mixed consortia* – with a significant presence of *Ruminococcus* – could use raw POME as substrate, thus eliminating the need of applying expensive pretreatments. Species such as *Ruminococcus albus* can produce biohydrogen from lignocellulosic substrates (Ntaikou et al. 2008).

Mañunga et al. (2019) reported a pH value between 5.0 and 5.4 units as a suitable range for the fermentation, when treating a carbohydrate-rich wastewater under mesophilic conditions. However, higher pH values were reported as optimised conditions, as demonstrated by Kamyab et al. (2019), giving a value of 6.2. Biohydrogen fermentation must cope with the use of cheap organic sources and attain a low energy demand if it is expected to become a feasible competitor in the biofuel market. These requirements translate in the capacity of the system for keeping healthy hydrogen-producing biomass and eliminate the application of sterile conditions, although being proposed by some researchers (Wongthanate et al. 2014; Xiao and Liu 2009). Reactor operating parameters must be such that allows the survival of the desirable microflora and avoidance of hydrogen consumers. However, keeping low pH values in the anaerobic reactor is not enough to prevent the growth of methanogens. Other factors as organic loading rate (OLR) and nitrogen content of the feed may cause the predominance of undesirable organisms (Cuetos et al. 2007). For this reason, short residence times are usually applied, allowing enough time for hydrolysis and partial acidification of the substrate, but preventing the proliferation of methanogens with lower growth rate.

Substrate concentration is also a relevant factor; organic content of the feed should be high enough to avoid shifts in microbial populations. Fernández et al. (2014) reported a decrease in hydrogen yields when decreasing substrate concentration in the feed. These authors studied the performance of a packed bed reactor using cheese whey permeate as substrate operating under non-sterile conditions. The prevalence of non-hydrogen producers was experienced when reducing OLR, indicating the dominance of *Sporolactobacillus* sp. and *Prevotella*. Similarly was the report of Jing et al. (2020) who studied the simultaneous saccharification and fermentation for producing hydrogen. These authors indicated that substrate concentration had the greatest influence on hydrogen production.

Based on previous difficulties and the imperious need to approach process feasibility at large-scale, recent research has focused on using mixed microflora, avoiding costly pretreatment techniques and applying real wastes under non-sterile conditions. Co-fermentation has also been proposed as a way for improving performance and balancing nutrients (Gómez et al. 2006; Kamyab et al. 2019; Sethupathy et al. 2019). However, having available the optimised proportion of substrates encounters several logistic difficulties since large-scale plants need to deal with those resources available in the vicinity all year round (Sevillano et al. 2021). In addition, keeping an active microflora is necessary, along with pH control of the fermentation.

Some authors have attained this double objective by coupling hydrogen and methane production using a recirculating stream. Alkalinity is recovered from the second stage and acceptable hydrogen yields are still obtained despite the cross circulation of microflora (Redondas et al. 2015; Algapani et al. 2019; Wang et al. 2020). Successful performance is attained by adjusting the recirculation ratio (Qin et al. 2019a).

Yang and Wang (2018) studied the co-fermentation of sewage sludge and flower waste, reporting that the presence of sludge in the fermentation enriched the system in *Clostridium* sp. and *Enterococcus* sp. This finding corroborates the possibility of attaining stable performance under conditions close to those found when dealing with real wastes. Qin et al. (2019b) evaluated a two-stage process for continuous production of hydrogen and methane. The first stage worked under thermophilic conditions, and the second one at mesophilic regimen. The two reactors were connected by a recirculation stream, thus continuously providing cellulose degrading bacteria which could only grow at mesophilic conditions. The recirculation stream also added the alkalinity necessary for keeping pH at optimum values in the hydrogen phase. The identification of microorganisms resulted interesting, it was performed at the family level, reporting a greater diversity in the mesophilic reactor dedicated to methane production. The major families of bacteria found in the first stage were affiliated to Ruminococcaceae, Clostridiaceae, Lactobacillaceae and Tissierellaceae. However, in the second stage Lactobacillaceae lost its predominance, indicating that the recirculation stream helped controlling the competition with this organism which is frequently found in hydrogen producing systems, and capable of outcompeting hydrogen producing microflora.

After the description of some of the current research regarding fermentative hydrogen production, it is reasonable to consider that the simple digestion of wastes may be transformed into a bioengineered process where different end-products are derived depending on the particular interest of the economy. Therefore, anaerobic digestion can become a process where the partial degradation of organic materials generates VFAs or hydrogen based on the operating conditions set to the reactor (Wainaina et al. 2019). The complete degradation would be attained in the final stage to accomplish the stabilisation of organics and methane production. Acid intermediaries and hydrogen can be easily produced and awake a greater interest in the future circular economy model. However, before this could be the case, major challenges need an urgent solution as stated in the study performed by the members of the Latin American Biohydrogen Network. This group analysed and discussed the different causes of instabilities reported when running dark fermentation systems. Reactor configuration and operating conditions must give satisfactory results to problems dealing with microbial community shifts, accumulation of fermentation products and the excessive growth of methanogens (Castelló et al. 2020).

The direct use of lignocellulosic material is another field where high level of improvement is expected. *Caldicellulosiruptor saccharolyticus* is an extreme thermophilic bacterium that belongs to the Clostridia class. This organism can grow on complex lignocellulosic organics and presents a hydrogen yield close to the theoretical value (4 mol H_2/mol hexose) when using carbohydrate-rich substrates (Willquist et al. 2010). Byrne et al. (2021) tested five members of the genus *Caldicellulosiruptor* (*C. owensensis*, *C. kronotskyensis*, *C. bescii*, *C. acetigenus* and *C. kristjanssonii*) with the aim to increase its tolerance to higher sugar concentrations and enhance H_2 reactor productivity. Although these authors reported that the application of adaptive laboratory evolution increased sugar tolerance, the development of reactor systems where biofilm formation is allowed seems necessary to keep acceptable performance under continuous operation.

Members of *Caldicellulosiruptor* genus have the ability to grow on sugar mixtures but with a preference for xylose and cellobiose, instead of glucose. This feature is of particular interest when dealing with lignocellulosic hydrolysates as substrates. *C. kronotskyensis* does not present a diauxic-like growth pattern and in contrast to *C. saccharolyticus*, *C. kronotskyensis* does not possess a second uptake system for glucose, showing a clear preference for cellobiose assimilation (Vongkampang et al. 2021). Therefore, using co-cultures with these organisms or applying multistage fermentation are reasonable options for attaining a complete conversion of complex substrates and increasing H_2 volumetric productivity.

7.4   ACETONE–BUTANOL–ETHANOL FERMENTATION

The ABE fermentation produces a mixture of solvents and acids (acetic and butyric) through the fermentation of sugars using solventogenic *Clostridium* species such as *C. acetobutylicum*, *C. beijerinckii* and *Clostridium saccharoperbutylacetonicum*, which are frequently cultivated with industrial purposes (Dürre 2008). The interest in this process has awakened once again because butanol is considered a drop-in fuel for gasoline, having more similar physical and chemical properties to gasoline than ethanol, the leader liquid biofuel (Lee et al. 2008). However, this process is still characterised by low productivity (Maddox 1989) and high operating costs associated with the recovery and separation of solvents, explaining thus the current null industrial application of this technology.

Butanol as a fermentation product presents high toxicity, inhibiting nutrient transport, glucose assimilation and membrane-bound ATPase activity (Bowles and Ellefson 1985). Most of the *Clostridia* reported in the literature cannot stand fermentation broths containing more than 2% solvent concentration (Lee et al. 2008). Therefore, this type of fermentation rarely produces more than 13 g/L of butanol and still suffers from the same limitations that caused the shot down of industrial fermentation plants for solvent production around 1930–1960, that is, high costs of fermentable sugars, high energy demand due to sterilisation needs and distillation of low solvent containing liqueur to separate butanol and acetone, along with high treatment costs of residual streams (Jones and Woods 1986).

ABE fermentation is a complex biphasic process with two distinct stages. A first acidogenesis phase where sugars are converted into acetic and butyric acid along with a decrease in culture pH, then a second stage takes place which is possibly triggered by enzyme regulation when a threshold acid value is reached (Grimmler et al. 2011). The transformation of intermediaries into butanol, acetone and ethanol occurs in this second phase, now accompanied by a pH increase (Maddox et al. 2000). Low pH values are necessary (around 4.5–5.0) to achieve high solvent yields, whereas higher pH values promote culture growth and acid production (Jones and Woods 1986; Guo et al. 2012; Jiang et al. 2014). These stages are critically affected by the intra and extracellular environment and pH conditions (Maddox et al. 2000). Besides, *Clostridium* is a strict anaerobe organism. Oxygen adversely affects cell growth, sugar consumption and protein synthesis. Thus, continuous bubbling of nitrogen or the use of reducing agents is needed to completely remove oxygen from the medium, increasing the operational costs (Wu et al. 2016).

The limits set by product toxicity also create constraints in the sugar content of the feeding stream and the type of operation. For this reason, the fermentation is started with a relatively low value of sugars if performed under batch conditions to limit final product concentration in the reactor liqueur. Under continuous conditions or fed-batch fermentation, the substrate concentration should be limited to keep product titre below the inhibitory level. Under these operating modes, the main component present in the reactor is the product. The low tolerance of wild-type strains to high levels of butanol in the fermenting liqueur explains the interest in genetically manipulated strains with higher butanol productivity (Tashiro et al. 2013).

Tomas et al. (2003) generated *C. acetobutylicum* ATCC 824 (pGROE1) to improve by 32% the production of butanol. Shin et al. (2021) isolated a new solventogenic *Clostridium* and applied simple genetic modification to attain a 31% increase in solvent production from a mixture of glucose and xylose as substrate. This strain was denoted as *C. beijerinckii* GSC1_R1 (gene-modified strain), reaching a productivity of 3.6 g/L h. Other authors obtained butanol titres close to 20 g/L by genetic manipulation, but reactor productivity is still extremely low (around 0.2–0.4 g/L h) (Formanek et al. 1997; Xue et al. 2012; Du et al. 2022).

Lignocellulosic by-products and residues from agriculture agro-industries have attracted attention as a suitable feedstock alternative for butanol production. Raw materials such as switchgrass, corn stover, apple pomace, tomato wastes or even prune vine shoots have been successfully employed using wild *Clostridium* strains, achieving butanol and ABE solvent titres above 9 g/L and 15 g /L, respectively (Paniagua-García et al. 2018; Hijosa-Valsero et al. 2017, 2020). Even though, important challenges need to be overcome. The pretreatment is a costly and complex process. The deconstruction and enzymatic hydrolysis of lignocellulosic biomass to fermentable sugars account for more than 50% of the total operating cost of upstream operations (Ibrahim et al. 2018) and product titres need significant improvements to keep distillation costs in a suitable range, thus the interest in obtaining engineered strains capable of treating hydrolysates or with the capacity of fermenting lignocellulosic substrates.

Zhiqiang et al. (2022) obtained an engineered strain from *C. cellulovorans* DSM 743B capable of producing butanol from lignocellulosic biomass. The authors reported a butanol titre of 3.47 g/L, a value that is far from being considered as optimum but still results interesting since it opens new opportunities for generating modified strains with the ability to use low-cost agricultural substrates and wastes. However, genetic modification may not necessarily be the answer. Co-cultivation may be easier to implement and develop at an industrial scale. Sequential fermentation and co-culturing of different strains take advantage of the ability of different *Clostridia* to transform wastes and cellulosic biomass in a co-fermentation system where other species use derived sugars and acids as substrates to complete the conversion process till obtaining the desired solvents.

Ozturk et al. (2020) evaluated a two-stage fermentation of thrown-away rice using an amylase-producing *Aspergillus oryzae*, followed by solvent-producing *Clostridium acetobutylicum* YM1 under non-sterile conditions. They obtained 10.91 g/L of butanol and 16.68 g/L ABE solvents. Wen et al. (2014) evaluated the fermentation of alkali extracted deshelled corn cobs as an example of lignocellulosic substrates. They used *Clostridium cellulovorans* 743B to saccharify this material and produce butyric acid, which was subsequently used by *C. beijerinckii* NCIMB 8052 to produce solvents (obtaining a butanol titre of 8.3 g/L). Zhang et al. (2021a,b) evaluated the use of food wastes as substrates using a mixture of *C. acetobutylicum* and *C. pasteurianum*. These authors applied enzymatic pretreatment to hydrolyse the waste material and then tested different co-culturing mixtures for these strains founding a proportion of 7/3 (*C. acetobutylicum* /*C. pasteurianum*) as the best conditions. As a result, a butanol concentration of 13.2 g/L in a 96 hours batch fermentation test was obtained, which is a much higher value but still too low to translate into a significant energy decrease in the subsequent distillation step, which is an unavoidable separation stage in the industrial implementation of this process.

Co-culture has also been proposed to include bacteria capable of degrading lignocellulosic substrates, but the implementation of this option may require a multistage approach due to different optimal growth temperatures, as these microorganisms are generally thermophilic. Other approaches for improving fermentation performance are the use of microorganisms to remove inhibitory compounds. The use of lignocellulosic material may aid in lowering raw material costs, but pretreatments applied to this type of substrate usually provoke the appearance of phenolic compounds that result toxic to the fermenting biomass. Detoxification of the hydrolysed substrate becomes necessary, thus incurring additional treatment costs unless microbial biomass carries out this process. This is the idea explored by Theiri et al. (2019) where a detoxification technology was developed using flocculation and biodetoxification. A co-culture composed of *Ureibacillus thermosphaericus* and *Cupriavidus taiwanensis* was used for removing phenolic compounds, attaining values that will not inhibit butanol fermentation in a second stage with *C. acetobutylicum* ATCC 824.

Other authors, such as Cui et al. (2020), tested co-culturing strategies for oxygen removal using *Bacillus* and *Clostridium* species. Bacillus genus is a fast-oxygen consumer bacterium because of its rapid growth, producing fewer by-products during ABE fermentation. These authors obtained a solvent production of 1.7 g/L of acetone, 4.8 g/L of butanol and 0.9 g/L of isopropanol – from 60 g/L of glucose – in a symbiotic system of engineered *B. subtilis* 1A1 strain (BsADH2) with *C. beijerinckii* G117. Another potential co-culture for ABE fermentation involves the use of *Saccharomyces cerevisiae*, which secretes amino acids under alcohol stress conditions that are assimilated by *Clostridium* species, improving alcohol resistance. Luo et al. (2015) increased by 16.6% the production of biobutanol (15.74 g/L) co-culturing *C. acetobutylicum* with *S. cerevisiae* thanks to the beneficial contribution of yeast secreting at least four amino acids (methionine, phenylalanine, tyrosine and lysine) in fermenting liqueur. Similar behaviour was observed in co-cultured systems with *S. cerevisiae* and *C. beijerinckii* with increments in butanol concentration of about 250% (12.76 g/L) compared to the control (Wu et al. 2019).

However, the major problem of low titre is still pending a solution. If this fermentation is expected to outcompete butanol production from the traditional chemical process, greater improvements are needed than those obtained by strain manipulation and co-culturing. A significant increase in butanol concentration is needed prior to distillation, along with improvements in product separation (Abo et al. 2019). Changes in operating conditions are also urgent, such as implementing continuous operation, multistage configuration or cell immobilisation. This latter type of reactor can use either a carrier or a retention barrier. The high energy demand is the other critical factor to overcome. Evident solutions are the application of *in situ* product recovery (ISPR) processes that would allow the use of a higher substrate concentration and avoid the effect of reaching toxic conditions (Jiménez-Bonilla and Wang 2018). Among these ISPR processes, gas-stripping, pervaporation, adsorption or liquid-liquid extraction can be highlighted for the easiness of being integrated into the fermentation process to achieve higher butanol productivity (S. Li et al. 2020).

The use of mixed microflora capable of operating under continuous conditions, creating a consortium with a wider capacity for degrading substrates and eliminating sterilisation requirements would be the most desirable objective. Carbohydrate-rich substrates may be scarce due to their use for animal feeding. Thus, lignocellulosic biomass which is abundant, is a more suitable candidate for different fermentations, including ABE. The conversion of lignocellulosic material implies, as the first stage, the release of fermentable sugars by different pretreatments either mechanical/chemical or biochemical conversions (Pinto et al. 2021). However, pretreatments applied for substrate hydrolysis not only raise the energy demand and operating costs but also bring, as a consequence, the presence of toxic compounds when high temperature and pressure conditions are applied. These inhibitory compounds compromise the subsequent fermentation steps and add complexities because of the extra equipment needed for their removal.

The revival of this fermentation should be linked to a biorefinery scheme where different fermentation stages and energy recovery can be integrated to optimise the use of hydrolysed sugars and recirculation of intermediaries. Valdez-Vazquez and Sanchez (2018) evaluated a biorefinery dedicated to the transformation of wheat straw, considering H_2, solvents and methane production to optimise the energy demand and operating costs. The scheme proposed by these authors displayed lower energy consumption and environmental impacts than conventional second-generation lignocellulosic biofuels biorefineries.

7.5   SYNGAS FERMENTATION

Given the difficulties in the pretreatment of biomass and requirements of strict sterile conditions in butanol production, the fermentation of syngas seems more attractive because of the advantage of reducing biomass pre-processing by gasifying the whole material to produce syngas or integrating this fermentation into the steel manufacturing process using waste gases emitted from the blast furnace. The use of microorganisms as catalysts to transform mainly CO and H_2 into other byproducts avoids the use of expensive metal catalysts prone to poisoning. In addition, the requirements of mild conditions to attain this conversion is another advantage offered which should not be disregarded (Munasinghe and Khanal, 2010). However, other factors may be carefully evaluated as the limited mass transfer of the gaseous substrate, which may translate into excessive reactor volumes, and the low productivity of this fermentation which leads to high production costs (Sun et al. 2019a). Nevertheless, syngas fermentation is a promising technology not only to produce ethanol but also to obtain superior alcohols such as butanol or hexanol.

Organisms such as *Clostridium ljungdahlii*, *C. autoethanogenum*, *Acetobacterium woodii*, *C. carboxidivorans* and *Butyribacterium methylotrophicum* have been studied for their abilities in using small gaseous molecules as substrates to produce acetate, ethanol, butyrate and butanol (Heiskanen et al. 2007; Johannes and Volker, 2015; Richter et al. 2016a; Lanzillo et al. 2020 ; Thi et al. 2020). The fermentation needs to be coupled to another process that provides the H_2 and CO/CO_2 gas stream. The gasification of biomass can produce in an efficient manner the full conversion of lignocellulosic material into syngas. However, the process still presents several challenges starting with profitability due to high installation costs (Sansaniwal et al. 2017), following with high transport costs of raw materials which increase with plant scale (Li et al. 2018) and high N_2 content in syngas when the process is carried out in the presence of air, instead of pure oxygen or steam. Despite these limitations, different acetogens are used as commercial production strains for industrial syngas fermentations in pilot or demonstration plants (Coskata, INEOS Bio, LanzaTech). Figure 7.3 presents a schematisation of the process considering a first gasification stage.

Efforts are being made at quasi-commercial scale to solve inhibitory problems associated with biomass derived syngas to provide efficient removal of hydrogen cyanide and tars (Bengelsdorf et al. 2013; Lane 2014). On the contrary, the use of

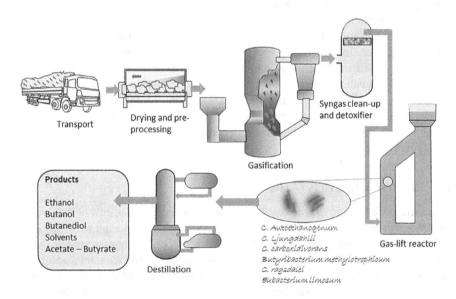

FIGURE 7.3   Schematisation of biomass gasification and syngas fermentation. Some of the organisms available for carrying out syngas transformation are exemplified in the scheme.

steel mill waste gases presented a better outcome. The LanzaTech's technology has been demonstrated at five industrial sites with over 40,000 hours of operation (BlueScope Steel, NZ; Shougang Steel, CN; BaoSteel, CN; China Steel, TW) and approximately 30,000 hours using syngas from industrial municipal solid waste gasification (Sekisui, JPN) (www.Basf.com). The first commercial ethanol plant Beijing Shougang LanzaTech New Energy Technology Co., Ltd. has earned Roundtable on Sustainable Biomaterials (RSB) certification. The plant is located at the Jingtang Steel Mill in Caofeidian in Hebei Province (bioenergyinternational.com).

Acetogenic bacteria use the Wood–Ljungdahl pathway (also known as acetyl-CoA biochemical pathway) and metalloenzymes when growing on C1 molecules and hydrogen. Nutrient-limited conditions (e.g. phosphate-limited media) promote solventogenesis; that is, ethanol is rather produced than acetate. This type of fermentation must cope with initially favouring microbial biomass and acid production and then shifting conditions to obtain solvents (Kennes et al. 2016). The production of ethanol and higher alcohols such as butanol and hexanol from CO/CO_2 is only limited to a few autotrophic anaerobic bacteria, standing out the strain *C. carboxidivorans* (Phillips et al. 2015). However, ethanol is the dominant alcohol in C1 fermentation while butanol is the major solvent. In HBE (hexane–butanol–ethanol) fermentation, final butanol concentrations hardly reach 1 g/L without strict pH control. Other authors reported higher values when applying pH regulation, but titres are extremely low, butanol values of 2.66 g/L from syngas were reported from CO fermentation using wild strains of *C. carboxidivorans* (Fernández-Naveira et al. 2016).

C. ljungdahlii is one of the most frequently used microorganisms in syngas fermentation to produce ethanol. Klasson et al. (1993) reported the highest ethanol concentration ever recorded (48 g/L) using *C. ljungdahlii* at a pH of 4.0–4.5 in a completely stirred tank reactor under nutrient-limited conditions during 560 hours of fermentation. Other studies focus on in increasing butanol production by genetic manipulation but butanol values achieved in the final fermenting liqueur are too low to result of any industrial interest (Köpke et al. 2010; Dürre 2016). Another strategy to overcome the metabolic pathway limitation of *C. ljungdahlii* was its co-culture with *C. kluyveri* (Richter et al. 2016b). However, significant improvements on HBE fermentations are still essential for process feasibility.

Clostridium strain P11[T] was tested at a semi-pilot scale using a biomass gasifier and a 100 L reactor for the fermentation of syngas. The main products derived from the fermentation were ethanol, produced in the solventogenesis phase, and acetic acid and 2-propanol, which were produced at the initial stage of the process when acid production was dominating (Kundiyana et al. 2010). Ethanol reached values of 25.3 g/L, but if compared with conventional ethanol fermentation, these values are too far from becoming competitive. In a different study, Liakakou et al. (2021) evaluated the coupling of biomass gasification and syngas fermentation using *Clostridium ljungdahlii* to understand the effect of inhibitory substances as tar components and assess the efficacy of the cleaning processes before the fermentation. The final ethanol concentration attained was 2.2 g/L, a value that may seem low but it is still promising given the great difficulties in removing toxic components of syngas derived from biomass gasification. Table 7.2 presents a list of different studies reported in the literature dedicated to improving product yields and lowering fermentation medium costs when fermenting syngas. Although efforts are being made, syngas clean-up has demonstrated challenging, causing a significant delay in semi-industrial operations when coupling syngas fermentation with biomass gasification.

This fermentation, just as any other type, has several requirements regarding the availability of nutrients for sustaining biomass in the reactor and the presence of buffering agents to regulate system pH. Therefore, it seems reasonable to attempt the use of a conventional fermentation system as a basic platform for attaining syngas conversion into solvents. This is the proposal developed by Richter et al. (2013), who used corn beer from a conventional yeast fermentation in the corn kernel-to-ethanol industry to produce a variety of solvents using carboxylic acid precursors and syngas. The outcome of this process is a fermenting liqueur requiring further distillation to recover solvents. The low titre of this fermentation translates into high energy demand unless a previous liquid extraction operation is implemented.

The transformation of syngas into methane rather than solvents may be of greater interest, since methane is a gaseous molecule that is already present in a different phase, avoiding the requirement of extra product recovery units. The biomethanation of syngas is a synergistic process where mixed microbial consortia use syngas as carbon and energy source producing methane and carbon dioxide. Mixtures of H_2 and CO can be transformed into methane both directly and stepwise with intermediary compounds mediating the process and comprising a complex network of biochemical reactions based on the water-gas shift reaction, acetogenesis, hydrogenotrophic methanation, carboxydotrophic methanation and acetoclastic methanation (Grimalt-Alemany et al. 2018).

Luo et al. (2013) demonstrated good performance of a thermophilic digester treating sewage sludge where a hollow fibre membrane module was introduced for feeding CO. The authors proved that the simultaneous treatment of sewage sludge and CO conversion could be attained. Species close to *Methanosarcina barkeri* and *Methanothermobacter thermautotrophicus* were the two main archaeal species involved in CO biomethanation. The conversion of H_2 into methane using a mesophilic digester treating sewage sludge was tested by Martínez et al. (2019b) using direct gas injection into the system and obtained a 12% increase in methane production. The limited hydrogen conversion was explained by mass transfer limitations and the lack of a gas recirculating system. In this case, the conversion into methane was not a direct process. Acetate was firstly produced as an intermediary, and then, methane was derived from the conversion of acetic. The authors reported the predominance of Methanoregulaceae and Methanobacteriaceae (both hydrogenotrophic archaea) and Methanosaetaceae, which follows the acetoclastic pathway. Therefore, the CO_2 composition in this reactor was kept at around 40%. This is contrary to results obtained by Luo et al. (2012), who also attained H_2 conversion but in this case with concomitant CO_2 assimilation, thus attaining an in-situ up-grading of biogas to biomethane, decreasing CO_2 concentration to 15%. In a later study under thermophilic conditions, the direct biomethanation was explained by the enrichment of the digestion system with Methanobacteriales, capable of mediating hydrogenotrophic methanogenesis (Luo and Angelidaki 2012).

Considering the capacity of conventional digestion to ferment a wide variety of substrates, the subsequent reasonable stage is to co-ferment syngas in an anaerobic

TABLE 7.2
Studies reported in the literature for syngas fermentation

Microorganisms	Product (g/L)	Aim	Reference
C. autoethanogenum	Ethanol: 2.24–3.76	Substituting yeast extract by malt and vegetable extracts	Thi et al. (2020)
	Ethanol: 4.23–4.57	Tryptone and peptone supplementation	Im et al. (2021)
	Ethanol: 3.45	Yeast extract optimisation	Xu et al. (2017)
C. ljungdahlii	Ethanol: 0.3	Nano particle addition to enhance syngas solubility	Kim et al. (2014)
	Ethanol: 2.0	pH regulation	Infantes et al. (2020)
	Ethanol: 1.09	Improving bioreactor productivity by the use of hollow fibre system	Anggraini et al. (2019)
	Ethanol: 0.34 2,3-Butanediol: 0.47	Optimising H_2/CO ratio	Jack et al. (2019)
	Ethanol: 0.85–3.75	Evaluating gas flow rate, media and effluent flow rate, pH level and stirrer speed	Acharya et al. (2019)
C. carboxidivorans P7	Ethanol: 2.0 Butanol: 1.0	Optimising medium trace metal composition	Han et al. (2020)
	Ethanol: 1.7–3.23 Butanol: 0.09–1.02	Effect of NH_3, H_2S and NO_x	Rückel et al. (2021)
	Ethanol: 3.64 Butanol: 1.35 Hexanol: 0.66	Temperature optimisation (two-steps: 37–25 °C)	Shen et al. (2020)
	Ethanol: 23.93	Improving bioreactor productivity by the use of hollow fibre system	Shen et al. (2014)
Butyribacterium methylotrophicum	Acetate: 2.3 Butyrate: 0.027	Evaluating process performance when using lignin derived syngas	Pacheco et al. (2021)
	Butanol: 2.7	Demonstration of growing on CO	Grethlein et al. (1991)
C. ragsdalei	Ethanol: 14.92	Implementation of genome shuffling technique for enhancing metabolite production	Patankar et al. (2021)
	Ethanol: 2.01 g/L	Key nutrients modulation: calcium pantothenate, vitamin B12, cobalt chloride ($CoCl_2$)	Kundiyana et al. (2011)
	Ethanol: 11–13.2	Addition of poultry litter biochar as substitute of costly buffer components	Sun et al. (2019H2 and CO2b)
Eubacterium limosum	Solvent product no reported Cell growth assessment	Use of methanol as a component for mixotrophy in syngas fermentation	Kim et al. (2021)

reactor. This strategy was studied by Yang et al. (2020) using a mixture of H_2:CO (5:4 ratio) under mesophilic and thermophilic conditions in reactors treating food wastes. These authors proved the feasibility of adapting existing reactors to syngas co-fermentation. Authors reported that under thermophilic syngas co-digestion, the order *Methanosarcinales* was dominant with the genus *Methanosarcina* (versatile aceticlastic methanogen) and *Methanobacterium* mainly present in the liquid samples and *Methanothermobacter* being abundant in the biofilm formed on the gas induction basket. In mesophilic reactors *Methanosaeta was dominant when the syngas flow was increased, thus indicating that CO and H_2 were easier to convert into acetate and then to CH_4*

7.6 BIOELECTROCHEMICAL SYSTEMS

A great variety of microorganisms can transfer electrons to a surface, with this feature being the main characteristic of BES. These organisms are often iron-reducing type bacteria, such as *Geobacter sulfurreducens*, producing high power densities at moderate temperatures (Logan et al. 2019). BES are in general, biological systems that use microorganisms in one or both electrodes (either anode or cathode). The reaction catalysed by microorganisms takes place at the electrode surface. The organic substrate is oxidised and electrons released are transferred to the cathode, where a reduction reaction occurs (Rozendal et al. 2008; Rabaey 2009; Sleutels et al. 2012).

BES, where chemical energy is converted into electrical energy, are called MFC, but if electricity is supplied to drive the reaction that does not otherwise occur spontaneously in any other case, then the system is known as MEC (Hamelers et al. 2010; Dey et al. 2022). In both types (MFC and MEC), the biological oxidative reaction generally occurs in the anodic compartment. Nevertheless, if a reduction reaction is stimulated by a small energy input in which a molecule is produced in the cathode compartment, the system is called microbial electrosynthesis (MES) (Zhang et al. 2021a,b). Figure 7.4 shows a scheme of the different configurations for microbial electrolysis systems.

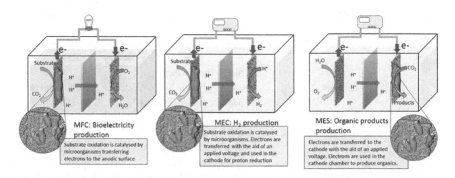

FIGURE 7.4 Microbial electrolysis cell representation. Different configurations are available where a reaction is catalysed by microorganisms. Electrons are transferred to an anodic surface and used to produce electricity or participate in cathodic reaction.

There is extensive research regarding the performance of BES, but scaling up the process demonstrated to be challenging given that it is a surface mediated process, and therefore, mass transfer limitations plays a relevant role in the final yield. It should also be added, that as in any other biological process, substrate characteristics and operating conditions greatly affect the final behaviour of the biofilm. Therefore, long-term performance needs to be carefully evaluated to guarantee that microbial shifts will not adversely affect the productivity of the reactor.

Since the appearance of the first bioelectrochemical reactor, several advancements have been materialised regarding to: (i) reactor design, (ii) chamber configuration, (iii) the reduction of electrode costs and materials and (iv) the development of membrane-less reactors (Tartakovsky et al. 2009; Ye et al. 2010; Khan et al. 2017; Jayabalan et al. 2019; González-Pabón et al. 2021 ; Zhang et al. 2021a,b). There is a better understanding of the microbiology associated with the biofilm responsible for electron transport and system performance. However, applying this technology at an industrial scale is still far from becoming a reality. There is an urgent need to demonstrate that BES can work at a larger scale and under realistic conditions after so many years of basic researched being founded by different institutions. This technology should reach a change in status and become a practical solution to real environmental problems moving away from a mere laboratory entertainment (Heidrich et al. 2013). Large pilot scale studies in this subject have been rather scarce. One of the main reasons is the high costs of the technology and the difficulties for keeping an active exoelectrogenic population when real wastewater and organic substrates are tested. Engineering strategies and different microbiological approaches are necessary to improve performance at a large scale to benefit from the promising potential of MFC technology (Yousaf et al. 2017).

Pure cultures within Geobacteraceae family, such as *Geobacter sulfurreducens*, *G. metallireducens* and *Desulfuromonas acetoxidans*, can support growth using organic compounds as substrate and transfer electrons produced from the oxidation to a graphite electrode (Bond et al. 2002; Bond and Lovley 2003). The mechanism of DIET was demonstrated by Summers et al. (2010) in a microbial co-culture of *Geobacter metallireducens* and *Geobacter sulfurreducens* producing an electrically conductive aggregate. This mechanism brings several advantages to anaerobic processes because enhances degradation under severe conditions where traditional digestion would be inhibited. Summers et al. (2010) isolated an exoelectrogenic bacterium identified as a representative of a novel species of the genus *Geobacter. Geobacter anodireducens* **sp.** is a strain that uses soluble or insoluble Fe(III) as the sole electron acceptor coupled with the oxidation of acetate but no fumarate. *Shewanella putrefaciens* grown under anaerobic conditions without nitrate is another organism characterised by electrochemical activity (Kim et al. 1999). *Desulfovibrio* is also an exoelectrogenic microorganisms capable of catalysing hydrogen production. This electroactive genus has been found in biocathodes involved in the bioelectrosynthesis reaction of methane from CO_2, when using mixed cultures as inoculum (Mateos et al. 2020). *Desulfovibrio* can also act as an acetogen and produce acetate when low sulphate levels are kept in an H_2/CO_2 environment (May et al. 2016; Mateos et al. 2018).

There are several organisms identified as having exolectrogenic properties. Table 7.3 shows a list of some of them with application in BES for electricity generation

(MFC), producing organic compounds by the oxidation of molecules in MEC, or simply electrogenic microorganisms involved in the reduction of inorganic compounds to obtain valuable organics (MES). The use of mixed cultures offers a more stable environment capable of tolerating perturbations and with a better adaptive capacity without sacrificing performance. However, the evaluation of BES with the use of pure cultures allows a better understanding of the process and offers solutions to mass transfer limitations and biofilm long-term performance problems (Cao et al. 2019).

Many BES systems are started-up after a long adaptation protocol to develop an active electrogenic microflora. The use of pure strains is evidently avoided, given that the main intended application of the technology is the treatment of waste streams. Thus, sterile conditions are senseless since microorganisms need to proliferate and compete in adverse environments where reactor conditions must be set to favour the culture of interest; otherwise, microbial shifts would become the dominant problem.

The literature reports the successful operation of MFC inoculated with mixed cultures, but many studies use synthetic wastes as feeding streams (Sharma et al. 2021; Hidayat et al. 2022). Sasaki et al. (2019) reported the presence of *Desulfovibrio*, *Clostridium*, *Lactococcus*, *Paludibacter* and *Bacteroides* in the anaerobic biofilm of an MFC, along with *Geobacter* sp., which was the dominant species. However, the MFC reporting a lower electricity production presented higher levels of methanogenic archaea (*Methanosarcina*, *Methanospirillum* and *Methanobacterium* spp.).

Rago et al. (2015) studied a MEC membrane-less reactor for four months, reporting unsuccessful performance due to methanogenesis build-up. The *Archaea* population dominating the cathodic and anodic biofilm was the hydrogen-oxidising genus *Methanobrevibacter*, of the Methanobacteriales order. The dominant population in the anode was *Geobacter*. The system was treated with 2-bromoethanesulphonate, but the addition of this compound did not prevent the reactor from producing methane. A similar result was previously reported by Cusick et al. (2011) using a 1-m³ cell for treating winery wastewater for 100 days. Results indicated that methane produced in the reactor was too high to be derived exclusively from the electrogenic activity, suspecting the growth of undesirable methanogens. Heidrich et al. (2014) tested a 100-L MEC treating wastewater for a whole year period considering seasonal variations in temperature. Contrary to other studies, these authors reported no methane production during the experiment, but the system performance was poor (0.007 m³ H_2/m³ reactor).

The methanogenic process also has a strong impact on MES systems, although the field has been less explored due to limited information and complexity. Several authors have focused on valorising carbon dioxide by reducing it into methane using microorganisms attached to the cathode electrode surface (Beese-Vasbender et al. 2015; Das et al. 2018; Ragab et al. 2020). This is a promising strategy for storing renewable energy. However, there is a great need to gain insight in microbe-electrode transfer mechanisms to increase conversion efficiency and better understand methane electrosynthesis by methanogenic Archaea. *Methanobacterium*, *Methanosarcina*, *Methanocorpusculum* and *Methanobrevibacter* genus have been reported on cathodic surfaces involved in the development of high conductive biofilm and electromethanogenic production (Z. Li et al. 2019; Bai et al. 2020; Izadi et al. 2020; J. Li et al. 2020; Pelaz et al. 2022).

TABLE 7.3
Electrogenic microorganisms used in MFC, MEC and MES

BES	Configuration	Organisms	Substrate	Reference
MFC	Electricity production of a dual-chamber mediatorless working in fed-batch mode	*Rhodoferax ferrireducens.* Performance comparison when MFC inoculation with waste activated sludge	Wastewater streams of a glutamate plant	Liu and Li (2007)
	Dual chamber. Limiting cell growth via removal of ingredients in anode media to improve coulumbic yields	*Rhodoferax ferrireducens.* comparison with mixed culture	Glucose	Liu et al. (2007)
	'Ministack' fuel cell	*Geobacter sulfurreducens* comparison with mixed culture	Acetate-fumarate medium	Nevin et al. (2008)
	One cassette-electrode comprising an air–cathode, a separator and an anode	*Geobacter* sp. Inoculation with activated sludge	Synthetic wastewater	Sasaki et al. (2019)
	Interactive relationship between bioelectricity production and anodophilic microbial community characteristics in MFC	*Pseudomonas, Desulfovibrio, Phyllobacterium, Desulfuromonas, Chelatococcus* and *Aminivibrio* inoculated with mixed carbon sources.	Activated sludge	Xin et al. (2019)
	Lamellar-type reactor with two end plates and three paired anode-cathode compartments	Inoculation with mixed microflora *Citrobacter* sp. substituted by *Pectinatus*-related population and subsequently by *Firmicutes* populations	Glycerol	Dennis et al. (2013)
	Dual compartment separated by cation exchange membrane	Inoculation with sewage sludge fermenter. Initial dominance of *Citrobacter* but decreasing with long-term operation	Glycerol	Zhou et al. (2015)
	Single-chamber air-cathode MFCs using PCR-DGGE and clone libraries to survey the microbial diversity	*Geobacter sulfurreducens, Clostridium* sp. Inoculated with wastewater	Glucose	Xing et al. (2009)
	Study of anode bacterial communities and their evolution using a two-chamber MFC.	*Geobacter sulfurreducens* inoculated with anaerobic sludge	Acetate	Jung and Regan (2007)
	Single-chamber air-cathode MFC. Study of the community structure with carbon brush as anode	*Shewanella putrefacies, Geobacter sulfurreducen* and *Desulfitobacterium hafniense* inoculated with anaerobic sludge.	Rice straw hydrolysate	Wang et al. (2014)

(Continued)

TABLE 7.3 (*Continued*)

Electrogenic microorganisms used in MFC, MEC and MES

BES	Configuration	Organisms	Substrate	Reference
	Interaction of MFC-*Acidiphilium* cells with carbon electrodes and the influence of O_2.	Inoculated with *Acidiphilium* sp. isolated from the Tinto River	Ferric iron – Ferrous iron	Malki et al. (2008)
MEC	The community structure was examined using a MEC configuration and sealing the air cathode.	*Thauera aromatica*, *Clostridium* sp. and *Geobacter* sp. (Wastewater served as both inoculum and substrate)	Dairy manure wastewater	Kiely et al. (2011)
	Investigate the performance of *Geobacter* species in MEC for hydrogen use	*Geobacter sulfurreducens* and *Geobacter metallireducens* with a mixed consortium	Acetate	Call et al. (2009)
	MEC with continuous flow operation to study the effect of anodic potential on biofilm microbial diversity	*Geobacter sulfurreducens*, *Pelobacter propionicus* with wastewater activated sludge as inoculum	Acetate	Torres et al. (2009)
	H-type MEC reactor with an anion exchange membrane to study the microbial community structure	*Pelobacter propionicus*, *Geobacter sulfurreducens* (anaerobic digested sludge as inoculum)	Ethanol	Parameswaran et al. (2010)
	Single-chamber MEC reactor to recover energy directly as electricity or hydrogen	*Geobacter* sp. with wastewater served as both inoculum and substrate	Domestic and winery wastewater	Cusick et al. (2010)
	Biofilm behaviour and external power supply for methane production in a single-chamber MEC	*Geobacter* sp. from anaerobic sludge inoculation	Volatile fatty acids	Luo et al. (2018)
MES	H-type microbial electrosynthesis system reactor separated using a proton exchange membrane	*Rhodobacter sphaeroides* for the production of hydrogen under illumination.	Carbon dioxide	S. Li et al. (2021)
	Dual-chamber MES reactor separated by a Nafion membrane.	Graphite rod electrode inoculated with *Sporomusa ovata*	Carbon dioxide	Bajracharya et al. (2022)
	Evaluate the effects of power outages in an acetogenic MES using a two-chamber reactor.	*Sporomusa*, *Clostridium* and *Desulfovibrio* inoculated from the effluent of an H-Cell MES reactor producing acetate from sodium bicarbonate.	Carbon dioxide	del Pilar Anzola Rojas et al. (2018)
	Three-electrode MES system for butyrate enhancement using a nickel ferrite-coated biocathode.	*Ochrobactrum* with EET capability and phylum Proteobacteria using a mixed microbial culture	Carbon dioxide	Tahir et al. (2021)

EET: Extracellular electron transfer.

7.7 PHOTO-FERMENTATION BY PURPLE NON-SULPHUR BACTERIA

Purple non-sulphur bacteria (PNSB) are a type of photosynthetic organism that do not release oxygen during the photosynthetic process and are usually red- or purple-coloured, belonging to the Rhodospirillaceae family (Lu et al. 2021). Cultures of *Rhodobacter sphaeroides, Rhodobacter capsulatus, Rhodopseudomonas palustris* and *Rhodospirillum rubrum* have been studied under laboratory conditions for different authors (Monroy et al. 2013; Kao et al. 2016; Akroum-Amrouche et al. 2019; Rodríguez et al. 2021; Ross and Pott 2022). *Rhodospirillum rubrum* has the interesting feature of also transforming CO into H_2 through biological water gas-shift reaction, representing another route for transforming syngas.

PNSB presents different metabolic pathways that enable them to grow in the presence of light or darkness, but it is the Glyoxylic acid cycle, the metabolic route unique to PNSB that provides them with the strong ability to assimilate short-chain fatty acids (Lu et al. 2021; Wu et al. 2021). Therefore, this type of photofermentation is an effective process for coupling with dark fermentation to attain the total conversion of organics.

These bacteria use sulphide as electron donor when growing photoautotrophically but at a much lower concentration than that required by sulphur bacteria, and thus the denomination of "non-sulphur" (Basak et al. 2014). Nitrogenase is the active enzyme involved in hydrogen production by PNSB under anoxic and nitrogen/N_2 deficient conditions – allocating all protons as H^{2+}, using sunlight as energy source and small carbon molecules as carbon source (Eroglu and Melis 2011). These bacteria also contain hydrogenases which participate in the production and assimilation of hydrogen in a single direction, or both directions catalysing up-take and production in a bidirectional way (Gabrielyan et al. 2015).

Photofermentation using PNSB bacteria has been extensively studied, with recent research focusing on the application of these organisms for the treatment of wastewater, the recovery of nutrients, CO_2 capture and production of valuable organics in addition to hydrogen production (Cerruti et al. 2020; Lu et al. 2021). However, experiences at a large scale are still null. The technology may be categorised as promising, but extensive research regarding the stability of the process under long-term operation and performance using real waste stream is urgently needed. This type of fermentation is full of challenges, and the fact that unsteady lights irradiation adversely affect hydrogen evolution (Abdelsalam et al. 2021) just add an extra complication to fermenter design, which already has to deal with light penetration problems, cell harvesting of low-density cultures, costly substrate pretreatments (Budiman et al. 2015; Sali and Mackey 2021), microbial contamination and having enough land availability for installing reactor units with high superficial area requirements.

Turon et al. (2021) tested a novel photo-bioreactor for the growth of *Rhodobacter capsulatus*, obtaining a hydrogen production of 157.7 ± 9.3 ml H_2/L/h, which is the highest value achieved up today. The reactor was a plate type with interconnected meandering channels designed to provide mixing characteristics. However, if H_2 production could be attained in the absence of light, then reactor design and operation could be highly simplified. This is the idea of the experimental work carried out by Rodríguez et al. (2021), who evaluated the use of *R. rubrum* to produce H_2 and polyhydroxybutyrate

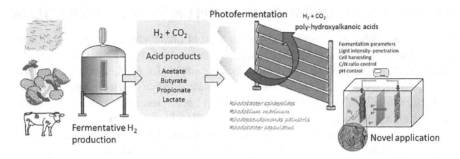

FIGURE 7.5 Schematic representation of photofermentation using PNSB for treating acidified effluents derived from fermentative hydrogen production.

using only CO/N_2 mixture in the dark. These are interesting efforts in the development of this technology, allowing to close the gap between laboratory scale and industrial application. Other efforts consider the application of PNSB exoelectrogenic ability in the degradation of recalcitrant molecules in photoMEC. Sogani et al. (2021) developed a hybrid system using a culture of *Rhodopseudomonas palustris* ATCC® 17007™ strain ATH 2.1.37 (NCIMB 11774). The reactor was a hybrid photo-assisted MFC containing two sections, one dedicated to the degradation of organic substances and the other to produce bioelectricity from hydrogen released during organic oxidation. However, this type of configuration must deal with scaling problems associated with MEC, yet to be solved and PNSB photofermentation stability. The future industrial development of this process integration seems to be far and current research is limited to small experimental assays under controlled laboratory conditions. Figure 7.5 presents a schematic diagram where this fermentation is integrated as a sequential stage in the treatment of biowastes. A first fermentation for producing H_2 and organic acids is considered, followed by a subsequent stage where photofermentation is introduced for assimilating soluble molecules and short chain fatty acids.

7.8 CONCLUSIONS

The different processes dedicated to biofuels and bioenergy production were reviewed, considering the anaerobic microbial communities involved. The application of these processes at a commercial scale is currently dominated by anaerobic digestion, an ancient process where a vast experience has already been gained by the spread of this technology worldwide. The solutions already available for different industrial microbial processes should be extrapolated to the treatment alternatives currently subject to extensive research. Dark fermentative hydrogen production keeps major similarities with the traditional digestion process. Therefore, this technology may probably find the fastest route to a large-scale application once aspects such as microflora stability and hydrogen storage are solved. Other processes like BES and photofermentation using NPSB still have serious challenges to overcome. Efforts should focus in increasing reactor scale and evaluate performance under long-term operations with conditions closer to those expected at a large scale.

ABE fermentation presents a glorious past, but constraints associated with operating costs, energy demand and logistics of raw materials relegated this technology to a stand-by point waiting for practical solutions to be developed. This process is still pending significant improvements in the field of reactor productivity and solvent extraction. Syngas fermentation presents current large-scale projects under development thanks to the ease in scaling-up. However, solutions are still needed to apply this process further to use syngas derived from biomass gasification. Research is needed in the removal of toxic compounds and syngas clean-up operations before this fermentation can find a niche in biomass valorisation technologies.

REFERENCES

Abdelsalam EM, Samer M, Moselhy MA, Arisha AH, Abdelqader AA, Attia YA (2021) Effects of He–Ne red and green laser irradiation on purple non-sulfur bacteria for biohydrogen production from food wastes. Biomass Convers Biorefinery. https://doi.org/10.1007/s13399-021-02084-7

Abo BO, Gao M, Wang Y, Wu C, Wang Q, Ma H (2019) Production of butanol from biomass: Recent advances and future prospects. *Environ Sci Pollut Res* 26(20):20164–20182. https://doi.org/10.1007/s11356-019-05437-y

Acharya B, Dutta A, Basu P (2019) Ethanol production by syngas fermentation in a continuous stirred tank bioreactor using *Clostridium ljungdahlii*. *Biofuels* 10(2):221–237. https://doi.org/10.1080/17597269.2017.1316143

Akroum-Amrouche D, Akroum H, Lounici H (2019) Green hydrogen production by Rhodobacter sphaeroides. *Energy Sources Part A Recover Util Environ Eff* 1–19. https://doi.org/10.1080/15567036.2019.1666190

Akturk AS, Demirer GN (2020) Improved food waste stabilization and valorization by anaerobic digestion through supplementation of conductive materials and trace elements. *Sustainability* 12(12):5222. https://doi.org/10.3390/su12125222

Algapani DE, Qiao W, Ricci M, Bianchi D, Wandera SM, Adani F, Dong R (2019) Bio-hydrogen and bio-methane production from food waste in a two-stage anaerobic digestion process with digestate recirculation. *Renew Energy* 130:1108–1115. https://doi.org/https://doi.org/10.1016/j.renene.2018.08.079

Angelidaki I, Batstone DJ (2010) Anaerobic digestion: Process. *Solid Waste Technol Manag* 583–600. https://doi.org/https://doi.org/10.1002/9780470666883.ch37

Anggraini ID, Keryanti, Kresnowati MTAP, Purwadi R, Noda R, Watanabe T, Setiadi T (2019) Bioethanol production via syngas fermentation of *Clostridium ljungdahlii* in a hollow fiber membrane supported bioreactor. *Int J Technol* 10(3):291–319. https://doi.org/https://doi.org/10.14716/ijtech.v10i3.2913

Attwood GT, Reilly K, Patel BKC (1996) *Clostridium proteoclasticum* sp. nov., a novel proteolytic bacterium from the bovine rumen. *Int J Syst Evol Microbiol* 46(3):753–758. https://doi.org/https://doi.org/10.1099/00207713-46-3-753

Bai Y, Zhou L, Irfan M, Liang TT, Cheng L, Liu YF, Liu JF, Yang SZ, Sand W, Gu JD, Mu BZ (2020) Bioelectrochemical methane production from CO_2 by *Methanosarcina barkeri* via direct and H_2-mediated indirect electron transfer. *Energy* 210:118445. https://doi.org/10.1016/J.ENERGY.2020.118445

Bajpai P (2017) Basics of anaerobic digestion process, in: Bajpai P (Ed.), *Anaerobic Technology in Pulp and Paper Industry*. Springer, Singapore, pp. 7–12. https://doi.org/10.1007/978-981-10-4130-3_2

Bajracharya S, Krige A, Matsakas L, Rova U, Christakopoulos P (2022) Dual cathode configuration and headspace gas recirculation for enhancing microbial electrosynthesis using *Sporomusa ovata*. *Chemosphere* 287:132188. https://doi.org/10.1016/J. CHEMOSPHERE.2021.132188

Bao Y, Dolfing J, Wang B, Chen R, Huang M, Li Z, Lin X, Feng Y (2019) Bacterial communities involved directly or indirectly in the anaerobic degradation of cellulose. *Biol Fertil Soils* 55(3):201–211. https://doi.org/10.1007/s00374-019-01342-1

Basak N, Jana AK, Das D, Saikia D (2014) Photofermentative molecular biohydrogen production by purple-non-sulfur (PNS) bacteria in various modes: The present progress and future perspective. *Int J Hydrogen Energy* 39(13):6853–6871. https://doi.org/https://doi. org/10.1016/j.ijhydene.2014.02.093

Basf.com. https://www.basf.com/global/en/who-we-are/organization/group-companies/BASF_ Venture-Capital/portfolio/LanzaTech-Inc.html [Accessed on 28-2-22].

Beese-Vasbender PF, Grote JP, Garrelfs J, Stratmann M, Mayrhofer KJJ (2015) Selective microbial electrosynthesis of methane by a pure culture of a marine lithoautotrophic archaeon. *Bioelectrochemistry* 102:50–55. https://doi.org/10.1016/J.BIOELECHEM.2014.11.004

Bengelsdorf FR, Straub M, Dürre P (2013) Bacterial synthesis gas (syngas) fermentation. *Environ Technol* 34(13–14):1639–1651. https://doi.org/10.1080/09593330.2013.827747

Bioenergy International. https://bioenergyinternational.com/biofuels-oils/beijing-shougang-lanzatech-new-energy-technology-ccu-plant-achieves-rsb-global-advanced-products-standard [Accessed on 8-3-22]

Bond DR, Holmes DE, Tender LM, Lovley DR (2002) Electrode-reducing microorganisms that harvest energy from marine sediments. *Science* 295(5554):483–485. https://doi. org/10.1126/SCIENCE.1066771

Bond DR, Lovley DR (2003) Electricity production by *Geobacter sulfurreducens* attached to electrodes. *Appl. Environ Microbiol* 69(3):1548–1555. https://doi.org/10.1128/ AEM.69.3.1548-1555.2003

Bowles LK, Ellefson WL (1985) Effects of butanol on *Clostridium acetobutylicum*. *Appl Environ Microbiol* 50(5):1165–1170. https://doi.org/10.1128/aem.50.5.1165-1170.1985

Budiman PM, Wu TY, Ramanan RN, Md. Jahim J (2015) Improvement of biohydrogen production through combined reuses of palm oil mill effluent together with pulp and paper mill effluent in photofermentation. *Energy Fuels* 29(9):5816–5824. https://doi.org/10.1021/ acs.energyfuels.5b01078

Byrne E, Björkmalm J, Bostick JP, Sreenivas K, Willquist K, van Niel EWJ (2021). Characterization and adaptation of *Caldicellulosiruptor* strains to higher sugar concentrations, targeting enhanced hydrogen production from lignocellulosic hydrolysates. *Biotechnol Biofuels* 14(1):210. https://doi.org/10.1186/s13068-021-02058-x

Call DF, Wagner RC, Logan BE (2009) Hydrogen production by *Geobacter* species and a mixed consortium in a microbial electrolysis cell. *Appl Environ Microbiol* 75(24):7579–7587. https://doi.org/10.1128/AEM.01760-09

Cao Y, Mu H, Liu W, Zhang R, Guo J, Xian M, Liu H (2019) Electricigens in the anode of microbial fuel cells: pure cultures versus mixed communities. *Microb Cell Factories* 18(1):1–14. https://doi.org/10.1186/S12934-019-1087-Z

Carillo P, Carotenuto C, Di Cristofaro F, Kafantaris I, Lubritto C, Minale M, Morrone B, Papa S, Woodrow P (2012) DGGE analysis of buffalo manure eubacteria for hydrogen production: effect of pH, temperature and pretreatments. *Mol Biol Rep* 39(12):10193–10200. https://doi.org/10.1007/s11033-012-1894-3

Castelló E, Nunes Ferraz-Junior AD, Andreani C, Anzola-Rojas M del P, Borzacconi L, Buitrón G, Carrillo-Reyes J, Gomes SD, Maintinguer SI, Moreno-Andrade I, Palomo-Briones R, Razo-Flores E, Schiappacasse-Dasati M, Tapia-Venegas E, Valdez-Vázquez I, Vesga-Baron A, Zaiat M, Etchebehere C (2020) Stability problems in the hydrogen production by dark fermentation: Possible causes and solutions. *Renew Sustain Energy Rev* 119:109602. https://doi.org/https://doi.org/10.1016/j.rser.2019.109602

Cerruti M, Stevens B, Ebrahimi S, Alloul A, Vlaeminck SE, Weissbrodt DG (2020) Enrichment and aggregation of purple non-sulfur bacteria in a mixed-culture sequencing-batch photobioreactor for biological nutrient removal from wastewater. *Front Bioeng Biotechnol* 8. https://doi.org/10.3389/fbioe.2020.557234

Cirne DG, Lehtomäki A, Björnsson L, Blackall LL (2007) Hydrolysis and microbial community analyses in two-stage anaerobic digestion of energy crops. *J Appl Microbiol* 103(3):516–527. https://doi.org/https://doi.org/10.1111/j.1365-2672.2006.03270.x

Cuetos MJ, Gómez X, Escapa A, Morán A (2007) Evaluation and simultaneous optimization of bio-hydrogen production using 3^2 factorial design and the desirability function. *J Power Sources* 169(1):131–139. https://doi.org/https://doi.org/10.1016/j.jpowsour.2007.01.050

Cuetos MJ, Martinez EJ, Moreno R, Gonzalez R, Otero M, Gomez X (2017) Enhancing anaerobic digestion of poultry blood using activated carbon. *J Adv Res* 8(3):297–307. https://doi.org/https://doi.org/10.1016/j.jare.2016.12.004

Cui Y, He J, Yang K-L, Zhou K (2020) Aerobic acetone-butanol-isopropanol (ABI) fermentation through a co-culture of *Clostridium beijerinckii* G117 and recombinant Bacillus subtilis 1A1. *Metab Eng Commun* 11:e00137. https://doi.org/https://doi.org/10.1016/j.mec.2020.e00137

Cusick RD, Bryan B, Parker DS, Merrill MD, Mehanna M, Kiely PD, Liu G, Logan BE (2011) Performance of a pilot-scale continuous flow microbial electrolysis cell fed winery wastewater. *Appl Microbiol Biotechnol* 89(6):2053–2063. https://doi.org/10.1007/S00253-011-3130-9

Cusick RD, Kiely PD, Logan BE (2010) A monetary comparison of energy recovered from microbial fuel cells and microbial electrolysis cells fed winery or domestic wastewaters. *Int J Hydrogen Energy* 35(17):8855–8861. https://doi.org/10.1016/J.IJHYDENE.2010.06.077

Cuzin N, Ouattara AS, Labat M, Garcia JL (2001) *Methanobacterium congolense* sp. nov., from a methanogenic fermentation of cassava peel. *Int J Syst Evol Microbiol* 51(2):489–493. https://doi.org/https://doi.org/10.1099/00207713-51-2-489

Das S, Chatterjee P, Ghangrekar MM (2018) Increasing methane content in biogas and simultaneous value added product recovery using microbial electrosynthesis. *Water Sci Technol* 77(5):1293–1302. https://doi.org/10.2166/WST.2018.002

del Pilar Anzola Rojas M, Mateos R, Sotres A, Zaiat M, Gonzalez ER, Escapa A, De Wever H, Pant D (2018) Microbial electrosynthesis (MES) from CO_2 is resilient to fluctuations in renewable energy supply. *Energy Convers Manag* 177(August):272–279. https://doi.org/10.1016/j.enconman.2018.09.064

Demirel B, Scherer P (2008) The roles of acetotrophic and hydrogenotrophic methanogens during anaerobic conversion of biomass to methane: a review. *Rev Environ Sci Bio/Technology* 7(2):173–190. https://doi.org/10.1007/s11157-008-9131-1

Dennis PG, Harnisch F, Yeoh YK, Tyson GW, Rabaey K (2013) Dynamics of cathode-associated microbial communities and metabolite profiles in a glycerol-fed bioelectrochemical system. *Appl Environ Microbiol* 79(13):4008. https://doi.org/10.1128/AEM.00569-13

Dey R, Maarisetty D, Baral SS (2022) A comparative study of bioelectrochemical systems with established anaerobic/aerobic processes. *Biomass Convers Biorefinery* 1:1–16. https://doi.org/10.1007/S13399-021-02258-3

Du G, Wu Y, Kang W, Xu Y, Li S, Xue C (2022) Enhanced butanol production in Clostridium acetobutylicum by manipulating metabolic pathway genes. *Process Biochem* 114:134–138. https://doi.org/https://doi.org/10.1016/j.procbio.2022.01.021

Dürre P (2008) Fermentative butanol production. *Ann N Y Acad Sci* 1125(1):353–362. https://doi.org/https://doi.org/10.1196/annals.1419.009

Dürre P (2016) Butanol formation from gaseous substrates. *FEMS Microbiol Lett* 363(6):fnw040. https://doi.org/10.1093/femsle/fnw040

El-Bery H, Tawfik A, Kumari S, Bux F (2013) Effect of thermal pre-treatment on inoculum sludge to enhance bio-hydrogen production from alkali hydrolysed rice straw in a mesophilic anaerobic baffled reactor. *Environ Technol* 34(13–14):1965–1972. https://doi.org/10.1080/09593330.2013.824013

Eroglu E, Melis A (2011) Photobiological hydrogen production: Recent advances and state of the art. *Bioresour Technol* 102(18):8403–8413. https://doi.org/https://doi.org/10.1016/j.biortech.2011.03.026

Etchebehere C, Castelló E, Wenzel J, del Pilar Anzola-Rojas M, Borzacconi L, Buitrón G, Cabrol L, Carminato VM, Carrillo-Reyes J, Cisneros-Pérez C, Fuentes L, Moreno-Andrade I, Razo-Flores E, Filippi GR, Tapia-Venegas E, Toledo-Alarcón J, Zaiat M (2016) Microbial communities from 20 different hydrogen-producing reactors studied by 454 pyrosequencing. *Appl Microbiol Biotechnol* 100(7):3371–3384. https://doi.org/10.1007/s00253-016-7325-y

Fernández C, Carracedo B, Martínez EJ, Gómez X, Morán A (2014) Application of a packed bed reactor for the production of hydrogen from cheese whey permeate: Effect of organic loading rate. *J Environ Sci Heal Part A* 49(2), 210–217. https://doi.org/10.1080/10934529.2013.838885

Fernández-Naveira Á, Abubackar HN, Veiga MC, Kennes C (2016) Efficient butanol-ethanol (B-E) production from carbon monoxide fermentation by *Clostridium carboxidivorans*. *Appl Microbiol Biotechnol* 100(7):3361–3370. https://doi.org/10.1007/s00253-015-7238-1

Fonknechten N, Chaussonnerie S, Tricot S, Lajus A, Andreesen JR, Perchat N, Pelletier E, Gouyvenoux M, Barbe V, Salanoubat M, Le Paslier D, Weissenbach J, Cohen GN, Kreimeyer A (2010) *Clostridium sticklandii*, a specialist in amino acid degradation: Revisiting its metabolism through its genome sequence. *BMC Genomics* 11(1):555. https://doi.org/10.1186/1471-2164-11-555

Formanek J, Mackie R, Blaschek HP (1997) Enhanced butanol production by *Clostridium beijerinckii* BA101 grown in semidefined P2 medium containing 6 percent maltodextrin or glucose. *Appl Environ Microbiol* 63(6):2306–2310. https://doi.org/10.1128/aem.63.6.2306-2310.1997

Gabrielyan L, Sargsyan H, Trchounian A (2015) Novel properties of photofermentative bio-hydrogen production by purple bacteria *Rhodobacter sphaeroides*: Effects of protonophores and inhibitors of responsible enzymes. *Microb Cell Fact* 14(1):131. https://doi.org/10.1186/s12934-015-0324-3

Gómez X, Cuetos MJ, Prieto JI, Morán A (2009) Bio-hydrogen production from waste fermentation: Mixing and static conditions. *Renew Energy* 34(4):970–975. https://doi.org/https://doi.org/10.1016/j.renene.2008.08.011

Gómez X, Morán A, Cuetos MJ, Sánchez ME (2006) The production of hydrogen by dark fermentation of municipal solid wastes and slaughterhouse waste: A two-phase process. *J Power Sources* 157(2):727–732. https://doi.org/https://doi.org/10.1016/j.jpowsour.2006.01.006

González J, Sánchez M, Gómez X (2018) Enhancing anaerobic digestion: The effect of carbon conductive materials. *C* 4(4):59. https://doi.org/10.3390/c4040059

González-Pabón MJ, Cardeña R, Cortón E, Buitrón G (2021) Hydrogen production in two-chamber MEC using a low-cost and biodegradable poly(vinyl) alcohol/chitosan membrane. *Bioresour Technol* 319. https://doi.org/10.1016/J.BIORTECH.2020.124168

Grethlein AJ, Worden RM, Jain MK, Datta R (1991) Evidence for production of n-butanol from carbon monoxide by *Butyribacterium methylotrophicum*. *J Ferment Bioeng* 72(1):58–60. https://doi.org/https://doi.org/10.1016/0922-338X(91)90147-9

Grimalt-Alemany A, Skiadas I V, Gavala HN (2018) Syngas biomethanation: State-of-the-art review and perspectives. *Biofuels Bioprod Biorefining* 12(1):139–158. https://doi.org/https://doi.org/10.1002/bbb.1826

Grimmler C, Janssen H, Krauße D, Fischer R-J, Bahl H, Dürre P, Liebl W, Ehrenreich A (2011) Genome-wide gene expression analysis of the switch between acidogenesis and solventogenesis in continuous cultures of *Clostridium acetobutylicum*. *Microb Physiol* 20(1):1–15. https://doi.org/10.1159/000320973

Guo H, Li S, Yang Z, Su X, Zhao S, Song B, Shi S (2022) Effect of continuous micro-aeration on hydrogen production by coal bio-fermentation. *Energy Sources Part A Recover Util Environ Eff* 1–11. https://doi.org/10.1080/15567036.2021.2024920

Guo T, Sun B, Jiang M, Wu H, Du T, Tang Y, Wei P, Ouyang P (2012) Enhancement of butanol production and reducing power using a two-stage controlled-pH strategy in batch culture of *Clostridium acetobutylicum* XY16. *World J Microbiol Biotechnol* 28(7):2551–2558. https://doi.org/10.1007/s11274-012-1063-9

Guo X, Wang C, Sun F, Zhu W, Wu W (2014) A comparison of microbial characteristics between the thermophilic and mesophilic anaerobic digesters exposed to elevated food waste loadings. *Bioresour Technol* 152:420–428. https://doi.org/https://doi.org/10.1016/j.biortech.2013.11.012

Hahnke S, Langer T, Klocke M (2018) *Proteiniborus indolifex* sp. nov., isolated from a thermophilic industrial-scale biogas plant. *Int J Syst Evol Microbiol* 68(3):824–828. https://doi.org/https://doi.org/10.1099/ijsem.0.002591

Hahnke S, Striesow J, Elvert M, Mollar XP, Klocke M (2014) *Clostridium bornimense* sp. nov., isolated from a mesophilic, two-phase, laboratory-scale biogas reactor. *Int J Syst Evol Microbiol* 64(Pt_8):2792–2797. https://doi.org/https://doi.org/10.1099/ijs.0.059691-0

Hallenbeck PC (2005) Fundamentals of the fermentative production of hydrogen. *Water Sci Technol* 52(1–2):21–29. https://doi.org/10.2166/wst.2005.0494

Hallenbeck PC (2009) Fermentative hydrogen production: Principles, progress, and prognosis. *Int J Hydrogen Energy* 34(17):7379–7389. https://doi.org/https://doi.org/10.1016/j.ijhydene.2008.12.080

Hamelers HVM, ter Heijne A, Sleutels THJA, Jeremiasse AW, Strik DPBTB, Buisman CJN (2010) New applications and performance of bioelectrochemical systems. *Appl Microbiol Biotechnol* 85(6):1673–1685. https://doi.org/10.1007/S00253-009-2357-1

Han Y-F, Xie B-T, Wu G, Guo Y-Q, Li D-M, Huang Z-Y (2020) Combination of trace metal to improve solventogenesis of *Clostridium carboxidivorans* P7 in syngas fermentation. *Front Microbiol* 11. https://doi.org/10.3389/fmicb.2020.577266

Hassa J, Maus I, Off S, Pühler A, Scherer P, Klocke M, Schlüter A (2018) Metagenome, metatranscriptome, and metaproteome approaches unraveled compositions and functional relationships of microbial communities residing in biogas plants. *Appl Microbiol Biotechnol* 102(12):5045–5063. https://doi.org/10.1007/s00253-018-8976-7

Heidrich ES, Dolfing J, Scott K, Edwards SR, Jones C, Curtis TP (2013) Production of hydrogen from domestic wastewater in a pilot-scale microbial electrolysis cell. *Appl Microbiol Biotechnol* 97(15):6979–6989. https://doi.org/10.1007/S00253-012-4456-7

Heidrich ES, Edwards SR, Dolfing J, Cotterill SE, Curtis TP (2014) Performance of a pilot scale microbial electrolysis cell fed on domestic wastewater at ambient temperatures for a 12 month period. *Bioresour Technol* 173:87–95. https://doi.org/10.1016/J.BIORTECH.2014.09.083

Heiskanen H, Virkajärvi I, Viikari L (2007) The effect of syngas composition on the growth and product formation of *Butyribacterium methylotrophicum*. *Enzyme Microb Technol* 41(3):362–367. https://doi.org/https://doi.org/10.1016/j.enzmictec.2007.03.004

Hidayat ARP, Widyanto AR, Asranudin A, Ediati R, Sulistiono DO, Putro HS, Sugiarso D, Prasetyoko D, Purnomo AS, Bahruji H, Ali BTI, Caralin I.S (2022) Recent development of double chamber microbial fuel cell for hexavalent chromium waste removal. *J Environ Chem Eng* 10(3):107505. https://doi.org/10.1016/J.JECE.2022.107505

Hijosa-Valsero M, Garita-Cambronero J, Paniagua-García AI, Díez-Antolínez R (2020) A global approach to obtain biobutanol from corn stover. *Renew Energy* 148:223–233. https://doi.org/https://doi.org/10.1016/j.renene.2019.12.026

Hijosa-Valsero M, Paniagua-García AI, Díez-Antolínez R (2017) Biobutanol production from apple pomace: The importance of pretreatment methods on the fermentability of lignocellulosic agro-food wastes. *Appl Microbiol Biotechnol* 101(21):8041–8052. https://doi.org/10.1007/s00253-017-8522-z

Ibrahim MF, Kim SW, Abd-Aziz S (2018) Advanced bioprocessing strategies for biobutanol production from biomass. *Renew Sustain Energy Rev* 91:1192–1204. https://doi.org/https://doi.org/10.1016/j.rser.2018.04.060

Im H, An T, Kwon R, Park S, Kim Y-K (2021) Effect of organic nitrogen supplements on syngas fermentation using *Clostridium autoethanogenum*. *Biotechnol Bioprocess Eng* 26(3):476–482. https://doi.org/10.1007/s12257-020-0221-4

Imachi H, Sakai S, Ohashi A, Harada H, Hanada S, Kamagata Y, Sekiguchi Y (2007) *Pelotomaculum propionicicum* sp. nov., an anaerobic, mesophilic, obligately syntrophic, propionate-oxidizing bacterium. *Int J Syst Evol Microbiol* 57(7):1487–1492. https://doi.org/https://doi.org/10.1099/ijs.0.64925-0

Infantes A, Kugel M, Neumann A (2020) Evaluation of media components and process parameters in a sensitive and robust fed-batch syngas fermentation system with *Clostridium ljungdahlii*. *Ferment* 6(2):61. https://doi.org/10.3390/fermentation6020061

Izadi P, Fontmorin JM, Godain A, Yu EH, Head IM (2020) Parameters influencing the development of highly conductive and efficient biofilm during microbial electrosynthesis: The importance of applied potential and inorganic carbon source. *npj Biofilms Microbiomes* 6(1):1–15. https://doi.org/10.1038/s41522-020-00151-x

Jack J, Lo J, Maness P-C, Ren ZJ (2019) Directing *Clostridium ljungdahlii* fermentation products via hydrogen to carbon monoxide ratio in syngas. *Biomass Bioenergy* 124:95–101. https://doi.org/https://doi.org/10.1016/j.biombioe.2019.03.011

Jang HM, Cho HU, Park SK, Ha JH, Park JM (2014) Influence of thermophilic aerobic digestion as a sludge pre-treatment and solids retention time of mesophilic anaerobic digestion on the methane production, sludge digestion and microbial communities in a sequential digestion process. *Water Res* 48:1–14. https://doi.org/https://doi.org/10.1016/j.watres.2013.06.041

Jayabalan T, Matheswaran M, Naina Mohammed S (2019) Biohydrogen production from sugar industry effluents using nickel based electrode materials in microbial electrolysis cell. *Int J Hydrogen Energy* 44(32):17381–17388. https://doi.org/10.1016/J.IJHYDENE.2018.09.219

Jiang M, Chen J, He A, Wu H, Kong X, Liu J, Yin C, Chen W, Chen P (2014) Enhanced acetone/butanol/ethanol production by *Clostridium beijerinckii* IB4 using pH control strategy. *Process Biochem* 49(8):1238–1244. https://doi.org/https://doi.org/10.1016/j.procbio.2014.04.017

Jiménez-Bonilla P, Wang Y (2018) In situ biobutanol recovery from clostridial fermentations: A critical review. *Crit Rev Biotechnol* 38(3):469–482. https://doi.org/10.1080/07388551.2017.1376308

Jing Y, Li F, Li Y, Jin P, Zhu S, He C, Zhao J, Zhang Z, Zhang Q (2020) Statistical optimization of simultaneous saccharification fermentative hydrogen production from corn stover. *Bioengineered* 11(1):428–438. https://doi.org/10.1080/21655979.2020.1739405

Johannes B, Volker M (2015) CO metabolism in the acetogen *Acetobacterium woodii*. *Appl Environ Microbiol* 81(17):5949–5956. https://doi.org/10.1128/AEM.01772-15

Jones DT, Woods D (1986) Acetone-butanol fermentation revisited. *Microbiol Rev* 50:484–524.

Jung K, Kim W, Park GW, Seo C, Chang HN, Kim Y-C (2015) Optimization of volatile fatty acids and hydrogen production from *Saccharina japonica*: Acidogenesis and molecular analysis of the resulting microbial communities. *Appl Microbiol Biotechnol* 99(7):3327–3337. https://doi.org/10.1007/s00253-015-6419-2

Jung S, Regan JM (2007) Comparison of anode bacterial communities and performance in microbial fuel cells with different electron donors. *Appl Microbiol Biotechnol* 77(2): 393–402. https://doi.org/10.1007/S00253-007-1162-Y/FIGURES/6

Kamyab S, Ataei SA, Tabatabaee M, Mirhosseinei SA (2019) Optimization of bio-hydrogen production in dark fermentation using activated sludge and date syrup as inexpensive substrate. *Int J Green Energy* 16(10):763–769. https://doi.org/10.1080/15435075.2019.1631828

Kao P-M, Hsu B-M, Chang T-Y, Chiu Y-C, Tsai S-H, Huang Y-L, Chang C-M (2016) Biohydrogen production by *Clostridium butyricum* and *Rhodopseudomonas palustris* in co-cultures. *Int J Green Energy* 13(7):715–719. https://doi.org/10.1080/15435075.2015.1088443

Kennes D, Abubackar HN, Diaz M, Veiga MC, Kennes C (2016) Bioethanol production from biomass: Carbohydrate vs syngas fermentation. *J Chem Technol Biotechnol* 91(2): 304–317. https://doi.org/https://doi.org/10.1002/jctb.4842

Kern T, Linge M, Rother M (2015) *Methanobacterium aggregans* sp. nov., a hydrogenotrophic methanogenic archaeon isolated from an anaerobic digester. *Int J Syst Evol Microbiol* 65(Pt_6):1975–1980. https://doi.org/https://doi.org/10.1099/ijs.0.000210

Khan MD, Khan N, Sultana S, Joshi R, Ahmed S, Yu E, Scott K, Ahmad A, Khan MZ (2017) Bioelectrochemical conversion of waste to energy using microbial fuel cell technology. *Process Biochem* 57:141–158. https://doi.org/10.1016/J.PROCBIO.2017.04.001

Kiely PD, Cusick R, Call DF, Selembo PA, Regan JM, Logan BE (2011) Anode microbial communities produced by changing from microbial fuel cell to microbial electrolysis cell operation using two different wastewaters. *Bioresour Technol* 102(1):388–394. https://doi.org/10.1016/J.BIORTECH.2010.05.019

Kim BH, Ikeda T, Park HS, Kim HJ, Hyun MS, Kano K, Takagi K, Tatsumi H (1999) Electrochemical activity of an Fe(III)-reducing bacterium, *Shewanella putrefaciens* IR-1, in the presence of alternative electron acceptors. *Biotechnol Tech* 13(7):475–478. https://doi.org/10.1023/A:1008993029309

Kim J, Park C, Kim T-H, Lee M, Kim S, Kim S-W, Lee J (2003) Effects of various pretreatments for enhanced anaerobic digestion with waste activated sludge. *J Biosci Bioeng* 95(3):271–275. https://doi.org/https://doi.org/10.1016/S1389-1723(03)80028-2

Kim J, Shin SG, Han G, O'Flaherty V, Lee C, Hwang S (2011) Common key acidogen populations in anaerobic reactors treating different wastewaters: Molecular identification and quantitative monitoring. *Water Res* 45(8):2539–2549. https://doi.org/https://doi.org/10.1016/j.watres.2011.02.004

Kim J-Y, Park S, Jeong J, Lee M, Kang B, Jang SH, Jeon J, Jang N, Oh S, Park Z-Y, Chang IS (2021) Methanol supply speeds up synthesis gas fermentation by methylotrophic-acetogenic bacterium, *Eubacterium limosum* KIST612. *Bioresour Technol* 321:124521. https://doi.org/https://doi.org/10.1016/j.biortech.2020.124521

Kim M, Gomec CY, Ahn Y, Speece RE (2003) Hydrolysis and acidogenesis of particulate organic material in mesophilic and thermophilic anaerobic digestion. *Environ Technol* 24(9):1183–1190. https://doi.org/10.1080/09593330309385659

Kim Y-K, Park SE, Lee H, Yun JY (2014) Enhancement of bioethanol production in syngas fermentation with *Clostridium ljungdahlii* using nanoparticles. *Bioresour Technol* 159:446–450. https://doi.org/https://doi.org/10.1016/j.biortech.2014.03.046

Klasson KT, Ackerson MD, Clausen EC, Gaddy JL (1993) Biological conversion of coal and coal-derived synthesis gas. *Fuel* 72(12):1673–1678. https://doi.org/https://doi.org/10.1016/0016-2361(93)90354-5

Köpke M, Held C, Hujer S, Liesegang H, Wiezer A, Wollherr A, Ehrenreich A, Liebl W, Gottschalk G, Dürre P (2010) *Clostridium ljungdahlii* represents a microbial production platform based on syngas. *Proc Natl Acad Sci* 107(29):13087–13092. https://doi.org/10.1073/pnas.1004716107

Kundiyana DK, Huhnke RL, Wilkins MR (2010) Syngas fermentation in a 100-L pilot scale fermentor: Design and process considerations. *J Biosci Bioeng* 109(5):492–498. https://doi.org/https://doi.org/10.1016/j.jbiosc.2009.10.022

Kundiyana DK, Huhnke RL, Wilkins MR (2011) Effect of nutrient limitation and two-stage continuous fermentor design on productivities during *"Clostridium ragsdalei"* syngas fermentation. *Bioresour Technol* 102(10):6058–6064. https://doi.org/https://doi. org/10.1016/j.biortech.2011.03.020

Lamed R, Naimark J, Morgenstern E, Bayer EA (1987) Specialized cell surface structures in cellulolytic bacteria. *J Bacteriol* 169(8):3792–3800. https://doi.org/10.1128/ jb.169.8.3792-3800.1987

Lane J (2014) On the mend: Why INEOS Bio isn't producing ethanol in Florida. *Biofuels Dig.* http://www.biofuelsdigest.com/bdigest/2014/09/05/on-the-mend-why-ineos-bio-isnt-reporting-much-ethanol-production/ [Accessed on 28-2-22].

Lanzillo F, Ruggiero G, Raganati F, Russo ME, Marzocchella A (2020) Batch syngas fermentation by *Clostridium carboxidivorans* for production of acids and alcohols. *Process* 8(9):1075. https://doi.org/10.3390/pr8091075

Lee SY, Park JH, Jang SH, Nielsen LK, Kim J, Jung KS (2008) Fermentative butanol production by clostridia. *Biotechnol Bioeng* 101(2):209–228. https://doi.org/https://doi. org/10.1002/bit.22003

Li J, Li Z, Xiao S, Fu Q, Kobayashi H, Zhang L, Liao Q, Zhu X (2020) Startup cathode potentials determine electron transfer behaviours of biocathodes catalysing CO_2 reduction to CH_4 in microbial electrosynthesis. *J CO2 Util* 35(June):169–175. https://doi. org/10.1016/j.jcou.2019.09.013

Li L, Zhang X, Zhu P, Yong X, Wang Y, An W, Jia H, Zhou J (2021) Enhancing biomethane production and pyrene biodegradation by addition of bio-nano FeS or magnetic carbon during sludge anaerobic digestion. *Environ Technol* 42(22):3496–3507. https://doi.org/1 0.1080/09593330.2020.1733674

Li S, Huang L, Ke C, Pang Z, Liu L (2020) Pathway dissection, regulation, engineering and application: Lessons learned from biobutanol production by solventogenic clostridia. *Biotechnol Biofuels* 13(1):39. https://doi.org/10.1186/s13068-020-01674-3

Li S, Sakuntala M, Song YE, Heo J ook, Kim M, Lee SY, Kim MS, Oh YK, Kim JR (2021) Photoautotrophic hydrogen production of *Rhodobacter sphaeroides* in a microbial electrosynthesis cell. *Bioresour Technol* 320:124333. https://doi.org/10.1016/J. BIORTECH.2020.124333

Li Y, Zhang Y, Yang Y, Quan X, Zhao Z (2017) Potentially direct interspecies electron transfer of methanogenesis for syntrophic metabolism under sulfate reducing conditions with stainless steel. *Bioresour Technol* 234:303–309. https://doi.org/https://doi.org/10.1016/j. biortech.2017.03.054

Li Z, Fu Q, Kobayashi H, Xiao S, Li J, Zhang L, Liao Q, Zhu X (2019) Polarity reversal facilitates the development of biocathodes in microbial electrosynthesis systems for biogas production. *Int J Hydrogen Energy* 44(48):26226–26236. https://doi.org/10.1016/j. ijhydene.2019.08.117

Li Z, Han C, Gu T (2018) Economics of biomass gasification: A review of the current status. *Energy Sources Part B Econ Plan Policy* 13(2):137–140. https://doi.org/10.1080/15567 249.2017.1410593

Liakakou ET, Infantes A, Neumann A, Vreugdenhil BJ (2021) Connecting gasification with syngas fermentation: Comparison of the performance of lignin and beech wood. *Fuel* 290:120054. https://doi.org/https://doi.org/10.1016/j.fuel.2020.120054

Lim JW, Park T, Tong YW, Yu Z (2020) The microbiome driving anaerobic digestion and microbial analysis, in: Li Y, Khanal SK (Eds.), *Advances in Bioenergy*. Elsevier, pp. 1–61. https://doi.org/10.1016/bs.aibe.2020.04.001

Lin C-Y, Wu C-C, Hung C-H (2008) Temperature effects on fermentative hydrogen production from xylose using mixed anaerobic cultures. *Int J Hydrogen Energy* 33(1):43–50. https:// doi.org/https://doi.org/10.1016/j.ijhydene.2007.09.001

Liu X, Zhu Y, Yang S-T (2006) Construction and characterization of ack deleted mutant of *Clostridium tyrobutyricum* for enhanced butyric acid and hydrogen production. *Biotechnol Prog* 22(5):1265–1275. https://doi.org/https://doi.org/10.1021/bp060082g

Liu Y, Balkwill DL, Aldrich HC, Drake GR, Boone DR (1999) Characterization of the anaerobic propionate-degrading syntrophs *Smithella propionica* gen. nov., sp. nov. and *Syntrophobacter wolinii*. *Int J Syst Evol Microbiol* 49(2):545–556. https://doi.org/https://doi.org/10.1099/00207713-49-2-545

Liu ZD, Du ZW, Lian J, Zhu XY, Li SH, Li HR (2007) Improving energy accumulation of microbial fuel cells by metabolism regulation using *Rhodoferax ferrireducens* as biocatalyst. *Lett Appl Microbiol* 44(4):393–398. https://doi.org/10.1111/J.1472-765X.2006.02088.X

Liu ZD, Li HR (2007) Effects of bio- and abio-factors on electricity production in a mediatorless microbial fuel cell. *Biochem Eng J* 36(3):209–214. https://doi.org/10.1016/J.BEJ.2007.02.021

Logan BE, Oh S-E, Kim IS, Van Ginkel S (2002) Biological hydrogen production measured in batch anaerobic respirometers. *Environ Sci Technol* 36(11):2530–2535. https://doi.org/10.1021/es015783i

Logan BE, Rossi R, Ragab A, Saikaly PE (2019) Electroactive microorganisms in bioelectrochemical systems. *Nat Rev Microbiol* 17(5):307–319. https://doi.org/10.1038/s41579-019-0173-x

Lu H, Zhang G, He S, Zhao R, Zhu D (2021) Purple non-sulfur bacteria technology: A promising and potential approach for wastewater treatment and bioresources recovery. *World J Microbiol Biotechnol* 37(9):161. https://doi.org/10.1007/s11274-021-03133-z

Luo G, Angelidaki I (2012) Integrated biogas upgrading and hydrogen utilization in an anaerobic reactor containing enriched hydrogenotrophic methanogenic culture. *Biotechnol Bioeng* 109(11):2729–2736. https://doi.org/https://doi.org/10.1002/bit.24557

Luo G, Johansson S, Boe K, Xie L, Zhou Q, Angelidaki I (2012) Simultaneous hydrogen utilization and in situ biogas upgrading in an anaerobic reactor. *Biotechnol Bioeng* 109(4):1088–1094. https://doi.org/https://doi.org/10.1002/bit.24360

Luo G, Wang W, Angelidaki I (2013) Anaerobic digestion for simultaneous sewage sludge treatment and CO biomethanation: Process performance and microbial ecology. *Environ Sci Technol* 47(18):10685–10693. https://doi.org/10.1021/es401018d

Luo H, Ge L, Zhang J, Zhao Y, Ding J, Li Z, He Z, Chen R, Shi Z (2015) Enhancing butanol production under the stress environments of co-culturing *Clostridium acetobutylicum/Saccharomyces cerevisiae* integrated with exogenous butyrate addition. *PLoS One* 10(10):e0141160. https://doi.org/10.1371/journal.pone.0141160

Luo L, Xu S, Jin Y, Han R, Liu H, Lü F (2018) Evaluation of methanogenic microbial electrolysis cells under closed/open circuit operations. *Environ Technol (UK)* 39(6):739–748. https://doi.org/10.1080/09593330.2017.1310934

Lynd LR, Weimer PJ, van Zyl WH, Pretorius IS (2002) Microbial cellulose utilization: Fundamentals and biotechnology. *Microbiol Mol Biol Rev* 66(3):506–577. https://doi.org/10.1128/MMBR.66.3.506-577.2002

Ma K, Liu X, Dong X (2005) *Methanobacterium beijingense* sp. nov., a novel methanogen isolated from anaerobic digesters. *Int J Syst Evol Microbiol* 55(1):325–329. https://doi.org/https://doi.org/10.1099/ijs.0.63254-0

Maddox IS (1989) The acetone-butanol-ethanol fermentation: Recent progress in technology. *Biotechnol Genet Eng Rev* 7(1):189–220.

Maddox IS, Steiner E, Hirsch S, Wessner S, Gutierrez NA, Gapes JR, Schuster KC (2000) The cause of "acid crash" and "acidogenic fermentations" during the batch acetone-butanol-ethanol (ABE-) fermentation process. *J Mol Microbiol Biotechnol* 2(1):95–100.

Malki M, De Lacey AL, Rodríguez N, Amils R, Fernandez VM (2008) Preferential Use of an Anode as an Electron Acceptor by an Acidophilic Bacterium in the Presence of Oxygen. *Appl Environ Microbiol* 74(14), 4472–4476. https://doi.org/10.1128/AEM.00209-08

Mañunga T, Barrios-Pérez JD, Zaiat M, Rodríguez-Victoria JA (2019) Evaluation of pretreatment methods and initial pH on mixed inoculum for fermentative hydrogen production from cassava wastewater. *Biofuels* 1–8. https://doi.org/10.1080/17597269.2019 .1680041

Martínez EJ, Blanco D, Gómez X (2019a) Two-stage process to enhance bio-hydrogen production, in: Treichel H, Fongaro G (Eds.), *Improving Biogas Production*. Springer International Publishing, Cham, pp. 149–179. https://doi.org/10.1007/978-3-030-10516-7_7

Martínez EJ, Sotres A, Arenas CB, Blanco D, Martínez O, Gómez X (2019b) Improving anaerobic digestion of sewage sludge by hydrogen addition: Analysis of microbial populations and process performance. *Energies* 12(7):1128. https://doi.org/10.3390/en12071228

Mateos R, Escapa A, San-Martín MI, De Wever H, Sotres A, Pant D (2020) Long-term open circuit microbial electrosynthesis system promotes methanogenesis. *J Energy Chem* 41:3–6. https://doi.org/10.1016/j.jechem.2019.04.020

Mateos R, Sotres A, Alonso RM, Escapa A, Morán A (2018) Impact of the start-up process on the microbial communities in biocathodes for electrosynthesis. *Bioelectrochemistry* 121:27–37. https://doi.org/10.1016/J.BIOELECHEM.2018.01.002

May HD, Evans PJ, LaBelle EV (2016) The bioelectrosynthesis of acetate. *Curr Opin Biotechnol* 42:225–233. https://doi.org/10.1016/J.COPBIO.2016.09.004

Mayer F, Coughlan MP, Mori Y, Ljungdahl LG (1987) Macromolecular organization of the cellulolytic enzyme complex of *Clostridium thermocellum* as revealed by electron microscopy. *Appl Environ Microbiol* 53(12):2785–2792. https://doi.org/10.1128/ aem.53.12.2785-2792.1987

Milán Z, Montalvo S, Ruiz-Tagle N, Urrutia H, Chamy R, Sánchez E, Borja R (2010) Influence of heavy metal supplementation on specific methanogenic activity and microbial communities detected in batch anaerobic digesters. *J Environ Sci Heal Part A* 45(11): 1307–1314. https://doi.org/10.1080/10934529.2010.500878

Monroy CI, Zlatev R, Stoytcheva M, González RER, Valdez B, Gochev V (2013) Light spectra and luminosity influence on photosynthetic hydrogen production by *Rhodobacter Capsulatus*. *Biotechnol Biotechnol Equip* 27(1):3513–3517. https://doi.org/10.5504/ BBEQ.2012.0130

Moreno R, Fierro J, Fernández C, Cuetos MJ, Gómez X (2015) Biohydrogen production from lactose: Influence of substrate and nitrogen concentration. *Environ Technol* 36(19): 2401–2409. https://doi.org/10.1080/09593330.2015.1032365

Moreno R, Gómez X (2012) Dark fermentative H$_2$ production from wastes: Effect of operating conditions. *J Environ Sci Eng A* 1(7A):936.

Munasinghe PC, Khanal SK (2010) Biomass-derived syngas fermentation into biofuels: Opportunities and challenges. *Bioresour Technol* 101(13):5013–5022. https://doi.org/ https://doi.org/10.1016/j.biortech.2009.12.098

Nevin KP, Richter H, Covalla SF, Johnson JP, Woodard TL, Orloff AL, Jia H, Zhang M, Lovley DR (2008) Power output and columbic efficiencies from biofilms of *Geobacter sulfurreducens* comparable to mixed community microbial fuel cells. *Environ Microbiol* 10(10):2505–2514. https://doi.org/10.1111/J.1462-2920.2008.01675.X

Nissilä ME, Tähti HP, Rintala JA, Puhakka JA (2011) Thermophilic hydrogen production from cellulose with rumen fluid enrichment cultures: Effects of different heat treatments. *Int J Hydrogen Energy* 36(2):1482–1490. https://doi.org/https://doi.org/10.1016/j. ijhydene.2010.11.010

Ntaikou I, Gavala HN, Kornaros M, Lyberatos G (2008) Hydrogen production from sugars and sweet sorghum biomass using *Ruminococcus albus*. *Int J Hydrogen Energy* 33(4): 1153–1163. https://doi.org/https://doi.org/10.1016/j.ijhydene.2007.10.053

Osman AI, Deka TJ, Baruah DC, Rooney DW (2020) Critical challenges in biohydrogen production processes from the organic feedstocks. *Biomass Convers Biorefinery* 1–19. https://doi.org/10.1007/s13399-020-00965-x

Ozturk AB, Al-Shorgani NKN, Cheng S, Arasoglu T, Gulen J, Habaki H, Egashira R, Kalil MS, Yusoff WMW, Cross JS (2020) Two-step fermentation of cooked rice with *Aspergillus oryzae* and *Clostridium acetobutylicum* YM1 for biobutanol production. *Biofuels* 1–7. https://doi.org/10.1080/17597269.2020.1813000

Pachapur VL, Sarma SJ, Brar SK, Le Bihan Y, Buelna G, Verma M (2015) Biological hydrogen production using co-culture versus mono-culture system. *Environ Technol Rev* 4(1): 55–70. https://doi.org/10.1080/21622515.2015.1068381

Pacheco M, Pinto F, Ortigueira J, Silva C, Gírio F, Moura P (2021) Lignin syngas bioconversion by *Butyribacterium methylotrophicum*: Advancing towards an integrated biorefinery. *Energies* 14(21):7124. https://doi.org/10.3390/en14217124

Paniagua-García AI, Hijosa-Valsero M, Díez-Antolínez R, Sánchez ME, Coca M (2018) Enzymatic hydrolysis and detoxification of lignocellulosic biomass are not always necessary for ABE fermentation: The case of Panicum virgatum. *Biomass Bioenergy* 116:131–139. https://doi.org/https://doi.org/10.1016/j.biombioe.2018.06.006

Parameswaran P, Zhang H, Torres CI, Rittmann BE, Krajmalnik-Brown R (2010) Microbial community structure in a biofilm anode fed with a fermentable substrate: The significance of hydrogen scavengers. *Biotechnol Bioeng* 105(1):69–78. https://doi.org/10.1002/BIT.22508

Park S, Baker JO, Himmel ME, Parilla PA, Johnson DK (2010) Cellulose crystallinity index: Measurement techniques and their impact on interpreting cellulase performance. *Biotechnol Biofuels* 3(1):10. https://doi.org/10.1186/1754-6834-3-10

Patankar S, Dudhane A, Paradh AD, Patil S (2021) Improved bioethanol production using genome-shuffled *Clostridium ragsdalei* (DSM 15248) strains through syngas fermentation. *Biofuels* 12(1):81–89. https://doi.org/10.1080/17597269.2018.1457313

Pecorini I, Peruzzi E, Albini E, Doni S, Macci C, Masciandaro G, Iannelli R (2020) Evaluation of MSW compost and digestate mixtures for a circular economy application. *Sustain* 12(7):3042. https://doi.org/10.3390/su12073042

Pelaz G, Carrillo-Peña D, Morán A, Escapa A (2022) Electromethanogenesis at medium-low temperatures: Impact on performance and sources of variability. *Fuel* 310:122336. https://doi.org/10.1016/J.FUEL.2021.122336

Phillips JR, Atiyeh HK, Tanner RS, Torres JR, Saxena J, Wilkins MR, Huhnke RL (2015) Butanol and hexanol production in *Clostridium carboxidivorans* syngas fermentation: Medium development and culture techniques. *Bioresour Technol* 190:114–121. https://doi.org/https://doi.org/10.1016/j.biortech.2015.04.043

Pinto T, Flores-Alsina X, Gernaey K V, Junicke H (2021) Alone or together? A review on pure and mixed microbial cultures for butanol production. *Renew Sustain Energy Rev* 147:111244. https://doi.org/https://doi.org/10.1016/j.rser.2021.111244

Qin Y, Li L, Wu J, Xiao B, Hojo T, Kubota K, Cheng J, Li Y-Y (2019b) Co-production of biohydrogen and biomethane from food waste and paper waste via recirculated two-phase anaerobic digestion process: Bioenergy yields and metabolic distribution. *Bioresour Technol* 276:325–334. https://doi.org/https://doi.org/10.1016/j.biortech.2019.01.004

Qin Y, Wu J, Xiao B, Cong M, Hojo T, Cheng J, Li Y-Y (2019a) Strategy of adjusting recirculation ratio for biohythane production via recirculated temperature-phased anaerobic digestion of food waste. *Energy* 179:1235–1245. https://doi.org/https://doi.org/10.1016/j.energy.2019.04.182

Rabaey K (2009) Bioelectrochemical systems: From extracellular electron transfer to biotechnological application. *Water Intell*. Online 8. https://doi.org/10.2166/9781780401621

Rachman MA, Furutani Y, Nakashimada Y, Kakizono T, Nishio N (1997) Enhanced hydrogen production in altered mixed acid fermentation of glucose by *Enterobacter aerogenes*. *J Ferment Bioeng* 83(4):358–363. https://doi.org/https://doi.org/10.1016/S0922-338X(97)80142-0

Ragab A, Shaw DR, Katuri KP, Saikaly PE (2020) Effects of set cathode potentials on microbial electrosynthesis system performance and biocathode methanogen function at a metatranscriptional level. *Sci Rep* 10:1–15. https://doi.org/10.1038/s41598-020-76229-5

Rago L, Ruiz Y, Baeza JA, Guisasola A, Cortés P (2015) Microbial community analysis in a long-term membrane-less microbial electrolysis cell with hydrogen and methane production. *Bioelectrochemistry* 106:359–368. https://doi.org/https://doi.org/10.1016/j.bioelechem.2015.06.003

Redondas V, Moran A, Martínez JE, Fierro J, Gomez X (2015) Effect of methanogenic effluent recycling on continuous H$_2$ production from food waste. *Environ Prog Sustain Energy* 34(1):227–233. https://doi.org/https://doi.org/10.1002/ep.11980

Richter H, Loftus SE, Angenent LT (2013) Integrating syngas fermentation with the carboxylate platform and yeast fermentation to reduce medium cost and improve biofuel productivity. *Environ Technol* 34(13–14):1983–1994. https://doi.org/10.1080/09593330.2013.826255

Richter H, Molitor B, Diender M, Sousa DZ, Angenent LT (2016b) A narrow pH range supports butanol, hexanol, and octanol production from syngas in a continuous co-culture of *Clostridium ljungdahlii* and *Clostridium kluyveri* with in-line product extraction. *Front Microbiol* 7. https://www.frontiersin.org/article/10.3389/fmicb.2016.01773

Richter H, Molitor B, Wei H, Chen W, Aristilde L, Angenent LT (2016a) Ethanol production in syngas-fermenting *Clostridium ljungdahlii* is controlled by thermodynamics rather than by enzyme expression. *Energy Environ Sci* 9(7):2392–2399. https://doi.org/10.1039/C6EE01108J

Rodríguez A, Hernández-Herreros N, García JL, Auxiliadora Prieto M (2021) Enhancement of biohydrogen production rate in *Rhodospirillum rubrum* by a dynamic CO-feeding strategy using dark fermentation. *Biotechnol Biofuels* 14(1):168. https://doi.org/10.1186/s13068-021-02017-6

Rosa D, Medeiros ABP, Martinez-Burgos WJ, do Nascimento JR, de Carvalho JC, Sydney EB, Soccol CR (2020) Biological hydrogen production from palm oil mill effluent (POME) by anaerobic consortia and *Clostridium beijerinckii*. *J Biotechnol* 323:17–23. https://doi.org/https://doi.org/10.1016/j.jbiotec.2020.06.015

Ross BS, Pott RWM (2022) Investigating and modeling the effect of light intensity on *Rhodopseudomonas palustris* growth. *Biotechnol Bioeng* 119(3):907–921. https://doi.org/https://doi.org/10.1002/bit.28026

Rozendal RA, Jeremiasse AW, Hamelers HVM, Buisman CJN (2008) Hydrogen production with a microbial biocathode. *Envioromental Sci Technol* 42(2):629–634. https://doi.org/10.1021/es071720

Rückel A, Hannemann J, Maierhofer C, Fuchs A, Weuster-Botz D 2021. Studies on syngas fermentation with *Clostridium carboxidivorans* in stirred-tank reactors with defined gas impurities. *Front Microbiol* 12:655390. https://doi.org/10.3389/fmicb.2021.655390

Sali S, Mackey HR (2021) The application of purple non-sulfur bacteria for microbial mixed culture polyhydroxyalkanoates production. *Rev Environ Sci Bio/Technol* 20(4):959–983. https://doi.org/10.1007/s11157-021-09597-7

Sansaniwal SK, Pal K, Rosen MA, Tyagi SK (2017) Recent advances in the development of biomass gasification technology: A comprehensive review. *Renew Sustain Energy Rev* 72:363–384. https://doi.org/https://doi.org/10.1016/j.rser.2017.01.038

Sasaki D, Sasaki K, Tsuge Y, Kondo A (2019) Less biomass and intracellular glutamate in anodic biofilms lead to efficient electricity generation by microbial fuel cells. *Biotechnol Biofuels* 12(1):1–11. https://doi.org/10.1186/S13068-019-1414-Y/FIGURES/6

Schnürer A 2016. Biogas production: Microbiology and technology, in: Hatti-Kaul R, Mamo G, Mattiasson B (Eds.), *Anaerobes in Biotechnology*. Springer International Publishing, Cham, pp. 195–234. https://doi.org/10.1007/10_2016_5

Schwarz W (2001) The cellulosome and cellulose degradation by anaerobic bacteria. *Appl Microbiol Biotechnol* 56(5):634–649. https://doi.org/10.1007/s002530100710

Sethupathy A, Arun C, Ravi Teja G, Sivashanmugam P (2019) Enhancing hydrogen production through anaerobic co-digestion of fruit waste with biosolids. *J Environ Sci Heal Part A* 54(6):563–569. https://doi.org/10.1080/10934529.2019.1571320

Sevillano CA, Pesantes AA, Peña Carpio E, Martínez EJ, Gómez X (2021) Anaerobic digestion for producing renewable energy—The evolution of this technology in a new uncertain scenario. *Entropy* 23(2):145. https://doi.org/10.3390/e23020145

Sharma M, Das PP, Sood T, Chakraborty A, Purkait MK (2021) Ameliorated polyvinylidene fluoride based proton exchange membrane impregnated with graphene oxide, and cellulose acetate obtained from sugarcane bagasse for application in microbial fuel cell. *J Environ Chem Eng* 9(6):106681. https://doi.org/10.1016/J.JECE.2021.106681

Shen S, Wang G, Zhang M, Tang Y, Gu Y, Jiang W, Wang Y, Zhuang Y (2020) Effect of temperature and surfactant on biomass growth and higher-alcohol production during syngas fermentation by *Clostridium carboxidivorans* P7. *Bioresour Bioprocess* 7(1):56. https://doi.org/10.1186/s40643-020-00344-4

Shen Y, Brown R, Wen Z (2014) Syngas fermentation of *Clostridium carboxidivoran* P7 in a hollow fiber membrane biofilm reactor: Evaluating the mass transfer coefficient and ethanol production performance. *Biochem Eng J* 85:21–29. https://doi.org/https://doi.org/10.1016/j.bej.2014.01.010

Shin Y-A, Choi S, Han M (2021) Simultaneous fermentation of mixed sugar by a newly isolated *Clostridium beijerinckii* GSC1. *Biotechnol Bioprocess Eng* 26(1):137–144. https://doi.org/10.1007/s12257-020-0183-6

Sleutels THJA, Ter Heijne A, Buisman CJN, Hamelers HVM (2012) Bioelectrochemical systems: An outlook for practical applications. *ChemSusChem* 5(6):1012–1019. https://doi.org/10.1002/CSSC.201100732

Sogani M, Pankan AO, Dongre A, Yunus K, Fisher AC (2021) Augmenting the biodegradation of recalcitrant ethinylestradiol using *Rhodopseudomonas palustris* in a hybrid photo-assisted microbial fuel cell with enhanced bio-hydrogen production. *J Hazard Mater* 408:124421. https://doi.org/https://doi.org/10.1016/j.jhazmat.2020.124421

Sousa DZ, Smidt H, Alves MM, Stams AJM (2007) *Syntrophomonas zehnderi* sp. nov., an anaerobe that degrades long-chain fatty acids in co-culture with Methanobacterium formicicum. *Int J Syst Evol Microbiol* 57(3):609–615. https://doi.org/https://doi.org/10.1099/ijs.0.64734-0

Stolze Y, Bremges A, Rumming M, Henke C, Maus I, Pühler A, Sczyrba A, Schlüter A (2016) Identification and genome reconstruction of abundant distinct taxa in microbiomes from one thermophilic and three mesophilic production-scale biogas plants. *Biotechnol Biofuels* 9(1):156. https://doi.org/10.1186/s13068-016-0565-3

Sträuber H, Lucas R, Kleinsteuber S (2016) Metabolic and microbial community dynamics during the anaerobic digestion of maize silage in a two-phase process. *Appl Microbiol Biotechnol* 100(1):479–491. https://doi.org/10.1007/s00253-015-6996-0

Stürmer B, Pfundtner E, Kirchmeyr F, Uschnig S (2020) Legal requirements for digestate as fertilizer in Austria and the European Union compared to actual technical parameters. *J Environ Manage* 253 :109756. https://doi.org/https://doi.org/10.1016/j.jenvman.2019.109756

Summers ZM, Fogarty HE, Leang C, Franks AE, Malvankar NS, Lovley DR (2010) Direct exchange of electrons within aggregates of an evolved syntrophic coculture of anaerobic bacteria. *Science* 330(6009):1413–1415. https://doi.org/10.1126/SCIENCE.1196526

Sun X, Atiyeh HK, Huhnke RL, Tanner RS (2019a) Syngas fermentation process development for production of biofuels and chemicals: A review. *Bioresour Technol Reports* 7:100279. https://doi.org/https://doi.org/10.1016/j.biteb.2019.100279

Sun X, Atiyeh HK, Zhang H, Tanner RS, Huhnke RL (2019b) Enhanced ethanol production from syngas by *Clostridium ragsdalei* in continuous stirred tank reactor using medium with poultry litter biochar. *Appl Energy* 236:1269–1279. https://doi.org/https://doi.org/10.1016/j.apenergy.2018.12.010

Tahir K, Miran W, Jang J, Woo SH, Lee DS (2021) Enhanced product selectivity in the microbial electrosynthesis of butyrate using a nickel ferrite-coated biocathode. *Environ Res* 196:110907. https://doi.org/10.1016/J.ENVRES.2021.110907

Tartakovsky B, Manuel MF, Wang H, Guiot SR (2009) High rate membrane-less microbial electrolysis cell for continuous hydrogen production. *Int J Hydrogen Energy* 34(2): 672–677. https://doi.org/10.1016/J.IJHYDENE.2008.11.003

Tashiro Y, Yoshida T, Noguchi T, Sonomoto K (2013) Recent advances and future prospects for increased butanol production by acetone-butanol-ethanol fermentation. *Eng Life Sci* 13(5):432–445. https://doi.org/https://doi.org/10.1002/elsc.201200128

Theiri M, Chadjaa H, Marinova M, Jolicoeur M (2019) Combining chemical flocculation and bacterial co-culture of *Cupriavidus taiwanensis* and *Ureibacillus thermosphaericus* to detoxify a hardwood hemicelluloses hydrolysate and enable acetone–butanol–ethanol fermentation leading to butanol. *Biotechnol Prog* 35(2):e2753. https://doi.org/https://doi.org/10.1002/btpr.2753

Thi HN, Park S, Li H, Kim Y-K (2020) Medium compositions for the improvement of productivity in syngas fermentation with *Clostridium autoethanogenum*. *Biotechnol Bioprocess Eng* 25(3):493–501. https://doi.org/10.1007/s12257-019-0428-4

Tomas CA, Welker NE, Papoutsakis ET (2003) Overexpression of groESL in *Clostridium acetobutylicum* results in increased solvent production and tolerance, prolonged metabolism, and changes in the cell's transcriptional program. *Appl Environ Microbiol* 69(8):4951–4965. https://doi.org/10.1128/AEM.69.8.4951-4965.2003

Torres CI, Krajmalnik-Brown R, Parameswaran P, Marcus AK, Wanger G, Gorby YA, Rittmann BE (2009) Selecting anode-respiring bacteria based on anode potential: Phylogenetic, electrochemical, and microscopic characterization. *Environ Sci Technol* 43(24), 9519–9524. https://doi.org/10.1021/ES902165Y/SUPPL_FILE/ES902165Y_SI_001.PDF

Treu L, Campanaro S, Kougias PG, Zhu X, Angelidaki I (2016) Untangling the effect of fatty acid addition at species level revealed different transcriptional responses of the biogas microbial community members. *Environ Sci Technol* 50(11):6079–6090. https://doi.org/10.1021/acs.est.6b00296

Turon V, Ollivier S, Cwicklinski G, Willison JC, Anxionnaz-Minvielle Z (2021) H$_2$ production by photofermentation in an innovative plate-type photobioreactor with meandering channels. *Biotechnol Bioeng* 118(3):1342–1354. https://doi.org/https://doi.org/10.1002/bit.27656

Valdez-Vazquez I, Sanchez A (2018) Proposal for biorefineries based on mixed cultures for lignocellulosic biofuel production: A techno-economic analysis. *Biofuels Bioprod Biorefining* 12(1):56–67. https://doi.org/https://doi.org/10.1002/bbb.1828

Vavilin VA, Fernandez B, Palatsi J, Flotats X (2008) Hydrolysis kinetics in anaerobic degradation of particulate organic material: An overview. *Waste Manag* 28(6):939–951. https://doi.org/https://doi.org/10.1016/j.wasman.2007.03.028

Venkiteshwaran K, Bocher B, Maki J, Zitomer D (2015) Relating anaerobic digestion microbial community and process function: Supplementary issue: Water microbiology. *Microbiol Insights* 8s2:MBI.S33593. https://doi.org/10.4137/MBI.S33593

Viana MB, Dams RI, Pinheiro BM, Leitão RC, Santaella ST, dos Santos AB (2019) The source of inoculum and the method of methanogenesis inhibition can affect biological hydrogen production from crude glycerol. *BioEnergy Res* 12(3):733–742. https://doi.org/10.1007/s12155-019-09994-5

Vongkampang T, Sreenivas K, Engvall J, Grey C, van Niel EWJ (2021) Characterization of simultaneous uptake of xylose and glucose in *Caldicellulosiruptor kronotskyensis* for optimal hydrogen production. *Biotechnol Biofuels* 14(1):91. https://doi.org/10.1186/s13068-021-01938-6

Wainaina S, Lukitawesa, Kumar Awasthi M, Taherzadeh MJ (2019) Bioengineering of anaerobic digestion for volatile fatty acids, hydrogen or methane production: A critical review. *Bioengineered* 10(1):437–458. https://doi.org/10.1080/21655979.2019.1673937

Wang J, Yin Y (2019) Progress in microbiology for fermentative hydrogen production from organic wastes. *Crit Rev Environ Sci Technol* 49(10):825–865. https://doi.org/10.1080/10643389.2018.1487226

Wang Y, Wang Z, Zhang Q, Li G, Xia C (2020) Comparison of bio-hydrogen and bio-methane production performance in continuous two-phase anaerobic fermentation system between co-digestion and digestate recirculation. *Bioresour Technol* 318:124269. https://doi.org/https://doi.org/10.1016/j.biortech.2020.124269

Wang Z, Lee T, Lim B, Choi C, Park J (2014) Microbial community structures differentiated in a single-chamber air-cathode microbial fuel cell fueled with rice straw hydrolysate. *Biotechnol Biofuels* 7. https://doi.org/10.1186/1754-6834-7-9

Wen Z, Wu M, Lin Y, Yang L, Lin J, Cen P (2014) Artificial symbiosis for acetone-butanol-ethanol (ABE) fermentation from alkali extracted deshelled corn cobs by co-culture of *Clostridium beijerinckii* and *Clostridium cellulovorans*. *Microb Cell Fact* 13(1):92. https://doi.org/10.1186/s12934-014-0092-5

Willquist K, Zeidan AA, van Niel EWJ (2010) Physiological characteristics of the extreme thermophile *Caldicellulosiruptor saccharolyticus*: An efficient hydrogen cell factory. *Microb Cell Fact* 9(1):89. https://doi.org/10.1186/1475-2859-9-89

Wongthanate J, Chinnacotpong K, Khumpong M (2014) Impacts of pH, temperature, and pretreatment method on biohydrogen production from organic wastes by sewage microflora. *Int J Energy Environ Eng* 5(1):6. https://doi.org/10.1186/2251-6832-5-6

Wu J, Dong L, Zhou C, Liu B, Feng L, Wu C, Qi Z, Cao G (2019) Developing a coculture for enhanced butanol production by *Clostridium beijerinckii* and *Saccharomyces cerevisiae*. *Bioresour Technol Reports* 6:223–228. https://doi.org/https://doi.org/10.1016/j.biteb.2019.03.006

Wu P, Wang G, Wang G, Børresen BT, Liu H, Zhang J (2016) Butanol production under microaerobic conditions with a symbiotic system of *Clostridium acetobutylicum* and *Bacillus cereus*. *Microb Cell Fact* 15(1):8. https://doi.org/10.1186/s12934-016-0412-z

Wu X, Ma G, Liu C, Qiu X, Min L, Kuang J, Zhu L (2021) Biosynthesis of pinene in purple non-sulfur photosynthetic bacteria. *Microb Cell Fact* 20(1):101. https://doi.org/10.1186/s12934-021-01591-6

Xiao B, Liu J (2009) Biological hydrogen production from sterilized sewage sludge by anaerobic self-fermentation. *J Hazard Mater* 168(1):163–167. https://doi.org/https://doi.org/10.1016/j.jhazmat.2009.02.008

Xin X, Chen BY, Hong J (2019) Unraveling interactive characteristics of microbial community associated with bioelectric energy production in sludge fermentation fluid-fed microbial fuel cells. *Bioresour Technol* 289:121652. https://doi.org/10.1016/J.BIORTECH.2019.121652

Xing D, Cheng S, Regan JM, Logan BE (2009) Change in microbial communities in acetate- and glucose-fed microbial fuel cells in the presence of light. *Biosens Bioelectron* 25(1):105–111. https://doi.org/10.1016/J.BIOS.2009.06.013

Xing T, Kong X, Dong P, Zhen F, Sun Y (2020) Leachate recirculation effects on solid-state anaerobic digestion of *Pennisetum hybrid* and microbial community analysis. *J Chem Technol Biotechnol* 95(4):1216–1224. https://doi.org/https://doi.org/10.1002/jctb.6310

Xu H, Liang C, Yuan Z, Xu J, Hua Q, Guo Y (2017) A study of CO/syngas bioconversion by *Clostridium autoethanogenum* with a flexible gas-cultivation system. *Enzyme Microb Technol* 101 :24–29. https://doi.org/https://doi.org/10.1016/j.enzmictec.2017.03.002

Xue C, Zhao J, Lu C, Yang S-T, Bai F, Tang I-C (2012) High-titer n-butanol production by *Clostridium acetobutylicum* JB200 in fed-batch fermentation with intermittent gas stripping. *Biotechnol Bioeng* 109(11):2746–2756. https://doi.org/https://doi.org/10.1002/bit.24563

Yang G, Wang J (2018) Synergistic biohydrogen production from flower wastes and sewage sludge. *Energy Fuels* 32(6):6879–6886. https://doi.org/10.1021/acs.energyfuels.8b01122

Yang G, Wang J (2019) Changes in microbial community structure during dark fermentative hydrogen production. *Int J Hydrogen Energy* 44(47):25542–25550. https://doi.org/https://doi.org/10.1016/j.ijhydene.2019.08.039

Yang Z, Liu Y, Zhang J, Mao K, Kurbonova M, Liu G, Zhang R, Wang W (2020) Improvement of biofuel recovery from food waste by integration of anaerobic digestion, digestate pyrolysis and syngas biomethanation under mesophilic and thermophilic conditions. *J Clean Prod* 256:120594. https://doi.org/https://doi.org/10.1016/j.jclepro.2020.120594

Ye Y, Wang L, Chen Y, Zhu S, Shen S (2010) High yield hydrogen production in a single-chamber membrane-less microbial electrolysis cell. *Water Sci Technol* 61(3):721–727. https://doi.org/10.2166/WST.2010.900

Yi J, Dong B, Jin J, Dai X 2014. Effect of increasing total solids contents on anaerobic digestion of food waste under mesophilic conditions: Performance and microbial characteristics analysis. *PLoS One* 9(7):e102548. https://doi.org/10.1371/journal.pone.0102548

Yin Y, Hu J, Wang J (2014) Enriching hydrogen-producing bacteria from digested sludge by different pretreatment methods. *Int J Hydrogen Energy* 39(25):13550–13556. https://doi.org/https://doi.org/10.1016/j.ijhydene.2014.01.145

Yousaf S, Anam M, Saeed S, Ali N (2017) Electricigens: Source, enrichment and limitations. *Environ Technol Rev* 6(1):117–134. https://doi.org/10.1080/21622515.2017.1318182

Zhang B, Zhang L-L, Zhang S-C, Shi H-Z, Cai W-M (2005) The influence of pH on hydrolysis and acidogenesis of kitchen wastes in two-phase anaerobic digestion. *Environ Technol* 26(3):329–340. https://doi.org/10.1080/09593332608618563

Zhang C, Ling Z, Huo S (2021a) Anaerobic fermentation of pretreated food waste for butanol production by co-cultures assisted with in-situ extraction. *Bioresour Technol Reports* 16:100852. https://doi.org/https://doi.org/10.1016/j.biteb.2021.100852

Zhang S, Jiang J, Wang H, Li F, Hua T, Wang W (2021b) A review of microbial electrosynthesis applied to carbon dioxide capture and conversion: The basic principles, electrode materials, and bioproducts. *J CO2 Util* 51:101640. https://doi.org/10.1016/J.JCOU.2021.101640

Zhiqiang W, Rodrigo L-A, Jianping L, Yu J, Sheng Y, Haruyuki A (2022) Improved n-butanol production from *Clostridium cellulovorans* by integrated metabolic and evolutionary engineering. *Appl Environ Microbiol* 85(7):e02560-18. https://doi.org/10.1128/AEM.02560-18

Zhou J, Zhou Y, You X, Zhang H, Gong L, Wang J, Zuo T (2021) Potential promotion of activated carbon supported nano zero-valent iron on anaerobic digestion of waste activated sludge. *Environ Technol* 1–14. https://doi.org/10.1080/09593330.2021.1924290

Zhou M, Freguia S, Dennis PG, Keller J, Rabaey K (2015) Development of bioelectrocatalytic activity stimulates mixed-culture reduction of glycerol in a bioelectrochemical system. *Microb Biotechnol* 8(3):483. https://doi.org/10.1111/1751-7915.12240

8 Applications of microbially synthesised nanoparticles in food sciences

Srinivasan Kameswaran, Vijaya Sudhakara Rao Kola, and Bellamkonda Ramesh

CONTENTS

8.1 INTRODUCTION

Applications for nanoparticles in industry and medicine have grown recently. Nanoparticles are characterised as having a size between 10 and 1,000 nm (Arshad, 2017). However, due to easier penetration and similar sizes to biomolecules, it is generally believed that materials smaller than 100 nm are beneficial for applications. Nanomaterials' reduced size offers a wide range of biological research opportunities.

DOI: 10.1201/9781003394600-9

Nanomaterials can interact with complicated biological systems in novel ways because their dimensions are similar to those of biomolecules. Design and development of multifunctional nanoparticles to diagnose, target, and treat diseases like cancer have been made possible by this quickly developing discipline (Sardar et al, 2014; Pastorino et al, 2019). The blood-tissue barriers are easily crossed by nanoscale molecules, components, and devices since they are roughly the same size as biological entities. For targeted and regulated distribution to the precise place, new strategies including drug delivery by nanocarriers are being used. They aid in enhancing therapeutic efficacy and lowering drug toxicity in the treatment of disease (Blanco et al., 2015; Pastorino et al, 2019; Ahmad et al, 2021). Additionally, nanocarriers interact with proteins both inside and outside of the cell in ways that do not change their biological characteristics and activity (Pastorino et al, 2019; Gao et al, 2020; Stillman et al, 2020). Such easy access to a living cell's inside offers tremendous benefits for basic and clinical research. As a result of their distinctive optical properties, such as fluorescence and surface plasmon resonance, nanomaterials are now receiving more attention for biomedical applications (Wang et al, 2007; Boisselier and Astruc, 2009; Aminabad et al, 2019; Elahi et al, 2019), particularly in the development of optics-based analytical techniques used for bioimaging (Xia, 2008; Chisanga et al, 2019; Kumar et al, 2019; Noori et al, 2020; Celiksoy et al, 2020). A metal surface's biocompatibility is an important factor in such biomedical applications, and metal nanoparticles created using biological systems, deliver metal ions with good biocompatibility.

There are many ways to make nanoparticles, including physical, chemical, and biological processes (Chen and Mao, 2007; Ahmad et al, 2015; Khatoon et al, 2015; Mazumder et al, 2016; Abdulla et al, 2021). In comparison to chemical or physical processes, green synthesis techniques like biological ones offer a method for synthesising nanoparticles that is affordable, sustainable, and less abrasive. Biological synthesis additionally provides control over size and form for necessary applications. Many species can create inorganic compounds either intracellularly or extracellularly, as is now widely known (Senapati et al, 2004). As reducing or stabilising agents for the synthesis of metal nanoparticles like gold, silver, copper, cadmium, platinum, palladium, titanium, and zinc, which have a wide range of industrial and biomedical applications, organisms like bacteria, fungi, actinomycetes, yeasts, viruses, and algae are being investigated. As a result, the current review article concentrates on the synthesis of different nanoparticles by microbes and how they are used in a variety of industries, with a special emphasis on the biomedical and pharmaceutical sectors.

8.2 NANOPARTICLE SYNTHESIS VIA MICROBIOLOGICAL STRAINS

Physical, chemical, and biological processes are the three main ways to create nanoparticles. These three strategies for creating nanoparticles fit into one of two categories: top-down or bottom-up techniques. In the top-down strategy, materials are mechanically reduced in size by being gradually broken down into nanoscale structures. Contrarily, the bottom-up approach relies on the assembly of nanoscale atoms or molecules into the molecular structure. The creation of the nanoparticles' chemical and biological components is dependent on the bottom-up approach, whereas top-down techniques often refer to the physical or chemical path (Gan and Li, 2012; Lombardo et al, 2020). Metallic nanoparticles can be created physically using UV radiation, sonochemistry, radiolysis, and laser ablation (Kundu et al, 2008; Mohamed and Abu-Dief, 2018; Maric et al, 2019; Sadrolhosseini et al, 2019; Silva et al, 2019; Amulya et al, 2020). These methods do have certain drawbacks. Although high purity nanoparticles of the desired size have been produced using physical and chemical methods, these procedures are frequently expensive and demand for hazardous substances. These nanoparticles may interact directly with the human body, where the associated toxicity becomes significant. The chemical synthesis process may result in the existence of specific toxic chemical species becoming adsorbed on the surface of nanoparticles, which may have negative effects in medical applications. Consequently, developing an environmentally benign production technique that can produce nanoparticles with low toxicity is one of nanotechnology's main goals.

To accomplish this goal, a number of researchers have concentrated their attention on biological processes because they are quick, affordable, and environmentally friendly (Eftekhari et al. 2023). Because of this, a wide variety of natural species, including viruses, bacteria, fungi, algae, and plants, are involved in the biological synthesis of nanoparticles (using their enzymes, proteins, DNA, lipids, and carbohydrates, etc.) (Prasad et al. 2016). Bacteria that decrease metals are identified environmental-friendly catalysts for bioremediation as well as materials production. In reality, respiration processes carried out by bacteria may contribute to the creation of several metal oxides (Kim et al, 2018). Through microbial dissimilatory anaerobic respiration, electrons can be transferred from reduced organic to oxidised inorganic molecules, enabling the production of crystal/nanoparticles and bioremediation processes. The ability of the genus Shewanella to oxidise organic acids as electron donors and reduce inorganic metals as electron acceptors has a long history (Heidelberg et al, 2002; Harris et al, 2018). Additionally, bacteria and other microorganisms have developed a technique for detoxifying the immediate cell environment by turning hazardous metal species into metal nanoparticles (Deplanche and Macaskie, 2008; Murray et al, 2017). Additionally, biomolecules released by bacteria were employed as capping and stabilising agents during the creation of nanoparticles. The process by which metal ions are first trapped on the surface or within the microbial cells is typically how the nanoparticles are created. In the presence of enzymes, the trapped metal ions are then converted to nanoparticles (Figure 8.1).

Microorganisms generally have two different effects on the production of minerals. They can alter the solution's makeup to make it more or less supersaturated with regard to a certain phase than it was before. The creation of organic polymers by microbes is a second way they can affect mineral formation. These polymers can influence nucleation by favouring (or hindering) the stability of the first mineral seeds. The size and form of biological nanoparticles can potentially be controlled by microbes, which are thought of as powerful ecofriendly green nanofactories. Despite the fact that plant-extract-based nanoparticle synthesis is a well-known biological nanosynthesis platform, nanoparticles produced in this fashion may change in yield due to seasonal variations as well as becoming polydisperse in nature due to the presence of phytochemicals (Mishra et al, 2013, 2016; Ovais et al, 2018; Sadaf et al, 2020; Ahmad et al, 2021). These are the distinct benefits that bacteria have over plants in terms of producing nanoparticles. Several microbes are therefore believed to be potential candidates for the production of nanoparticles (Priyadarshini et al, 2013).

8.3 BIOSYNTHESIS OF NANOPARTICLES BY BACTERIA

Microbes' amazing capacity to adapt to stressful environmental situations is what leads to the production of reduced metal ions by them (Kulkarni et al, 2015). Supernatants from several bacteria, such as *Pseudomonas proteolytic, Pseudomonas meridiana, Pseudomonas Antarctica, Arthrobacter gangotriensis,* and *Arthrobacter kerguelensis,* serve as microbial cell factories and are employed as reducing agents in the manufacture of silver nanoparticles (AgNPs) (Shaligram et al, 2009; Singh et al, 2015). Recent research using *Bacillus* brevis to produce AgNPs against multidrug-resistant strains of *Staphylococcus aureus* and *Salmonella typhi* revealed outstanding antibacterial capabilities (Saravanan et al, 2018). Another bacterial strain known to accumulate AgNPs by an intracellular mechanism is *Pseudomonas stutzeri* (Klaus et al, 1999). AgNPs have also been created in the periplasmic region of *Bacillus species* (Pugazhenthiran et al, 2009). The inhabitants of gold mines would be better equipped to withstand the poisonous effects of soluble gold and manufacture gold nanoparticles (AuNPs) (Srinath et al, 2018). The colour of the AuNPs containing colloidal solution varied greatly when *Acinetobacter* sp. SW30 was incubated with varying concentrations of gold chloride and different cell densities, indicating change in size and shape. Surprisingly, monodispersed spherical AuNP of size 19 nm was seen at the lowest cell density and $HAuC_{14}$ salt concentration, whereas cell number increase led to polyhedral AuNP (39 nm) production. Amino acids are involved in the reduction of the gold salt, and amide groups help to stabilise the AuNP (Wadhwani et al, 2016). Additionally, silver, gold, and their alloy nanocrystals have been biosynthesised inside the lactic acid bacteria cells (Nair and Pradeep, 2002). Two distinct strains of *Pseudomonas aeruginosa* were utilised in one sample to produce AuNPs of various sizes in order to create AuNPs (Husseiny et al, 2007). The extracellular creation of AuNPs in a range of sizes and shapes was also discovered to be mediated by *Rhodopseudomonas capsulata*. The strain was utilised to produce triangular plate (50–400 nm) and spherical (10–50 nm) AuNPs (He et al, 2007). *Serratia ureilytica* was employed to create ZnO nanoflowers, which were then applied to cotton fabrics to provide antibacterial properties against *S. aureus* and *E. coli* (Dhandapani et al, 2014). ZnO nanoparticle biosynthesis by *Lactobacillus plantarum* has also been observed (Selvarajan and Mohanasrinivasan, 2013). For the production of ZnO nanoparticles with additional antibacterial properties, the Gram-negative bacterial strain *Aeromonas hydrophila* has been investigated. Utilising *Halomonas elongate*, triangular CuO nanoparticles with antibacterial activity against *S. aureus* and *E. coli* were created (Rad et al, 2018). Another study used the *Bacillus cereus* strain to produce super paramagnetic iron oxide nanoparticles with diameters of about 29.3 nm. As a usage, their dose-dependent anti-cancer activities were observed in MCF-7 (breast cancer) and 3T3 (mouse fibroblast) cell lines (Fatemi et al, 2018). *Streptomyces* sp. (intracellular pathway) has been used to reduce manganese sulphate and zinc sulphate to produce manganese and zinc nanoparticles quickly and easily. Manganese and zinc NPs were between 10 and 20 nm in size (Waghmare et al, 2011).

Surfactin, which was synthesised by the *Bacillus amyloliquifaciens* strain KSU-109, assisted in the manufacture of stable cadmium sulphide nanoparticles with an average size of 3–4 nm (Singh et al, 2011). According to research, *Escherichia coli* E-30 and *Klebsiella pneumoniae* K-6 produce cadmium sulphide nanoparticles with an average size of 3.2–44.9 nm. These two strains also exhibit the highest levels of antimicrobial activity against *G. candidum, A. fumigatus, S. aureus, B. subtilis*, and *E. coli* strains (Abd Elsalam et al, 2018). Antimony sulphide nanoparticles made by *Serratia marcescens* were reported to have a size range of 35 nm (Bahrami et al, 2012), whereas selenium nanoparticles made by *Pseudomonas aeruginosa* ATCC 27853 were reported to have a size of 96 nm (Kora and Rastogi, 2016). Lead nanoparticles made from *Cocos nucifera* were found to be 47 nm in size and to have excellent anti-*S. aureus* action (Elango and Roopan, 2015). The bacteria from Gabal El Sela in Egypt's Eastern Dessert were employed for the intracellular manufacture of uranium nanoparticles with sizes ranging from 2.9 to 21.13 nm.

Due to the existence of bioactive components that aid in stabilising and functionalising the nanoparticles and reduce the number of steps required for synthesis, cyanobacteria are a phylum of photosynthetic bacteria that have been extensively studied for their potential to manufacture nanoparticles. Their rapid growth also makes it possible to produce more biomass, which helps with nanosynthesis. The majority of the times, cyanobacterial biomass extracts free of cells are employed for nanosynthesis. The cyanobacterium *Oscillatoria limnetica*'s aqueous extracts have been helpful in stabilising and producing AgNPs through reduction. The nanoparticles' sizes ranged from 3.30 to 17.97 nm, and they exhibited anti-cancer and antimicrobial activities (Hamouda et al, 2019). Aqueous biomass extracts were used to further a similar AgNP synthesis by *Microchaete* sp. NCCU-342, and spherical, polydispersed nanoparticles of 60–80 nm size were produced (Husain et al, 2019). 4.5–26 nm *Desertifilum* sp. derived AgNPs shown antibacterial activity and cytotoxicity against HepG2, Caco-2, and MCF-7cancer cells (Hamida et al, 2020). *Scytonema* sp., *Nostoc* sp., and *Phormidium* sp. are some other cyanobacterial strains being studied for the production of nanoparticles (Al Rashed et al, 2018). A fascinating investigation utilised *Plectonema boryanum* (strain UTEX 485) filamentous cyanobacterium biomass that responded with AgNO3. It was discovered that cyanobacterium cells and their surface both precipitated AgNPs. The size of intracellular nanoparticles was discovered to be (10 nm), but extracellular nanoparticles showed size in the range of (1–200 nm) (Lengke et al, 2007a). In addition, *P. boryanum* has been shown to convert gold (III)-chloride solutions to create gold (I) sulphide intracellularly; this species is also known to create platinum and palladium NPs (Lengke et al, 2006, 2007b). So, cyanobacteria offer a potentially fruitful platform for biogenic nanosynthesis with a wide range of uses.

8.4 ACTINOMYCETES SYNTHESISE NANOPARTICLES

Due to their importance in the creation of metal nanoparticles despite being the least studied, actinomycetes have attracted a lot of attention (Golinska et al, 2014). Due to their synthesis of numerous bioactive components and extracellular enzymes through their saprophytic activity, actinomycetes are regarded as superior groupings among

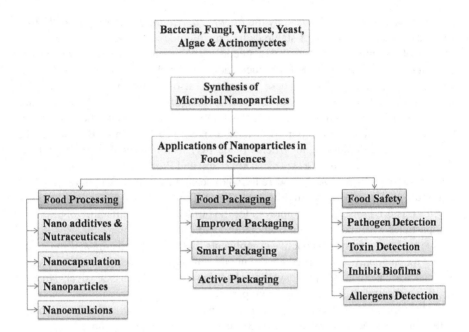

FIGURE 8.1   Microbial nanoparticle synthesis and applications in the food sciences.

microbial species of commercial value (Kumar et al, 2008). Only a handful of the actinomycete genera, including *Thermomonospora, Nocardia, Streptomyces*, and *Rhodococcus*, have been found for the biosynthesis and characterisation of AuNPs (El-Batal et al, 2015). The most competitive species for biosynthesis is thought to be the *Streptomyces* genus (Zonooz et al, 2012). Actinomycetes produce nanoparticles when metal ions undergo intracellular reduction on the surface of mycelia and cytoplasmic membranes (Ahmad et al, 2003b). According to some researchers, the electrostatic interactions between Ag+ and negatively charged carboxylate groups in mycelial cell wall enzymes may be the method by which metal nanoparticles are synthesised intracellularly. Enzymes in the cell wall that cause the formation of silver nuclei reduce the silver ions; this causes the nuclei to grow by further decreasing the silver ions and accumulating Ag+ ions (Abdeen et al, 2014). Sintubin et al proposed a novel process for the intracellular creation of AgNPs utilising lactic acid bacteria (2009). Additionally, a number of other teams have also reported using actinomycetes strains to synthesise metal nanoparticles intracellularly (Usha et al, 2010; Balagurunathan et al, 2011; Prakasham et al, 2012; Sukanya et al, 2013).

8.5   FUNGI-BASED NANOPARTICLE SYNTHESIS

The successful use of myconanotechnological techniques constitutes another biogenic avenue for the manufacture of different metal nanoparticles. Nanosynthesis can occur either extracellularly or intracellularly, just like in bacteria and cyanobacteria. Metal salts in the mycelia, which fungi can utilise, are transformed into a less hazardous form via the intracellular pathway (Molnar et al, 2018; Rajeshkumar and

Sivapriya, 2020). Extracellular biosynthesis is involved in the application of fungus extracts (Zhao et al, 2018; Rajeshkumar and Sivapriya, 2020). Due to their abundance of bioactive metabolites, high aggregation, and enhanced production, fungi are considerably more resourceful than bacteria in the manufacture of nanoparticles (Castro-Longoria et al, 2011; Alghuthaymi et al, 2015). AuNP biosynthesis has been found to be possible in a variety of filamentous fungus. Several techniques were used in this study to biosynthesise AuNPs. The scientists proposed that media elements and substances released by fungi may be employed to stabilise the nanoparticles (Molnar et al, 2018; Guilger-Casagrande and de Lima, 2019). Another team used the three distinct fungal strains *Fusarium oxysporum*, *Fusarium* sp., and *Aureobasidium pullulans* to biosynthesise the reported AuNPs. The authors proposed that sugar reduction was involved in shaping the structure of AuNPs and that biosynthesis took place inside fungal vacuoles. Additionally, the biomolecules or proteins that the fungus used to create the secondary metabolite contains serve as capping and stabilising agents (Zhang et al, 2011). In a different investigation, a number of *Fusarium oxysporum* strains were utilised to produce extracellular silver metal nanoparticles between 20 and 50 nm (Ahmad et al, 2003a). By analysing UV-Visible, fluorescence, and enzymatic activity, the metal ion reduction by nitrate dependent reductase and extracellular shuttle quinone was confirmed (Duran et al, 2005, 2007). Kumar and his team used the nitrate reductase enzyme obtained from *Fusarium oxysporum*, phytochelatinin, and 4-hydroxyquinoline to create in vitro AgNPs (10–25 nm) stabilised in the presence of reduced cofactor nicotinamide adenine dinucleotide phosphate (NADPH) by a capping peptide (Kumar et al, 2007). Although condition optimisation led to AgNPs of 2.86 nm, a different study found that *Rhizopus stolonifera* extracts were responsible for the production of monodispersed AgNPs of 9.4 nm size (Abdelrahim et al, 2017). AgNPs were produced extracellularly using *Candida glabrata*, which indicated that they have potent antibacterial activity (Jalal et al, 2018). Excellent antibacterial capability was shown by ZnO nanoparticles mediated by *Aspergillus niger*, and Bismarck brown dye deteriorated by up to 90% (Kalpana et al, 2018). *Aspergillus nidulans* has recently been used to create cobalt oxide nanoparticles (Vijayanandan and Balakrishnan, 2018). The biosynthesis of platinum nanoparticles with sizes between 100 and 180 nm by the fungus *Fusarium oxysporum* was observed (Riddin et al, 2006). The ability to create nanoparticles either extracellularly or intracellularly has been demonstrated by the fungi *Verticillium* sp., *Fusarium oxysporum*, and *Aspergillus flavus* (Mukherjee et al, 2002; Bhainsa and D'Souza, 2006). The switch from bacteria to fungi has the additional benefit that downstream biomass processing and handling may be made considerably easier when building natural nanofactories (Aziz et al. 2016, 2019; Prasad 2017) (Figure 8.1).

8.6   YEAST-BASED NANOPARTICLE SYNTHESIS

It is known that yeast strains from various genera use various processes for the synthesis of nanoparticles, leading to notable variances in size, particle location, monodispersity, and other features. According to one study, detoxification processes in yeast cells produce glutathione, two types of metal-binding ligands (metallothioneins and phytochelatins), and other compounds. Most of the examined yeast species depend on these molecules to stabilise the resultant complexes and determine the process for nanoparticle production (Hulkoti and Taranath, 2014). In the presence of harmful metals, yeast cells frequently convert the absorbed metal ions into complex polymer molecules that are non-toxic to the cell as a defence strategy. Typically, the term "semiconductor crystals" or "quantum semiconductor crystals" is used to describe these nanoparticles made in the yeast (Dameron et al, 1989). The ability of yeast cells to create semiconductor nanoparticles, especially those of cadmium sulphide, is particularly well recognised (CdS). There have been reports on yeasts such *Pichia capsulata* (Subramanian et al, 2010), *Candida guilliermondii* (Mishra et al, 2011), *Saccharomyces boulardii* (Kaler et al, 2013), *Kluyveromyces marxianus* (Ashour, 2014), *Candida utilis* (Waghmare et al, 2015), *Candida lusitaniae* (Eugeni et al, 2016), *Rhodotorula glutinis*, and *Rhodotorula mucilaginosa* (Cunha et al, 2018). AgNPs were made using the silver-tolerant yeast strain MKY3 (Kowshik et al, 2002).

8.7   ALGAE-BASED NANOPARTICLE SYNTHESIS

Algae are also increasingly being used for the production of nanoparticles. *Sargassum muticum* was utilised to create ZnO nanoparticles, and it was discovered to have anti-apoptotic and anti-angiogenesis actions in HepG2 cells (Sanaeimehr et al, 2018). *Sargassum crassifolium*, a macroalgae and seagrass, has been used in the production of AuNPs. Intriguingly, this study's UV absorption spectra showed a blue shift as *S. crassifolium* concentration was raised, which was explained by a reduction in nanoparticle size caused by an increase in nucleation sites in the reductant (Maceda et al, 2018). Using *Cystoseira trinodis*, CuO nanoparticles with dimensions of about 7 nm were produced biochemically. They were found to have enhanced antioxidant and antibacterial capabilities as well as methylene blue degradation potential (Gu et al, 2018). Aluminium oxide nanoparticles with a size of around 20 nm were created using *Sargassum ilicifolium* (Koopi and Buazar, 2018). There have been reports of the production of AuNPs by a number of algae strains, including *Turbinaria conoides, Laminaria japonica, Acanthophora spicifera*, and *Sargassum tenerrimum* (Ghodake and Lee, 2011; Swaminathan et al, 2011; Vijayaraghavan et al, 2011; Ramakrishna et al, 2016). The creation of unique core (Au)-shell (Ag) nanoparticles has also been studied using *Spirulina platensis* (Govindaraju et al, 2008).

8.8   VIRAL NANOPARTICLE SYNTHESIS

Due to their biocompatibility, biodegradability, capability for mass manufacturing, programmable scaffolds, and simplicity of genetic editing for desired features, viruses have emerged as intriguing candidates as nanoparticles for biomedical applications. Since viral bodies have a size between 20 and 500 nanometres, they are themselves naturally occurring nanoparticles. Biomedical applications have taken use of their resilience and capacity to release their genetic material in response to environmental changes. Viral nanoparticles (VNPs) have mostly been used in imaging, theranostics, vaccinations, immunotherapeutics, and gene and medication delivery. While bacteriophages and plant viruses have been investigated for medication delivery, vaccinations, and immunotherapeutics, most mammalian viruses are used in gene delivery.

Additionally, VNPs can be labelled with a variety of ligands to serve as imaging, therapeutic, or targeted agents for a variety of biomedical purposes (Steinmetz, 2010). Virus-like particles (VLPs), which are made from the proteins that cover viruses, belong to a related family of materials (Chung et al, 2020). These nanoparticles are dynamic, self-assembling moieties with symmetrical, monodisperse structures and can have viral, plant, or animal origins. The creation of VNPs entails production in a host organism (whether it a bacterium, an animal, or a plant), followed by chemical conjugation and tuning, in vitro and in vivo testing, and evaluation (Steinmetz, 2010). The toxicity of VNPs, particularly for human pathogens, is an important factor to take into account while employing them. In contrast to mammalian viruses like adenoviruses, bacteriophages and plant viruses are favoured. The immunogenicity of the viral particle also has an impact on how quickly it leaves the tissue and how much it accumulates there. Often, attachment to molecules like PEG aids in protecting particular biointeractions (Bruckman et al, 2008). Chemotherapeutic medicines have been delivered using a variety of VNPs and VLPs. In human hepatocellular carcinoma cells, VLPs modified with targeting peptide and a dose of doxorubicin, cisplatin, and 5-fluorouracil were successful (Ashley et al, 2011). In platinum-resistant ovarian cancer cells, VNPs generated from the tobacco mosaic virus have been employed to deliver the drug cisplatin (Franke et al, 2017). Better antibacterial action has been observed with chloramphenicol-loaded bacteriophage fd-based nanoparticles than with chloramphenicol alone against harmful bacteria like *Staphylococcus aureus* (Yacoby et al, 2006). Since VNPs have rigid structures and long rotational correlation times, which lead to high relaxivity, they are also used as MRI contrast agents. A large range of contrast agents, including gadolinium, can also be chelated to their inner or exterior surfaces due to their polyvalent nature (Steinmetz, 2010). The possibility of creating vaccinations against viruses like hepatitis B, HIV, and Neospora caninum using such nanoparticles has also been investigated (Oh and Han, 2020).

The handling of bacterial or viral strains that could be dangerous or pathogenic to humans is another crucial precaution that needs to be stressed. Thus, it is crucial to prioritise related biological safety issues in order to deploy microorganism-mediated nanosynthesis on a wide scale for commercial exploitation.

8.9 FOOD PROCESSING WITH NANOTECHNOLOGY

The food ingredients with nanostructures are being created with the promise of better consistency, taste, and texture (Cientifica Report, 2006). Nanotechnology helps reduce the amount of food wasted due to microbial infestation while also extending the shelf life of various types of food ingredients (Pradhan et al, 2015; Karlo et al. 2023). Presently, food additives are delivered in food products using nanocarriers without affecting their fundamental morphology. As it was shown that in some cell lines, only submicron nanoparticles can be absorbed efficiently but not the larger size micro-particles, particle size may directly alter the distribution of any bioactive molecule to various places inside the body (Ezhilarasi et al, 2013). The following characteristics of an ideal delivery system are desired: (i) the ability to deliver the active compound precisely at the desired location; (ii) assurance of availability at the desired time and rate; and (iii) effectiveness in maintaining the active compound at the desired level for extended periods of time (in storage condition). The creation of encapsulation, emulsions, biopolymer matrices, simple solutions, and association colloids using nanotechnology provides effective delivery methods with all the properties listed above. In food packaging, nano polymers are attempting to take the place of traditional materials. Food pollutants, mycotoxins, and microbes can all be detected using nanosensors (Bratovčić et al, 2015).

Traditional encapsulation techniques are less effective at encapsulating and releasing substances than nanoparticles. In addition to masking tastes or odours and regulating interactions between active ingredients and the food matrix, controlling the release of the active ingredients, ensuring availability at a predetermined time and rate, and shielding them from moisture, heat, chemical, or biological degradation during processing, storage, and use, nanoencapsulations also show compatibility with other substances in the system (Weiss et al, 2006). Additionally, these delivery methods can enter tissues deeply due to their tiny size, enabling effective administration of active substances to specific body locations (Lamprecht et al, 2004). Many encapsulating delivery techniques based on synthetic and natural polymers have been developed to increase the bioavailability and preservation of the active food ingredients. Additionally, the contribution of nanotechnology to the enhancement of food products in terms of (i) food texture, (ii) food appearance, (iii) food flavour, (iv) food nutritional content, and (v) food shelf-life can be used to assess the significance of nanotechnology in food processing. It is a truth that unexpectedly, nanotechnology not only affects all of the aforementioned factors but has also significantly changed food goods, giving them novel properties.

8.10   FOOD'S TEXTURE, TASTE, AND APPEARANCE

Nanotechnology offers a variety of choices to raise food quality and also contributes to improving food flavour. Techniques for nanoencapsulation have been widely employed to enhance taste release and retention and to provide culinary balance (Nakagawa, 2014). Anthocyanins, a highly reactive and unstable plant pigment with a variety of biological functions, were nanoencapsulated by Zhang et al (2014). The heat stability and photostability of apo recombinant soybean seed H-2 subunit ferritin (rH-2) were enhanced by encapsulating cyanidin-3-O-glucoside (C3G) molecules within the inner cavity. The creation of multifunctional nanocarriers for the delivery and protection of bioactive molecules. Rutin is a common dietary flavonoid with significant pharmacological activity, but its use in the food sector is restricted due to poor solubility. As compared to free rutin, the ferritin nanocages encapsulation improved the solubility, thermal stability, and UV radiation resistance of ferritin trapped rutin (Yang et al, 2015). Since they can be made with natural food ingredients using simple production techniques and may be tailored to improve water-dispersion and bioavailability, the use of nanoemulsions to deliver lipidsoluble bioactive chemicals is quite popular (Ozturk et al, 2015).

Due to their subcellular size, which results in a higher drug bioavailability, nanoparticles offer a promising method of improving the bioavailability of nutraceutical compounds when compared to larger particles, which typically release encapsulated compounds more slowly and over longer time periods. Numerous metallic oxides, including titanium dioxide and silicon dioxide (SiO_2), have historically been utilised in food products as colouring or flow agents (Ottaway, 2010). One of the most popular food nanomaterials for carrying tastes or scents in food products is silicon dioxide (SiO_2) (Dekkers et al, 2011).

8.11 NUTRITIONAL VALUE

The bulk of bioactive substances, including lipids, proteins, carbohydrates, and vitamins, are sensitive to the duodenum's and stomach's highly acidic environment and enzyme activity. These bioactive compounds can endure such harsh circumstances by being encapsulated, but they can also easily integrate into food products, which is difficult to do in non-capsulated form due to the low water solubility of these bioactive compounds. In order to optimise the distribution of medications, vitamins, or delicate micronutrients in daily diets, tiny edible capsules based on nanoparticles are being developed (Yan and Gilbert, 2004; Koo et al, 2005). In order to more effectively distribute nutrients like protein and antioxidants for precisely targeted nutritional and health benefits, numerous techniques like nanocomposite, nano-emulsification, and nanostructuration have been used. It has been discovered that polymeric nanoparticles are suitable for encasing bioactive substances (such as flavonoids and vitamins) and transporting such substances to their intended targets (Langer and Peppas, 2003).

8.12 THE SHELF-LIFE OR PRESERVATION

By slowing down the processes of degradation or stopping degradation until the product is delivered at the target site, nanoencapsulation of these bioactive components extends the shelf-life of functional foods where bioactive components frequently get degraded and ultimately led to inactivation due to the hostile environment. Additionally, the edible nano-coatings on different food components may operate as a barrier to moisture and gas exchange, transmit tastes, colours, enzymes, antioxidants, and anti-browning agents, and extend the shelf life of manufactured meals even after the packaging has been opened (Renton, 2006; Weiss et al, 2006). By manipulating the interfacial layer's characteristics around functional components, it is frequently possible to slow down chemical breakdown processes. For instance, curcumin (*Curcuma longa*), the most active and least stable bioactive component, displayed reduced antioxidant activity and was discovered to be stable to pasteurisation and at various ionic strengths upon encapsulation (Sari et al, 2015).

8.13 PACKAGING FOR FOOD USING NANOTECHNOLOGY

An ideal packing material must be strong, biodegradable, and gas and moisture permeable (Couch et al, 2016). Better packaging materials with improved mechanical strength, barrier properties, antimicrobial films, and nanosensing for pathogen detection and alerting consumers to the safety status of food are just a few of the advantages that nano-based "smart" and "active" food packagings have over conventional packaging methods (Mihindukulasuriya and Lim, 2014: Chausali et al., 2021).

To enhance food packaging, nanocomposites can be employed as an active material for packaging and material coating (Pinto et al, 2013). Numerous researchers were interested in investigating the antimicrobial capabilities of organic compounds such as bacteriocins, organic acids, and essential oils as well as their usage in polymeric matrices as antimicrobial packaging (Gálvez et al, 2007; Schirmer et al, 2009). However, due to their extreme sensitivity to these physical conditions, these compounds cannot be used in the several food processing procedures that call for high temperatures and pressures. An effective antibacterial activity can be obtained in small concentrations and greater stability in harsh environments using inorganic nanoparticles. As a result, there has been a lot of interest in employing these nanoparticles in food packaging that is antimicrobial in recent years. In order to prevent or deter any microbes' development on food surfaces, an antimicrobial package is essentially a sort of active packaging that comes into contact with the food item or the headspace inside (Soares et al, 2009). Numerous nanoparticles have been demonstrated to possess antibacterial characteristics, including metal oxide nanoparticles like silver, titanium oxide or zinc oxide, chitosan, copper, and nanoparticles of metal (Bradley et al, 2011; Tan et al, 2013).

Nanocomposite and nanolaminates have been actively employed in food packaging to provide a barrier from excessive heat and mechanical damage, extending food shelf-life. Nanoparticle applications are not just restricted to antimicrobial food packaging. By incorporating nanoparticles into packaging materials, quality food with a longer shelf life is made available. To have more mechanical and thermostable packaging materials, polymer composites are made. Improved polymer composites are produced by using a lot of inorganic or organic fillers. The development of more durable packaging materials at a lower cost has been made possible by the inclusion of nanoparticles into polymers (Sorrentino et al, 2007). Utilising innocuous nanoscale fillers like chitin or chitosan, silica (SiO_2) nanoparticles, clay and silicate nanoplatelets, or chitin or chitosan results in a polymer matrix that is lighter, stronger, more fire resistant, and has better thermal properties (Duncan, 2011; Othman, 2014). Due to their barrier and structural integrity, antimicrobial nanocomposite films made by impregnating fillers with at least one dimension in the nanometric range or nanoparticles into polymers have a dual benefit (Rhim and Ng, 2007).

8.14   NANOSENSORS FOR PATHOGEN DETECTION

Nanomaterials for use in biosensor construction provide a high level of sensitivity and other cutting-edge characteristics. In food microbiology, nanosensors or nano-biosensors are used to quantify the available food ingredients, detect infections in processing plants or in food material, and notify distributors and consumers of the food's safety status (Cheng et al, 2006; Helmke and Minerick, 2006). The nanosensor functions as an indicator that reacts to changes in environmental factors like humidity or temperature in storage rooms, microbial contamination, or the deterioration of items (Bouwmeester et al, 2009). Numerous nanostructures have been studied for potential use in biosensors, including thin films, nanorods, nanoparticles, and nanofibers (Jianrong et al, 2004). Rapid and extremely sensitive detection devices for microbial compounds or cells have been made possible by thin film-based optical immunosensors. These immunosensors encapsulate particular antibodies, antigens, or protein molecules on thin nanofilms or sensor chips, which generate signals when target molecules are detected (Subramanian, 2006). Escherichia coli O157:H7 and Staphylococcus aureus were quickly detected using an electrochemical impedance spectrum using a dimethylsiloxane microfluidic immunosensor combined with a particular antibody mounted on an alumina nanoporous membrane (Tan et al, 2011). In the tracking, tracing, and monitoring chain for food quality, nanotechnology can also help with the identification of pesticides (Liu et al, 2008), infections (Inbaraj and Chen, 2015), and toxins (Palchetti and Mascini, 2008).

Due to its quick detection, ease of use, and affordability, carbon nanotube-based biosensors have also attracted a lot of attention. They have also been effectively used to detect microbes, poisons, and other degradation products in food and beverages (Nachay, 2007). Toxin antibodies tied to these nanotubes alter their conductivity in a measurable way when bound to aquatic toxins, making them useful for the detection of waterborne toxins (Wang et al, 2009). Additionally, food condition is monitored by an array of nanosensors used in electronic tongues and noses that emit signals in response to aromas or gases emitted by food (Garcia et al, 2006). The interaction between different odorants and chemicals that have been coated on the crystal surface of the quartz crystal microbalance (QCM) can be detected by an electric nose that is based on the QCM. Quartz crystal surfaces that have been altered with various functional groups or biological molecules, such as amines, enzymes, lipids, and diverse polymers, have been employed in numerous investigations on small molecule detection (Kanazawa and Cho, 2009).

8.15 ASPECTS OF RELATED SAFETY CONCERNS, HEALTH RISKS, AND REGULATORY ASPECTS

Due to subsequent transfer of particle nanomaterials from the packaging into the food as a result of subpar packing performance, eating foods that have come into touch with nanopackaging may present an exposure route and constitute a serious health risk. This effect would be strongly influenced by the food's specific ingestion rate, degree of migration, packaging matrix, and nanomaterial's toxicity (Cushen et al, 2012). Nano-based goods have a negative impact on health and provide safety problems and health risks due to overconsumption, bioaccumulation, and increased activity (Cagri et al, 2004; Maynard 2007; Rasmussen et al, 2010; Cushen et al, 2012; Jovanovic, 2015). It is important to conduct an adequate risk assessment before exposing humans to higher concentrations of these materials through inhalation or skin contact because it could lead to serious safety concerns, particularly with regard to long-term toxicity (He and Hwang, 2016). According to some research, AgNPs contained in packaging may migrate into food and be consumed by people (Echegoyen and Nern, 2013); nevertheless, nothing is known regarding the toxicity of these particles. However, these nanoparticles may collect in several organs in animals, including the liver, spleen, stomach, small intestine, and kidneys (McClements and Xiao, 2017). Furthermore, a single oral intake of ZnO nanoparticles could cause issues like lung, renal, and liver harm (Esmaeillou et al, 2013). The use of titanium oxide and its final disposal can have an impact on both people and the environment, which raises the possibility of environmental and human health risks (Yang et al, 2014). Policymakers, conscientious consumers, international regulatory bodies, and numerous other stake holders are all paying attention to how nanotechnology is being applied. The knowledge of traceability and monitoring of the physical, chemical, and functional properties of nano-sized materials is complicated by the interplay between nano-bio-eco (He et al, 2018). Without sufficient scientific study and sequential case studies, it is very challenging to create any regulations or approve any legislation. Additionally, it is challenging since each nanomaterial behaves differently in different products under certain processing conditions (He et al, 2018). A large number of food products using nanotechnology are getting close to R&D trials and are waiting for legislation, however since the law was first suggested in 2003, there has been very little progress in the legislative process. However, some safety guidelines on the acceptable range or limits of nanomaterials employed in food applications must be established, as above-mentioned behaviour may change beyond a certain threshold. The use of nanoparticles in food packaging may be questioned by environmental authorities because it will ultimately be thrown away in some way. Chaudhry et al have collated information on laws pertaining to applications of nanotechnologies in food, Both Hodge et al (2014) and Gergely et al(2010). A global knowledge-sharing platform is required by food processors, academics, researchers, and consumers to debate and address all facets of application, consumption, disposal, and long-term impacts. It will improve this promising field's research and applications even further.

8.16   CONSTRAINTS IN TECHNOLOGY AND DIFFICULTIES

Although there are many obstacles, nanotechnology offers a significant potential to create novel products and procedures in the food industry. Producing edible delivery systems with cost-effective processing processes and safe formulations for human consumption is the main problem (Dupas and Lahmani, 2007). To maintain the wholesomeness of foods, it is extremely important to prevent the migration and leaching of nanoparticles from packing materials into food products. Sometimes the NSMs are separated after being introduced, either directly or indirectly, as a result of migration from other sources (Hannon et al, 2016). At the nanoscale, the behaviour of the materials is completely different, and our technical understanding of its analysis is still limited. The full comprehension of the functions at the nanoscale and the toxicity of nanomaterials will improve their practical application and safety standards. It is necessary to examine the effects of nanoparticles, potential risks, related toxicological concerns, and environmental difficulties. There have been reports of nanoparticles invading cells and organs after breaching the biological barrier (Su and Li, 2004). Different chemical processes used in the synthesis of nanoparticles have negative side effects and produce harmful, environmentally harmful byproducts that seriously pollute the environment (Cha and Chinnan, 2004; Singhal et al, 2011). Therefore, when manufacturing, packaging, and consuming nano-based food products by humans, a comprehensive risk assessment programme, regulatory policy, biosafety, and public concerns must be taken into account in addition to popularity and public demand (Davies, 2007; Yu et al, 2012; Shi et al, 2013; Bajpai et al, 2018). Prior to commercial application and for the production of antibacterial nanoparticles with environmentally benign materials, in vitro and in vivo investigations involving nanoparticle interactions with living things are required (Cha and Chinnan, 2004; Das et al, 2011).

8.17 COMMERCIALISATION POTENTIAL AND FUTURE OPPORTUNITIES

The applications of nanotechnology in food science and research have made amazing strides. Nanotechnology helps with toxin, pathogen, and pesticide detection as well as tracking, tracing, and monitoring to ensure food quality is maintained. The lack of qualified labour, the expense of analysis, and the expense of purchasing top-tier technical equipment do not constitute hurdles for nanotechnology. Some nano-systems, however, are still in their infancy or are being evolved into powerful nano-components. For widespread use, more in-depth research can be done in the following areas. Challenges and safety concerns might both be taken into account at once.

a. The revolutionary concepts of "packaging technologies," creating biomarkers that are particular to an antigen, and combining nanoparticles to create nanocomposite polymeric films are all slowly coming to fruition. Future expansion and industrial uses are subject to extensive research.

b. Nanocomposites are biodegradable, carbon-neutral compounds. As a result, their use as food packaging material may be tapped in the near future. Similarly, the economic potential of nanosilica as a surface coating material with improved barrier qualities might be investigated (Chaudhry and Castle, 2011).

c. Antigen-specific biomarkers typically help to detect the presence of the organism that causes food rotting (Cho et al, 2008). It would be simple and quick to harness its utility in food pathogen detection, including bacteria, viruses, and mycotoxins (rapid, accurate, and labour intensive).

d. Additionally relevant to consumers is the use of nanosensors in film packaging to detect gases produced as a result of food spoiling (Chaudhry and Castle, 2011; Das et al, 2011). Such sensors can identify food spoilage at every point in the supply chain, reducing overall food loss to the advantage of food producers, merchants, and customers.

e. Additionally, sensors are being incorporated into packaging materials using carbon nanotubes primarily for the detection of bacteria, hazardous chemicals, and food spoilage (Tully et al, 2006). Future packaging materials for smart and intelligent food packaging systems can be thoroughly researched (Wang and Irudayaraj, 2008).

f. In a similar vein, in-depth study insights may be obtained to explain the potential of nanofibers in the field of food packaging and with regard to the interactions of various food ingredients during nanoencapsulation.

In the coming years, the range of formulation and manufacture of functional foods is projected to be expanded thanks to foods developed from nanotechnology. If rules and regulations tailored to nanotechnology are established to address the many safety issues that are related to this technology, it may grow to dominate the whole food processing industry. According to recent forecasts, nanotechnology is expected to be the cutting-edge technology with infinite development rate by 2050 to solve most industrial and societal challenges because of its capacity to find cooperative solutions both at the micro and macro level (Aithal, 2016).

8.18 CONCLUSIONS

The scope of the study and ensuing commercial applications is progressively expanding from one spectrum to another. Nanotechnology offers a great deal of promise to advance food science over a broad spectrum, encompassing many different fields of study and encompassing many different facets of food processing. Nanotechnology has a bright future in the field of food technology, with applications ranging from increasing product shelf life to improving food storage to tracing and tracking contaminants/pollutants to introducing nutritional or health supplements into the body through food. The risk of foodborne infections, which are fatal even at very low contamination levels, can be eliminated by supporting the traditional microbe separation methods with nanotechnology-based technologies. Recent advances in the field of nanoscience-based applications offer chances to reorganise production processes, promote resource conservation, and even alter consumer eating habits. These technologies help agro-ecosystems develop, opening the way for sustainable agricultural growth. Nanotechnology is a crucial technique for overcoming the ongoing difficulties related to packaging. If a packing solution is both economically feasible and environmentally friendly, it may be relied upon. The quality, storability, safety, and security of food will ultimately be significantly improved by such advancements, which will help both food producers and consumers. However, more research is required in accordance with the precautionary principle, particularly in regards to the migratory patterns of NSMs in food matrix, the cytotoxicity of nanoparticles in humans, and their potential effects on consumer health and safety as well as the environment.

REFERENCES

Abd Elsalam, S. S., Taha, R. H., Tawfeik, A. M., El-Monem, A., Mohamed, O., and Mahmoud, H. A. (2018). Antimicrobial activity of bio and chemical synthesized cadmiumsulfide nanoparticles. *Egypt. J. Hosp. Med.* 70, 1494–1507.

Abdeen, S., Geo, S., Praseetha, P. K., and Dhanya, R. P. (2014). Biosynthesis of silver nanoparticles from actinomycetes for therapeutic applications. *Int. J. Nano Dimens.* 5, 155–162.

Abdelrahim, K., Mahmoud, S. Y., Ali, A. M., Almaary, K. S., Mustafa, A. E. Z. M. A., and Husseiny, S. M. (2017). Extracellular biosynthesis of silver nanoparticles using *Rhizopus stolonifer. Saudi J. Biol.* Sci. 24, 208–216.

Abdulla, N. K., Siddiqui, S. I., Fatima, B., Sultana, R., Tara, N., Hashmi, A. A., et al (2021). Silver based hybrid nanocomposite: a novel antibacterial material for water cleansing. *J. Clean. Prod.* 284, 124746.

Ahmad, A., Mukherjee, P., Senapati, S., Mandal, D., Khan, M. I., Kumar, R., et al (2003a). Extracellular biosynthesis of silver nanoparticles using the fungus *Fusarium oxysporum. Colloids Surf. B Biointerfaces* 28, 313–318.

Ahmad, A., Senapati, S., Khan, M. I., Kumar, R., Ramani, R., Srinivas, V., et al (2003b). Intracellular synthesis of gold nanoparticles by a novel alkalotolerant actinomycete, *Rhodococcus* species. *Nanotechnology* 14, 824.

Ahmad, R., Mohsin, M., Ahmad, T., and Sardar, M. (2015). Alpha amylase assisted synthesis of TiO2 nanoparticles: structural characterization and application as antibacterial agents. *J. Hazard. Mater.* 283, 171–177.

Ahmad, R., Srivastava, S., Ghosh, S., and Khare, S. K. (2021). Phytochemical delivery through nanocarriers: a review. *Colloids Surf. B Biointerfaces* 197, 111389.

Aithal, P. S. (2016). Nanotechnology innovations & business opportunities: a review. *International Journal of Management, IT and Engineering*, 6(1), 182– 204.

Alghuthaymi, M. A., Almoammar, H., Rai, M., Said-Galiev, E., and Abd-Elsalam, K. A. (2015). Myconanoparticles: synthesis and their role in phytopathogens management. *Biotechnol. Biotechnol. Equip.* 29, 221–236.

Al Rashed, S., Al Shehri, S., and Moubayed, N. M. S. (2018). Extracellular biosynthesis of silver nanoparticles from Cyanobacteria. *Biomed. Res.* 29, 2859–2862.

Aminabad, N. S., Farshbaf, M., and Akbarzadeh, A. (2019). Recent advances of gold nanoparticles in biomedical applications: state of the art. *Cell Biochem. Biophys.* 77, 123–137.

Amulya, M. A. S., Nagaswarupa, H. P., Kumar, M. R. A., Ravikumar, C. R., and Kusuma, K. B. (2020). Sonochemical synthesis of MnFe$_2$O$_4$ nanoparticles and their electrochemical and photocatalytic properties. *J. Phys. Chem. Solids* 148, 109661.

Arshad, A. (2017). Bacterial synthesis and applications of nanoparticles. *Nano Sci. Nano Technol.* 11, 119.

Ashley, C. E., Carnes, E. C., Phillips, G. K., Durfee, P. N., Buley, M. D., Lino, C. A., et al (2011). Cell-specific delivery of diverse cargos by bacteriophage MS2 virus-like particles. *ACS Nano* 5, 5729–5745.

Ashour, S. M. (2014). Silver nanoparticles as antimicrobial agent from *Kluyveromyces marxianus* and *Candida utilis*. *Int. J. Curr. Microbiol. Appl. Sci.* 3, 384–396.

Aziz, N., Faraz, M., Sherwani, M. A., Fatma, T., Prasad, R. (2019). Illuminating the anticancerous efficacy of a new fungal chassis for silver nanoparticle synthesis. *Front Chem* 7, 65. doi: 10.3389/fchem.2019.00065

Aziz, N., Pandey, R., Barman, I., Prasad, R. (2016). Leveraging the attributes of *Mucor hiemalis*-derived silver nanoparticles for a synergistic broad-spectrum antimicrobial platform. *Front Microbiol* 7, 1984. doi: 10.3389/fmicb.2016.01984

Bahrami, K., Nazari, P., Sepehrizadeh, Z., Zarea, B., and Shahverdi, A. R. (2012). Microbial synthesis of antimony sulfide nanoparticles and their characterization. *Ann. Microbiol.* 62, 1419–1425.

Bajpai, V. K., Kamle, M., Shukla, S., Mahato, D. K., Chandra, P., Hwang, S. K., … Han, Y. K. (2018). Prospects of using nanotechnology for food preservation, safety, and security. *J. Food Drug Anal.*, 26(4), 1201– 1214.

Balagurunathan, R., Radhakrishnan, M., Rajendran, R. B., and Velmurugan, D. (2011). Biosynthesis of gold nanoparticles by actinomycete *Streptomyces viridogens* strain HM10. *J. Biochem. Biophys.* 48, 331–335.

Bhainsa, K. C., and D'Souza, S. F. (2006). Extracellular biosynthesis of silver nanoparticles using the fungus *Aspergillus fumigatus*. *Colloids Surf. B Biointerfaces* 47, 160–164.

Blanco, E., Shen, H., and Ferrari, M. (2015). Principles of nanoparticle design for overcoming biological barriers to drug delivery. *Nat. Biotechnol.* 33, 941.

Boisselier, E., and Astruc, D. (2009). Gold nanoparticles in nanomedicine: preparations, imaging, diagnostics, therapies and toxicity. *Chem. Soc. Rev.* 38, 1759–1782.

Bouwmeester, H., Dekkers, S., Noordam, M. Y., Hagens, W. I., Bulder, A. S., Heer, C., et al (2009). Review of health safety aspects of nanotechnologies in food production. *Reg. Toxicol. Pharmacol.* 53, 52–62.

Bradley, E. L., Castle, L., and Chaudhry, Q. (2011). Applications of nanomaterials in food packaging with a consideration of opportunities for developing countries. *Trends Food Sci. Technol.* 22, 603–610.

Bratovčić, A., Odobašić, A., Ćatić, S., and Šestan, I. (2015). Application of polymer nanocomposite materials in food packaging. *Croat. J. Food Sci. Technol.* 7, 86–94.

Bruckman, M. A., Kaur, G., Lee, L. A., Xie, F., Sepulveda, J., Breitenkamp, R., et al (2008). Surface modification of tobacco mosaic virus with "click" chemistry. *Chembiochem* 9, 519–523.

Cagri, A., Ustunol, Z., and Ryser, E. T. (2004). Antimicrobial edible films and coatings. *J. Food Protect.* 67(4), 833–848.

Castro-Longoria, E., Vilchis-Nestor, A. R., and Avalos-Borja, M. (2011). Biosynthesis of silver, gold and bimetallic nanoparticles using the filamentous fungus *Neurospora crassa*. *Colloids Surf. B Biointerfaces* 83, 42–48.

Celiksoy, S., Ye, W., Wandner, K., Schlapp, F., Kaefer, K., Ahijado-Guzman, R., et al (2020). Plasmonic nanosensors for the label-free imaging of dynamic protein patterns. *J. Phys. Chem. Lett.* 11, 4554–4558.

Cha, D. S., and Chinnan, M. S. (2004). Biopolymer-based antimicrobial packaging: a review. *Crit. Rev. Food Sci. Nutr.* 44(4), 223– 237.

Chaudhry, Q., and Castle, L. (2011). Food applications of nanotechnologies: an overview of opportunities and challenges for developing countries. *Trends Food Sci. Technol* 22(11), 595–603.

Chausali, N., Saxena, J., Prasad, R. (2021) Recent trends in nanotechnology applications of bio-based packaging. *Journal of Agriculture and Food Research*, https://doi.org/10.1016/j.jafr.2021.100257

Chen, X., and Mao, S. S. (2007). Titanium dioxide nanomaterials: synthesis, properties, modifications, and applications. *Chem. Rev.* 107, 2891–2959.

Cheng, Q., Li, C., Pavlinek, V., Saha, P., and Wang, H. (2006). Surface-modified antibacterial TiO2/Ag+ nanoparticles: preparation and properties. *Appl. Surface Sci.* 252, 4154–4160.

Chisanga, M., Muhamadali, H., Ellis, D. I., and Goodacre, R. (2019). Enhancing disease diagnosis: biomedical applications of surface-enhanced Raman scattering. *Appl. Sci.* 9, 1163.

Cho, Y. J., Kim, C. J., Kim, N., Kim, C. T., and Park, B. (2008). Some cases in applications of nanotechnology to food and agricultural systems. *BioChip J.* 2(3), 183–185.

Chung, Y. H., Cai, H., and Steinmetz, N. F. (2020). Viral nanoparticles for drug delivery, imaging, immunotherapy, and theranostic applications. *Adv. Drug Deliv. Rev.* 156, 214–235.

Cientifica Report (2006). Nanotechnologies in the food industry, Published August 2006. Available at: http://www.cientifica.com/www/details.php?id47 [accessed October 24, 2006].

Couch, L. M., Wien, M., Brown, J. L., and Davidson, P. (2016). Food nanotechnology: proposed uses, safety concerns and regulations. *Agro. Food Ind. Hitech.* 27, 36–39.

Cunha, F. A., Da Cso Cunha, M., Da Frota, S. M., Mallmann, E. J. J., Freire, T. M., Costa, L. S., et al (2018). Biogenic synthesis of multifunctional silver nanoparticles from *Rhodotorula glutinis* and *Rhodotorula mucilaginosa*: antifungal, catalytic and cytotoxicity activities. *World J. Microbiol. Biotechnol.* 34, 127.

Cushen, M., Kerry, J., Morris, M., Cruz-Romero, M., and Cummins, E. (2012). Nanotechnologies in the food industry–recent developments, risks and regulation. *Trends Food Sci. Technol.* 24(1), 30–46.

Dameron, C. T., Reese, R. N., Mehra, R. K., Kortan, A. R., Carroll, P. J., Steigerwald, M. L., et al (1989). Biosynthesis of cadmium sulphide quantum semiconductor crystallites. *Nature* 338, 596–597.

Das, S., Jagan, L., Isiah, R., Rajesh, B., Backianathan, S., and Subhashini, J. (2011). Nanotechnology in oncology: characterization and in vitro release kinetics of cisplatin-loaded albumin nanoparticles: implications in anticancer drug delivery. *Indian J. Pharmacol.* 43(4), 409.

Davies, J. C. (2007). *EPA and Nanotechnology: Oversight for the 21st Century* (Washington, DC: Woodrow Wilson International Center for Scholars).

Dekkers, S., Krystek, P., Peters, R. J., Lankveld, D. X., Bokkers, B. G., van Hoeven-Arentzen, P. H., et al (2011). Presence and risks of nanosilica in food products. *Nanotoxicology* 5, 393–405.

Deplanche, K., and Macaskie, L. E. (2008). Biorecovery of gold by *Escherichia coli* and *Desulfovibrio desulfuricans*. *Biotechnol. Bioeng.* 99, 1055–1064.

Dhandapani, P., Siddarth, A. S., Kamalasekaran, S., Maruthamuthu, S., and Rajagopal, G. (2014). Bio-approach: ureolytic bacteria mediated synthesis of ZnO nanocrystals on cotton fabric and evaluation of their antibacterial properties. *Carbohydr. Polym.* 103, 448–455.

Duncan, T. V. (2011). Applications of nanotechnology in food packaging and food safety: barrier materials, antimicrobials and sensors. *J. Colloid Interface Sci.* 363, 1–24.

Dupas, C., and Lahmani, M. (Eds.). (2007). *Nanoscience: Nanotechnologies and Nanophysics* (Berlin: Springer Science & Business Media).

Duran, N., Marcato, P. D., Alves, O. L., De Souza, G. I. H., and Esposito, E. (2005). Mechanistic aspects of biosynthesis of silver nanoparticles by several *Fusarium oxysporum* strains. *J. Nanobiotechnol.* 3, 8.

Duran, N., Marcato, P. D., De Souza, G. I. H., Alves, O. L., and Esposito, E. (2007). Antibacterial effect of silver nanoparticles produced by fungal process on textile fabrics and their effluent treatment. *J. Biomed. Nanotechnol.* 3, 203–208.

Echegoyen, Y., and Nerín, C. (2013). Nanoparticle release from nano-silver antimicrobial food containers. *Food Chem. Toxicol.* 62, 16–22.

Eftekhari, A., Khalilov, R., Kavetskyy, T., Keskin, C., Prasad, R., Rosic, G. L. (2023). Biological/chemical-based metallic nanoparticles synthesis, characterization, and environmental applications. *Frontiers in Chemistry.* 11:1191659. doi: 10.3389/fchem.2023.1191659

Elahi, N., Kamali, M., and Baghersad, M. H. (2019). Recent biomedical applications of gold nanoparticles: a review. *Talanta* 184, 537–556.

Elango, G., and Roopan, S. M. (2015). Green synthesis, spectroscopic investigation and photocatalytic activity of lead nanoparticles. *Spectrochimi. Acta Part AMol. Biomol. Spectrosc.* 139, 367–373.

El-Batal, A., Mona, S., and Al-Tamie, M. (2015). Biosynthesis of gold nanoparticles using marine *Streptomyces cyaneus* and their antimicrobial, antioxidant and antitumor (in vitro) activities. *J. Chem. Pharm. Res.* 7, 1020–1036.

Esmaeillou, M., Moharamnejad, M., Hsankhani, R., Tehrani, A. A., and Maadi, H. (2013). Toxicity of ZnO nanoparticles in healthy adult mice. *Environ. Toxicol. Pharmacol.* 35(1), 67–71.

Eugenio, M., Muller, N., Frases, S., Almeida-Paes, R., Lima, L. M. T. R., Lemgruber, L., et al (2016). Yeast-derived biosynthesis of silver/silver chloride nanoparticles and their antiproliferative activity against bacteria. *RSC Adv.* 6, 9893–9904.

Ezhilarasi, P. N., Karthik, P., Chhanwal, N., and Anandharamakrishnan, C. (2013). Nanoencapsulation techniques for food bioactive components: a review. *Food Bioprocess Technol.* 6, 628–647.

Fatemi, M., Mollania, N., Momeni-Moghaddam, M., and Sadeghifar, F. (2018). Extracellular biosynthesis of magnetic iron oxide nanoparticles by *Bacillus cereus* strain HMH1: characterization and in vitro cytotoxicity analysis on MCF-7 and 3T3 cell lines. *J. Biotechnol.* 270, 1–11.

Franke, C. E., Czapar, A. E., Patel, R. B., and Steinmetz, N. F. (2017). Tobacco mosaic virus-delivered cisplatin restores efficacy in platinum-resistant ovarian cancer cells. *Mol. Pharm.* 15, 2922–2931.

Gálvez, A., Abriouel, H., López, R. L., and Omar, N. B. (2007). Bacteriocinbased strategies for food biopreservation. *Int. J. Food Microbiol.* 120, 51–70.

Gan, P. P., and Li, S. F. Y. (2012). Potential of plant as a biological factory to synthesize gold and silver nanoparticles and their applications. *Rev. Environ. Sci. Bio. Technol.* 11, 169–206.

Gao, C., Wang, Y., Ye, Z., Lin, Z., Ma, X., and He, Q. (2020). Biomedical micro/nanomotors: from overcoming biological barriers to in vivo imaging. *Adv. Mater.* 33, 2000512.

Garcia, M., Aleixandre, M., Gutiérrez, J., and Horrillo, M. C. (2006). Electronic nose for wine discrimination. *Sens. Actuat. B* 113, 911–916.

Gergely, G., Wéber, F., Lukács, I., Illés, L., Tóth, A., Horváth, Z., ... Balázsi, C. (2010). Nano-hydroxyapatite preparation from biogenic raw materials. *Open Chem.* 8(2), 375–381.

Ghodake, G., and Lee, D. S. (2011). Biological synthesis of gold nanoparticles using the aqueous extract of the brown algae *Laminaria japonica. J. Nanoelectr. Optoelectr.* 6, 268–271.

Golinska, P., Wypij, M., Ingle, A. P., Gupta, I., Dahm, H., and Rai, M. (2014). Biogenic synthesis of metal nanoparticles from actinomycetes: biomedical applications and cytotoxicity. *Appl. Microbiol. Biotechnol.* 98, 8083–8097.

Govindaraju, K., Basha, S. K., Kumar, V. G., and Singaravelu, G. (2008). Silver, gold and bimetallic nanoparticles production using single-cell protein (*Spirulina platensis*) Geitler. *J. Mater. Sci.* 43, 5115–5122.

Gu, H., Chen, X., Chen, F., Zhou, X., and Parsaee, Z. (2018). Ultrasound-assisted biosynthesis of CuO-NPs using brown alga *Cystoseira trinodis*: characterization, photocatalytic AOP, DPPH scavenging and antibacterial investigations. *Ultrason. Sonochem.* 41, 109–119.

Guilger-Casagrande, M., and de Lima, R. (2019). Synthesis of silver nanoparticles mediated by fungi: a review. *Front. Bioeng. Biotechnol.* 7, 287.

Hamida, R. S., Abdelmeguid, N. E., Ali, M. A., Bin-Meferij, M. M., and Khalil, M. I. (2020). Synthesis of silver nanoparticles using a novel cyanobacteria *Desertifilum* sp. extract: their antibacterial and cytotoxicity effects. *Int. J. Nanomed.* 15, 49.

Hamouda, R. A., Hussein, M. H., Abo-Elmagd, R. A., and Bawazir, S. S. (2019). Synthesis and biological characterization of silver nanoparticles derived from the cyanobacterium *Oscillatoria limnetica. Sci. Rep.* 9, 13071.

Hannon, J. C., Kerry, J. P., Cruz-Romero, M., Azlin-Hasim, S., Morris, M., and Cummins, E. (2016). Assessment of the migration potential of nanosilver from nanoparticle-coated low density polyethylene food packaging into food simulants. *Food Addit. Contam Part A* 33, 167–78.

Harris, H. W., Sanchez-Andrea, I., Mclean, J. S., Salas, E. C., Tran, W., El-Naggar, M. Y., et al (2018). Redox sensing within the genus *Shewanella. Front. Microbiol.* 8, 2568.

He, S., Guo, Z., Zhang, Y., Zhang, S., Wang, J., and Gu, N. (2007). Biosynthesis of gold nanoparticles using the bacteria *Rhodopseudomonas capsulata. Mater. Lett.* 61, 3984–3987.

He, X., Fu, P., Aker, W. G., and Hwang, H. M. (2018). Toxicity of engineered nanomaterials mediated by nano–bio–eco interactions. *J. Environ. Sci. Health, Part C* 36(1), 21–42.

He, X., and Hwang, H. M. (2016). Nanotechnology in food science: functionality, applicability, and safety assessment. *J. Food Drug Anal.* 24(4), 671–681.

Heidelberg, J. F., Paulsen, I. T., Nelson, K. E., Gaidos, E. J., Nelson, W. C., Read, T. D., et al (2002). Genome sequence of the dissimilatory metal ionreducing bacterium *Shewanella oneidensis. Nat. Biotechnol.* 20, 1118–1123.

Helmke, B. P., and Minerick, A. R. (2006). Designing a nano-interface in a microfluidic chip to probe living cells: challenges and perspectives. *Proc. Nat. Acad. Sci. U.S.A.* 103, 6419–6424.

Hodge, G. A., Maynard, A. D., and Bowman, D. M. (2014). Nanotechnology: rhetoric, risk and regulation. *Sci. Public Policy* 41(1), 1–14.

Hulkoti, N. I., and Taranath, T. C. (2014). Biosynthesis of nanoparticles using microbes-a review. *Colloids Surf. B Biointerfaces* 121, 474–483.

Husain, S., Afreen, S., Yasin, D., Afzal, B., and Fatma, T. (2019). Cyanobacteria as a bioreactor for synthesis of silver nanoparticles-an effect of different reaction conditions on the size of nanoparticles and their dye decolorization ability. *J. Microbiol. Methods* 162, 77–82.

Husseiny, M. I., Abd El-Aziz, M., Badr, Y., and Mahmoud, M. A. (2007). Biosynthesis of gold nanoparticles using *Pseudomonas aeruginosa*. Spectrochim. *Acta Part A Mol. Biomol. Spectrosc.* 67, 1003–1006.

Inbaraj, B. S., and Chen, B. H. (2015). Nanomaterial-based sensors for detection of foodborne bacterial pathogens and toxins as well as pork adulteration in meat products. *J. Food Drug Anal.* 24, 15–28.

Jalal, M., Ansari, M. A., Alzohairy, M. A., Ali, S. G., Khan, H. M., Almatroudi, A., et al (2018). Biosynthesis of silver nanoparticles from oropharyngeal *Candida glabrata* isolates and their antimicrobial activity against clinical strains of bacteria and fungi. *Nanomaterials* 8, 586.

Jianrong, C., Yuqing, M., Nongyue, H., Xiaohua, W., and Sijiao, L. (2004). Nanotechnology and biosensors. *Biotechnol. Adv.* 22, 505–518.

Jovanovic, B. (2015). Critical review of public health regulations of titanium dioxide, a human food additive. *Integrat. Environ. Assess. Manage,* 11, 10–20. https://doi.org/10.1002/ieam.1571

Kaler, A., Jain, S., and Banerjee, U. C. (2013). Green and rapid synthesis of anticancerous silver nanoparticles by *Saccharomyces boulardii* and insightinto mechanism of nanoparticle synthesis. *Biomed Res. Int.* 2013, 872940.

Kalpana, V. N., Kataru, B. A. S., Sravani, N., Vigneshwari, T., Panneerselvam, A., and Rajeswari, V. D. (2018). Biosynthesis of zinc oxide nanoparticles using culture filtrates of *Aspergillus niger*: antimicrobial textiles and dye degradation studies. *OpenNano* 3, 48–55.

Kanazawa, K., and Cho, N. J. (2009). Quartz crystal microbalance as a sensor to characterize macromolecular assembly dynamics. *J. Sens.* 6, 1–17.

Karlo, J., Prasad, R., Singh, S.P. (2023) Biophotonics in food technology: Quo vadis? *Journal of Agriculture and Food Research* https://doi.org/10.1016/j.jafr.2022.100482

Khatoon, N., Ahmad, R., and Sardar, M. (2015). Robust and fluorescent silver nanoparticles using *Artemisia annua*: biosynthesis, characterization and antibacterial activity. *Biochem. Eng. J.* 102, 91–97.

Kim, T.-Y., Kim, M. G., Lee, J.-H., and Hur, H.-G. (2018). Biosynthesis of nanomaterials by *Shewanella* species for application in lithium ion batteries. *Front. Microbiol.* 9, 2817.

Klaus, T., Joerger, R., Olsson, E., and Granqvist, C.-G. R. (1999). Silver-based crystalline nanoparticles, microbially fabricated. *Proc. Natl. Acad. Sci. U.S.A.* 96, 13611–13614.

Koo, O. M., Rubinstein, I., and Onyuksel, H. (2005). Role of nanotechnology in targeted drug delivery and imaging: a concise review. *Nanomed. Nanotechnol. Biol. Med.* 1, 193–212.

Koopi, H., and Buazar, F. (2018). A novel one-pot biosynthesis of pure alpha aluminum oxide nanoparticles using the macroalgae *Sargassum ilicifolium*: a green marine approach. *Ceram. Int.* 44, 8940–8945.

Kora, A. J., and Rastogi, L. (2016). Biomimetic synthesis of selenium nanoparticles by *Pseudomonas aeruginosa* ATCC 27853: an approach for conversion of selenite. *J. Environ. Manage.* 181, 231–236.

Kowshik, M., Ashtaputre, S., Kharrazi, S., Vogel, W., Urban, J., Kulkarni, S. K., et al (2002). Extracellular synthesis of silver nanoparticles by a silver-tolerant yeast strain MKY3. *Nanotechnology* 14, 95.

Kulkarni, R. R., Shaiwale, N. S., Deobagkar, D. N., and Deobagkar, D. D. (2015). Synthesis and extracellular accumulation of silver nanoparticles by employing radiation-resistant *Deinococcus radiodurans*, their characterization, and determination of bioactivity. *Int. J. Nanomed.* 10, 963.

Kumar, R. M. P., Venkatesh, A., and Moorthy, V. H. S. (2019). Nanopits based novel hybrid plasmonic nanosensor fabricated by a facile nanofabrication technique for biosensing. *Mater. Res. Express* 6, 1150b6.

Kumar, S. A., Abyaneh, M. K., Gosavi, S.W., Kulkarni, S. K., Pasricha, R., Ahmad, A., et al (2007). Nitrate reductase-mediated synthesis of silver nanoparticles from AgNO3. *Biotechnol. Lett.* 29, 439–445.

Kumar, S. A., Peter, Y. A., and Nadeau, J. L. (2008). Facile biosynthesis, separation and conjugation of gold nanoparticles to doxorubicin. *Nanotechnology* 19, 495101.

Kundu, S., Maheshwari, V., and Saraf, R. F. (2008). Photolytic metallization of Au nanoclusters and electrically conducting micrometer long nanostructures on a DNA scaffold. *Langmuir* 24, 551–555.

Lamprecht, A., Saumet, J. L., Roux, J., and Benoit, J. P. (2004). Lipid nanocarriers as drug delivery system for ibuprofen in pain treatment. *Int. J. Pharma.* 278, 407–414.

Langer, R., and Peppas, N. A. (2003). Advances in biomaterials, drug delivery, and bionano-technology. *AIChE J.* 49, 2990–3006.

Lengke, M. F., Fleet, M. E., and Southam, G. (2006). Synthesis of platinum nanoparticles by reaction of filamentous cyanobacteria with platinum (IV) – chloride complex. *Langmuir* 22, 7318–7323.

Lengke, M. F., Fleet, M. E., and Southam, G. (2007a). Biosynthesis of silver nanoparticles by filamentous cyanobacteria from a silver (I) nitrate complex. *Langmuir* 23, 2694–2699.

Lengke, M. F., Fleet, M. E., and Southam, G. (2007b). Synthesis of palladium nanoparticles by reaction of filamentous cyanobacterial biomass with a palladium (II) chloride complex. *Langmuir* 23, 8982–8987.

Liu, S., Yuan, L., Yue, X., Zheng, Z., and Tang, Z. (2008). Recent advances in nanosensors for organophosphate pesticide detection. *Adv. Powder. Technol.* 19, 419–441.

Lombardo, D., Calandra, P., Pasqua, L., and Magazu, S. (2020). Self-assembly of organic nanomaterials and biomaterials: the bottom-up approach for functional nanostructures formation and advanced applications. *Materials* 13, 1048.

Maceda, A. F., Ouano, J. J. S., Que, M. C. O., Basilia, B. A., Potestas, M. J., and Alguno, A. C. (2018). Controlling the absorption of gold nanoparticles via green synthesis using *Sargassum crassifolium* extract. *Key Eng. Mater.* 765, 44–48.

Maric, I., Stefanic, G., Gotic, M., and Jurkin, T. (2019). The impact of dextran sulfate on the radiolytic synthesis of magnetic iron oxide nanoparticles. *J. Mol. Struct.* 1183, 126–136.

Maynard, A. D. (2007). Nanotechnology: the next big thing, or much ado about nothing? *Ann. Occup. Hyg.* 51(1), 1–12.

Mazumder, J. A., Ahmad, R., and Sardar, M. (2016). Reusable magnetic nanobiocatalyst for synthesis of silver and gold nanoparticles. *Int. J. Biol. Macromol.* 93, 66–74.

McClements, D. J., and Xiao, H. (2017). Is nano safe in foods? Establishing the factors impacting the gastrointestinal fate and toxicity of organic and inorganic food-grade nanoparticles. *npj Sci. Food* 1(1), 1–13.

Mihindukulasuriya, S. D. F., and Lim, L. T. (2014). Nanotechnology development in food packaging: a review. *Trends Food Sci. Technol.* 40, 149–167.

Mishra, A., Ahmad, R., Perwez, M., and Sardar, M. (2016). Reusable green synthesized bio-mimetic magnetic nanoparticles for glucose and H2O2 detection. *Bionanoscience* 6, 93–102.

Mishra, A., Ahmad, R., Singh, V., Gupta, M. N., and Sardar, M. (2013). Preparation, characterization and biocatalytic activity of a nanoconjugate of alpha amylase and silver nanoparticles. *J. Nanosci. Nanotechnol.* 13, 5028–5033.

Mishra, A., Tripathy, S. K., and Yun, S.-I. (2011). Bio-synthesis of gold and silver nanoparticles from *Candida guilliermondii* and their antimicrobial effect against pathogenic bacteria. *J. Nanosci. Nanotechnol.* 11, 243–248.

Mohamed, W. S., and Abu-Dief, A. M. (2018). Synthesis, characterization and photocatalysis enhancement of Eu2O3-ZnO mixed oxide nanoparticles. *J. Phys. Chem. Solids* 116, 375–385.

Molnar, Z., Bodai, V., Szakacs, G., Erdelyi, B., Fogarassy, Z., Safran, G., et al (2018). Green synthesis of gold nanoparticles by thermophilic filamentous fungi. *Sci. Rep.* 8, 3943.

Mukherjee, P., Senapati, S., Mandal, D., Ahmad, A., Khan, M. I., Kumar, R., et al (2002). Extracellular synthesis of gold nanoparticles by the fungus *Fusarium oxysporum*. *ChemBioChem* 3, 461–463.

Murray, A. J., Zhu, J., Wood, J., and Macaskie, L. E. (2017). A novel biorefinery: biorecovery of precious metals from spent automotive catalyst leachates into new catalysts effective in metal reduction and in the hydrogenation of 2-pentyne. *Miner. Eng.* 113, 102–108.

Nachay, K. (2007). Analyzing nanotechnology. *Food Technol.* 1, 34–36.

Nair, B., and Pradeep, T. (2002). Coalescence of nanoclusters and formation of submicron crystallites assisted by *Lactobacillus* strains. *Crystal Growth Design* 2, 293–298.

Nakagawa, K. (2014). Nano- and micro-encapsulation of flavor in food systems. In *Nano-and Microencapsulation for Foods*, Chapter 10, ed. H.-S. Kwak (Oxford: John Wiley & Sons), 249–272.

Noori, R., Ahmad, R., and Sardar, M. (2020). Nanobiosensor in health sector: the milestones achieved and future prospects. In *Nanobiosensors for Agricultural, Medical and Environmental Applications*, eds M. Mohsin, R. Naz, and A. Ahmad (Gateway East: Springer), 63–90.

Oh, J.-W., and Han, D.-W. (2020). Virus-based nanomaterials and nanostructures. *Nanomaterials* 10, 567.

Othman, S. H. (2014). Bio-nanocomposite materials for food packaging applications: types of biopolymer and nano-sized filler. *Agric. Agric. Sci. Proc.* 2, 296–303.

Ottaway, P. B. (2010). Nanotechnology in supplements and foods – EU concerns. Available at: http://www.accessmylibrary.com/coms2/summary_ 0286-37130259_ITM [accessed February 26, 2010].

Ovais, M., Khalil, A. T., Ayaz, M., Ahmad, I., Nethi, S. K., and Mukherjee, S. (2018). Biosynthesis of metal nanoparticles via microbial enzymes: a mechanistic approach. *Int. J. Mol. Sci.* 19, 4100.

Ozturk, A. B., Argin, S., Ozilgen, M., and McClements, D. J. (2015). Formation and stabilization of nanoemulsion-based vitamin E delivery systems using natural biopolymers: whey protein isolate and gum. *Food Chem.* 188, 256–263.

Palchetti, I., and Mascini, M. (2008). Electroanalytical biosensors and their potential for food pathogen and toxin detection. *Anal. Bioanal. Chem.* 391, 455–471.

Pastorino, F., Brignole, C., Di Paolo, D., Perri, P., Curnis, F., Corti, A., et al (2019). Overcoming biological barriers in neuroblastoma therapy: the vascular targeting approach with liposomal drug nanocarriers. *Small* 15, 1804591.

Pinto, R. J. B., Daina, S., Sadocco, P., Neto, C. P., and Trindade, T. (2013). Antibacterial activity of nanocomposites of copper and cellulose. *BioMed Res. Int.* 6, 280512.

Pradhan, N., Singh, S., Ojha, N., Srivastava, A., Barla, A., Rai, V., et al (2015). Facets of nanotechnology as seen in food processing, packaging, and preservation industry. *BioMed Res. Int.* 2015, 365672.

Prakasham, R. S., Kumar, B. S., Kumar, Y. S., and Shankar, G. G. (2012). Characterization of silver nanoparticles synthesized by using marine isolate *Streptomyces albidoflavus*. J. *Microbiol. Biotechnol.* 22, 614–621.

Prasad, R., Pandey, R., Barman, I. (2016) Engineering tailored nanoparticles with microbes: quo vadis. *WIREs Nanomed Nanobiotechnol* 8:316–330. doi: 10.1002/wnan.1363

Prasad, R. (2017). *Fungal Nanotechnology: Applications in Agriculture, Industry, and Medicine.* Springer Nature Singapore Pte Ltd. (ISBN 978-3-319-68423-9)

Priyadarshini, S., Gopinath, V., Priyadharsshini, N. M., Mubarakali, D., and Velusamy, P. (2013). Synthesis of anisotropic silver nanoparticles using novel strain, *Bacillus flexus* and its biomedical application. *Colloids Surf. B Biointerfaces* 102, 232–237.

Pugazhenthiran, N., Anandan, S., Kathiravan, G., Prakash, N. K. U., Crawford, S., and Ashokkumar, M. (2009). Microbial synthesis of silver nanoparticles by *Bacillus* sp. *J. Nanopart. Res.* 11, 1811.

Rad, M., Taran, M., and Alavi, M. (2018). Effect of incubation time, CuSO4 and glucose concentrations on biosynthesis of copper oxide (CuO) nanoparticles with rectangular shape and antibacterial activity: taguchi method approach. *Nano Biomed. Eng.* 10, 25–33.

Rajeshkumar, S., and Sivapriya, D. (2020). Fungus-mediated nanoparticles: characterization and biomedical advances. In *Nanoparticles in Medicine*, ed. A. Shukla (Gateway East: Springer), 185–199.

Ramakrishna, M., Babu, D. R., Gengan, R. M., Chandra, S., and Rao, G. N. (2016). Green synthesis of gold nanoparticles using marine algae and evaluation of their catalytic activity. *J. Nanostruct. Chem.* 6, 1–13.

Rasmussen, J. W., Martinez, E., Louka, P., and Wingett, D. G. (2010). Zinc oxide nanoparticles for selective destruction of tumor cells and potential for drug delivery applications. *Expert Opin. Drug Deliv.* 7(9), 1063– 1077.

Renton, A. (2006). Welcome to the World of Nano Foods. Available at: http: //observer.guardian.co.uk/foodmonthly/futureoffood/story/0,1971266,00.html [accessed January 17, 2008].

Rhim, J. W., and Ng, P. K. (2007). Natural biopolymer-based nanocomposite films for packaging applications. *Crit. Rev. Food Sci. Nutri.* 47, 411–433.

Riddin, T. L., Gericke, M., and Whiteley, C. G. (2006). Analysis of the inter-and extracellular formation of platinum nanoparticles by *Fusarium oxysporum* f. sp. *lycopersici* using response surface methodology. *Nanotechnology* 17, 3482.

Sadaf, A., Ahmad, R., Ghorbal, A., Elfalleh, W., and Khare, S. K. (2020). Synthesis of cost-effective magnetic nano-biocomposites mimicking peroxidase activity for remediation of dyes. *Environ. Sci. Pollut. Res.* 27, 27211–27220.

Sadrolhosseini, A. R., Rashid, S. A., Shafie, S., and Soleimani, H. (2019). Laser ablation synthesis of Ag nanoparticles in graphene quantum dots aqueous solution and optical properties of nanocomposite. *Appl. Phys. A* 125, 82.

Sanaeimehr, Z., Javadi, I., and Namvar, F. (2018). Antiangiogenic and antiapoptotic effects of green-synthesized zinc oxide nanoparticles using *Sargassum muticum* algae extraction. *Cancer Nanotechnology*, 9, 1–16.

Saravanan, M., Barik, S. K., Mubarakali, D., Prakash, P., and Pugazhendhi, A. (2018). Synthesis of silver nanoparticles from *Bacillus brevis* (NCIM 2533) and their antibacterial activity against pathogenic bacteria. *Microb. Pathog.* 116, 221–226.

Sardar, M., Mishra, A., and Ahmad, R. (2014). Biosynthesis of metal nanoparticles and their applications.. In *Biosensors and Nanotechnology*, eds A. Tiwari and A. P. F. Turner (Beverly, MA: Scrivener Publishing), 239–266.

Sari, P., Mann, B., Kumar, R., Singh, R. R. B., Sharma, R., Bhardwaj, M., et al (2015). Preparation and characterization of nanoemulsion encapsulating curcumin. *Food Hydrocol.* 43, 540–546.

Schirmer, B. C., Heiberg, R., Eie, T., Møretrø, T., Maugesten, T., and Carlehøg, M. (2009). A novel packaging method with a dissolving CO2 headspace combined with organic acids prolongs the shelf life of fresh salmon. *Int. J. Food Microbiol.* 133, 154–160.

Selvarajan, E., and Mohanasrinivasan, V. (2013). Biosynthesis and characterization of ZnO nanoparticles using *Lactobacillus plantarum* VITES07. *Mater. Lett.* 112, 180–182.

Senapati, S., Mandal, D., Ahmad, A., Khan, M. I., Sastry, M., and Kumar, R. (2004). Fungus mediated synthesis of silver nanoparticles: a novel biological approach. *Indian J. Phys.* 78, 101–105.

Shaligram, N. S., Bule, M., Bhambure, R., Singhal, R. S., Singh, S. K., Szakacs, G., et al (2009). Biosynthesis of silver nanoparticles using aqueous extract from the compactin producing fungal strain. *Proc. Biochem.* 44, 939–943.

Shi, S., Wang, W., Liu, L., Wu, S., Wei, Y., and Li, W. (2013). Effect of chitosan/ nano-silica coating on the physicochemical characteristics of longan fruit under ambient temperature. *J. Food Eng.*, 118, 125–131.

Silva, N., Ramirez, S., Diaz, I., Garcia, A., and Hassan, N. (2019). Easy, quick, and reproducible sonochemical synthesis of CuO nanoparticles. *Materials* 12, 804.

Singh, B. R., Dwivedi, S., Al-Khedhairy, A. A., and Musarrat, J. (2011). Synthesis of stable cadmium sulfide nanoparticles using surfactin produced by *Bacillus amyloliquifaciens* strain KSU-109. *Colloids Surf. B Biointerfaces* 85, 207–213.

Singh, R., Shedbalkar, U. U., Wadhwani, S. A., and Chopade, B. A. (2015). Bacteriagenic silver nanoparticles: synthesis, mechanism, and applications. *Appl. Microbiol. Biotechnol.* 99, 4579–4593.

Singhal, G., Bhavesh, R., Kasariya, K., Sharma, A. R., and Singh, R. P. (2011). Biosynthesis of silver nanoparticles using *Ocimum sanctum* (Tulsi) leaf extract and screening its antimicrobial activity. *J. Nanopart. Res.*, 13(7), 2981–2988.

Sintubin, L., De Windt, W., Dick, J., Mast, J., Van Der Ha, D., Verstraete, W., et al (2009). Lactic acid bacteria as reducing and capping agent for the fast and efficient production of silver nanoparticles. *Appl. Microbiol. Biotechnol.* 84, 741–749.

Soares, N. F. F., Silva, C. A. S., Santiago-Silva, P., Espitia, P. J. P., Gonçalves, M. P. J. C., Lopez, M. J. G., et al (2009). Active and intelligent packaging for milk and milk products. In *Engineering Aspects of Milk and Dairy Products*, eds J. S. R. Coimbra and J. A. Teixeira (New York: CRC Press), 155–174.

Sorrentino, A., Gorrasi, G., and Vittoria, V. (2007). Potential perspectives of bionanocomposites for food packaging applications. *Trends Food Sci. Technol.* 18, 84–95.

Srinath, B. S., Namratha, K., and Byrappa, K. (2018). Eco-friendly synthesis of gold nanoparticles by *Bacillus subtilis* and their environmental applications. *Adv. Sci. Lett.* 24, 5942–5946.

Steinmetz, N. F. (2010). Viral nanoparticles as platforms for next-generation therapeutics and imaging devices. *Nanomed. Nanotechnol. Biol. Med.* 6, 634–641.

Stillman, N. R., Kovacevic, M., Igor, B., and Sabine, H. (2020). In silico modelling of cancer nanomedicine, across scales and transport barriers. *NPJ Comput. Mater.* 6, 1–10.

Su, S. L., and Li, Y. (2004). Quantum dot biolabeling coupled with immunomagnetic separation for detection of *Escherichia coli* O157:H7. *Analyt. Chem.* 76(16), 4806–4810.

Subramanian, A. (2006). A mixed self-assembled monolayer-based surface Plasmon immunosensor for detection of *E. coli* O157H7. *Biosens. Bioelectron.* 7, 998–1006.

Subramanian, M., Alikunhi, N. M., and Kandasamy, K. (2010). In vitro synthesis of silver nanoparticles by marine yeasts from coastal mangrove sediment. *Adv. Sci. Lett.* 3, 428–433.

Sukanya, M. K., Saju, K. A., Praseetha, P. K., and Sakthivel, G. (2013). Therapeutic potential of biologically reduced silver nanoparticles from actinomycete cultures. *J. Nanosci.* 2013, 940719.

Swaminathan, S., Murugesan, S., Damodarkumar, S., Dhamotharan, R., and Bhuvaneshwari, S. (2011). Synthesis and characterization of gold nanoparticles from alga *Acanthophora spicifera* (VAHL) boergesen. *Int. J. Nanosci. Nanotechnol.* 2, 85–94.

Tan, F., Leung, P. H. M., Liud, Z., Zhang, Y., Xiao, L., Ye, W., et al (2011). Microfluidic impedance immunosensor for *E. coli* O157:H7 and *Staphylococcus aureus* detection via antibody-immobilized nanoporous membrane. *Sensor. Actuat. B. Chem.* 159, 328–335.

Tan, H., Ma, R., Lin, C., Liu, Z., and Tang, T. (2013). Quaternized chitosan as an antimicrobial agent: antimicrobial activity, mechanism of action and biomedical applications in orthopedics. *Int. J. Mol. Sci.* 14, 1854–1869.

Tully, E., Hearty, S., Leonard, P., and O'Kennedy, R. (2006). The development of rapid fluorescence-based immunoassays, using quantum dot-labelled antibodies for the detection of *L. monocytogenes* cell surface proteins. *Int. J. Biol. Macromolec.* 39, 127–134.

Usha, R., Prabu, E., Palaniswamy, M., Venil, C. K., and Rajendran, R. (2010). Synthesis of metal oxide nano particles by *Streptomyces* sp. for development of antimicrobial textiles. *Global J. Biotechnol. Biochem.* 5, 153–160.

Vijayanandan, A. S., and Balakrishnan, R. M. (2018). Biosynthesis of cobalt oxide nanoparticles using endophytic fungus *Aspergillus nidulans*. *J. Environ. Manage.* 218, 442–450.

Vijayaraghavan, K., Mahadevan, A., Sathishkumar, M., Pavagadhi, S., and Balasubramanian, R. (2011). Biosynthesis of Au (0) from Au (III) via biosorption and bioreduction using brown marine alga *Turbinaria conoides*. *Chem. Eng. J.* 167, 223–227.

Wadhwani, S. A., Shedbalkar, U. U., Singh, R., Vashisth, P., Pruthi, V., and Chopade, B. A. (2016). Kinetics of synthesis of gold nanoparticles by *Acinetobacter* sp. SW30 isolated from environment. *Indian J. Microbiol.* 56, 439–444.

Waghmare, S. R., Mulla, M. N., Marathe, S. R., and Sonawane, K. D. (2015). Ecofriendly production of silver nanoparticles using *Candida utilis* and its mechanistic action against pathogenic microorganisms. *3 Biotech* 5, 33–38.

Waghmare, S. S., Deshmukh, A. M., Kulkarni, S. W., and Oswaldo, L. A. (2011). Biosynthesis and characterization of manganese and zinc nanoparticles. *Univ. J. Environ. Res. Technol.* 1, 64–69.

Wang, C., and Irudayaraj, J. (2008). Gold nanorod probes for the detection of multiple pathogens. *Small* 4(12), 2204–2208.

Wang, H. U. I., Brandl, D. W., Nordlander, P., and Halas, N. J. (2007). Plasmonic nanostructures: artificial molecules. *Acc. Chem. Res.* 40, 53–62.

Wang, L., Chen, W., Xu, D., Shim, B. S., Zhu, Y., Sun, F., et al (2009). Simple, rapid, sensitive, and versatile SWNT-paper sensor for environmental toxin detection competitive with *ELISA. Nano Lett.* 9, 4147–4152.

Weiss, J., Takhistov, P., and McClements, J. (2006). Functional materials in food nanotechnology. *J. Food Sci.* 71, R107–R116.

Xia, Y. (2008). Nanomaterials at work in biomedical research. *Nat. Mater.* 7, 758–760.

Yacoby, I., Shamis, M., Bar, H., Shabat, D., and Benhar, I. (2006). Targeting antibacterial agents by using drug-carrying filamentous bacteriophages. *Antimicrob. Agents Chemother.* 50, 2087–2097.

Yan, S. S., and Gilbert, J. M. (2004). Antimicrobial drug delivery in food animals and microbial food safety concerns: an overview of in vitro and in vivo factors potentially affecting the animal gut microflora. *Adv. Drug Deliv. Rev.* 56, 1497–1521.

Yang, R., Zhou, Z., Sun, G., Gao, Y., Xu, J., Strappe, P., et al (2015). Synthesis of homogeneous protein-stabilized rutin nanodispersions by reversible assembly of soybean (*Glycine max*) seed ferritin. *RSC Adv.* 5, 31533–31540.

Yang, Y., Doudrick, K., Bi, X., Hristovski, K., Herckes, P., Westerhoff, P., and Kaegi, R. (2014). Characterization of food-grade titanium dioxide: the presence of nanosized particles. *Environ. Sci. Technol.*, 48(11), 6391–6400.

Yu, Y. W., Zhang, S. Y., Ren, Y. Z., Li, H., Zhang, X. N., and Di, J. H. (2012). Jujube preservation using chitosan film with nano-silicon dioxide. *J. Food Eng.* 113, 408–414.

Zhang, T., Lv, C., Chen, L., Bai, G., Zhao, G., and Xu, C. (2014). Encapsulation of anthocyanin molecules within a ferritin nanocage increases their stability and cell uptake efficiency. *Food Res. Int.* 62, 183–192.

Zhang, X., He, X., Wang, K., and Yang, X. (2011). Different active biomolecules involved in biosynthesis of gold nanoparticles by three fungus species. *J. Biomed. Nanotechnol.* 7, 245–254.

Zhao, X., Zhou, L., Riaz Rajoka, M. S., Yan, L., Jiang, C., Shao, D., et al (2018). Fungal silver nanoparticles: synthesis, application and challenges. *Crit. Rev. Biotechnol.* 38, 817–835.

Zonooz, N. F., Salouti, M., Shapouri, R., and Nasseryan, J. (2012). Biosynthesis of gold nanoparticles by *Streptomyces* sp. ERI-3 supernatant and process optimization for enhanced production. *J. Clust. Sci.* 23, 375–382.

Section II

Understanding microbiology for
environmental sustainability

9 Understanding the soil microbiome

Perspectives for environmental bioremediation

M. Subhosh Chandra and M. Srinivasulu

CONTENTS

9.1 INTRODUCTION

Weather alteration and land use exercises are probable to influence biogeochemical cycling by the soil microorganisms in the ecology. Comprehension of how soil microorganisms might react with these factors is enormously significant to alleviate the procedure of land deterioration. The incorporated study of shotgun metagenomics application attached to weather forecasting statistics was sign up to untie the effect of seasons and land exploit transform on soil microorganisms on conserved and experimental meadows in Caatinga drylands. Multivariate analysis recommended microorganisms of conserved soils with seasonal alteration were shaped mainly in water scarcity, with a sturdy augment of *Actinobacteria* and *Proteobacteria* species in arid and rainy seasons, correspondingly. Whereas nutritional accessibility remarkably played a decisive function in motivating the microbiome in agricultural soils. The sturdy enhancement of bacterial species fit in to the unfamiliar phylum *Acidobacteria* in soils with dry season influenced by ferti-irrigation methods assumes they survive

DOI: 10.1201/9781003394600-11

at nutrient rich environments, particularly carbon and environmental role in reducing the influence of chemical fertilizers.

Majority of current climatic projections recommend enhancement of 11–23% in degree of worldwide arid lands for period of 100 years (Huang and Guan 2016). An urgent necessity to comprehension land employ alters influence soil microbiota in sequence to foresee ecological balance for the growth of additional sustainability land managing in arid land. Seasonal tropical dry forests (STDF) were broadly scattered and the major endangered and less examined the forest ecosystem all over the world (Gillespie et al. 2012). "Caatinga," a Brazil microbiome compiled of biggest STDF in South America, is situated in semiarid north eastern Brazil (occupying approximately 11% of province) marina above 23 million people, suggesting more human operate pressure in normal atmosphere (Beuchle et al. 2015). Even though harboring huge diverseness and a greater height of endemic organisms, this ecology is tranquil less examined plus conserved (Santos et al. 2011). Anthropogenic methods (i.e., crop growing, farm animals, and predatory extractive) and aerial alterations were caused severe ecological spoil, i.e., speed up desertification and effects on carbon plus nitrogen cycles. Escalated attentiveness has been rewarded for farming, depiction, and biological prospective of Caatinga microorganisms because of their exclusive organic features that permit them to live on with harsh climatic circumstances (higher temperature, high UV light, and water scarcity) (Monteiro et al. 2009, Fernandes-Junior et al. 2015).

Metagenomics methods and 16S rRNA sequence were hunted to measure vast microbes assortment and focused on metabolic approaches that allow Caatinga microorganisms to endure in extreme situations (Pacchioni et al. 2014, Leite et al. 2017). Even though few reports have indicated soil microbiota were formed by seasons, this is yet to be apparent how microorganisms acclimatize to change of together Caatinga dry forest into farming lands, adding to essential influence in ecosystem performance. Gileno et al. (2019) studied along with Sao Francisco river valley and consists of conserved vicinity of Caatinga dry forest and other one contains Caatinga fragments encircled by farming lands and subsequently overdone with input of water and nutrients by cropped regions. These regions constitute usual ecological versions for analyzing seasons and anthropogenic stressors demarcation soil microbiome in semiarid circumstances. With this framework, they assumed that: (1) ecological alterations connected in seasonal sequence were key component delineation taxonomic and efficient characteristics of microorganisms in conserved soils from Caatinga dry forest and (2) Input of water and nutrient with land use methods influences usual composition and functional ability of soil microorganisms, by effective measures on biogeochemical ways and soil functionality.

Anthropogenic actions and its waste dumping is a worldwide concern, but specifically difficult in areas with permissive environmental policies and legislation. Waste water is very essential to the farmers in some developing countries, because of its higher nutrient levels. Though, the eternal appliance of waste water able to change physical, chemical, and biological characteristics of soil (Antil et al. 2007, Kharche et al. 2011) and leads maximum titers of heavy metals (Kharche et al. 2011) and dyes. Sewage waste appliance on soils, may not thoroughly practiced it contains highly disease causing organisms that create risk (Lapen et al. 2008, Edwards et al. 2009).

Usually highly depleted soils sludge can be a rich resource of nutrients. It possess N, P, and organic matter too, it also consists heavy metals and chlorinated hydrocarbons (Jang et al. 2010). Increasing heavy metal pollution globally is accredited to anthropogenic actions. Worldwide, released toxic wastes through industry and cities were contaminating soil, air, and water, making unfavorable for cultivation of crop in addition to human and animal welfare (Wahid et al. 2000).There is an imperative necessitate to clean factory wastes to confiscate pollutants before release into soil or rivers. Attentiveness was given to bioremediation schemes of contaminants because of their constant character and enhanced alertness among the worldwide (Ali 2010). Various physicochemical techniques were employed to treat factory wastes (Arslan-Alaton 2007). Though, techniques were pricey and not eco-friendly as they produce huge amount of mud, they need safe removal and also cause secondary contamination (Zhang et al. 2004).

9.2 ROLE OF MICROBES IN ENVIRONMENTAL REMEDIATION

Most microbes are used to eliminate toxic substances from environment and these microbes plays an important role in eliminating pollutants from soil and minimizing their toxic effect in their surrounding (EPA 2016). The microbes or their metabolites are employed to devastate, eradicate, or immobilize toxic substances in the atmosphere (Uqab et al. 2016). PGPR employed in enhancing plant growth using bioremediation of polluted and deteriorate soils and aquatic forms and eliminate pollutants in surroundings (Gouda et al. 2018). Microbes consist enzymes with prospective to detoxify industrial wastes in various ecological circumstances (Pandey et al. 2007). Bioremediation is mainly depends on metabolic activities of microorganism to detoxify aromatic compounds into less harmful substances (Xu and Zhou 2017). It has wide range of applications contrast to physical and chemical methods. For instance, biological method is potential to deteriorate constant and refractory substances. Various applications i.e., a decrease in waste accumulation, less handling, work over a broad array of temperatures, and are easy and simple to operate (Kulshrestha and Husain 2007). Spina et al. (2018) explained that significance of fungi in degradation of pesticides and contaminants. The approach of synthetic communities suggested long-term way to assist remediation as fungi emerged as efficient. Soil contamination is worldwide problem; there are approximately 342,000 polluted sites, this number could be increased over 5,00,000 in 2025 (Anon 2012). Polycyclic aromatic hydrocarbons (PAHs) are a type of toxic compounds, consists of two benzene rings organized in different conformances (Wickliffe et al. 2014). PAHs are discharged into the surroundings as partial burning of crude oil and are all over in nature, with soil is their key basin (Collins et al. 2013). Sixteen of these compounds are main contaminants due to carcinogenic and mutagenic nature, by the United States Environmental Protection Agency (USEPA) (Wang et al. 2010). Bacteria plays a crucial role in degradation of polluted chemicals like PAHs (Fernandez-Luqueno et al. 2011, Brussaard 2012). Due to hydrophobicity of PAHs makes them inadequate to microbes, leading to their perseverance in atmosphere (Posada-Baquero and Ortega-Calvo 2011). A variety of fungi and bacteria competent to degrade PAHs were isolated from polluted sites (Doyle et al. 2008, Ghosal et al. 2016), by Pseudomonas,

Sphingomonas, and Mycobacterium spp. among majority PAH deteriorate bacteria (Bastiaens et al. 2000, Johnsen and Karlson 2005). *Proteobacteria, Actinobacteria,* and *Firmicutes* mainly enhanced the degradation of PAHs and petroleum hydrocarbons by culture independent analyses (Fuentes et al. 2014).

However, remediation methods were effectively useful to PAH polluted sites in the laboratory and field as well (Sayara et al. 2011, Lors et al. 2012), this practice remains badly unstated, generally in requisites of the formation and purpose of microbial grouping concerned (Yang et al. 2015). In global ecosystem, soil microorganisms play an imperative function in sustained structure and biogeochemical cycles, appropriate carbon and modest climate (Bardgett and Van Der Putten 2014). Enhancing urbanization and human-driven practices leads to climate alteration and global warming, it is crucial to comprehend systems that assist soil microbiome to manage by outer factors, and influences on input of biological cycles for microbial process (Andrew et al. 2012, Shade et al. 2013). One gram of surface soil consists billions of microorganisms (Hughes et al. 2001). Recent past, developments in culture independent methods and DNA sequence methods were given a profound perceptive of biotic and abiotic factors disturbing soil microbial consortium; conquering the restrictions of culture-based methods. On the whole in dry and half-dry atmospheres, the availability of temperature and water were use over control on soil microbiome (Nielsen and Ball 2015, Armstrong et al. 2016, Zhao et al. 2016a,b). Short term water grade is an important component adaptable microbial action and microbiota consortium in rice fields differing in pH, organic matter, and soil texture (Liao et al. 2018). The influence of land employ and administration on soil microorganisms were intensively investigated in forests and grassland environments (Rasche et al. 2011, Thapa et al. 2018). In dryland sites, crop production is main disturbing land employ kind nourishing one-third of worldwide people, and notably affect soil healthiness and feature (United Nations Environmental Management Group 2011). The farming supplements, includes manure, herbicide, and irrigation impact on soil diversity of microorganisms and ecology performance. Because of water constrained ecological units occupy approximately 41.5% of earth's exterior (Sorensen 2007) and indisputable significance within biological series (Maestre et al. 2016), reports observed influences of ecological factors i.e., type of weather, land use type, and administration in soil microbial communities purposes (Tian et al. 2017, Pajares et al. 2018).

Anthropogenic activities usually weakening of ecosystems all over the world make them unsuitable for endurance of individual bio-entity. This might be the reason to fast diminishing natural resources. From the past few decades environmental pollution has been a main anxiety, upsetting quality of life. Extensive industrialization, inapt agricultural approaches, released of unchecked contaminants into land and aquatic systems has severely polluted the environment. This has evolved utilization of insufficient normal resources, augment in infertile lands, loss of diverseness, difficulty of drinking water and vast financial losses which are highly hard to remove. Artificial substances were enhancing ever and most of them are recalcitrant and majority are xenobiotic. Based on the record, ten million tons of chemicals were entered into ecosystem all over the world every year. Because of toxic chemicals, i.e., PAHs, polychlorinated biphenyls (PCBs), discharged into the soil and water bodies turned to polluted. These contaminants were toxic and constant, caused severely damage to all forms of life in the

TABLE 9.1

Various types of bacteria and fungi used for bioremediation of pesticides/pollutants

S. no.	Name of the bacteria	Name of the pollutant	Reference
1.	*Pseudomonas, Sphingomonas, Mycobacterium* spp.	PAHs	Bastiaens et al. (2000), Johnsen and Karlson (2005)
2.	*Proteobacteria, Actinobacteria,* and *Firmicutes*	PAHs	Fuentes et al. (2014)
3.	*Alphaproteobacteria*	PAHs	Kuppusamy et al. (2016)
4.	*Pseudomonas, Bacillus, Sphingomonas*	PAHs	Wang et al. (2016)
5.	*Dehalococcoides, Dehalobium*	PCBs	Cutter et al. (2001), Fennell et al. (2004)
6.	*Pseudomonas, Burkholderia, Comamonas, Cupriavidus, Sphingo-monas, Acidovorax, Rhodococcus, Corneybacterium,* and *Bacillus*	PCBs	Furukawa and Fujihara (2008), Seeger and Pieper (2010)
7.	*Pseudomonas aeruginosa*	PCBs	Hatamian-Zarmi et al. (2009)
8.	*Burkholderia xenovorans*	PCBs	Rodrigues et al. (2006)
9.	*Arthrobacter* sp. strain B1B and *Ralstonia eutrophus* H850	PCBs	Gilbert and Crowley (1997)
10.	*Rhodococcus* sp. strain RHA1	PCBs	Iwasaki et al. (2007)
Name of the fungi			
11.	*P. chrysosporium, Lentinus edodes,* and *Phlebia brevispora*	PCBs	Kamei et al. (2006a,b)
12.	*Trametes versicolor*	PCBs	Cloete and Celliers (1999)
13.	*Irpex lacteus, Bjerkanderaadusta, Pycnoporus cinnabarinus,* and *Phanerochaete magnoliae*	PCBs	Cvancarov et al. (2012)
14.	*Pleurotus ostreatus*	PCBs	Kubatova et al. (2001), Cvancarov et al. (2012)

environment (Naveen 2018). Different types of bacteria and fungi used for bioremediation of pesticides/pollutants are presented in Table 9.1.

9.2.1 Role of bacteria in remediation of polycyclic aromatic compounds

Several studies were reported PAH degradation by bacteria with various molecular approaches (Festa et al. 2016, Kuppusamy et al. 2016). Particularly, independent molecular procedures of PAH degradation earmarked bioremediation circumstances. For instance, Jiao et al. (2017) noticed a different types of contaminants were reliable, mostly distinct influences on soil microbiome, chiefly suppressing or triggering definite microbial species. Zhao et al. (2016) observed Mycobacterium donate ring-hydroxylating dioxygenases concerned in early measures of fluoranthene disintegration, though an additional varied group of bacteria given to the metabolism of downstream fluoranthene breakdown products. Kuppusamy et al. (2016) reported

PAH-degrading *Alphaproteobacteria* able to persevere for widespread in soils polluted by heavy metals and PAHs contrast to *Actinobacteria*. Certain modifications can enhance remediation efficacy, for instance Wang et al. (2016) reported surfactants enhanced PAH elimination from soil and increased prevalence of Pseudomonas, Bacillus, and Sphingomonas. Similarly, growth of PAH breaking consortium into contaminated soil too increase remediation and improve the plenty of genes connected by formation of PAH disintegration products can be subsequently metabolized in TCA cycle (Zafra et al. 2016). Festa et al. (2016) noticed that growth of PAH-degrading Sphingobium enhances phenanthrene degradation in soil but had no output in breakdown of PAHs in a constantly polluted soil. Some reports observed less associations amid soil supplements, soil microbial communities and PAH degradation. For instance, Thomas and Cébron (2016) reported phenanthrene alteration in plants did not shown maximum elimination of phenanthrene in bulk soil, although enhances in plenty of PAH-ring-hydroxylating genes. Likewise, Delgado-Balbuena et al. (2016) not shown sturdy associations amid soil microbiome structure and anthracene exclusion in polluted soil. Though these investigations suggest precious path into molecular ecology of PAH degradation in soils, they differ significantly in their goals, target habitat, PAH type, and investigational plan.

Present, literature shows the influences a single PAH in indigenous soil microorganism with no confusing results of several treatments and pollutants were scarce. PAH pollution influences physiologic pressure and trigger detoxification and stress resistance on soil microbial species (De Menezes et al. 2012). So, PAH contact not only enhances the growth of PAH-degrading microbes, but it may also influence the ecological stabilization of soil microorganisms, in turn influencing variety of soil procedures (Griffiths and Philipot 2013).

Microbial environment was influenced with species richness, evenness, and composition, though big diverseness were not essentially a signal of an ecosystem (Griffiths and Philipot 2013, Shade 2017). As amid interruption theory suggested to describe normal surveillance ecology of higher diverseness stages at moderate stages of interruption, the applicable of this assumption to microbial ecosystems is less assured (Gibbons et al. 2016). Bacterial community in soil enhanced three-ring PAH phenanthrene has contrasted with unamended soil in phenanthrene degradation by 16S rRNA sequence. The soil employed in this experimentation had earlier record to PAH contamination because of the closed timber treatment service, though at sampling PAH titers in the soil were same in uncontaminated soils. De Menezes et al. (2012) observed gene expression in a phenanthrene-polluted soil, one time was investigated and the influence of PAH in soil microbial community structure was not determined. Sean et al. reported dominant phenanthrene responsible bacteria proved selfish characteristics, enhancing significantly in large quantity in the supplemented samples but lesser level in control, as a result of their ability to acclimatize and grow in ecologically troubled environments.

9.2.2 Role of Fungi in Remedy of Polycyclic Aromatic Compounds

PAHs are a kind of xenobiotic composites, these are discharged into soil because of inadequate incineration of organic materials Viz., oil, petroleum gas, wood, municipal, and urban waste (Ijoma and Tekere 2017, Kadri et al. 2017). PAHs are greatly bioaccumulative compounds in soil and sediments due to their higher hydrophobicity and lower volatility (Singh 2006). Higher concentration of PAHs can also cause kidney and liver damage, skin inflammation, repress immune reactions, an embryotoxic effect in pregnancy and show genotoxic, carcinogenic, mutagenic, and teratogenic effect (Rostami and Juhasz 2011, Rengarajan et al. 2015). Hence, it is very crucial to transform PAH compounds in soil. The USEPA was identified sixteen PAHs as priority contaminants, such as phenanthrene and pyrene (Jin et al. 2007, Wu et al. 2016). It is a cheap method so it can be directly employed in site (Boopathy 2000a,b). Number of bacterial species were capable to breakdown PAHs (Singh 2006), but present scientists focus on breakdown of PAHs in land and water ecosystems with fungi, by using co-metabolic pathway fungi eliminate PAHs. Fungi did not utilize PAHs as a carbon and energy resource (Casillas et al. 1996). Fungi plays an imperative role in ecosystem with controlling the movement of supplements and energy with mycelia (Tisma et al. 2010). Particularly white rot fungi plays an important role in remediation (Domínguez et al. 2005, Wong 2009). Lignolytic enzymes mainly secreted by white rot fungi i.e., laccase, lignin peroxidase and manganese peroxidase (Kitamura et al. 2005). These lignolytic enzymes play an imperative role in degradation and mineralization of different organic contaminants (Wang et al. 2009).

9.2.3 Effect of Bacteria in Remedy of Polychlorinated Biphenyl

Biotransformation of PCBs is mainly performed by aerobic and anaerobic bacteria. Bacterial cultures of *Dehalococcoides* and *Dehalobium* were connected to halogenations of PCBs (Cutter et al. 2001, Fennell et al. 2004). Various aerobic bacteria able to oxidize PCBs were noticed (Pieper and Seeger 2008). Bacterial species of *Pseudomonas*, *Burkholderia*, *Comamonas*, *Cupriavidus*, *Sphingomonas*, *Acidovorax*, *Rhodococcus*, *Corneybacterium*, and *Bacillus* genera were identified (Furukawa and Fujihara 2008, Seeger and Pieper 2010).

Many studies on the chemical (Wang et al. 2014), physical (Nollet et al. 2003), and biological (Borja et al. 2005) degradation of PCBs were performed to transform these dangerous pollutants. Of which discarding methods, biological PCB breakdown is precious and widely examined. Harkness et al. (1993) reported that PCBs degradation from 37% to 55% at 73 days of incubation by a native aerobic microorganism from Hudson river. Many other aerobic microorganisms with ability to degrade PCBs were noticed, such as a strain of *Pseudomonas aeruginosa* (Hatamian-Zarmi et al. 2009), *Burkholderia xenovorans* (Rodrigues et al. 2006), *Arthrobacter* sp. strain *B1B*, *Ralstonia eutrophus* H850 (Gilbert and Crowley 1997), and *Rhodococcus* sp. strain RHA1 (Iwasaki et al. 2007). Generally, biodegradation was concentrated on use of bacteria and fungi (Pieper and Seeger 2008).

9.2.4 Influence of Fungi in Remediation of Polychlorinated Biphenyl

The procedure of PCB degradation by fungi was also well recognized. Some of the studies developed the flourishing fungal transformation ability in soils (Yadav et al. 1995, Siracusa et al. 2017). Not like bacteria, lignolytic organisms of *Phanerochaete chrysosporium*, a white rot fungus, can calcify tetrachloro and hexachloro alternative to PCB congeners as well as Aroclor 1254 (Pieper et al. 1992, Thomas et al. 1992). Other reports have explained that *P. chrysosporium* breakdown at higher intensity (10 ppm) of Aroclor 1242, 1254, and 1260 (Yadav et al. 1995). PCB degradation by wood-degrading basidiomycetes fungi was well studied (Stella et al. 2015). White rot fungi efficiently breakdown lignin to CO_2 in plants (Boominathan and Reddy 1992). According to Bumpus et al. (1985) and Eaton (1985) white rot fungi *P. chrysosporium* breakdown dioxins, PCBs and other chloro-organics. *P. chrysosporium* was most elaboratively studied lignolytic white rot fungi calcify xenobiotics (Hammel 1992, Bumpus 1993). Number of white rot fungi were assessed for the capability to decompose PCBs (Covino et al. 2016).

Many reports were proved that white rot fungi i.e., *P. chrysosporium* (Kamei et al. 2006a,b), *Trametes versicolor* (Cloete and Celliers 1999), *Lentinus edodes*, *Phlebia brevispora* (Kamei et al. 2006a,b), *Irpex lacteus*, *Bjerkanderaadusta*, *Pycnoporus cinnabarinus*, *Phanerochaete magnoliae* (Cvancarov et al. 2012), and specifically *Pleurotus ostreatus* (Kubatova et al. 2001, Cvancarov et al. 2012) could successfully remove PCB. However, few white rot fungi were examined on PCB contaminated soil (Federici et al. 2012), though *P. ostreatus* was, thus far, likely very efficient PCB-degrading microbe (Kubatova et al. 2001). Oyster mushrooms were observed highly asset to industries like fungal remediation uses. *P. ostreatus* was known for more capable for elimination of PCBs by earlier researchers. Zeddel et al. (1993) explained *P. ostreatus* particularly eliminated PCBs with soil mixed by wood chips. *P. ostreatus* was effectively used at pollutant assemblage from 100 to 650 ppm for single isomers and 2,500 ppm for all PCBs.

9.3 DEGRADATION OF ORGANOPHOSPHATE PESTICIDES BY BACTERIA

Some organophosphorus pesticides (OPs) are very toxic and are still they are using for control of insect. Wide exercise of OPs causes severe problems over food safety and environmental contamination. In the recent past, different bacterial species were isolated and recorded for breakdown of OPs. *Serratia* sp. SPL-2 breakdown methidathion (Li et al. 2013). *Pseudomonas aeruginosa* IS-6 breakdown acephate, methamidophos, methyl parathion, dimethoate, and malathion (Ramu and Seetharaman 2014). Biodegradation of chlorpyrifos using soil bacteria's consists seven species of *Pseudomonas*, *Agrobacterium*, and *Bacillus* were examined (Maya et al. 2011). *Bacillus cereus*, *Bacillus subtilis*, *Brucella melitensis*, *Klebsiella* cultures, *Pseudomonas aeruginosa*, *Pseudomonas fluorescence*, and *Serratia marcescens* were breakdown from 46% to 72% of chlorpyrifos as carbon source at 20 days of incubation (Lakshmi et al. 2008). Several factors influence bacterial degradation of OPs, includes water holding capacity, temperature, pH, and even aging of pesticides in soils prior to inoculation. In a report of *Enterobacter* sp., aging of chlorpyrifos noticed about 10% of it persistent in soil (Singh et al. 2006). The diazinon degrading *Serratia marcescens* was degrade chlorpyrifos, fenitrothion, and parathion (Cycon et al. 2013). *Stenotrophomonas* sp. SMSP-1 was fully disintegrate methyl parathion, fenitrothion, and ethyl parathion and degrade 37.8% of fenthion and 59.7% of phoxim, but cannot degrade chlorpyrifos at 20 mg/L (Shen et al. 2010). *Stenotrophomonas* sp. PF32 was effectively disintegrate methyl parathion, phoxim, fenthion, sumithion, triazophos, and chlorpyrifos (Xu et al. 2009). *Stenotrophomonas maltophilia* MHF ENV20 was degrade chlorpyrifos by a half-life of 96 hours plus its metabolite trichlorophenol and diethyl thiophosphate salt. *Stenotrophomonas* sp. YC-1 was also use chlorpyrifos as carbon source, but cannot subsequently degrade its metabolite, trichlorophenol YC-1 can fully degrade 100 mg/L of chlorpyrifos by 24 hours (Yang et al. 2006). *Stenotrophomonas* sp. DSP-4 can totally degrade 100 mg/L chlorpyrifos in 24 hours (Li et al. 2008).

To date, the hydrolase was found in methyl parathion breakdown *Pseudaminobacter salicylatoxidans* MP-1, *Achromobacter xylosoxidans* MP-2, *Ochrobactrum tritici* MP-3, *B. melitensis* MP-7, *Plesiomonas* sp. M6, *Sphingopyxis* sp. DLP-2, *Pseudomonas stutzeri* HS-D36 (Zhang et al. 2005, Sun 2011), chlorpyrifos-degrading *Cupriavidus* sp. DT-1 (Lu et al. 2013), and phoxim-degrading *Delftia* sp. XSP-1 (Shen et al. 2007). Methyl parathion is the optimum substrate for methyl parathion hydrolase (MPH) from Pseudomonas sp. WBC-3, which degrade chlorpyrifos, ethyl parathion, and sumithion (Chu et al. 2003, Liu 2003). Burkholderia sp. FDS-1 degrade methyl parathion, parathion, and sumithion, but not chlorpyrifos, methamidophos, phoxim, and triazophos (Zhang 2005). Methyl parathion hydrolase was an intracellular enzyme in some species (Xu 2005, Zhang 2005) and cell membrane bound in the other species (Wang 2007). Various types of bacteria used for bioremediation of OPs are presented in Table 9.2.

TABLE 9.2
Various types of bacteria used for bioremediation of OPs

S.no.	Name of the bacteria	Name of the OP	Reference
1.	*Serratia* sp. SPL-2	Methidathion	Li et al. (2013)
2.	*Pseudomonas aeruginosa* IS-6	Acephate, methamidophos, methyl parathion, dimethoate, and malathion	Ramu and Seetharaman (2014)
3.	*Pseudomonas, Agrobacterium, Bacillus*	Chlorpyrifos	Maya et al. (2011)
4.	*Bacillus cereus, Bacillus subtilis, Brucella melitensis, Klebsiella* cultures, *Pseudomonas aeruginosa, Pseudomonas fluorescence, Serratia marcescens*	Chlorpyrifos	Lakshmi et al. (2008)
5.	*Enterobacter* sp.	Chlorpyrifos	Singh et al. (2006)
6.	*Serratia marcescens*	Chlorpyrifos, fenitrothion, parathion	Cycon et al. (2013)
7.	*Pseudaminobacter salicylatoxidans* MP-1, *Achromobacter xylosoxidans* MP-2, *Ochrobactrum tritici* MP-3, *B. melitensis* MP-7, *Plesiomonas* sp. M6, *Sphingopyxis* sp. DLP-2, *Pseudomonas stutzeri* HS-D36	Methyl parathion	Wang et al. (2008); Sun (2011), Zhang et al. (2005)

9.4   DEGRADATION OF ORGANOPHOSPHATE PESTICIDES BY FUNGI

In handling and appliance of organophosphorous insecticides, terrestrial ecosystems, particularly soil and water, obtain huge amounts of pesticides (Singh et al. 2004). Because of the extent of this difficulty and lack of a logical answer, rapid, efficient, and ecologically responsible cleaning up method is greatly required to remove poisonous pollutants (Boopathy 2000a,b). Usually, physical and chemical cleanup methods are expensive and not much effective. But, biological methods such as remediation by microorganisms were proven very effective (Schoefs et al. 2004). Organophosphorus insecticides disintegration in soil was widely reported; but, the influence of OPs on soil microbes was obtained less reflection. *A. flavus* and *A. sydowii* phosphatases effectively degrades pesticides at 300–1,000 ppm in soil (Hasan 1999). Various PTEs were identified, i.e., organophosphate hydrolase, MPH, organophosphorus acid anhydrolase, diisopropylfluorophosphatase, and paraoxonase 1, carboxylesterases (Bigley and Raushel 2013).

9.5   CONCLUSIONS

Environmental changes, agricultural practices and land use activities severely affect the soil microbiome. It is indispensable to realize how land use practices persuade the soil microbial ecological systems to visualize the ecosystem strength for evolution of more sustained soil management in dryland ecosystems. Expand attentiveness was recently been paid to the isolation, identification, characterization, and applications of microorgnisms because of their exceptional biological characteristics that tolerate to live under extreme circumstances, i.e., higher temperature, higher pH, high UV light, water, and nutrient scarcity. Microorganisms have vast array of enzyme systems with great potential to calcify industrial wastes with diverse environmental circumstances. Bioremediation is a plan of action that depends on metabolic capability of microorganisms to modify aromatic compounds into non-toxic compounds. Understanding "the soil microbiome interactions with pollutants" is very essential in order to achieve effective environmental bioremediation. Therefore, further in-depth research is needed on soil microbiome in view of environmental cleanup.

REFERENCES

Ahmad M, Pataczek L, Hilger TH et al. (2018) Perspectives of microbial inoculation for sustainable development and environmental management. *Front Microbiol* 9: 2992. https://doi.org/10.3389/fmicb.2018.02992

Ali H (2010) Biodegradation of synthetic dyes – a review. *Water Air Soil Pollut* 213(1–4): 251–273.

Andrew DR, Fitak RR, Munguia-Vega A et al. (2012) Abiotic factors shape microbial diversity in Sonoran Desert soils. *Appl Environ Microbiol* 78(21): 7527–7537. https://doi.org/10.1128/AEM.01459-12

Anon (2012) *Soil Contamination: A Severe Risk for the Environment and Human Health.*

Antil RS, Dinesh, Dahiya SS (2007) Utilization of sewer water and its significance in INM. *Proceedings of ICAR sponsored Winter School on Integrated Nutrient Management* 79–83. https://doi.org/10.1111/j.1365-2427.2009.02299.x

Armstrong A, Valverde A, Ramond JB et al. (2016) Temporal dynamics of hot desert microbial communities reveal structural and functional responses to water input. *Sci Report* 6: 34434. https://doi.org/10.1038/srep34434 1

Arslan-Alaton I (2007) Degradation of a commercial textile biocide with advanced oxidation processes and ozone. *J Environ Manag* 82: 145–154. https://doi.org/10.1016/j.jenvman.2005.12.021

Bardgett RD, Van Der Putten WH (2014) Belowground biodiversity and ecosystem functioning. *Nature* 515: 505–511. https://doi.org/10.1038/nature

Bastiaens L, Springael D, Wattiau P et al. (2000) Isolation of adherent polycyclic aromatic hydrocarbon (PAH)-degrading bacteria using PAH-sorbing carriers. *Appl Environ Microbiol* 66: 1834–1843. https://doi.org/10.1128/aem.66.5.1834-1843.2000

Beuchle R, Grecchi RC, Shimabukuro YE et al. (2015) Land cover changes in the Brazilian Cerrado and Caatinga biomes from 1990 to 2010 based on a systematic remote sensing sampling approach. *Appl Geogr* 58: 116–127. http://dx.doi.org/10.1016/j.apgeog.2015.01.017

Bigley AN, Raushel FM (2013) Catalytic mechanisms for phosphotriesterases. *Biochim Biophys Acta (BBA)-Proteins Proteom* 1834: 443–453. https://doi.org/10.1016/j.bbapap.2012.04.004

Boominathan K, Reddy CA (1992) Fungal degradation of lignin: biotechnological applications. In *Handbook of Applied Mycology,* Arora DK, Elander RP, Mukerji KG, Eds., Marcel Dekker Inc.: New York, Volume 4, pp. 763–782.

Boopathy R (2000a) Bioremediation of explosives contaminated soil. *Inter Biodeter Biodegra* 46: 29–36.

Boopathy R (2000b) Factors limiting bioremediation technologies. *Bioresour Technol* 74: 63–67. https://doi.org/10.1016/S0960-8524(99)00144-3

Borja J, Taleon DM, Auresenia J, Gallardo S (2005) Polychlorinated biphenyls and their biodegradation. *Process Biochem* 40: 1999–2013. https://doi.org/10.1016/j.procbio.2004.08.006

Brussaard L (2012) Ecosystem services provided by the soil biota. *Soil Ecol Ecosys Ser* 45–58. https://doi.org/10.1093/acprof:oso/9780199575923.003.0005

Bumpus JA (1993) White-rot fungi and their potential use in soil bioremediation *processes. Soil Biochem* 8: 65–100.

Bumpus JA, Tien M, Wright D, Aust SD (1985) Oxidation of persistent environmental pollutants by a white rot fungus. *Science* 228: 1434–1436. https://doi.org/10.1126/science.3925550.

Casillas RP, Crow SA, Heinze TM et al. (1996) Initial oxidative and subsequent conjugative metabolites produced during the metabolism of phenanthrene by fungi. *J Indust Microbiol* 16: 205–215.

Chu X, Zhang X, Chen Y et al. (2003) Study on the properties of methyl parathion hydrolase from *Pseudomonas* sp. WBC-3. *Wei sheng wu xue bao= Acta Microbiol Sin* 43: 453–459.

Cloete TE, Celliers L (1999) Removal of Aroclor 1254 by the white rot fungus *Coriolus versicolor* in the presence of different concentrations of Mn (IV) oxide. *Inter Biodeter Biodeg* 44: 243–253.

Collins CD, Mosquera-Vazquez M, Gomez-Eyles J et al. (2013) Is there sufficient 'sink' in current bioaccessibility determinations of organic pollutants in soils? *Environ Pollut* 181: 128–132. https://doi.org/10.1016/j.envpol.2013.05.053

Covino S, Stella T, Cajthaml T (2016) Mycoremediation of organic pollutants: principles, opportunities, and pitfalls. In *Fungal Applications in Sustainable Environmental Biotechnology*, Purchase, D, Ed., Springer: Cham, pp. 185–231.

Cutter LA, Watts JE, Sowers KR, May HD (2001) Identification of a microorganism that links its growth to the reductive dechlorination of 2, 3, 5, 6-chlorobiphenyl. *Environ Microbiol* 3: 699–709.

Cycon M, Zmijowska A, Wojcik M, Piotrowska-Seget Z (2013) Biodegradation and bioremediation potential of diazinon-degrading *Serratia marcescens* to remove other organophosphorus pesticides from soils. *J Environ Manage* 117: 7–16. https://doi.org/10.1016/j.jenvman.2012.12.031

De Menezes A, Clipson N, Doyle E (2012) Comparative metatranscriptomics reveals widespread community responses during phenanthrene degradation in soil. *Environ Microbiol* 14: 2577–2588. https://doi.org/10.1111/j.1462-2920.2012.02781.x

Delgado-Balbuena L, Bello-Lopez JM, Navarro-Noya YE, Rodriguez-Valentin A, Luna-Guido ML, Dendooven L (2016) Changes in the bacterial community structure of remediated anthracene-contaminated soils. *PLoS One* 11: e0160991. https://*doi*.org/10.1371/journal.pone.0160991

Domínguez A, Couto SR, Sanroman MA (2005) Dye decolorization by *Trametes hirsuta* immobilized into alginate beads. *World J Microbiol Biotechnol* 21: 405–409.

Doyle E, Muckian L, Hickey AM, Clipson N (2008) Microbial PAH degradation. *Adv Appl Microbiol* 65: 27–66. https://doi.org/10.1016/S0065-2164(08)00602-3

Eaton DC (1985) Mineralization of polychlorinated biphenyls by *Phanerochaete chrysosporium*: a ligninolytic fungus. *Enzyme Microb Technol* 7: 194–196.

Edwards M, Topp E, Metcalfe CD et al. (2009) Pharmaceutical and personal care products in tile drainage following surface spreading and injection of dewatered municipal biosolids to an agricultural field. *Sci Total Environ* 407: 4220–4230. https://doi.org/10.1016/j.scitotenv.2009.02.028

EPA (2016) *Greenhouse Gas Emissions, US Environmental Protection Agency.* Available online at: https://www.epa.gov/ghgemissions/overview-greenhouse-gases/ (Accessed November 24, 2016).

Federici E, Giubilei M, Santi G et al. (2012) Bioaugmentation of a historically contaminated soil by polychlorinated biphenyls with *Lentinus tigrinus. Microb Cell Fact* 11: 35.

Fennell DE, Nijenhuis I, Wilson SF et al. (2004) *Dehalococcoides ethenogenes* strain 195 reductively dechlorinates diverse chlorinated aromatic pollutants. *Environ Sci Technol* 38: 2075–2081. https://doi.org/10.1021/es034989b

Fernandes-Junior PI, Aidar SDT, Morgante CV et al. (2015) The resurrection plant *Tripogon spicatus* (Poaceae) harbors a diversity of plant growth promoting bacteria in northeastern Brazilian Caatinga. *Revista Brasileira de Ciencia do Solo* 39: 993–1002. https://doi.org/10.1590/01000683rbcs20140646

Fernandez-Luqueno F, Valenzuela-Encinas C, Marsch R et al. (2011) Microbial communities to mitigate contamination of PAHs in soil – possibilities and challenges: a review. *Environ Sci Pollut Res* 18: 12–30. https://doi.org/10.1007/s11356-010-0371-6

Festa S, Macchi M, Cortes F et al. (2016) Monitoring the impact of bioaugmentation with a PAH-degrading strain on different soil microbiomes using pyrosequencing. *FEMS Microbiol Ecol* 92. https://doi.org/10.1093/femsec/fiw125

Fuentes S, Mendez V, Aguila P, Seeger RM (2014) Bioremediation of petroleum hydrocarbons: catabolic genes, microbial communities, and applications. *Appl Microbiol Biotechnol* 98: 4781–4794. https://doi.org/10.1007/s00253-014-5684-9

Furukawa K, Fujihara H (2008) Microbial degradation of polychlorinated biphenyls: biochemical and molecular features. *J Biosci Bioeng* 105: 433–449. https://doi.org/10.1263/jbb.105.433

Ghosal D, Ghosh S, Dutta TK, Ahn Y (2016) Current state of knowledge in microbial degradation of polycyclic aromatic hydrocarbons (PAHs): a review. *Front Microbiol* 7: 1369. https://doi.org/10.3389/fmicb.2016.01369

Gibbons SM, Scholz M, Hutchison AL et al. (2016) Disturbance regimes predictably alter diversity in an ecologically complex bacterial system. *Mbio* 7(6), e01372-16. https://doi.org/10.1128/mBio.01372-16

Gilbert ES, Crowley DE (1997) Plant compounds that induce polychlorinated biphenyl biodegradation by *Arthrobacter* sp. strain B1B. *Appl Environ Microbiol* 63: 1933–1938.

Gillespie TW, Lipkin B, Sullivan L et al. (2012) The rarest and least protected forests in biodiversity hotspots. *Biodivers Conser* 21: 3597–3611. https://doi.org/10.1007/s10531-012-0384-1

Gouda S, Kerry RG, Das G et al. (2018) Revitalization of plant growth promoting rhizobacteria for sustainable development in agriculture. *Microbiol Res* 206: 131–140. https://doi.org/10.1016/j.micres.2017.08.016

Griffiths BS, Philippot L (2013) Insights into the resistance and resilience of the soil microbial community. *FEMS Microbiol Rev* 37: 112–129. https://doi.org/10.1111/j.1574-6976.2012.00343.x

Hammel KE (1992) Oxidation of aromatic pollutants by lignin-degrading fungi and their extracellular peroxidases. *Metal Ion Biol Sys* 28: 41–60.

Harkness MR, McDermott JB, Abramowicz DA et al. (1993) In situ stimulation of aerobic PCB biodegradation in Hudson River sediments. *Science* 259: 503–507. https://doi.org/10.1126/science.8424172

Hasan HAH (1999) Fungal utilization of organophosphate pesticides and their degradation byAspergillus flavus and *A. sydowii* in soil. *Folia Microbiol* 44: 77.

Hatamian-Zarmi A, Shojaosadati SA, Vasheghani-Farahani E et al. (2009) Extensive biodegradation of highly chlorinated biphenyl and Aroclor 1242 by *Pseudomonas aeruginosa* TMU56 isolated from contaminated soils. *Int Biodeter Biodeg* 63: 788–794. https://doi.org/10.1016/j.ibiod.2009.06.009

Huang JPY, Guan XD (2016) Accelerated dryland expansion under climate change. *Nat Climate Change* 6: 166–171. https://doi.org/10.1038/nclimate2837

Hughes JB, Hellmann JJ, Ricketts TH, Bohannan BJ (2001) Counting the uncountable: statistical approaches to estimating microbial diversity. *Appl Environ Microbiol* 67: 4399–4406. https://doi.org/10.1128/aem.67.10.4399-4406.2001

Ijoma GN, Tekere M (2017) Potential microbial applications of co-cultures involving ligninolytic fungi in the bioremediation of recalcitrant xenobiotic compounds. *Int J Environ Sci Technol* 14: 1787–1806.

Iwasaki T, Takeda H, Miyauchi K et al. (2007) Characterization of two biphenyl dioxygenases for biphenyl/PCB degradation in a PCB degrader, *Rhodococcus* sp. strain RHA1. *Biosci Biotechnol Biochem* 71: 993–1002.

Jang HN, Park SB, Kim JH, Seo YC (2010) Combustion characteristics and emission of hazardous air pollutants in commercial fluidized bed combustors for sewage sludge. The 13th International Conference on Fluidization - New Paradigm in Fluidization Engineering, Art. 77, May 16-21, 2010, Gyeong-ju, Korea

Jiao S, Zhang Z, Yang F et al. (2017) Temporal dynamics of microbial communities in microcosms in response to pollutants. *Molec Ecol* 26: 923–936. https://doi.org/10.1111/mec.13978

Jin D, Jiang X, Jing X, Ou Z (2007) Effects of concentration, head group, and structure of surfactants on the degradation of phenanthrene. *J Hazard Material* 144: 215–221. https://doi.org/10.1016/j.jhazmat.2006.10.012 PMID: 17113708

Johnsen AR, Karlson U (2005) PAH degradation capacity of soil microbial communities—does it depend on PAH exposure? *Microb Ecol* 50: 488–495. https://doi.org/10.1007/s00248-005-0022-5

Kadri T, Rouissi T, Brar SK et al. (2017) Biodegradation of polycyclic aromatic hydrocarbons (PAHs) by fungal enzymes: a review. *J Environ Sci* 51: 52–74.

Kamei I, Kogura R, Kondo R (2006a) Metabolism of 4, 4′-dichlorobiphenyl by white-rot fungi *Phanerochaete chrysosporium* and *Phanerochaete* sp. MZ142. *Appl Microbiol Biotechnol* 72: 566–575.

Kamei I, Sonoki S, Haraguchi K, Kondo R (2006b) Fungal bioconversion of toxic polychlorinated biphenyls by white-rot fungus, *Phlebia brevispora*. *Appl Microbiol Biotechnol* 73: 932–940.

Kharche VK, Desai VN, Pharande AL (2011) Effect of sewage irrigation on soil properties, essential nutrient and pollutant element status of soils and plants in a vegetable growing area around Ahmednagar city in Maharashtra. *J Ind Soc Soil Sci* 59: 177–184.

Kitamura S, Suzuki T, Sanoh S et al. (2005) Comparative study of the endocrine-disrupting activity of bisphenol A and 19 related compounds. *Toxicol Sci* 84: 249–259. https://doi.org/10.1093/toxsci/kfi074

Kubatova A, Erbanova P, Eichlerova I et al. (2001) PCB congener selective biodegradation by the white rot fungus *Pleurotus ostreatus* in contaminated soil. *Chemosphere* 43: 207–215. https://doi.org/10.1016/s0045-6535(00)00154-5

Kulshrestha Y, Husain Q (2007) Decolorization and degradation of acid dyes mediated by salt fractionated turnip (*Brassica rapa*) peroxidases. *Toxicol Environ Chem* 89: 255–267. https://doi.org/10.1080/02772240601081692

Kuppusamy S, Thavamani P, Megharaj M et al. (2016) Pyrosequencing analysis of bacterial diversity in soils contaminated long-term with PAHs and heavy metals: implications to bioremediation. *J Hazard Material* 317: 169–179. https://doi.org/10.1016/j.jhazmat.2016.05.066

Lakshmi CV, Kumar M, Khanna S (2008) Biotransformation of chlorpyrifos and bioremediation of contaminated soil. *Int Biodet Biodegr* 62: 204–209.

Lapen DR, Topp E, Edwards M et al. (2008) Effect of liquid municipal biosolid application method on tile and ground water quality. *J Environ Qual* 37: 925–936. https://doi.org/10.2134/jeq2006.0486

Leite J, Fischer D, Rouws LF et al. (2017) Cowpea nodules harbor non-rhizobial bacterial communities that are shaped by soil type rather than plant genotype. *Front Plant Sci* 7: 2064. https://doi.org/10.3389/fpls.2016.02064

Li C, Lan Y, Zhang J et al. (2013) Biodegradation of methidathion by *Serratia* sp. in pure cultures using an orthogonal experiment design, and its application in detoxification of the insecticide on crops. *Ann Microbiol* 63: 451–459.

Li X, Jiang J, Gu L et al. (2008) Diversity of chlorpyrifos-degrading bacteria isolated from chlorpyrifos-contaminated samples. *Int Biodeterior Biodegr* 62: 331–335. https://doi.org/10.1016/j.ibiod.2008.03.001

Liao H, Chapman SJ, Li Y, Yao H (2018) Dynamics of microbial biomass and community composition after short-term water status change in Chinese paddy soils. *Environ Sci Pollut Res* 25: 2932–2941.

Liu LL (2003) *Research of Methyl Parathion Hydrolase*, Wuhan Institute of Virology, CAS, Beijing, China.

Lors C, Damidot D, Ponge JF, Perie, F (2012) Comparison of a bioremediation process of PAHs in a PAH-contaminated soil at field and laboratory scales. *Environ Pollut* 165: 11–17. https://doi.org/10.1016/j.envpol.2012.02.004

Lu P, Li Q, Liu H et al. (2013) Biodegradation of chlorpyrifos and 3, 5, 6-trichloro-2-pyridinol by *Cupriavidus* sp. DT-1. *Bioresour Technol* 127: 337–342. https://doi.org/10.1016/j.biortech.2012.09.116

Maestre FT, Eldridge DJ, Soliveres S et al. (2016) Structure and functioning of dryland ecosystems in a changing world. *Annu Rev Ecol Evol Syst* 47: 215–237. https://doi.org/10.1146/annurev-ecolsys-121415-032311

Maya K, Singh RS, Upadhyay SN, Dubey SK (2011) Kinetic analysis reveals bacterial efficacy for biodegradation of chlorpyrifos and its hydrolyzing metabolite TCP. *Proc Biochem* 46: 2130–2136. https://doi.org/10.1016/j.procbio.2011.08.012

Monteiro JM, Vollu RE, Coelho MRR et al. (2009) Comparison of the bacterial community and characterization of plant growth-promoting rhizobacteria from different genotypes of *Chrysopogon zizanioides* (L.) Roberty (vetiver) rhizospheres. *J Microbiol* 47: 363–370. https://doi.org/10.1007/s12275-009-0048-3

Naveen KA (2018) Bioremediation: a green approach for restoration of polluted ecosystems. *Environ Sustain* 1: 305–307. https://doi.org/10.1007/s42398-018-00036-y

Nielsen UN, Ball BA (2015) Impacts of altered precipitation regimes on soil communities and biogeochemistry in arid and semiarid ecosystems. *Global Change Biol* 21: 1407–1421.

Nollet H, Roels M, Lutgen P et al. (2003) Removal of PCBs from wastewater using fly ash. *Chemosphere* 53:655–665. https://doi.org/10.1016/S0045-6535(03)00517-4

Pacchioni RG, Carvalho FM, Thompson CE (2014) Taxonomic and functional profiles of soil samples from Atlantic forest and Caatinga biomes in northeastern Brazil. *Microbiol Open* 3: 299–315. https://doi.org/10.1002/mbo3.169

Pajares S, Campo J, Bohannan BJ, Etchevers JD (2018) Environmental controls on soil microbial communities in a seasonally dry tropical forest. *Appl Environ Microbiol* 84: e00342–18. https://doi.org/10.1128/AEM.00342-18

Pandey A, Singh P, Iyengar L (2007) Bacterial decolorization and degradation of azo dyes. *Int Biodeter Biodegr* 59:73–84. https://doi.org/10.1016/j.ibiod.2006.08.006

Pieper DH, Seeger M (2008) Bacterial metabolism of polychlorinated biphenyls. *J Molecular Microbiol Biotechnol* 15: 121–138. https://doi.org/10.1159/000121325

Pieper DH, Winkler R, Sandermann JRH (1992) Formation of a toxic dimerization product of 3, 4 dichloroaniline by lignin peroxidase from *Phanerochaete chrysosporium*. *Angew Chem Int Ed Engl* 31: 68–70.

Posada-Baquero R, Ortega-Calvo JJ (2011) Recalcitrance of polycyclic aromatic hydrocarbons in soil contributes to background pollution. *Environ Poll* 159: 3692–3699. https://doi.org/10.1016/j.envpol.2011.07.012

Ramu S, Seetharaman B (2014) Biodegradation of acephate and methamidophos by a soil bacterium *Pseudomonas aeruginosa* strain IS-6. *J Environ Sci Health Part B* 49: 23–34. https://doi.org/10.1080/03601234.2013.836868

Rasche F, Knapp D, Kaiser C (2011) Seasonality and resource availability control bacterial and archaeal communities in soils of a temperate beech forest. *The ISME J* 5: 389–402. https://doi.org/10.1038/ismej.2010.138

Rengarajan T, Rajendran P, Nandakumar N et al. (2015) Exposure to polycyclic aromatic hydrocarbons with special focus on cancer. *Asian Pacific J Tropical Biomed* 5: 182–189. https://doi.org/10.1016/S2221-1691(15)30003-4Rodrigues JL, Kachel CA, Aiello MR et al. (2006) Degradation of Aroclor 1242 dechlorination products in sediments by *Burkholderia xenovorans* LB400 (ohb) and *Rhodococcus* sp. strain RHA1 (fcb). *Appl Environ Microbiol* 72: 2476–2482. https://doi.org/10.1128/AEM.72.4.2476-2482.2006

Rostami I, Juhasz AL (2011) Assessment of persistent organic pollutant (POP) bioavailability and bioaccessibility for human health exposure assessment: a critical review. *Crit Rev Environ Sci Technol.* 41: 623–656. https://doi.org/10.1080/10643380903044178

Santos MG, Oliveira MT, Figueiredo KV et al. (2014) Caatinga, the Brazilian dry tropical forest: can it tolerate climate changes? *Theor Exp Plant Physiol* 26:83–99. https://doi.org/10.1007/s40626-014-0008-0

Sayara T, Borràs E, Caminal G, Sarrà M, Sánchez A (2011) Bioremediation of PAHs-contaminated soil through composting: influence of bioaugmentation and biostimulation on contaminant biodegradation. *Int Biodeter Biodegr* 65: 859–865. https://doi.org/10.1016/j.ibiod.2011.05.006

Schoefs O, Perrier M, Samson R (2004) Estimation of contaminant depletion in unsaturated soils using a reduced-order biodegradation model and carbon dioxide measurement. *Appl Microbiol Biotechnol* 64: 53–61. https://doi.org/10.1007/s00253-003-1423-3

Seeger M, Pieper DH (2010) Genetics of biphenyl biodegradation and co -metabolism of PCBs. In *Handbook of Hydrocarbon and Lipid Microbiology*, Timmis, KN, Ed., Springer: Heidelberg, pp. 1179–1199.

Shade A (2017) Diversity is the question, not the answer. *ISME J* 11: 1–6. https://doi.org/10.1038/ismej.2016.118

Shade A, Caporaso JG, Handelsman J, Knight R, Fierer N (2013) A meta-analysis of changes in bacterial and archaeal communities with time. *The ISME J* 7: 1493. https://doi. org/10.1038/ismej.2013.54

Shen YJ, Hong YF, Hong Q, Jiang X, Li SP (2007) Isolation, identification and characteristics of a phoxim-degrading bacterium XSP-1. *Huan jing ke xue Huanjing kexue* 28: 2833–2837.

Shen YJ, Lu P, Mei H, Yu HJ, Hong Q, Li SP (2010) Isolation of a methyl parathion-degrading strain *Stenotrophomonas* sp. SMSP-1 and cloning of the ophc2 gene. *Biodegradation* 21: 785–792. https://doi.org/10.1007/s10532-010-9343-2

Singh BK, Walker A, Morgan JAW, Wright DJ (2004) Biodegradation of chlorpyrifos by Enterobacter strain B-14 and its use in bioremediation of contaminated soils. *Appl Environ Microbiol* 70:4855–4863. https://doi.org/10.1128/AEM.70.8.4855-4863.2004

Singh BK, Walker A, Wright DJ (2006) Bioremedial potential of fenamiphos and chlorpyrifos degrading isolates: influence of different environmental conditions. *Soil Biol Biochem* 38: 2682–2693. https://doi.org/10.1016/j.soilbio.2006.04.019

Singh H (2006) *Mycoremediation: Fungal Bioremediation*, John Wiley & Sons: New York, USA.

Siracusa G, Becarelli S, Lorenzi R (2017) PCB in the environment: bio-based processes for soil decontamination and management of waste from the industrial production of *Pleurotus ostreatus*. *New Biotechnol* 39: 232–239. https://doi.org/10.1016/j.nbt.2017.08.011.

Sorensen L (2007) *A Spatial Analysis Approach to the Global Delineation of Dryland Areas of Relevance to the CBD Programme of Work on Dry and Sub-Humid Lands*. Cambridge: United Nations Environment Programme World Conservation Monitoring Centre.

Spina F, Cecchi G, Landinez-Torres A et al. (2018) Fungi as a toolbox for sustainable bioremediation of pesticides in soil and water. *Plant Biosyst Int J Deal All Aspects Plant Biol* 152: 474–488. https://doi.org/10.1080/11263504.2018.1445130

Stella T, Covino S, Burianová E et al. (2015) Chemical and microbiological characterization of an aged PCB-contaminated soil. *Sci Total Environ* 533: 177–186.

Sun JJ (2011) Characterization of a Mp-degrading strain Sphingopyxissp. In *DLP-2, Cloning and Functional Verification of Tnmpd*, Nanjing Agricultural University, Nanjing, China.

Thapa VR, Ghimire R, Mikha MM, Idowu OJ, Marsalis MA (2018) Land use effects on soil health in semiarid drylands. *Agric Environ Lett* 3(1). https://doi.org/10.2134/ ael2018.05.0022

Thomas DR, Carswell KS, Georgiou G (1992) Mineralization of biphenyl and PCBs by the white rot fungus *Phanerochaete chrysosporium*. *Biotechnol Bioeng* 40: 1395–1402.

Thomas F, Cébron A (2016) Short-term rhizosphere effect on available carbon sources, phenanthrene degradation, and active microbiome in an aged-contaminated industrial soil. *Fron Microbiol* 7: 92. https://doi.org/10.3389/fmicb.2016.00092

Tian Q, Taniguchi T, Shi WY, Li G, Yamanaka N, Du S (2017) Land-use types and soil chemical properties influence soil microbial communities in the semiarid Loess Plateau region in China. *Sci Rep* 7: 45289. https://doi.org/10.1038/srep45289

Tisma M, Sudar M, Vasic-Racki D, Zelic B (2010) Mathematical model for *Trametes versicolor* growth in submerged cultivation. *Bioproc Biosys Eng* 33: 749–758. https://doi. org/10.1007/s00449-009-0398-6

United Nations Environmental Management Group (2011) Global drylands: a UN system-wide response. http://www.zaragoza.es/contenidos/medioambiente/onu//issue07/1107-eng.pdf

Uqab B, Mudasir S, Nazir R (2016) Review on bioremediation of pesticides. *J Biorem Biodeg* 7: 2. https://doi.org/10.4172/2155-6199.1000343

Wahid A, Nasir MGA, Ahmad SS (2000) Effects of water pollution on growth and yield of soybean. *Acta Scient* 10: 51–58

Wang B, Xiong L, Zheng Y et al. (2008) Cloning and expression of the mpd gene from a newly isolated methylparathion-degrading strain of bacteria. *Acta Scientiae Circumstantiae* 10: 1969–1974

Wang C, Sun H, Li J et al. (2009) Enzyme activities during degradation of polycyclic aromatic hydrocarbons by white rot fungus *Phanerochaete chrysosporium* in soils. *Chemosphere* 77: 733–738. https://doi.org/10.1016/j.chemosphere.2009.08.028

Wang LW, Li F, Zhan Y, Zhu L (2016) Shifts in microbial community structure during in situ surfactant-enhanced bioremediation of polycyclic aromatic hydrocarbon-contaminated soil. *Environ Sci Pollut Res* 23: 14451–14461. https://doi.org/10.1007/s11356-016-6630-4

Wang S (2007) *Isolation and Identification of Methyl Parathion Degrading Bacteria and Cloning of Hydrolase Gene*, Shandong Agricultural University, Shandong, China.

Wang S, Hao C, Gao Z et al. (2014) Theoretical investigation on photodechlorination mechanism of polychlorinated biphenyls. *Chemosphere* 95: 200–205. https://doi.org/10.1016/j.chemosphere.2013.08.066

Wang W, Simonich SLM, Xue M et al. (2010) Concentrations, sources and spatial distribution of polycyclic aromatic hydrocarbons in soils from Beijing, Tianjin and surrounding areas, *North China*. *Environ Poll* 158: 1245–1251. https://doi.org/10.1016/j.envpol.2010.01.021

Wickliffe J, Overton E, Frickel S et al. (2014) Evaluation of polycyclic aromatic hydrocarbons using analytical methods, toxicology, and risk assessment research: seafood safety after a petroleum spill as an example. *Environ Health Perspect* 122: 6–9. https://doi.org/10.1289/ehp.1306724

Wong DW (2009) Structure and action mechanism of ligninolytic enzymes. *Appl Biochem Biotechnol* 157: 174–209. https://doi.org/10.1007/s12010-008-8279-z

Wu M, Xu Y, Ding W, Li Y, Xu H (2016) Mycoremediation of manganese and phenanthrene by *Pleurotus eryngii* mycelium enhanced by Tween 80 and saponin. *Appl Microbiol Biotechnol* 100: 7249–7261. https://doi.org/10.1007/s00253-016-7551-3

Xu W (2005) *Purification and Characterization of Methyl Parathion Hydrolase from Plesiomonas sp. M6*, Nanjing Agricultural University, Nanjing, China.

Xu Y, Zhou NY (2017) Microbial remediation of aromatics-contaminated soil. *Front Environ Sci Eng* 11: 1.

Xu YX, Feng ZZ, Lu P et al. (2009) Isolation and characterization of capable of degrading parathion-methyl bacterium *Stenotrophomonas* sp. PF32. *Chin J Pesticide Sci* 3: 329–334.

Yadav JS, Quensen JF, Tiedje JM, Reddy CA (1995) Degradation of polychlorinated biphenyl mixtures (Aroclors 1242, 1254, and 1260) by the white rot fungus *Phanerochaete chrysosporium* as evidenced by congener-specific analysis. *Appl Environ Microbiol* 61: 2560–2565. https://doi.org/10.1128/AEM.61.7.2560-2565.1995

Yang C, Liu N, Guo X, Qiao C (2006) Cloning of mpd gene from a chlorpyrifos-degrading bacterium and use of this strain in bioremediation of contaminated soil. *FEMS Microbiol Lett* 265: 118–125. https://doi.org/10.1111/j.1574-6968.2006.00478.x

Yang Y, Wang J, Liao J et al. (2015) Abundance and diversity of soil petroleum hydrocarbon-degrading microbial communities in oil exploring areas. *Appl Microbiol Biotechnol* 99: 1935–1946. https://doi.org/10.1007/s00253-014-6074-z

Zafra G, Taylor TD, Absalon AE, Cortes-Espinosa DV (2016) Comparative metagenomic analysis of PAH degradation in soil by a mixed microbial consortium. *J Hazard Mater* 318: 702–710. https://doi.org/10.1016/j.jhazmat.2016.07.060

Zeddel A, Majcherczyk A, Hüttermann A (1993) Degradation of polychlorinated biphenyls by whiterot fungi *Pleurotus ostreatus* and *Trametes versicolor* in a solid state system. *Toxicol Environ Chem* 40: 255–266. https://doi.org/10.1080/02772249309357947

Zhang F, Yediler A, Liang X, Kettrup A (2004) Effects of dye additives on the ozonation process and oxidation by-products: a comparative study using hydrolyzed CI Reactive Red 120. *Dyes Pigment* 60: 1–7.

Zhang R, Cui Z, Jiang J et al. (2005) Diversity of organophosphorus pesticide-degrading bacteria in a polluted soil and conservation of their organophosphorus hydrolase genes. *Can J Microbiol* 51: 337–343. https://doi.org/10.1139/w05-010

Zhang Z (2005) *Isolation and Characterization of Fenitrothioni-degrading Strain FDS-l (Burkhohelria sp.), Cloning and Expression ofmpd Gene*, Nanjing Agricultural University.

Zhao C, Miao Y, Yu C et al. (2016) Soil microbial community composition and respiration along an experimental precipitation gradient in a semiarid steppe. *Sci Rep* 6: 1–9.

Zhao JK, Li XM, Ai GM et al. (2016) Reconstruction of metabolic networks in a fluoranthene-degrading enrichments from polycyclic aromatic hydrocarbon polluted soil. *J Hazar Mater* 318: 90–98. https://doi.org/10.1016/j.jhazmat.2016.06.055

10 Sensory mechanism in bacteria for xenobiotics utilization

Oluwafemi Adebayo Oyewole,
Muhammed Muhammed Saidu,
Japhet Gaius Yakubu, and Mordecai Gana

CONTENTS

10.1 INTRODUCTION

Before the various sensory mechanisms in bacteria and their applications for xenobiotics utilization are discussed, it is important to understand and define the terminology "xenobiotics". The term xenobiotics in this context refer to compounds that are alien to a living organism, often of abiotic or man-made origin, that have propensity to accumulate and are present in the environment in greater magnitude than would be expected to occur (Park *et al.*, 2014; Bharadwaj, 2018). The accumulation of wide range of toxic xenobiotics in the environment resulting from both natural and human activities has thus caused a global concern (Singh, 2017). Industrialization has caused a lot of positive improvements to the standards of human living. However, it comes with various prices; some affects the life style of man while others are detrimental to man's health. However, the ecosystem at large is at the receiving end of various xenobiotic pollutions arising from industrialization and other human

DOI: 10.1201/9781003394600-12

253

activities. In order to avoid the accumulation of xenobiotic compounds, measures need to be put in place to detect and neutralize the harmful effects caused by xenobiotics. Xenobiotics include industrially synthesized chemicals such as synthetic chemicals, chlorinated and nitroaromatic compounds, pesticides, dyes and pharmaceuticals products. Other chemicals of natural origin that are introduced into the environment through human activities, these include various components of crude oil and other organic compounds and their breakdown products are also considered xenobiotics (Parales *et al.*, 2015). Certain recalcitrant xenobiotics that amass in the environment and have become a serious problem as a result of their accumulation and toxicity include polycyclic aromatic hydrocarbons (PAHs), trichloroethylene (TCE) and polychlorinated biphenyls (PCBs). To a great extent, these compounds accumulate specifically in the subsurface environment and water sources. Xenobiotic pollutants are mainly introduced into the environment from industries such as pharmaceuticals, fuels, paper, polymers and textile and agricultural inputs. For instance, they may occur as synthetic chlorocarbon such as pesticides and plastics, organic chemicals such hydrocarbons and some fractions of coal and crude oil (Kumar *et al.*, 2017). Various examples of xenobiotics include: synthetic dyes, phenolic products, paint, petro products, fertilizers and plastic material (Kumar *et al.*, 2017).

Xenobiotics have been continually released into the environment over the last hundred years as a result of human activities, such as industrial and agricultural processes, and they have adverse effects on the pristine nature of the environment causing serious problems such as environmental pollution due to their recalcitrant nature to degradation (Nagata, 2020). The continual amassing of recalcitrant xenobiotic compounds in the environment has caused a serious problem globally. Xenobiotics of industrial source include paper and pulp residues, chemical and pharmaceutical wastes, plastic wastes, paint and dye effluents and solid waste residues (dumps) (Kumar *et al.*, 2017). Xenobiotics are capable of causing cancer (carcinogenic) and mutation (mutagenic), and they are also capable of causing defect to a foetus, i.e. they have teratogenic effect and continue to remain in the environment over a very long period of time (Janssen and Stucki, 2020). Thus, it is important for early detection and possible eradication of xenobiotics from the environment.

10.2   BACTERIAL SENSORY MECHANISMS FOR XENOBIOTICS

With the help of biotechnology and advanced tools in microbiology, bacterial species have been utilized in assessing the health of the environment with the help of sensory cells. Bacteria face constant challenges in their changing environment. An intricate sensory system has been developed in bacteria to respond to the vicissitude nature of their immediate environment. Most of these sensory systems are found in other prokaryotes and some eukaryotes. Most of the sensory mechanisms have evolved in bacteria for responding to various changes in the environment (Alvarado *et al.*, 2020).

Sensory mechanisms in bacteria comprise small set of protein domain through which a bacterium cell detect many different sensory signals such as light, nutrients, antibiotics, redox potential, xenobiotics and other environmental stresses (Alvarado *et al.*, 2020). Internal and external stimuli are sensed bringing about changes in protein sequence, which is either inferred in enzymatic reactions, transport mechanism or other vital cellular processes. These activities can then translate into pathways of characterized kinases, which transmit the information to the deoxyribonucleic acid (DNA) or other response units (Alvarado *et al.*, 2020).

Certain exceptional abilities have been developed by the bacteria to control various aspects of their reaction in response to signals within the cell and outside the cell environment. Important cell components with unique abilities of detecting changes through immediate interaction with the physical and chemical stimuli are the various macromolecules (proteins or RNA) (Aravind *et al.*, 2015). Bacteria use the two-component systems to detect various signals which bring about multitude of downstream effects, thereby affecting gene expression (Alvarado *et al.*, 2020).

Numerous sensory mechanisms are available to bacteria for perceiving environmental stimuli, through cell's detection and relay of environmental and intercellular stimuli to activate the appropriate response to them. The sensing of stimuli in bacteria occurs at the molecular level using small set of protein domain. Protein conformation analysis with experimental findings formed the basis of characterizing these protein domains. Vast array of sensory stimuli such as chemicals (xenobiotics), light and redox potential are detected by the protein domains (Aravind *et al.*, 2015).

Forays into the sensory mechanisms of bacteria began with the operon hypothesis of Jacob and Monod in 1959 (Aravind *et al.*, 2015). By the second half of the 1990s, this picture was to undergo a major modification, thanks to the rise of genomics and the expansion of studies on signaling in new prokaryotic models (Aravind *et al.*, 2015). From the earliest days, computational analysis of protein sequence plays a major role in dissection of signaling system at the molecular level (Somavanshi *et al.*, 2016). There have been enormous improvement over the last decades in the understanding of the structural complexity and functional roles of these signal receptors, and how their interaction with various signals results in changes in their activity and downstream processes, and the understanding of the receptors for some systems extends to the atomic level. Recent genome analysis of bacteria strain that sense xenobiotics have implied that they indeed evolved relatively recently by isolating genes for the detection and degradation of xenobiotics and mobile genetic elements played significant roles in the recruitment of genetically engineered genes (Nagata *et al.*, 2019). A number of bacterial strains with inherent capacity of detecting and subsequently degrading a wide range of xenobiotics have been isolated and studied; these strains of bacteria exhibit tendency to move in response to compounds they possess the capacity of utilizing (Luu *et al.*, 2015). Although the origin of the gene and the evolutionary trends of the strains are mysteries yet to be unraveled, current studies on the comprehensive analyses of genome and metagenome of the strains aimed at providing some clues for understanding the mysteries, and the genes for the signaling of xenobiotics can be used for exploring new mechanisms for the evolution of bacteria (Nagata *et al.*, 2015).

A number of degradative genes responsible for the metabolisms of xenobiotics are found on the plasmids, transposons or collectively found on chromosomes. This provides an insight into the evolution of pathways of xenobiotics metabolisms which ease the genetic manipulation process such that strains with enhanced degradative capacity for pollutants degradation are developed (Monzón *et al.*, 2018).

10.3   CLASSES OF SENSORY MECHANISMS IN BACTERIA FOR DETECTING XENOBIOTICS

Bacteria have evolved a well-developed sensory system, which regulate response to various sensory stimuli. Two-component systems, which are characterized by the kinase enzymes play a major role in sensing and signaling in bacteria and they consist of a sensor phosphokinase or multi-kinase networks and a corresponding signaling protein (Groisman, 2016; Zschiedrich *et al.*, 2016; Willett and Crosson, 2017; Francis and Porter, 2019). There are certain distinct sensory mechanisms in bacteria through which bacteria detect and relay environmental and intracellular stimuli to activate the appropriate response to them. They allow bacteria to sense xenobiotics resulting in a multitude of downstream processes which have effect on the gene expression (Desai and Kenney, 2017). There are two distinct classes of sensory mechanisms in bacteria for xenobiotics, these are; the canonical sensory mechanism and non-canonical sensory mechanisms.

10.4   CANONICAL SENSORY MECHANISM IN BACTERIA

Canonical sensory mechanism comprises a single or many regulatory phosphokinase enzymes and a signaling protein, which is phosphorylated upon activation of sensor by external stimuli such as xenobiotics (Groisman, 2016; Desai and Kenney, 2017). The canonical sensors are found on the cell membrane, which is the case of many sensor kinases or could also be found in the periplasm or in the cytoplasm (Wiech *et al.*, 2015; Desai and Kenney, 2017; Matson *et al.*, 2017; Galperin, 2018; Masilamani *et al.*, 2018; May *et al.*, 2019; Osman *et al.*, 2019).

Bacterial chemotactic sensory module which governs bacteria chemotaxis to xenobiotics is a common example of the canonical sensory mechanism; it consists

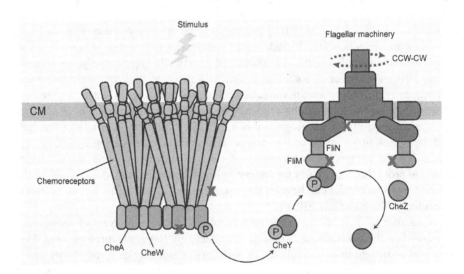

FIGURE 10.1   Canonical sensory mechanism (Alvarado et al., 2020; Copyright © 2020 Alvarado, Behrens and Josenhans).

of trans-membrane sensing unit, response regulator CheY and the histidine kinase CheA (Parkinson *et al.*, 2015). This system controls the movement of bacteria by regulating flagellar rotation thereby directing the swimming of the bacteria brought about by the signal from specialized taxis sensors in response to the stimulus as shown in Figure 10.1 (Terahara *et al.*, 2018; Alvarado *et al.*, 2020). There are receptors present in the cytoplasm for receiving signals which are closely associated with the adaptor protein CheW and the histidine kinase CheA and upon encounter with xenobiotics, the histidine kinase CheA is phosphorylated thereby activating the signaling protein CheY and the signal is transmitted to the flagella motor proteins which induces a change in the direction of flagella rotation toward the xenobiotics to metabolize it as source of carbon and energy, the dephosphorylation of the phosphorylated CheY by certain proteins in this case CheZ (phosphatase of CheY) depicts signal termination at the motor switch (FliM, FliN and FliG). P in the diagram depicting phosphoryl group, cytoplasmic membrane indicated as CM, clockwise as CW and counter-clockwise as CCW (Terahara *et al.*, 2018; Alvarado *et al.*, 2020).

10.5   NON-CANONICAL SENSORY MECHANISM IN BACTERIA

Non-canonical sensory mechanism in bacteria involves specific protein–protein interaction resulting in conformational changes in protein sequence which affects the downstream relay of signals and often translates into kinase pathway (Tsang *et al.*, 2015). Proteins interaction of the non-canonical sensory processes can be linked to the canonical sensory modules in regulating some vital events such as chemotaxis of bacteria to xenobiotics, however, in some cases, the non-canonical sensory mechanism act entirely independently of canonical sensory modules (Alvarado *et al.*, 2020). Non-canonical sensory processes regulate metabolism of xenobiotics and motility in bacteria (Tsang *et al.*, 2015). Various proteins interact in the non-canonical sensory mechanism in flagellar assembly regulation (Bi and Sourjik, 2018). Six chemotaxis proteins interact to modulate the direction of flagellar motor rotation in response to xenobiotics concentration gradient. Three dimers of different paralogous receptors bind to an adaptor protein CheW forming a complex sensory protein. The complex is associated with CheA, a kinase with autophosphorylation activity protein. The response regulator CheY and also the methylesterase CheB receive a phosphoryl group from CheY i.e. they become phosphorylated. The methyltransferase CheR and the methylesterase CheB mediate adaptation of the receptors to a constant concentration by maintaining the methylation level, and CheZ dephosphorylates CheY acting as the phosphatase of CheY (Bi and Sourjik, 2018). When the receptors present on the outer membrane of the cell detect changes in the concentration of the xenobiotics, they transmit the signal to the kinase with autophosphorylation ability CheA, which phosphorylates the signaling protein CheY. The phosphorylated CheY thereby induces a clockwise turn in the direction of the flagellar motor rotation

10.6   XENOBIOTICS RECEPTORS IN BACTERIA

When humans perceive a smell or taste a food, drugs or any chemical substance, receptors present on the tissues of the sense organ detect the chemical and relay the information to the brain through the neurons, where many cells interact to process the information and generate appropriate response for the signal received (Machuca *et al.*, 2016). Bacteria on the other hand, being unicellular in nature must make use of various sensory receptors to sense and response to every change around it. Bacteria possess a distinct ability of detecting changes in the concentration of a chemical substance as small as 0.1% in their environment, tantamount to a drop diluted in a pool of 1,000 drops (Machuca *et al.*, 2016). Biologically, researchers are yet to discover a system with sensitivity over such a wide range. Bacteria have developed a complex sensory system that enables them to detect xenobiotics in the environment and subsequently metabolize them as carbon and energy source. In bacteria, the sensory receptors are located on the outer membrane of the cell which, are responsible for receiving signals from the outside environment, similar to the human body, the organs eyes, nose and ears that receive signals are located on the body surface (Irazoki *et al.*, 2016). Xenobiotics receptors detect a wide range of xenobiotics and translate the information into the chemosensory pathways that show a major mode of signal transformation (Ortega *et al.*, 2017).

The receptors are highly specific in the chemicals they detect similar to the sensors of higher system and they respond to changes in the environment that are crucial to the bacteria in a like manner the humans respond to various changes in their environment necessary for their survival (Jones and Armitage, 2017). Reports have showed that chemosensory pathways mediate bacterial chemotaxis to xenobiotics and also control other vital cellular processes (Upadhyay *et al.*, 2016; Bardy *et al.*, 2017). Key to these systems is the complex sensory proteins formed by dimers of different receptors and the coupling protein CheW closely linked to the kinase protein CheA (Parkinson *et al.*, 2015). When bacteria encounter xenobiotics, the complex sensory proteins made of receptors and coupling protein transmit the signal to CheA which translates in directional change of flagellar motor rotation toward the xenobiotics (Parkinson *et al.*, 2015; Bi and Sourjik, 2018; Alvarado *et al.*, 2020). Receptors are usually composed of two main modules; input and output module. A single globular domain usually makes up the input module although some receptors have the input module composed of two or more domains (Sampedro *et al.*, 2015). The output module is composed of two symmetrical antiparallel coiled coils which constitutes the cytoplasmic signaling domain. Receptors detect xenobiotics and receive signals from the environment via the input module (Rico-Jimenez *et al.*, 2016).

10.6.1 CHARACTERIZATION OF SENSORY SIGNALS

Responses to the signals are broadly characterized as; direct rapid responses, filtered responses, signal amplification, altering the shape of a response and negative regulation, allosteric and feedback regulation and memory. Direct rapid responses are responses to stimuli (xenobiotics) mediated by one-component system where sensory domains are usually fused to DNA-binding domains, to direct transcription of that particular target gene and rapidly alter the transcriptional state of the cell (Aravind *et al.*, 2015).

> **Filtered responses**: In this case a sensor domain might be combined to a catalytic effector domain via signal transmitter elements. These elements act as a preliminary filter for the propagation of the stimuli received by the sensor

domains and thus ensure its controlled transmission. Additional control steps are present downstream and receiver domains before the signal is converted to a transcriptional response. These systems allow a controlled response that allow sensing of thresholds and may be contrasted with the more continuous and rapid responses afforded by the one-component system (Aravind *et al.*, 2015).

Signal amplification: Reports has been made on bacteria amplification of stimuli generated from xenobiotics. There usually use to be amplification of the initial stimulus sensed by the sensory domain is useful in directing global state changes in the cell. These are usually mediated by enzymes that sense the signal via a sensory domain and amplify it by generating a second messenger that is further sensed by specific domains (Aravind *et al.*, 2015).

Altering the shape of a response and negative regulation: When bacteria sense xenobiotics in the environment, certain enzymes such as phosphatases are being secreted to function downstream of sensor domains and create specifically shaped responses such as sharp peak or a shutdown of an on-going responses (Aravind *et al.*, 2015).

Allosteric and feedback regulation: In this case, sensory domains combine with catalytic domains can often serve to regulate the action of an enzyme in an allosteric manner or sense a feedback from a downstream process (Aravind *et al.*, 2015).

Memory: In bacteria, the alteration of responses to stimuli subsequent to the initial stimulus might be termed memory in a signaling system. This is seen in the form of covalent modification of signaling domains (Aravind *et al.*, 2015).

10.7   METABOLISM OF THE TARGET XENOBIOTICS

Recent scientific research has discovered variety of bacterial strains that possess the inherent capacity of degrading diverse range of xenobiotic pollutants. Studies of pure isolate and mixed cultures of bacterial strains under aerobic and anaerobic conditions revealed the various products of both aerobic and anaerobic biodegradation pathways with bacterial strains associated with each pathway (Nagata, 2020).

The researchers suggest that when a receptor detects xenobiotics, transmission of signals between closely linked receptors result in the rearrangement of the receptors, this can be likened to when water freezes, there is an orderly rearrangement of the water molecules to form a new structure. Upon rearrangement, the signal is amplified by the array of receptors indicating that certain chemical have been sensed. The signal is transmitted to bring about the activation of the kinase resulting in a series of reactions to generate the appropriate response which is translated into directional change in the flagella spin; this enables the bacterium to swim toward the xenobiotics (Alvarado *et al.*, 2020). This is preceded by the secretion of organic acid or enzymes by the bacteria for gradual metabolism of the target xenobiotics and utilization as carbon and energy source thereby eliminating the xenobiotics accumulated as toxic compounds in the environment (Chong, 2015). Several enzymes capable of metabolizing xenobiotics have been isolated from various xenobiotics degrading bacterial strains and the factors influencing their activity studied (Monzón *et al.*,

2018). Currently, bacteria are isolated from polluted soils, waste water and residual sites which are thought to be resistant to wide range of toxic xenobiotics, which is as a result of their high tolerance even at higher xenobiotics concentration (Yeo *et al.*, 2015). The isolated and studied tolerant bacteria are used for degrading toxic xenobiotic pollutants, heavy metals and other solid waste effluent in the environment (Yeo *et al.*, 2015; Gomathi *et al.*, 2020). Naturally isolated strains from the site of contamination can be used for the treatment of industrial wastes or genetically modified strains may be used to enhance the degradation process (Farber *et al.*, 2019; Tusher *et al.*, 2020).

10.8   APPLICATIONS OF SENSORY MECHANISMS IN BACTERIA FOR XENOBIOTICS

Sensory mechanisms in bacteria for xenobiotics are of paramount significance and have contributed immensely to the field of biotechnology and environmental microbiology. Dissection of the molecular mechanisms of sensing and signaling in bacteria has been a major facet of modern biochemistry and molecular biology in the alleviation of pollutions related to the environment (Aravind *et al.*, 2015). In achieving this, the bacterial strains involved needs to be able to detect xenobiotic compounds first before metabolizing it.

10.9   DETECTION OF XENOBIOTIC COMPOUNDS

Sensory systems in bacteria enable the detection of xenobiotic compounds in the environment. Xenobiotics are constantly discharged into the ecosystem causing various environmental problems such as pollution (Nagata, 2020). The continual amassing of recalcitrant xenobiotics in the environment discharged from various sources such as homes or industries has caused a serious global concern. Some microorganisms that detect those xenobiotic chemicals and adapt to such environmental conditions have been isolated and employed as bio-indicators of xenobiotics in the soil. They have also been used to assess the quality of soil for the presence of xenobiotics.

10.10   ANALYSIS OF CHEMOTAXIS OF BACTERIA TO XENOBIOTICS

Sensory mechanisms in bacteria play a major role in regulating the movement of bacteria within the environment in response to xenobiotic chemicals. Bacteria exhibit chemotaxis to xenobiotic chemicals as they tend to move toward higher concentration of xenobiotic chemicals to ultimately utilize them for energy and carbon (Nishikino *et al.*, 2018). Xenobiotic are detected by receptors present on the cell membrane that bring about the phosphorylation of the complex sensory proteins and the signaling protein resulting in change of direction of the flagellar rotation thereby regulating the swimming behavior (Parales *et al.*, 2015; Nishikino *et al.*, 2018).

There are reported pathways of bacterial chemotaxis to xenobiotics (Parales *et al.*, 2015). The concentric circles often indicate an increasing chemoattractant concentration with the darkest green-shaded circle having the highest concentration. Various receptors present on the membranes are responsible for the sensing of xenobiotics. Xenobiotics that found their way into the periplasm are detected by the classical methyl-accepting chemotaxis proteins and appropriate response generated. Upon detection of xenobiotics, the substrates are efficiently transported across the membrane for metabolism of the target chemicals. The transfer of catabolic plasmid is likely to increase by the chemotaxis of bacteria to xenobiotics. The range of xenobiotics sensed as chemoattractant, the chemicals a bacterium possesses and the capacity to degrade are ultimately enhanced by the horizontal transfer of catabolic plasmids (Parales *et al.*, 2015).

Xenobiotics biodegradation is a metabolic process that employs the use of degradative enzymes to completely breakdown toxic xenobiotics in the environment. The use of bacteria or microorganisms in general is proven to be a more economical and environmental friendly way of eliminating pollutants from the environment, a process referred to as bioremediation.

Recent studies on the sensory mechanisms in bacteria for xenobiotics are widely used in environmental microbiology and biotechnology for the detection and degradation of xenobiotic (Mpofu *et al.*, 2020). Bacteria that degrade xenobiotics have been isolated and studied (Narwal and Gupta, 2017). Majority of the aerobic bacterial strains that degrade xenobiotics possess various degradation pathways and thus, serve as excellent model for analyzing ways through which bacteria strive and adapt to various environmental conditions (Nagata, 2020).

The biodegradation pathway of xenobiotic occurs aerobically and anaerobically (Zhang *et al.*, 2017). The products of the aerobic and anaerobic biodegradation pathways and various xenobiotics suitable for each pathways are released, the aerobic biodegradation yields water, carbon dioxide, biomass and residues as products while the anaerobic pathway yields methane in addition to the products of the aerobic biodegradation pathway (Zhang *et al.*, 2017).

There is an increasing interest in the ability of bacteria to detect and appropriately metabolize xenobiotics as a result of their persistency and toxicity. The use of microorganisms in the treatment of pollutants, a process known as bioremediation is considered a more economical way or removing toxic xenobiotics from the environment. Research has shown that diverse group of bacterial strains are capable of metabolizing wide range of xenobiotics (Zhang *et al.*, 2017). Furthermore, bacteria play key role in biogeochemical cycles and in sustainable development of the ecosystem. The bacterial genera recognized for their degradative abilities includes both aerobic such as *Alcaligenes, Bacillus, Pseudomonas, Escherichia, Sphingomonas, Sphingobium, Pandoraea, Rhodococcus, Mycobacterium, Gordonia, Moraxella, Micrococcus* and anaerobic types including; *Methanococcus, Pelatomaculum, Methanosaeta, Methanobacterium, Desulfotomaculum, Syntrophosomonas, Syntrophobacter, Syntrophus, Desulphovibrio* and *Methanospirillum* (Singh, 2017). Aerobic bacteria have been found to degrade pesticides, alkanes and PAHs as energy and carbon sources. In the same vein, anaerobic bacteria are used in the breakdown of PCBs as treatment of river sediments and in dechlorination of the toxic solvent TCE and chloroform (Zhang *et al.*, 2017).

10.11   PROGNOSIS OF THE EVOLUTION OF BACTERIA

The fundamental studies on the sensory mechanisms in bacteria for xenobiotics metabolism have shown that the bacterial gene capable of sensing xenobiotics can be employed as model for investigating the progressional changes of bacteria in the environment with regards to xenobiotics degradation (Nagata, 2020). On-going studies on the varieties of bacterial strains that sense xenobiotics, having being isolated and studied revealed that such strains can be used as probes for exploring various ways through which bacteria adapt to xenobiotic chemicals and their evolutionary changes in the environment. Many chemotaxis pathways for xenobiotics sensing have been explored due to evolution of bacteria and these involves physiological processes for direct sensing of xenobiotics, active transport and metabolism of the target xenobiotic chemicals. The pathways stressed the need for sensing in determining the roles of the bacteria in xenobiotics degradation under the constantly changing nature of the soil environment (Parales *et al.*, 2015).

10.12   CONCLUSION

Bacterial species are important bio-indicators in the environment. They possess various receptors and sensory mechanisms they use in detecting an environment polluted by xenobiotics. They have evolved various pathways they use in metabolizing xenobiotics into less harmful forms that other living things can utilize.

Bacterial pathways for sensing and detecting xenobiotics should be harnessed to detect even the smallest amount of xenobiotic compounds from the environment as well as to be able to transform them into less toxic form, as this can save the planet a lot of troubles arising from bioaccumulation of xenobiotics in the environment. Furthermore, bacterial genes responsible for chemotactic response be enhanced to facilitate bacterial use as an effective tool for sensing and metabolism of xenobiotics.

REFERENCES

Alvarado, A., Behrens, W., & Josenhans, C. (2020). Protein activity sensing in bacteria in regulating metabolism and motility. *Frontiers in Microbiology,* 10, 3055. https://doi.org/10.3389/fmicb.2019.03055

Aravind, L., Iyer, L. M., & Anantharaman, V. (2015). Natural history of sensor domains in bacterial signaling system. *Current Opinion in Microbiology,* 6, 490–497.

Bardy, S. L., Briegel, A., Rainville, S., & Krell, T. (2017). Recent advances and future prospects in bacterial and archaeal locomotion and signal transduction. *Journal of Bacteriology,* 10, e00230-17. https://doi.org/10.1128/JB.00203-17

Bharadwaj, A. (2018). Bioremediation of xenobiotics: An eco-friendly cleanup approach. In: Parmar V., Malhotra P., Mathur D. (Eds), *Green Chemistry in Environmental Sustainability and Chemical Education* (pp. 1–13). Springer, Singapore.

Bi, S., & Sourjik, V. (2018). Stimulus sensing and signal processing in bacterial chemotaxis. *Current Opinion in Microbiology,* 45, 22–29.

Chong, N. (2015). Model development with defined biological mechanisms for xenobiotic treatment activated sludge at steady state. *Environmental Science and Pollution Research,* 11, 8567–8575.

Desai, S. K., & Kenney, L. J. (2017). To approximately P or not to approximately P? Non-canonical activation by two-component response regulators. *Molecular Microbiology,* 103, 203–213. https://doi.org/10.1111/mmi.13532

Farber, R., Rosenberg, A., Rozenfeld, S., Banet, G., & Cahan, R. (2019). Bioremediation of artificial diesel-contaminated soil using bacterial consortium immobilized to plasma-pretreated wood waste. *Microorganisms,* 7, 497.

Francis, V. I., & Porter, S. L. (2019). Multikinase networks: Two-component signaling networks integrating multiple stimuli. *Annual Review on Microbiology,* 73, 199–223. https://doi.org/10.1146/annurev-micro-020518-115846

Galperin, M. Y. (2018). What bacteria want. *Environmental Microbiology,* 20, 4221–4229. https://doi.org/10.1111/1462-2920.14398

Gomathi, T., Saranya, M., Radha, E., Vijayalakshmi, K., Supriya, P. P., & Sudha, P.N. (2020). Bioremediation. *Encyclopedia of Marine Biotechnology,* 10, 3139–3172.

Groisman, E. A. (2016). Feedback control of two-component regulatory systems. *Annual Reviews of Microbiology,* 70, 103–124. https://doi.org/10.1146/annurev-micro-102215-095331

Irazoki, O., Mayola, A., Campoy, S., & Barbe, J. (2016). SOS system induction inhibits the assembly of chemoreceptor signaling clusters in *Salmonella enterica. PLoS One.* https://doi.org/10.1371/journal.pone.0146685

Janssen, D. B., & Stucki, G. (2020). Perspectives of genetically engineered microbes for groundwater bioremediation. *Environmental Sciences: Processes Impacts,* 22, 487–499.

Jones, C. W., & Armitage, J. P. (2017). Essential role of the cytoplasmic chemoreceptor TlpT in the de novo formation of chemosensory complexes in *Rhodobacter sphaeroides. Journal of Bacteriology.* https://doi.org/10.1128/JB.00366-17

Kumar, M., Prasad, R., Goyal, P., Teotia, P., Tuteja, N., Varma, A., & Kumar, V. (2017). Environmental biodegradation of xenobiotics: Role of potential microflora. In: Hashmi M., Kumar V., Varma A. (Eds), *Xenobiotics in the Soil Environment. Soil Biology* (vol. 49, pp. 319–334). Springer, Cham. https://doi.org/10.1007/978-3-319-47744-2_21

Luu, R. A., Kootstra, J., Brunton, C., Nesteryuk, V., Parales, J. V., Ditty, J. L., & Parales, R. E. (2015). Integration of chemotaxis, transport, and catabolism in *Pseudomonas putida* and identification of the aromatic acid chemoreceptor PcaY. *Molecular Microbiology,* 96, 134–147.

Machuca, M. A., Liu, Y. C., Beckham, S. A., Gunzburg, M. J., & Roujeinikova, A. (2016). The crystal structure of the tandem-PAS sensing domain of *Campylobacter jejuni* chemoreceptor Tlp1 suggests indirect mechanism of ligand recognition. *Journal of Structural Biology,* 194, 205–213. https://doi.org/10.1016/j.jsb.2016.02.019

Masilamani, R., Cian, M. B., & Dalebroux, Z. D. (2018). *Salmonella* Tol-Pal reduces outer membrane glycerophospholipid levels for envelope homeostasis and survival during bacteremia. *Infection and Immunity,* 86, e00173-18. https://doi.org/10.1128/IAI. 00173-18

Matson, J. S., Livny, J., & Dirita, V. J. (2017). A putative *Vibrio cholerae* two-component system controls a conserved periplasmic protein in response to the antimicrobial peptide polymyxin B. *PLoS One,* 12, e0186199. https://doi.org/10.1371/journal.pone.0186199

May, K. L., Lehman, K. M., Mitchell, A. M., & Grabowicz, M. (2019). A stress response monitoring lipoprotein trafficking to the outer membrane. *mBio,* 10, e00618-19. https://doi.org/10.1128/mBio.00618-19

Monzón, G. C., Nisenbaum, M., Seitz, K. H., & Murialdo, S. E. (2018). New findings on aromatic compounds' degradation and their metabolic pathways, the biosurfactant production and motility of the halophilic bacterium *Halomonads sp.* KHS3. *Current Microbiology,* 75(8), 1108–1118.

Mpofu, E., Chakraborty, J., Suzuki-Minakuchi, C., Okada, K., Kimura, T., & Nojiri, H. (2020). Biotransformation of monocyclic phenolic compounds by *Bacillus licheniformis* TAB7. *Microorganisms,* 8, 26.

Nagata, Y. (2020). Microbial degradation of xenobiotics. *Microorganisms, 8*, 487. https://doi.org/10.3390/microorganisms8040487

Nagata, Y., Kato, H., Ohtsubo, Y., & Tsuda, M. (2019). Lessons from the genomes of lindane-degrading sphingomonads. *Environmental Microbiology Republican, 11*, 630–644.

Nagata, Y., Ohtsubo, Y., & Tsuda, M. (2015). Properties and biotechnological applications of natural and engineered haloalkane dehalogenases. *Applied Microbiological Biotechnology, 99*, 9865–9881.

Narwal, S. K., & Gupta, R. (2017). Biodegradation of xenobiotic compounds: An overview. In: Bhakta, J. N. (Ed.), *Handbook of Research on Inventive Bioremediation Techniques* (pp. 186–212). IGI Global. https://doi.org/10.4018/978-1-5225-2325-3.ch008

Nishikino, T., Hijikata, A., Miyanoiri, Y., Onoue, Y., Kojima, S., & Shirai, T. (2018). Rotational direction of flagellar motor from the conformation of FliG middle domain in marine Vibrio. *Science Republican, 8*, 17793. https://doi.org/10.1038/s41598-018-35902-6

Ortega, Á., Zhulin, I. B., & Krell, T. (2017). Sensory repertoire of bacterial chemoreceptors. *Microbiology and Molecular Biology Review.* https://doi.org/10.1128/MMBR00033-17

Osman, D., Martini, M. A., Foster, A. W., Chen, J., Scott, A. J. P., & Morton, R. J. (2019). Bacterial sensors define intracellular free energies for correct enzyme metalation. *Nature Chemical Biology, 15*, 241–249. https://doi.org/10.1038/s41589-018-0211-4

Parales, J. E., Luu, R. A., Hughes, J. G., & Ditty, J. L. (2015). Bacterial chemotaxis to xenobiotic chemicals and naturally-occurring analogs. *Current Opinion in Biotechnology, 33*, 318–326. http://dx.doi.org/10.1016/j.copbio.2015.03.017

Park, Y. C., Lee, S., & Cho, M. H. (2014). The simplest flowchart stating the mechanisms for organic xenobiotics-induced toxicity: Can it possibly be accepted as a "central dogma" for toxic mechanisms? *Toxicology Research.* https://doi.org/10.5487/TR.2014.30.3.179F

Parkinson, J. S., Hazelbauer, G. L., & Falke, J. J. (2015). Signaling and sensory adaptation in *Escherichia coli* chemoreceptors. *Trends Microbiology, 23*, 257–266. https://doi.org/10.1016/j.tim.2015.03.003

Rico-Jimenez, M., Reyes-Darias, J. A., Ortega, A., Diez-Pena, A. I., Morel, B., & Krell, T. (2016). Two different mechanisms mediate chemotaxis to inorganic phosphate in *Pseudomonas aeruginosa. Science Republican, 6*, 28967. https://doi.org/10.1038/srep28967

Sampedro, I., Parales, R. E., Krell, T., & Hill, J. E. (2015). *Pseudomonas* chemotaxis. *FEMS Microbiology Review, 39*, 17–46.

Singh, R. (2017). Biodegradation of xenobiotics- a way for environmental detoxification. *International Journal of Development Research, 7*(07), 14082–14087.

Somavanshi, R., Ghosh, B., & Sourjik, V. (2016). Sugar influx sensing by the phosphotransferase system of *Escherichia coli. PLoS Biology, 14*, e2000074. https://doi.org/10.1371/journal.pbio.2000074

Terahara, N., Inoue, Y., Kodera, N., Morimoto, Y. V., Uchihashi, T., & Imada, K. (2018). Insight into structural remodeling of the FlhA ring responsible for bacterial flagellar type III protein export. *Science Advances, 4*, eaao7054. https://doi.org/10.1126/sciadv.aao7054

Tsang, J., Hirano, T., Hoover, T. R., & Mcmurry, J. L. (2015). *Helicobacter pylori* FlhA binds the sensor kinase and flagellar gene regulatory protein FlgS with high affinity. *Journal of Bacteriology, 197*, 1886–1892. https://doi.org/10.1128/JB.02610-14

Tusher, T. R., Shimizu, T., Inoue, C., & Chien, M. F. (2020). Enrichment and analysis of stable 1,4-dioxane-degrading microbial consortia consisting of novel dioxane-degraders. *Microorganisms, 8*, 50.

Upadhyay, A. A., Fleetwood, A. D., Adebali, O., Finn, R. D., & Zhulin, I. B. (2016). Cache domains that are homologous to, but different from PAS domains comprise the largest superfamily of extracellular sensors in prokaryotes. *PLoS Computational Biology, 12*, e1004862. https://doi.org/10.1371/journal.pcbi.1004862

Wiech, E. M., Cheng, H.-P., & Singh, S. M. (2015). Molecular modeling and computational analyses suggests that the *Sinorhizobium meliloti* periplasmic regulator protein ExoR adopts a superhelical fold and is controlled by a unique mechanism of proteolysis. *Protein Sciences*, 24, 319–327. https://doi.org/10.1002/pro.2616

Willett, J. W., & Crosson, S. (2017). Atypical modes of bacterial histidine kinase signaling. *Molecular Microbiology,* 103, 197–202. https://doi.org/10.1111/mmi.13525

Yeo, B. J., Goh, S., Zhang, J., Livingston, A. G., & Fane, A. G. (2015). Novels MBRs for the removal of organic priority pollutants from industrial wastewaters. *Journal of Chemical Technology & Biotechnology*, 11, 1949–1967.

Zhang, S., Gedalanga, P. B., & Mahendra, S. (2017). Advances in bioremediation of 1,4-dioxane contaminated waters. *Journal of Environmental Management,* 204, 765–774.

Zschiedrich, C. P., Keidel, V., & Szurmant, H. (2016). Molecular mechanisms of two-component signal transduction. *Journal of Molecular Biology,* 428, 3752–3775. https://doi.org/10.1016/j.jmb.2016.08.003

11 Biofilms

Recent advances in bioremediation

Iqra Bano, Syed Shams ul Hassan,
Shireen Aziz, and Muhammad Ahmer Raza

CONTENTS

11.1 INTRODUCTION

A biofilm is a collection of microbial cells that are surface-associated and encased in a matrix of extracellular polymeric material (Donlan, 2002). The conditions of the environment in which the biofilm forms determine the kind of materials to which the microorganisms can cling. Generally, the microorganisms have a potential to attach themselves to non-cellular material like as particles of corrosion, clay or split particles, and mineral crystals (Patra et al., 2018). The first time that bacteria were seen on tooth surfaces was by Van Leeuwenhoek, who is also credited with the discovery of microbial biofilms, using his crude microscopes. Microorganisms have been classified as planktonic, freely suspended cells since the beginning of microbiology, and their morphological, physiological, and growth characteristics in nutrient-rich culture conditions have been used to describe them (Singh et al., 2006). The initial connection of microorganisms is weak and reversible, but if the colonies remain attached, irreversible attachment takes place, and integrins and proteins on the cell surface may result in persistent

DOI: 10.1201/9781003394600-13

adhesion (Jiang et al., 2021). There are microbial biofilms all over the natural world mostly on plants, on soil, on rocks, on human or animal tissue, on medical indwelling devices, and on bodily implants (Sentenac et al., 2022). Furthermore, the aquatic invertebrates like many fish eat graze on the biofilms, which are crucial parts of the food web in rivers and streams. Extreme settings, such as the hot, acidic springs in the Yellowstone National Park as well as on glaciers in Antarctica, may be conducive to the growth of biofilms (Proal, 2008). However, in the vast majority of natural settings, microorganisms are found attached to surfaces and interfaces as multicellular aggregates held together by the slime they secrete. To put it simply, biofilms are communities of microorganisms that have colonized a particular surface. They populate practically every wet setting with adequate nutrition supply and the ability to cling to surfaces (Sharma, 2022). One species of bacteria can be responsible for the formation of a biofilm (Figure 11.1); however, biofilms can also be composed of several other types of bacteria, fungi, algae, and protozoa (De Carvalho, 2007). About 97% of the biofilms are either water, which is linked to the shells of microbial cells or solvents, the physical qualities of that which (such as viscosity) are governed by the solutes that are dissolved in it. In addition to water as well as microbial cells, the film matrix is composed of a variety of extruded polymers, absorbed nutrients and metabolites, products of cell lysis, and even particulate matter and detritus from the environment immediately surrounding the biofilm (Weitere et al., 2018). Moreover, within a biofilm environment, it is possible to find all of the major classes of macromolecules, including proteins, polysaccharides, DNA, and RNA, in addition to peptidoglycan, lipids, and phospholipids, as well as other components of cells (Figure 11.2). Previously some scientists have revealed that the existence of uronic acids in biofilms, such as D-glucuronic, D-galacturonic, and mannuronic acids, as well as ketal-linked pyruvates, is what gives biofilms their anionic feature (Sutherland, 2001). The rapid expansion of chemical companies over the last several decades has led to the poisoning of the environment as a result of the hazardous waste effluents produced by these businesses. The continued presence of chemical pollutants and the subsequent environmental concerns have prompted the general population to become more aware of the prospect of catastrophic long-term environmental effects. As a result of this, several different methods are currently being established, and additional research is currently being carried out, to discover means of sustaining the ecosystem (Polprasert & Liyanage, 1996). The use of microorganisms in the process of bioremediation is a relatively new in situ approach to the elimination of environmental toxins (Perelo, 2010). The significance of biofilm communities for bioremediation processes has now been realized, and it is clear that biological processes are superior to chemical and physical approaches for treating hazardous effluents in terms of efficiency and economics (Saeed et al., 2021). Since cells in a biofilm have a greater chance of survival and adaptation (particularly during periods of stress) as they are sheltered within the matrix of a biofilm, biofilm-mediated bioremediation offers a more efficient and safer alternative to bioremediation utilizing planktonic microbes (Shukla et al., 2017). In this article of a review series, the applications of biofilm-mediated bioremediation processes are discussed.

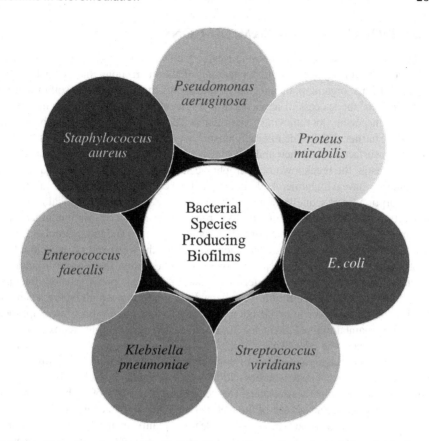

FIGURE 11.1 The demonstration of biofilm producing bacteria including both Gram-positive as well as Gram-negative species.

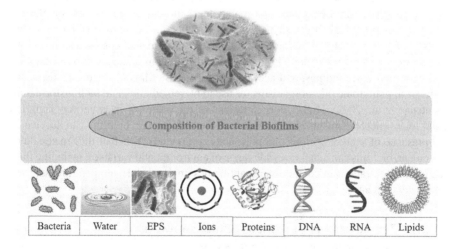

FIGURE 11.2 The demonstration of composition of bacterial biofilm, showing different important components of biofilm.

11.2 BIOFILMS AND BIOREMEDIATIONS

Bioremediation is the use of microorganisms to remove hazardous pollutants from soil, water, and the atmosphere. An understanding of the interactions that occur between microorganisms, organic pollutants, and the elements that make up the soil or aquifer is required for the successful application of a bioremediation procedure. This under-standing is necessary for removing contaminants from the environment using biological methods (Sharma, 2022). The microorganisms' physiological features, such as the cre-ation of biosurfactants and their ability to engage in chemotaxis, increase bioavailability and, as a result, the breakdown of hydrophobic substances (Yao & Habimana, 2019b). Because they have a high microbial biomass and the capacity to immobilize substances via biosorption (inert sequestration in addition to connections with biological material), biosorption (enhanced concentration of microbes under influence), and biomineraliza-tion, microscopic organisms that normally produce polymers and bioaccumulate on the hydrocarbon substrate are particularly appropriate for the treatment of recalcitrant oil (Mohapatra et al., 2019). Biosorption is passive sequestration that occurs in addition to interactions with biological matter (formation of insoluble precipitates by interact-ing with biological matter). Biofilms are capable of supporting a high biomass density, which makes it easier for mineralization processes to take place. This is accomplished by biofilms maintaining pH levels, localized solute concentrations, and redox potentials that are ideal near the cells. This is made possible by the one-of-a-kind design of the biofilm as well as the regulated flow of fluids throughout its interior (Muhammad et al., 2020). The treatment of large volumes of dilute aqueous solutions, such as those found in industrial and municipal wastewaters, frequently makes use of biofilm-based reactors. Bioremediation takes advantage of the fact that many naturally occurring microorgan-isms, including bacteria and fungi, can degrade the contaminant in which they are found. Even better, the microorganisms in question are highly amenable to genetic engineering to facilitate the breakdown of target contaminants (Kokare et al., 2009). Many distinct bacterial species may be used in a bioremediation project, each with its own unique set of metabolic pathways, enzymes, and metabolites. In comparison to their planktonic counterparts, microbial biofilms enjoy several benefits, including the facilitation of the transfer of genetic material, resistance to adverse environmental conditions, the ability to remain in a variety of metabolic states, the ability to communicate with one another and with their surroundings, and access to nutrients. Biofilms of microorganisms are superior to planktonic microorganisms for use in bioremediation and biotransformation (Sentenac et al., 2022). Higher concentrations of the chemical in issue may suppress planktonic microorganisms in contaminated areas. However, biofilms can survive in the presence of such dangerous and poisonous compounds. Whether they're sessile or flocculent, biofilms can adapt to a wide range of environmental variables, including fluc-tuating concentrations of contaminants, antibiotics, minerals, pH, temperature, salinity, and water content (Singh et al., 2006).

11.2.1 THE IMPORTANCE OF BIOFILMS IN THE REMOVAL OF HEAVY METALS FROM THE ENVIRONMENT

Some heavy metals, including mercury (Hg), lead (Pb), cadmium (Cd), copper (Cu), nickel (Ni), and cobalt (Co), have the potential to be hazardous to humans, whereas others, including iron (Fe), manganese (Mn), selenium (Se), and zinc (Zn) can pose a role as micronutrients. Heavy metal pollution of freshwater environments has emerged as one of the most pressing environmental concerns in recent years as a direct result of the discharge of significant quantities of wastewaters that were previously contaminated with metals (Teitzel & Parsek, 2003). The sectors that use heavy metals like Cd, Se, Cu, Ni, Pb, and Zn are by far the most dangerous of the chemical-intensive industries since these metals have a high level of toxicity and are easily soluble in aquatic environments. Heavy metals, once they have entered the food chain, are capable of being absorbed by living creatures (Haque et al., 2021). Due to the recalcitrance of heavy metals in the environment, high quantities of heavy metals can accumulate not only in aquatic biotopes but also in the human body. Metal and radionuclide bioremediation relies heavily on the dispersion and diversity of the bacteria that live in contaminated areas, as well as the genes that encode the phenotypes responsible for interactions between metals and microbes (Patra et al., 2018). Bioremediation of heavy metals can be accomplished through immobilization, concentration, and partitioning of an environmental compartment. This reduces the potential for the risks that were previously expected. As a result of the physiological reaction biofilms exhibit during the process of absorbing water and inorganic or organic solutes, biofilms have the potential to also influence the destinies of other substances in their immediate environment (Rummel et al., 2017). Recently, Valls and Lorenzo have outlined several different scenarios in which a cellular property that is already present in certain strains can be coupled with another strain's phenotype using genetic engineering, hence improving the trait (Lorenzo & Valls, 2002). Molecular techniques enable the construction of better strains with particular metal-binding capabilities by improving metal precipitation processes, expressing metal-chelating peptides and proteins, and introducing metal transformation activities in tough environmental strains. Then, these strains can be applied to cleaning up the environment (Muhammad et al., 2020).

11.2.2 The importance of biofilms in the removal of hydrocarbons from the environment

Recalcitrant chemicals known as chlorinated compounds are found in the effluents of several different chemical industries and can move fast over soils. Despite being present in such vast amounts, they are considered one of the most pervasive soil and groundwater contaminants and can cause cancer even at very low concentrations (Chattopadhyay et al., 2022). Nitroaromatic compounds are indeed an additional class of xenobiotics that have been utilized in the production of a variety of useful products, including foams, medicines, pesticides, and explosives. These compounds are resistant to biodegradation because they contain nitro groups, and the conversion of these compounds by microbes frequently results in the creation of toxic metabolites (Ju & Parales, 2010). There is a possibility that biofilm-mediated bioremediation might be improved even further by altering the metabolic processes and enzymes involved in the degradation process, as well as by increasing the copy number of genes that are engaged in the process, as well as by upgrading the strains. For example, a chemotactic strain that has been modified to have catabolic genes would be highly effective at biodegradation. Strains that are capable of both chemotaxis and biodegradation would be ideal for biofilm development. Biofilm formation would be advantageous for strains that are capable of both biodegradation and chemotaxis (Muhammad et al., 2020). The existence of such strains would make the production of biofilm easier to accomplish. Co-adhesion and synergistic interaction with species that form biofilms have been suggested as a potential alternate mechanism for the persistence and spread of other strains in several papers. The combination of genetically modified microbes, adjustment of physio-chemical properties, and substrate concentration in bioreactors should be given top priority when developing bioremediation solutions for this reason (Donlan & Costerton, 2002).

11.2.3 The importance of chemotaxis in both the process of biodegradation and the creation of biofilm

The migration of organisms that takes place in response to a chemical nutrient or a chemical gradient is referred to as chemotaxis. In addition to being an essential component of the biodegradation process, it aids in the discovery by bacteria of the optimal conditions for their growth and survival (Yaryura et al., 2008). Chemotaxis is likely chosen as a beneficial behavior in bacteria in situations of restricted carbon and energy sources. Additionally, it is feasible that xenobiotic breakdown skills are selected as a favorable behavior in bacteria after exposure to such chemicals. The use of bacteria that show chemotaxis toward contaminants has gotten less attention even though the bacterial breakdown of pollutants is effective for the bioremediation of contaminated locations in several different instances (Pandey & Jain, 2002). The bioaccumulation of the material to the microbial species is the initial thing that must be addressed throughout bioremediation. One of the main obstacles to the successful bioremediation of polluted locations is the bioavailability of organic contaminants, which can be overcome by using chemotactic bacteria. Chemotactic cells can detect substances that specialize in adhering to dirt particles and swim toward them; this will allow you to circumvent mass-transfer constraints that hinder the bioremediation process (Donlan & Costerton, 2002). To put it another way: when the cells are placed near a surface, a chain reaction of biofilm growth and surfactant generation begins, resulting in increased bioavailability and biodegradation. The process of chemotaxis plays a significant part in the formation of biofilm in a variety of microorganisms. This process directs bacteria to swim forward into nutrient content (hydrophobic toxins) that are adsorbed to an exterior, which is then followed by the attachment of the bacterial flagellum to the surface (Kostakioti et al., 2013). Adhesion to abiotic surfaces requires flagella, and flagella also play an important role in the early stages of biofilm development. In addition, chemotaxis and/or motility may be necessary for bacteria that are part of a forming biofilm for them to migrate along the surface to proliferate and spread (Dang & Lovell, 2016).

11.2.4 The importance of biofilms in field of agriculture

The use of pesticides to manage phytopathogens and chemical fertilizers in agriculture has increased soil contamination and harmful chemical buildup in the soil. Additionally, the spread of dangerous chemicals could contaminate the environment and cause diseases in humans. Biological control may be a different approach to eco-friendly agricultural practices (Velmourougane et al., 2017). A biological control, sometimes known as a "biocontrol," is a type of bacterium that stops the growth of pathogens or makes substances that gives protection and encourage their growth. Numerous researchers have recently emphasized the advantages of biocontrol agent biofilms and their capacity to generate biofilms. As well as soil particles, mushrooms, and organic waste, bacteria can colonize plants and produce biofilms on their stems, leaves, and rhizosphere. The following phases are involved in the production of bio-film on plants (Harms, 2011). (1) Free-floating planktonic bacteria must first adhere to one another. Bacteria now can move around. (2) The loss of locomotor organs and bacterial adherence to the substrate are the second stages. (3) Following that, germs begin to multiply. (4) Next, exo-polysaccharides are produced as a result of the spatial arrangement of cells and the formation of the biofilm. (5) Regulated bio-film dispersion occurs as a result of biofilm aging or unfavorable environmental cir-cumstances for maintaining the biofilm. Plants are also known to harbor pathogens that produce biofilms (Velmourougane et al., 2017). For instance, the Gram-negative bacteria Dickeya dadantii causes soft rot diseases in a variety of plant species. On chicory leaves, the bacteria invade and create biofilms, which lead to the develop-ment of disease-causing enzymes. However, Pseudoalteromonas tunicate, a marine Gram-negative bacterium, forms biofilms when exposed to the green macroalga Ulvalactuca (P. tunicate). P. tunicate is an endophytic bacterium that makes anti-fouling substances that prevent biofilm and colonization (Solanki et al., 2020). Similar to this, Pseudomonas chlororaphis forms biofilms on wheat roots to guard against fungus-related illness. Beside this in the field of land irrigation the manual irriga-tion, sub-irrigation, surface irrigation, localized irrigation, drip irrigation, sprinkler irrigation, center pivot irrigation, and lateral move irrigation are some of the many types that can be utilized for a wide variety of irrigation tasks (Figure 11.3). The vast majority of these utilize pipes, drippers, or sprinklers, which provide the essential surface to enable the establishment of biofilms (Pachepsky et al., 2012). Microscopy techniques of the most recent generation could be utilized in order to characterize

FIGURE 11.3   The land irrigation the manual irrigation, sub-irrigation, surface irrigation, drip irrigation, and sprinkler irrigation utilizing biofilms.

the structural features of biofilms at various stages of biofilm formation. It is possible to use omics technology in conjunction with computational methods and fluorescence in situ hybridization procedures in order to establish the microbial makeup of an irrigation water distribution system (IWDS) biofilm as well as the geographic distribution of various microbial groups or people within that biofilm (Alam, 2014). A potential risk to food safety should be taken into consideration since biofilms in IWDSs are crucial in the movement of diseases, chemicals, and environmental contaminants into downstream irrigated crops (Yao & Habimana, 2019b). In terms of the function of biofilms, researchers should concentrate their efforts on investigating the interspecific and intraspecific transfer of specific genes that occur through biofilms, such as the transfer of genes that cause antibiotic resistance through horizontal gene transfer (Yao & Habimana, 2019a).

11.3   CONCLUSION

A microbial aggregation or association known as the biofilm is characterized by its ability to attach to biotic or abiotic structures or environments. According to the review, the current level of phytoremediation for heavy metals contains a great deal of potential for the adsorption and purification of heavy metals, notably from biofilm bacteria. The formation of bacterial biofilms takes occurs in a systematic and well-ordered set of processes, and the biofilm lifestyle is the most prevalent form of bacterial existence in both natural and man-made environments. Microorganisms that are connected with biofilms play a significant role in the elimination of chemicals as well as the transformation of hazardous pollutants into harmless compounds. The moment has come to extend the capabilities of gene delivery within biofilms for the treatment of heavy metals. In the bio sorbent cell wall, which functions as an active binding site for increased metal concentrations, polysaccharides, and peptidoglycan can be discovered. A method like this is helpful not only to the environment but also to the bottom line because it has advantages like accelerated kinetics and enhanced metal-binding throughout a broad temperature and pH range. The bioremediation process stands to gain from a deeper understanding of the role that microbiological processes play in the toleration and degradation of pollutants. The mechanisms and genes that are involved in the creation of biofilms can be of assistance in the discovery of new bioremediation techniques.

REFERENCES

Alam, M. (2014). Microbial status of irrigation water for vegetables as affected by cultural practices. *In Faculty of Landscape Architecture, Horticulture and Crop Production Science Department of Biosystems and Technology Alnarp Doctoral*. Doctoral Thesis. ISSN 1652-6880 https://pub.epsilon.slu.se/10986/1/alam_m_140124.pdf

Chattopadhyay, I., Rajesh Banu, J., Usman, T. M. M., & Varjani, S. (2022). Exploring the role of microbial biofilm for industrial effluents treatment. *Bioengineered, 13*(3), 6420–6440. https://doi.org/10.1080/21655979.2022.2044250

Dang, H., & Lovell, C. R. (2016). Microbial surface colonization and biofilm development in marine environments. *Microbiology and Molecular Biology Reviews : MMBR, 80*(1), 91–138. https://doi.org/10.1128/MMBR.00037-15

De Carvalho, C. C. C. R. (2007). Biofilms: Recent developments on an old battle. *Recent Patents on Biotechnology*, *1*(1), 49–57.

Donlan, R. M. (2002). Biofilms: Microbial life on surfaces. *Emerging Infectious Diseases*, *8*(9), 881.

Donlan, R. M., & Costerton, J. W. (2002). Biofilms: Survival mechanisms of clinically relevant microorganisms. *Clinical Microbiology Reviews*, *15*(2), 167–193. https://doi.org/10.1128/CMR.15.2.167-193.2002

Haque, M. M., Mosharaf, M. K., Haque, M. A., Tanvir, M. Z. H., & Alam, M. K. (2021). Biofilm formation, production of matrix compounds and biosorption of copper, nickel and lead by different bacterial strains. *Frontiers in Microbiology*, *12*(June), 1–19. https://doi.org/10.3389/fmicb.2021.615113

Harms, H. (2011). Bioavailability and bioaccessibility as key factors in bioremediation. In: E. Moo-Young (ed.) *Comprehensive Biotechnology*, pp. 83–94. Academic Press. https://doi.org/https://doi.org/10.1016/B978-0-08-088504-9.00367-6

Jiang, Z., Nero, T., Mukherjee, S., Olson, R., & Yan, J. (2021). Searching for the secret of stickiness: how biofilms adhere to surfaces. *Frontiers in Microbiology*, *12*, 686–793.

Ju, K.-S., & Parales, R. E. (2010). Nitroaromatic compounds, from synthesis to biodegradation. *Microbiology and Molecular Biology Reviews : MMBR*, *74*(2), 250–272. https://doi.org/10.1128/MMBR.00006-10

Kokare, C. R., Chakraborty, S., Khopade, A. N., & Mahadik, K. R. (2009). Biofilm: importance and applications. *Indian Journal of Biotechnology*, *8*, 159–168.

Kostakioti, M., Hadjifrangiskou, M., & Hultgren, S. J. (2013). Bacterial biofilms: development, dispersal, and therapeutic strategies in the dawn of the postantibiotic era. *Cold Spring Harbor Perspectives in Medicine*, *3*(4), a010306. https://doi.org/10.1101/cshperspect.a010306

Lorenzo, D., & Valls, M. (2002). Exploiting the genetic and biochemical capacities of bacteria for the remediation of heavy metal pollution. *FEMS Microbiology Reviews*, 2002 Nov; *26*(4), 327–338.

Mohapatra, R. K., Behera, S. S., Patra, J. K., Thatoi, H., & Parhi, P. K. (2019). Potential application of bacterial biofilm for bioremediation of toxic heavy metals and dye-contaminated environments. *New and Future Developments in Microbial Biotechnology and Bioengineering: Microbial Biofilms Current Research and Future Trends in Microbial Biofilms*, May 2021, 267–281. https://doi.org/10.1016/B978-0-444-64279-0.00017-7

Muhammad, M. H., Idris, A. L., Fan, X., Guo, Y., Yu, Y., Jin, X., Qiu, J., Guan, X., & Huang, T. (2020). Beyond risk: bacterial biofilms and their regulating approaches. *Frontiers in Microbiology*, *11*(May), 1–20. https://doi.org/10.3389/fmicb.2020.00928

Pachepsky, Y., Morrow, J., Guber, A., Shelton, D., Rowland, R., & Davies, G. (2012). Effect of biofilm in irrigation pipes on microbial quality of irrigation water. *Letters in Applied Microbiology*, *54*(3), 217–224. https://doi.org/10.1111/j.1472-765X.2011.03192.x

Pandey, G., & Jain, R. K. (2002). Bacterial chemotaxis toward environmental pollutants: role in bioremediation. *Applied and Environmental Microbiology*, *68*(12), 5789–5795. https://doi.org/10.1128/AEM.68.12.5789-5795.2002

Patra, J. K., Vishnuprasad, C. N., & Das, G. (2018). Microbial biotechnology. *Microbial Biotechnology*, *1*(August), 1–479. https://doi.org/10.1007/978-981-10-6847-8

Perelo, L. W. (2010). In situ and bioremediation of organic pollutants in aquatic sediments. *Journal of Hazardous Materials*, *177*(1–3), 81–89.

Polprasert, C., & Liyanage, L. R. J. (1996). Hazardous waste generation and processing. *Resources, Conservation and Recycling*, *16*(1–4), 213–226.

Proal, A. (2008). Understanding biofilms. *Bacteriality—Exploring Chronic Disease*, *26*. http://bacteriality.com/2008/05/26/biofilm

Rummel, C. D., Jahnke, A., Gorokhova, E., Kühnel, D., & Schmitt-Jansen, M. (2017). Impacts of biofilm formation on the fate and potential effects of microplastic in the aquatic environment. *Environmental Science and Technology Letters*, *4*(7), 258–267. https://doi. org/10.1021/acs.estlett.7b00164

Saeed, M. U., Hussain, N., Sumrin, A., Shahbaz, A., Noor, S., Bilal, M., Aleya, L., & Iqbal, H. M. N. (2021). Microbial bioremediation strategies with wastewater treatment potentialities–A review. *Science of the Total Environment*, 2022 Apr 20; *818*, 151754. https://doi.org/10.1016/j.scitotenv.2021.151754. Epub 2021 Nov 17. PMID: 34800451.

Sentenac, H., Loyau, A., Leflaive, J., & Schmeller, D. S. (2022). The significance of biofilms to human, animal, plant and ecosystem health. *Functional Ecology*, *36*(2), 294–313.

Sharma, P. (2022). Role and significance of biofilm-forming microbes in phytoremediation - a review. *Environmental Technology and Innovation*, *25*, 102–182. https://doi. org/10.1016/j.eti.2021.102182

Shukla, S. K., Mangwani, N., Karley, D., & Rao, T. S. (2017). Bacterial biofilms and genetic regulation for metal detoxification. In: Surajit Das, Hirak Ranjan Dash (eds.) *Handbook of Metal-Microbe Interactions and Bioremediation* (1st edition), pp. 317–332. CRC Press. https://doi.org/10.1201/9781315153353

Singh, R., Paul, D., & Jain, R. K. (2006). Biofilms: implications in bioremediation. *Trends in Microbiology*, *14*(9), 389–397. https://doi.org/10.1016/j.tim.2006.07.001

Solanki, M. K., Solanki, A. C., Kumari, B., Kashyap, B. K., & Singh, R. K. (2020). Chapter 12 - Plant and soil-associated biofilm-forming bacteria: their role in green agriculture. In: M. K. Yadav & B. P. Singh (eds.) *New and Future Developments in Microbial Biotechnology and Bioengineering*, pp. 151–164. Elsevier. https://doi.org/https://doi. org/10.1016/B978-0-444-64279-0.00012-8

Sutherland, I. W. (2001). The biofilm matrix–an immobilized but dynamic microbial environment. *Trends in Microbiology*, *9*(5), 222–227.

Teitzel, G. M., & Parsek, M. R. (2003). Heavy metal resistance of biofilm and planktonic Pseudomonas aeruginosa. *Applied and Environmental Microbiology*, *69*(4), 2313–2320. https://doi.org/10.1128/AEM.69.4.2313-2320.2003

Velmourougane, K., Prasanna, R., & Saxena, A. K. (2017). Agriculturally important microbial biofilms: present status and future prospects. *Journal of Basic Microbiology*, *57*(7), 548–573. https://doi.org/10.1002/jobm.201700046

Weitere, M., Erken, M., Majdi, N., Arndt, H., Norf, H., Reinshagen, M., Traunspurger, W., Walterscheid, A., & Wey, J. K. (2018). The food web perspective on aquatic biofilms. *Ecological Monographs*, *88*(4), 543–559.

Yao, Y., & Habimana, O. (2019a). Biofilm research within irrigation water distribution systems: trends, knowledge gaps, and future perspectives. *Science of the Total Environment*, *673*, 254–265. https://doi.org/10.1016/j.scitotenv.2019.03.464

Yao, Y., & Habimana, O. (2019b). Biofilm research within irrigation water distribution systems: trends, knowledge gaps, and future perspectives. *The Science of the Total Environment*, *673*, 254–265. https://doi.org/10.1016/j.scitotenv.2019.03.464

Yaryura, P. M., León, M., Correa, O. S., Kerber, N. L., Pucheu, N. L., & García, A. F. (2008). Assessment of the role of chemotaxis and biofilm formation as requirements for colonization of roots and seeds of soybean plants by Bacillus amyloliquefaciens BNM339. *Current Microbiology*, *56*(6), 625–632. https://doi.org/10.1007/s00284-008-9137-5

12 Extracellular enzymatic activity of bacteria in aquatic ecosystems

Gabriel Gbenga Babaniyi, Olaniran Victor Olagoke, and Sesan Abiodun Aransiola

CONTENTS

12.1 INTRODUCTION

Extracellular enzymes are key functional components of marine environments because they initiate the breakdown of organic macromolecules. Extracellular enzyme activity (EEA) measurements in seawater can give crucial information about the biogeochemical cycling of organic materials in the ocean. An exoenzyme, or extracellular enzyme, on the other hand, is an enzyme secreted by a cell that acts outside of that cell. Exoenzymes are produced by both prokaryotic and eukaryotic cells and have been found to be an important part of a variety of biological activities (Li et al., 2019). Microorganisms also play an important part in the remineralization of organic matter in the water. The microbial loop transforms, repackages, and respires an estimated 50% of primary production in surface water (Azam et al., 1983; Arnosti et al., 2005; Azam & Malfatti, 2007). The majority of organic matter in the ocean

DOI: 10.1201/9781003394600-14

is in the form of chemically complicated macromolecules that are too massive to breach the cytoplasmic membrane. Extracellular enzymes must first hydrolyze them into small molecules (600 Da) before they can be taken up by the microbial cell. As a result, extracellular enzymes are important players in the organic matter cycle in marine ecosystems (Arnosti, 2011; Orsi et al., 2018).

Extracellular enzymes from marine bacteria are either cell-associated or dissolved in the water column. Cell-associated enzymes are bound to the cell surface, or localized in the periplasmic space (Li et al., 2019). For free-living microorganisms, these enzymes present a cost-effective solution. They can assist the cell in preferentially accessing dissolved organic materials (DOM) due to the dilute nature of DOM. The substrate, on the other hand, must either penetrate the cell wall or be physically present around the cell (Allison et al., 2012). Some polysaccharide substrates could be taken up directly into the periplasm of "selfish" organisms without the need for extracellular hydrolysis products (Hehemann et al., 2019; Reintjes et al., 2019). Active secretion by cells, bacterial starvation, and changes in cell permeability are all possible sources of dissolved enzymes, which belong to a kind of "living dead" realm (Li et al., 2019). Furthermore, they can be created during the grazing process on bacterial colonies and released following viral lysis (Baltar, 2018). These enzymes can hydrolyze faraway substrates as they spread, but the hydrolysis products may not be collected by the parent cell. Dissolved enzymes perform their crucial job away from the cell due to their lengthy longevity, especially in deep waters (Li et al., 2019). In several situations, dissolved EEA could account for a significant fraction (up to 100%) of total marine EEA, indicating a mismatch between marine microorganisms and enzymatic activity (Baltar et al., 2010, 2019; D'ambrosio et al., 2014). Because of the high nutrient concentrations on particles and the relaxed necessity for cellular interaction with particle organic matter, a "secreting dissolved enzyme" method for particle-associated bacteria may be beneficial (Baltar et al., 2019). The substrate has a loose hydrolysis-uptake connection with these bacteria. The "secreting dissolved enzyme" method might sometimes benefit free-living bacteria through cooperative efforts, however it is likely to be expensive (Pai et al., 2012; Celiker & Gore, 2013).

However, because most marine microbes cannot be grown and because genomic and in situ gene investigations are questionable, there is insufficient information to reflect EEA in marine ecosystems or element recycling on a global scale. Thus, in situ EEA measurement is critical for gaining a fundamental understanding of the biogeochemical cycle of organic matter, as evidenced by field investigations (Baltar, 2018). However, most enzyme activity tests infer biopolymer enzyme activity using low molecular weight substrate approximations. These proxies lack the three-dimensional structure of biopolymers in solution and so are unable to accurately depict polymer breakdown (Li et al., 2019). As a result, little is known about the rates of hydrolysis of actual polymers. Furthermore, the substrate proxies do not represent the activity of endo-acting enzymes that cleave to the inner of polymer chains. Although several fluoresceinamine labeled polysaccharides have been used to measure polysaccharide degrading enzymes in the marine ecosystem (Reintjes et al., 2017, 2019), EEA on proteinaceous polymers, which make up a large portion of organic matter in primary production, has not been measured. In addition, very little is known about the differences in hydrolysis rates between polymers and their oligomers. Heterotrophic bacteria, according to Kalwasińska and Brzezinska (2013), play an important part in the cycling and mineralization of organic materials in freshwater. So, the majority of organic compounds in aquatic ecosystems are macromolecular and thus inaccessible to bacterial cells. As a result, high molecular weight macromolecules must be broken into simple compounds by extracellular enzymes before being taken up by microbial cells and served as a source of carbon, nitrogen, and energy. Chrost (1991) distinguished between enzymes that remain connected to their producers (ectoenzymes) and those that are dissolved in water or adsorbed to particles (endoenzymes) (extracellular).

12.1.1 DIFFERENCE BETWEEN INTRACELLULAR AND EXTRACELLULAR ENZYMES

The main difference between intracellular and extracellular enzymes, according to Lakna (2019), is that intracellular enzymes, also known as endoenzymes, work within the cell to aid intracellular digestion, whereas extracellular enzymes, also known as exoenzymes, work outside the cell to aid extracellular digestion. Furthermore, intracellular enzymes make up the majority of enzymes, while external enzymes make up the minority. Extracellular enzymes act on the polymer's end to break down its monomers one at a time, while intracellular enzymes break down big polymers into smaller chains of monomers. In a cell, there are two types of digestive enzymes: intracellular and external enzymes. In general, they are classified according to where the action takes place.

Intracellular enzymes, also known as endoenzymes: are enzymes that work within the cell. Inside the cell of both eukaryotes and prokaryotes, they are responsible for millions of metabolic events. Intracellular enzymes are therefore responsible for both photosynthesis and cellular respiration within the cell. Furthermore, these enzymes are in charge of DNA replication, protein synthesis, and other processes. In unicellular organisms, intracellular enzymes are also responsible for food digestion within food vacuoles. Intracellular digestion is the term for this process. These intracellular enzymes are usually found in lysosomes (Lakna, 2019; Li et al., 2019). Furthermore, lysosomes contain digestive enzymes that are responsible for the destruction of aged cells. Intracellular enzymes also break down big polymers into smaller monomer chains. The enzyme endoamylase, for example, breaks down big amylose molecules into shorter dextrin chains. Exoenzymes, on the other hand, start at the end and break down monomer subunits of huge polymers (Baltar, 2018).

Extracellular enzymes: also known as exoenzymes, are enzymes that function outside of the cell. The number of external enzymes is often lower than the amount of intracellular enzymes. Furthermore, they are in charge of extracellular digestion, which occurs in animals' alimentary canals. Different types of auxiliary organs emit digestive enzymes into the alimentary canal's lumen, which the food passes through. Carbohydrates, proteins, lipids, and nucleic acids in meals are digested into monomer units known as monosaccharides and disaccharides, amino acids, fatty acids, and nucleotides, respectively, by mixing with these enzymes (Orsi et al., 2018). Moreover, the digestion of decaying organic waste is carried out by extracellular enzymes produced by decomposers into the environment. Decomposers also play an important function in ecosystems by recycling nutrients. Furthermore, these organisms can receive nutrients via their cell walls, which are the byproducts of extracellular digestion. Plants and other species can also receive these nutrients through their roots (Lakna, 2019; Li et al., 2019). However, extracellular enzymes that can breakdown a wide variety of natural polymers to monomers or oligomers are synthesized and regulated by many heterotrophic bacteria. Extracellular enzymes, according to Priest (1984), are those that are found outside of the cytoplasmic membrane. Bacteria create the majority of extracellular enzymes found in water basins. They may also come from autolytic mechanisms and other species like fungus or phytoplankton to a lesser extent. Peptidase, endo- and exonucleases, 50 nucleotidase, lipase, a- and b-glucosidases, and alkaline phosphatase are among the ecto- and extracellular enzymes involved in the degradation of organic materials (Kalwasińska & Brzezinska, 2013).

12.1.2 Similarities and difference between intracellular and extracellular enzymes

Similarities: Below are similarities between intracellular and extracellular according to Lakna (2019) and Hehemann et al. (2019):

- Intracellular and extracellular enzymes are the two types of digestive enzymes that occur in cells.
- Both occur in eukaryotes as well as prokaryotes.
- They differ by their location of the action.
- Based on their action, they have different important functions in the cell.
- However, their main function is to undergo digestion of food particles.
- Both are protein molecules made up of chains of amino acids.

Differences: According to Lakna (2019) and Orsi et al. (2018), below are the differences between intracellular and extracellular:

Intracellular enzymes are the enzymes that function inside the cell. Most enzymes are intracellular enzymes, breaking down large polymers into small chains of

Intracellular enzymes	Extracellular enzymes
It refer to the enzymes which act inside the cell	It refer to the enzymes made by the cell but, work in the outside of the cell
They are also known as endoenzymes	They are known as exoenzymes
They account for the majority of enzymes	They account for the minority
They breakdown large polymers into smaller chains of monomers	They act on the end of the polymer to breakdown its monomers one at a time
They undergo intracellular digestion	They undergo extracellular digestion
They are responsible for the digestion of food particles inside the cytoplasm of unicellular organisms	They are responsible for the digestion of food inside the alimentary canal of higher animals and the extracellular digestion in decomposers such as fungi and bacteria

Source: Authors compilation, 2022.

monomers. Generally, intracellular enzymes occur in unicellular organisms that undergo intracellular digestion of food particles. On the other hand, extracellular enzymes are the minor group of enzymes, functioning outside the cell. In contrast, they breakdown large polymers into monomers starting from the ends. Basically, extracellular enzymes are responsible for the digestion inside the alimentary canal in higher animals as well as the extracellular digestion in decomposers including fungi and bacteria. Therefore, the main difference between intracellular and extracellular enzymes is the location of action and importance. However, enzyme action transforms the nonfluorescent substrate into a fluorescent product. In aquatic habitats, the ecological importance of these strategies has been debated (Chrost 1991). Therefore, current understanding of the mechanisms of organic matter colonization and enzymatic breakdown by bacteria is still incomplete. Because, artificial substrates containing a monomer covalently bonded to a fluorescent molecule, such as 4-methylumbelliferone (MUF) or 7-amino-4-methylcoumarin, are available for each class of hydrolytic enzyme (MCA).

12.2 EXTRACELLULAR ENZYMATIC AND ACTIVITY

Enzymes are frequently used as catalysts in chemical reactions in biology. As a result, enzymes regulate physiological responses by either accelerating or triggering them. In this context, the importance of enzymes in biochemical processes cannot be emphasized. Controlling the enzyme's level of substrate amounts regulates the enzyme's activity (Lakna, 2019). So, extracellular enzymes play an important function in bacterial metabolism by helping bacteria in the digestion and use of organic compounds. Extracellular polymeric compounds are an example of these substances (extracellular polymeric substances; EPS). Biosynthesized polymers, or EPS, boost microbial activity. However, microbial enzymatic activity has been linked to indices of water eutrophication and/or the trophic status of aquatic environments in several studies (Kalwasińska & Brzezinska, 2013). According to Siuda and Chróst (2002), there are a few enzymatic microbial activities (such as alkaline phosphatase, esterase, and notably aminopeptidase activity) that are very useful for quickly determining the current trophic condition of lakes. These enzymes' activities grow exponentially along a trophic gradient and are strongly correlated with the trophic state index of lakes.

Nevertheless, extracellular enzymes and EPS are important for microbial growth, survival, and activity in the ocean. Because EPS is amphiphilic, it can operate as a biological surfactant in the event of an oil leak. Extracellular enzymes aid microorganisms in digesting and using organic matter components, such as EPS, which can promote growth and microbial activity. However, given that phytoplankton production of EPS can range from 3 to 40% of total primary productivity, and that primary productivity accounts for 45–50 Pg C yr^{-1} in the ocean, the quantity of carbon released as EPS might be anywhere between 1.5 and 20 Pg C yr^{-1}. Furthermore, bacteria are known to create EPS, which has been linked to their ability to guard against harmful environmental conditions (Manivasagan & Kim, 2014). The composition of EPS is diverse, consisting primarily of carbohydrates, proteins, sugar monomers, amino acids, and uronic acids. EPS can be used as a carbon, nitrogen, and phosphorus substrate to help bacteria and mixotrophic phytoplankton grow, hence increasing microbial activity (Quigg et al., 2016). Extracellular enzymes accelerate the breakdown of complex polymers in EPS to less complex compounds, which heterotrophic microorganisms would otherwise be unable to utilize, according to several studies. Extracellular enzymes (exoenzymes) such as – and – glucosidase, alkaline phosphatase, leucine aminopeptidase, and lipase are secreted by heterotrophic bacteria to help them degrade the various components of EPS (Kamalanathan et al., 2018). Because enzymes are specific to the substrates they operate on, and bacteria differ in their ability to create different types of enzymes, changes in enzyme activity can be utilized as an indirect indicator of microbial functional diversity and/ or nutritional content in a system (Caldwell, 2005). During the oil spill, the oil and Corexit may have served as a source of carbon to the microorganisms, thereby affecting EPS generation and enzyme activity. Oil and Corexit have also been observed to alter the microbial ecology in favor of hydrocarbon degraders (Bacosa et al., 2015; Kleindienst et al., 2015; Doyle et al., 2018). As a result, it's important to look into how adding oil and Corexit affects EPS production, extracellular enzyme production, the microbial ecology, and aggregate formation.

Furthermore, extracellular enzymes are produced by microbial communities in soils to obtain energy and resources from complex macromolecules in the soil environment. These enzymes are vital in the carbon (C), nitrogen (N), and phosphorus (P) cycles, and they catalyze critical transformations in the environment (Wallenstein and Burns, 2011). Despite the fact that soil microbes influence biogeochemical processes across the soil profile, most investigations of microbial populations and their related enzymes have been limited to the upper 15 cm of the soil (Fierer et al., 2003a). Tropical forest soils are generally several meters deep, and tropical subsoils retain a significant amount of carbon, accounting for around half of the carbon deposited at depths greater than one meter (Jobbágy and Jackson, 2000). As a result, it's acceptable to assume that microbial communities play a role in nitrogen cycling in deeper regions of tropical soil profiles than previously measured. Carbon availability and composition, pH, temperature, redox state, texture, and mineralogy are all factors that influence microbial communities (Stone et al., 2014). All of these characteristics can alter dramatically with depth, some by orders of magnitude. The quantity, composition, and functions of soil microbial communities are influenced by the environmental gradient reflected by depth profiles. Microbial biomass and substrate pools, for example, tend to decrease as soil depth increases (Fierer et al., 2003a; Fang & Moncrieff, 2005; Kramer et al., 2013). Similarly, the organization of microbial communities alters with depth, presumably reflecting an increasing dominance of organisms that can maintain basal metabolism in low-energy environments (Hoehler and Jorgensen, 2013). Several investigations have discovered that extracellular enzymes in soils influence the decomposition of organic matter and catalyze critical transformations in carbon, nitrogen, and phosphorus cycle, and that potential enzyme activity decrease with depth. The majority of EEA research has, however, been limited to relatively carbon and nutrient-rich surface soils. Several centimeters of nutrient-rich surface soil can be found on top of meters of resource-poor subsoil, whose microbial ecology is little understood. However, in the vast majority of cases, specific enzyme activities (activity normalized to biomass or substrate availability) have not been measured, making it difficult to distinguish physiologically adaptive changes in enzyme activities from changes caused solely by differences in biomass and substrate quantities (Trasar-Cepeda et al., 2008). The few studies that have standardized microbial assimilation or mineralization activities to the size of the microbial biomass have found that specific activity increases with depth or has similar values throughout the soil profile (Gelsomino & Azzellino, 2011; Kramer et al., 2013). Furthermore, higher mineral association in subsoils might lead to better organic material stabilization (Eusterhues et al., 2003; Rasse et al., 2005). Changes in the percentage of enzymes bound to clay minerals or organoemineral complexes as a function of depth can affect both prospective enzyme activity and enzyme turnover rates (Alasalvar et al., 2002; Allison, 2006).

12.2.1 FACTORS INFLUENCING EXTRACELLULAR ENZYME ACTIVITY

Extracellular enzyme synthesis is dependent on nutrient availability and environmental conditions, and it supplements bacteria' direct intake of nutrients. To access the carbon and nutrients trapped in detritus, a system of extracellular enzymes is required due to the diverse chemical composition of organic matter. The ability of microorganisms to break down these various substrates varies, and only a few species have the ability to breakdown all of the plant cell wall components available (Allison et al., 2009). Some exoenzymes are created constitutively at low levels to detect the presence of complex polymers, and their expression is increased when the substrate is abundant (Klonowska et al., 2002). Fungi can respond dynamically to changing availability of specific resources due to their sensitivity to the presence of varied amounts of substrate. Because the enzymes are prone to denature, degrade, or diffuse away from the producing cell, the benefits of exoenzyme production may be lost after secretion. Enzyme production and secretion, on the other hand, is an energy-intensive activity that uses resources that could otherwise be used for reproduction. As a result, there is evolutionary pressure to conserve those resources by limiting output (Schimel, 2003; Allison et al., 2010). While most microbes can assimilate simple monomers, polymer degradation is specialized, with just a few organisms capable of degrading resistant polymers like cellulose and lignin (Baldrian et al., 2012). Each microbial species has its own set of extracellular enzyme genes and is tailored to breakdown certain substrates (Allison et al., 2009). Furthermore, the availability of a given substrate regulates the expression of genes that encode for enzymes. In the presence of a low-molecular-weight soluble substrate like glucose, for example, enzyme synthesis is inhibited by inhibiting the transcription of cellulose-degrading enzymes (Hanif et al., 2004).

Exoenzyme expression and activity can be influenced by environmental factors such as soil pH, soil temperature, moisture content, and plant litter type and quality. Variations in seasonal temperatures can cause microbe metabolic needs to fluctuate in lockstep with changes in plant nutritional requirements (Online-Wikipedia, 2020). Agricultural methods such as fertilizer additions and tillage can modify the spatial distribution of resources in the soil profile, resulting in altered exoenzyme activity (Poll et al., 2003). Moisture enhances the loss of soluble monomers by diffusion and exposes soil organic matter to enzyme catalysis. Furthermore, osmotic stress caused by changes in water potential can affect enzyme activity because bacteria divert energy from enzyme production to the manufacture of osmolytes in order to preserve cellular structures. In natural aquatic environments, there are three general pathways for organic matter decomposition. Predation, particle feeding, and dissolved organic matter intake are all factors because it is a main carbon and energy source in many aquatic environments. Bacteria have a role in the latter two because they can hydrolyze nonliving particles, competing with particle feeders, and take up tiny organic molecules, which is their sole domain. Extracellular enzymes in the intestines of mammals and the enzymatic activity of associated bacteria are involved in particle breakdown. As a result, successful organic matter competition among tropic levels is also a matter of extracellular enzymatic efficiency. The enzymes of free-living bacteria are primarily responsible for the degradation of dissolved organic macromolecules, which are then incorporated into the tiny molecules produced by enzymatic hydrolysis. Bacterial activity thus has a significant impact on the concentration and speciation of dissolved organic compounds in water. Extracellular hydrolysis is a rather sluggish process compared to the absorption of low-molecular-weight organic matter, hence dissolved macromolecules are likely to make up a large portion of the DOM pool in the water (LMWOM). Depending on a variety of conditions, the effectiveness of animals feeding on particles might vary significantly. The chemical content and size of the particles will have a significant impact on microbial particle hydrolysis. The sluggish but constant microbial component, on the one hand, and the pulse-feeding actions of animals, on the other hand, will influence competition for organic particles between animals and microorganisms (Hoppe, 1991).

Exposed seafloor basalts do, however, make up a 600,000-km^2 continuous undersea habitat. Endolithic basalt microorganisms are diverse and abundant, with certain bacterial clades appearing to be more common on basalts than elsewhere. Bacterial phyla show abundance trends that correspond to rock geochemistry, demonstrating that microbial communities are strongly influenced by rock composition (Jacobson Meyers et al., 2014). Methanogenesis, nitrogen fixation, anaerobic ammonium oxidation, denitrification, Fe reduction, and dissimilatory sulfate reduction genes have been found in basalt microbial communities, demonstrating that basalts have the potential for a wide range of biogeochemical changes. However, no information on the metabolic activity rates of basaltic microorganisms is currently available. In harsh settings in general, and basalts in particular, the relationship between the existence of prokaryotes and the activity and function of enzymes is understudied. Existing data sets from various marine ecosystems can be compared with information from freshly explored locations, and hydrolytic extracellular enzymes have been found to be indications of metabolically active bacteria. Furthermore, because organic matter is more refractory in deep sea environments, extracellular enzyme hydrolases should play a key role in the start of organic matter recycling. Recent research has revealed that the most prevalent Archaea in marine sediments develops unique extracellular enzymes for protein breakdown (Cycling, 2014).

12.2.2 EXTRACELLULAR ENZYME ACTIVITY IN FUNGI DURING PLANT DECOMPOSITION

Fungi are thought to be responsible for the majority of extracellular enzymes involved in polymer breakdown in leaf litter and soil. Fungi produce a mixture of oxidative and hydrolytic enzymes to efficiently break down lignocelluloses like wood by adjusting their metabolism to the availability of varied amounts of carbon and nitrogen in the environment. Plant litter degradation begins with the destruction of cellulose and other labile substrates, followed by lignin depolymerization, which results in increased oxidative enzyme activity and changes in microbial community structure. Similarly, cellulose and hemicellulose are embedded in a pectin scaffold in plant cell walls, which needs pectin degrading enzymes like polygalacturonases and pectin lyases to weaken the plant cell wall and expose hemicellulose and cellulose to enzymatic destruction (Ridley et al., 2001; Lagaert et al., 2009). Enzymes that oxidize aromatic compounds, such as phenol oxidases, peroxidases, and laccases, catalyze the degradation of lignin. Several genes encoding lignin-degrading exoenzymes are found in many fungi (Courty et al., 2009). Saprotrophic ascomycetes and basidiomycetes are the most efficient wood degraders. Based on the appearance of the decaying material, these fungi are traditionally classed as brown rot (Ascomycota and Basidiomycota), white rot (Basidiomycota), and soft rot (Ascomycota) (Burns et al., 2013). White rot fungi breakdown cellulose and lignin, while brown rot fungus prefer to attack cellulose and hemicellulose. Basidiomycetes use hydrolytic enzymes such endoglucanases, cellobiohydrolase, and -glucosidase to destroy cellulose (Baldrian & Valášková, 2008). Endoglucanases are extensively produced by fungi, and cellobiohydrolases have been recovered from a variety of white-rot fungus and plant diseases. Many wood-rotting fungus, including white and brown rot fungi, mycorrhizal fungi, and plant diseases, produce -glucosidases. -glucosidases can cleave xylose, mannose, and galactose in addition to cellulose (Burns et al., 2013).

Expression of manganese-peroxidase is triggered by the presence of manganese, hydrogen peroxide, and lignin in white-rot fungi like *Phanerochaete chrysosporium*, whereas laccase is activated by the availability of phenolic substances. Basidiomycetes produce lignin-peroxidase and manganese-peroxidase, which is commonly used to assess basidiomycete activity, notably in biotechnology applications. Laccase, a copper-containing enzyme that destroys polymeric lignin and humic compounds, is produced by the majority of white-rot species (Burns et al., 2013). Brown-rot basidiomycetes are most frequent in coniferous woods, and they get their name from the fact that they degrade wood into a brown residue that crumbles easily. These fungi prefer to attack hemicellulose in wood before moving on to cellulose, leaving lignin generally unaffected. Soft-rot Ascomycetes decaying wood is dark and soft. *Trichoderma reesei*, a soft-rot *Ascomycete*, is widely employed in industrial applications as a source of cellulases and hemicellulases. *T. reesei*, *Aspergillus* species, and freshwater ascomycetes have all been found to have laccase activity (Baldrian & Valášková, 2008).

Meanwhile, sample harvesting before analysis, mixing of samples with buffers, and the use of substrate are all ways for assessing soil enzyme activity. Sample transit from the field, storage procedures, assay pH conditions, substrate concentrations, assay temperature, sample mixing and preparation can all affect results (German et al., 2011). Colorimetric assays using a *p*-nitrophenol-linked substrate or fluorometric assays using a MUF-linked substrate are required for hydrolytic enzymes. Lignin breakdown and humification are mediated by oxidative enzymes such as phenol oxidase and peroxidase. The oxidation of L-3,4-dihydoxyphenylalanine (L-DOPA), pyrogallol (1,2,3-trihydroxybenzene), or 2,2′-azino-bisphenol S is used to measure phenol oxidase activity (3-ethylbenzothiazoline-6-sulfonic acid). The phenol oxidase experiment is done in parallel with another assay in which L-DOPA and hydrogen peroxide (H_2O_2) are added to each sample to determine peroxidase activity (Sinsabaugh, 2010). Peroxidase activity is indicated by the difference in results between the two assays. In most enzyme tests, proxies are used to reveal enzymes' exo-acting activity. Exo-acting enzymes hydrolyze substrates toward the end of their life cycle. Endo-acting enzymes that break down polymers in the middle chain must be represented by additional substrate proxies. New enzyme assays strive to capture the diversity of enzymes and more clearly determine their potential activity (Sinsabaugh, 2010; German et al., 2011; Burns et al., 2013). Molecular approaches to estimate the number of enzyme-coding genes are being utilized to correlate enzymes with their makers in soil environments, thanks to improved technologies. Proteomic approaches can disclose the existence of enzymes in the environment and link them to the species that produce them, while transcriptome analyses are now used to investigate genetic regulation of enzyme expression (Sinsabaugh, 2010).

12.3   NATURES OF EXTRACELLULAR ENZYMES/ ENZYMATIC ACTIVITY

In terrestrial, freshwater, and marine ecosystems, the majority of primary production enters detrital food webs and is eaten by heterotrophic microbes. The majority of macromolecular detritus must first be digested into assimilable substrates by enzymes secreted into the environment, and to some extent also discharged via cell lysis (Arnosti et al., 2014; Münster, 1992). The autochthonous generation of DOM by autotrophs in humic waters, however, may not be sufficient to meet the carbon and energy needs of bacteria (Münster, 1992). In comparison to pure water systems, humic fluids may be able to transport more microorganisms. This greater carrying capacity is associated with external allochthonous carbon, energy, and nutrition supplies. Furthermore, particulate organic matter (POM) and DOM's carbon content may differ by a factor of 10–100 (Münster, 1992). In humic systems compared to clear water systems, this ratio is substantially more prominent. The primary substrate pool for microheterotrophs in these conditions is made up of DOM. The most effective user of this DOM among them is bacteria (Arnosti et al., 2014). They are essential to the microbial food chain.

12.3.1   Abiotic drivers

Terrestrial, freshwater, and marine systems are diverse habitats with resources, microbes, and biological processes dispersed in a non-uniform way. Surface interactions as controls on enzymatic activity at micrometer scales (Arnosti et al., 2014). The quantity of solid surface areas within a given volume, which distinguishes ecosystems ranging from ocean waters to soils, has a significant impact on this microscale heterogeneity. Surface contacts have an impact on enzymes and their functions. In contrast to pelagic seas, mineral particle size, distribution, content, and density greatly control enzyme activity in soils and sediments. In addition to being protected from deterioration for a lot longer than liquid enzymes, sorbed enzymes may become more active following desorption (Münster, 1992). In turn, these interactions control the size of the active enzyme pool, access to organic substrates, and turnover rates. Enzyme–surface interactions are governed by mineralogy, the composition of related organic compounds, and thermodynamic parameters (Wallenstein & Weintraub, 2008; Sinsabaugh, 2010). The source and vicinity of terrestrial runoff, phytoplankton productivity, turbulence, and flocculation processes all affect particle density and composition in pelagic systems (aquatic habitats excluding bottom and coastline). According to Simon et al. (2002), DOM can create ephemeral surfaces like marine snow, lake snow, or river snow that increase accessible surface area and serve as "hotspots" for biogeochemical reactions and increased enzyme activity (Smith et al., 1992; Grossart & Simon, 1998; Ziervogel et al., 2010). Small-scale (spatial and temporal) temperature and pH fluctuations' effects on enzyme activity The most important environmental factors influencing enzyme kinetics and substrate binding are temperature and pH. The rate and size of spatial and temporal temperature and pH fluctuations are also influenced by water volume and hydraulic residence duration. For instance, dramatic temperature changes in marine systems occur over far larger spatial and temporal scales than they do in freshwater and terrestrial ecosystems. Due to the enormous carbonate concentrations in the ocean, pH fluctuations are very minor, and salinity variations are typically small over very broad spatial scales. When compared to the ocean, most freshwater systems often exhibit a significantly larger range of salinities, pH, and temperature excursions. This is because they typically lack a carbonate buffer system with capacity equivalent to the ocean. Soil pH fluctuates by orders of magnitude at large sizes, such as across ecosystems. Different parent materials, vegetation, and weathering may be reflected in smaller scale vertical gradients within soil ecosystems. pH gradients may develop inside soil aggregates at much finer sizes. The second axis of our conceptual design represents these spatial and/or temporal differences in temperature or pH, which significantly separate soil and freshwater ecosystems but not marine waters or sediments, with the exception of intertidal zones.

12.3.2 Biotic drivers

The low explanatory power of the first two physicochemical axes for variations in enzyme activities in marine waters, the effects of microbial community composition and capabilities on enzyme activities, and other factors suggest that, at the very least, a third axis is required to distinguish controls on enzyme activities across this range of systems (Ogunlade et al., 2021). As the third axis, we have selected biotic factor variations in the make-up and capacities of microbial communities, also known as community functional diversity. Since it is still not possible to quantify the precise degree to which community make-up and capacities influence enzyme activities in an ecosystem, the position of a given habitat in the diagram is arbitrary. However, incorporating the functional diversity of the microbial community allows us to conceptualize the cross-system influences that can influence the activity of enzymes (Wegner et al. 2013). A growing number of studies showing the significance of community functional diversity in regulating enzyme activity in marine environments lend credence to the choice of this factor. The bulk of active enzymes in marine waters appear to be more closely related to generating bacteria than enzymes in soil environments, where abiotic surfaces more strongly control enzyme activity and turnover rates. This is supported by the relative lack of solid surfaces in ocean waters. Individual organisms have been shown to have distinct enzymatic capabilities and substrate specialization by studies of organisms isolated from marine waters, genomic and biochemical analyses of cultured isolates, and genomic analyses of single cells directly sorted from ocean waters (Martinez-Garcia et al. 2012). It is yet unclear exactly how variances in microbial communities' composition and functions relate to one another. However, metagenomic analyses in ocean waters demonstrate regional and temporal variations in polysaccharide hydrolase genes linked with particular phylogenetic groupings, which extend to variations in enzymatic performance, supporting the existence of such variances in the ocean (Gomez-Pereira et al. 2010, 2012; Arnosti et al. 2012; Teeling et al. 2012). Terrestrial and freshwater ecosystems are divided by community functional diversity. Recent research demonstrates that the functional characteristics of the microbial population in soils are influenced by changes in the availability of nutrients and water. In terrestrial ecosystems, in situ enzyme activities are dynamic and responsive to changes in microbial biomass and community structure, or production. They decrease with stability (mineral and organic sorption) and degradation (Arnosti et al. 2012). However, not all variations in enzyme activity are connected to variations in the make-up of microbial communities. For instance, in a recent investigation of a freshwater system, considerable temporal fluctuations in enzyme activity along stream flow routes were discovered, but limited spatial variation and little correlation with bacterial community composition was seen (Frossard 2012). It is unclear to what degree variations in enzyme activity are caused by modifications in the make-up of microbial communities as opposed to modifications in the metabolic control of certain microorganisms.

12.3.3 FRESHWATER SYSTEMS

Because there is no water stress in freshwater ecosystems and nutrient availability is increased by wastewater inputs, agricultural runoff, and urban runoff, enzyme activity per unit organic matter are often higher in freshwater environments than in soils. In freshwater systems, analyses of biofilm, surface sediment, and hyporheic zone enzyme activity have traditionally concentrated on connections with microbial substrate consumption and production, or community composition. There have been less studies of enzyme activity in the water column since the majority of ecosystem metabolism of lotic (flowing water) systems is linked to sediment microbial populations (Teeling et al. 2012). Litter decomposition studies have been a part of inland water studies since measurements of cellulase, b-glucosidase, and phosphatase activities related to deciduous leaf litter decomposing in a woodland stream. Since allochthonous inputs of plant litter account for a significant portion of the organic matter input to many inland water systems (Sinsabaugh & Shah, 2012). It has been utilized to estimate the instantaneous decomposition rates for POM from enzyme measurements since subsequent studies often reveal positive correlations between lignocellulose-degrading enzyme activities and decomposition rates (Sinsabaugh et al., 1994; Jackson et al., 1995). Given the hydrodynamic interconnection of inland water ecosystems, it is considerably simpler in freshwater than in terrestrial or marine systems to establish statistical relationships between enzyme activity, microbial metabolism, resource availability, and landscape features. As a result, freshwater microbial communities' enzyme interactions are the subject of more synthetic investigations. In relation to decomposition, nutrient cycling, and detoxification, for instance, Krauss et al. (2011) review the ecology and physiology of aquatic fungi, and Sinsabaugh and Shah (2012) compare the activities and ratios of b-glucosidase, phosphatase, leucine aminopeptidase, and N-acetylglucosaminidase in freshwater sediments and terrestrial soils. One conclusion from the latter review is that freshwater sediments have mean ratios of b-glucosidase: phosphatase activity that are much higher than those of terrestrial soils, which is consistent with sediments having lower elemental C:P ratios. The earliest models for estimating decomposition rates from enzyme measurements and resource allocation models that link enzyme stoichiometry to nutrient availability were developed as a result of research on extracellular enzymatic activities connected to disintegrating POM (Sinsabaugh et al., 1994; Jackson et al., 1995). Planktonic and hyporheic studies connected investigations of microbial community composition to resource composition and metabolism. The development and testing of models that relate enzyme activity to ecological stoichiometry and the carbon use effectiveness of microbial communities followed the emergence of these linkages (Sinsabaugh & Shah, 2012). aquatic systems Large ocean volumes with low DOM concentrations and sparsely dispersed, transient patches of particles are included in the research of microbial extracellular enzymes in marine environments. Below these waters are compact sediments with a range of origins, compositions, and depths. Determining the activity of heterotrophic microbial communities in the context of the breakdown of complex organic substrates is the overarching theme of the majority of extracellular enzyme research conducted inside these environments. Our understanding of the distribution and use of microbial enzymes in marine systems is significantly limited by two considerations. Due to ship time restrictions and the difficulty of collecting samples, access to a large portion of the ocean is restricted. With few exceptions (Baltar et al., 2009, 2010, 2013), most analyses of enzyme activity in marine waters have therefore been conducted in shallow coastal areas or surface- and near-surface waters (upper 200 m of the ocean's average depth of 4,000 m). The second issue, which includes limitations in our ability to quantify enzyme activity accurately and issues with substrate sorption, contributes to the relative dearth of information on enzyme activities in marine sediments.

A growing number of studies have concentrated specifically on determining the activities of a wider range of enzymes, including those of endo-acting enzymes that hydrolyze polymers mid-chain, despite the fact that most investigations in marine systems have used a few substrate proxies to measure enzyme activities. These initiatives have made use of bigger peptides labeled with and without additional fluorophores, fluorescently labeled polysaccharides, plankton-derived extracts, or spin probes (Liu et al., 2010). These substrates show the fundamental significance of substrate structure in controlling hydrolysis rates: in coastal waters, larger peptides hydrolyzed more quickly than smaller peptides with the same chemical composition; hydrolysis rates among peptides with the same components in a different order were also observed to vary (Liu et al. 2010). These investigations also show that the polysaccharide substrates that are easily hydrolyzed in the underlying sediments are frequently only hydrolyzed by a portion of the microbial populations in ocean waters. The make-up of the microbial communities in the water column and the underlying sediments may vary, which could explain these discrepancies in enzymatic capacities (Durbin & Teske, 2011). The ability of heterotrophic microbial communities to hydrolyze various polysaccharide substrates also showed latitudinal gradients in a recent investigation (Arnosti et al., 2011). With increasing latitude, the range of substrates hydrolyzed narrows, mirroring a pattern of latitudinal shifts in community diversity and declining species richness at higher latitudes seen in other studies (Baldwin et al., 2005; Pommier et al., 2007; Fuhrman et al., 2008). It is yet unknown what causes the observed depth- and site-related variations in the enzyme activity of heterotrophic microbial communities in ocean waters. The choice of this factor as the third axis reflects the conclusion that variations in microbial community composition and large-scale patterns of microbial biogeography, which are now being observed in ocean environments, must be significant factors affecting the enzymatic capabilities of entire communities (Pommier et al., 2007; Zinger et al., 2011; Friedline et al., 2012). A well-established sequence of organic matter transformations during early diagenesis of plant litter is one of the aspects that contributes to understanding enzyme activity in freshwater and terrestrial environments. Contrarily, the difficulty of determining the chemical structure of marine POM and DOM (Hedges et al., 2000; Lee et al., 2004) at the level that is pertinent for considerations of the structural specificities of enzymes prevents the current state of the art in ocean waters and sediments from establishing a meaningful linkage between the presences of specific substrates and enzyme activities. These limitations severely restrict the creation of enzyme activity models that can be used in marine environments.

12.3.4 STRUCTURING FACTORS ACROSS ENVIRONMENTS: THE SAME OR DIFFERENT?

Extracellular enzymes work in all environments to provide vital supplies for heterotrophic bacteria. Therefore, extracellular enzymes play a key role in the cycling of nutrients and carbon. Even though the biogeochemical cycles in these ecosystems are interrelated, there haven't been many attempts to combine enzyme research from various environments. Due to variations in environmental conditions, community organization, and the number and character of substrates, marine and terrestrial systems in particular have been seen as being unsuitable for comparative investigations (Reintjes et al., 2019). As was mentioned above, environmental factors like surface area, temperature, pH, and moisture content have a significant impact on the activities of organisms and enzymes in terrestrial systems (and in many inland waters), but they have little predictive power for patterns of enzyme activity measured in marine systems, where biotic factors like microbial community composition and substrate diversity may have the biggest impact. The spatial links between microbial communities and enzyme activities in marine settings have received more attention in study, while the impact of solid surfaces and dynamic environmental features on their activities in terrestrial systems have received more attention. Numerous shared research requirements have been discovered despite historical disparities in study foci and real variances in environmental influences on enzyme activity among habitats. Researchers within areas would also gain from increased cross-disciplinary research and collaborations. For instance, aquatic researchers were the first to apply fluorescence-based enzyme assays in the environment, which are significantly more sensitive than the colorimetric assays often used in terrestrial investigations (Hehemann et al., 2019; Lakna, 2019). The use of fluorescence-based assays has made it possible for researchers studying soil to examine enzyme activity under temperature and pH settings that are more like those seen in the natural world. On the other hand, if fundamental substrate-enzyme dynamics can be discovered, enzyme models created in the context of freshwater and terrestrial systems may offer insight into the dynamics of marine extracellular enzymes. The underlying process is the same in terrestrial, freshwater, and marine environments, despite the fact that the relative importance of particular structuring factors (such as pH, the presence of surfaces, etc.) varies greatly. Microorganisms produce extracellular enzymes in order to gain a selective advantage; these enzymes then catalyze biogeochemical cycles. The ultimate regulation and biogeochemical effects of extracellular enzymes across settings should be better understood with increased cross-system collaboration among researchers.

12.4   AQUATIC BACTERIOLOGY

With an emphasis on freshwater, estuarine, and oceanic habitats, aquatic microbiology is committed to furthering the study of bacteria in aqueous environments. Aquatic microorganisms are essential to the biogeochemical cycles of the planet and serve a variety of roles in ecosystems. As a result, a number of variables, including water chemistry, nutrient runoff from nearby land, temperature, the amount of organic matter present, pH, oxygen concentration, light, and water movement, have a significant impact on the type and abundance of microbes found in freshwater ecosystems. Because aquatic microorganisms are an essential component of the food chain and web and play a crucial role in the cycling of nutrients in their environment. The breakdown of organic materials in dead plants and animals provides food for many microbes. However, the make-up of microbial communities can fluctuate significantly between different habitats in a same lake or stream, leading to different communities at varying depths or with various benthic substrate properties (Yang et al., 2020). The individual microbes present, as well as their overall and comparative abundance, will be greatly influenced by local conditions. While some microbes can survive in almost any aquatic environment, including anaerobic settings or regions with temperature extremes, this is not always the case. Similar to human communities, microbial communities respond fast to changes in their environment. When conditions are good, they reproduce quickly, and when they are unfavorable, they quickly disappear. These modifications could be seasonal or brought on by extreme weather, such as storms that provide higher-than-normal amounts of nutrient or chemical runoff from nearby land, changes in land use, or other environmental disturbances. The algal, plant, and animal species that depend on them either directly or because of their function in the ecosystem's nutrient cycling will be impacted as microbial populations alter in response to shifting environmental conditions (Grossart et al., 2020). A window into the overall health of the aquatic system is provided by tracking changes in the microbial populations that live there. However, aquatic microbes from the benthos and surface water, including bacterial, fungal, and archaeal species, should be collected (sediment-water interface). In all environments, microbes control the cycling of nutrients. The composition and activity of microbial communities are influenced by temperature, nutrition and carbon availability, physical dispersion in moving water, and competition. Due to these reasons, surface water chemistry and benthic algae samples should be taken at the same time as aquatic microorganism samples. Researchers will be able to evaluate changes in this crucial ecosystem group of species and gain vital insights into the general productivity and health of the aquatic environment by gathering simple measurements of biomass, enzyme activity, and DNA.

12.4.1 EFFECT OF ENZYMATIC ACTIVITY ON AQUATIC ECOSYSTEM

Temperature, biodiversity, community structure, and nutrient richness are just a few of the parameters that have an impact on how well wooded headwater stream habitats function. The activity of the microbes in these ecosystems, which serve as the building blocks of trophic webs, drive global carbon (C) and nutrient cycles, and alter the composition of the atmosphere, is particularly influenced by temperature (Bautz, 2019). Headwater forest streams play a crucial part in the global C cycle because they may sequester, process, and transfer large amounts of terrestrial carbon. Extracellular enzymes are used to catabolize organic materials in aquatic habitats, and this process is regarded as a rate-controlling stage in the overall C cycle (Arnosti et al., 2014). This is why a number of research have concentrated on how prospective climatic changes may affect aquatic habitats. By the end of the century, climate scientists predict that the average global surface temperature will have risen by 1–4°C (IPCC, 2014). Increases in the rate of microbially mediated litter decomposition may have significant effects on stream ecosystem processes under climate change because they may result in a lack of food for higher trophic levels in aquatic environments. In headwater forest streams, photosynthesis and subsequent primary production are constrained by the lack of light as a result of shadowing. Most of the species in these environments are heterotrophic, which means they depend on organic carbon sources (Ferreira & Chauvet, 2011; Sinsabaugh & Shah, 2012). Forest streams get significant allochthonous inputs of organic matter from the riparian zone, which includes leaves and twigs of riparian trees and shrubs, as a result of natural senescence and abscission. These inputs support the microbial metabolism of decomposers associated with trash in stream environments. Thus, in these ecosystems, microbes associated with trash are in charge of mediating the flow of energy and nutrients to higher trophic levels. In freshwater streams, fungus and bacteria are primarily responsible for the decomposition of organic materials (Gulis et al., 2019). Growing data points to fungi, also referred to as aquatic hyphomycetes, being the dominant microbial species in aerobically degrading plant litter. At the base of aquatic food webs, where it supports aerobic metabolism, this organic matter decomposes, whereas litter C is absorbed into fungal biomass and released as CO_2 through microbial respiration. For aquatic invertebrate shredders producing fine particle organic matter to be utilized downstream by filter-feeders and collectors, the process of fungal decomposition improves litter palatability (Gulis &

Bärlocher, 2017). Therefore, in forest stream ecosystems, aquatic hyphomycetes play a crucial intermediary role between decaying organic matter, secondary production (the accumulation of microbial biomass), and higher trophic levels. James H. Brown developed the Metabolic Theory of Ecology (MTE) in an effort to statistically define how ecological systems might react to temperature variations (Brown et al., 2004). A quantitative hypothesis known as MTE describes how metabolic rates would change as a function of body size and temperature within the range of normal biological temperatures (0–40°C) (Brown et al., 2004). According to Brown, metabolism is a wholly biological process that is subject to the principles of mass, energy, and thermodynamics (Brown et al., 2004). According to Brown, metabolic rate is the fundamental biological rate because it defines how quickly energy is taken in, transformed, and distributed throughout living things (Brown et al., 2004). According to Gulis and Bärlocher (2017), the metabolic processes of litter-associated decomposers typically involve the acquisition and incorporation of carbon (C) and other nutrients (like N and P) into microbial biomass (secondary production), as well as the release of CO_2 as a result of aerobic respiration. Individual biochemical reaction rates, metabolic rates, and practically all other rates of biological activity all scale exponentially with temperature, according to the MTE (Brown et al., 2004). Since aquatic microorganisms that are connected with litter derive their energy from the oxidation of organic substances, their metabolic rate and respiration rate are related.

Although temperature sensitivity may differ by metabolic activity, the apparent activation energy (Ea) of the respiratory complex, based on several experimental estimations, has been characterized as being around 0.65 eV (Gillooly et al., 2001; Yvon-Durocher et al., 2012). It is also possible to evaluate the kinetics of microbial enzymes involved in C sequestration using their apparent activation energy, which is defined by the structure and function of the participating enzymes in a metabolic process (Sierra 2012; Sinsabaugh & Shah 2012). To determine whether the responses adhere to the straightforward MTE prediction of monotonous (exponential) increase across wide temperature intervals, experiments are required to better understand how changes in temperature affect the extracellular enzymatic activity of aquatic litter-associated microorganisms. Stream communities of aquatic hyphomycetes have lignocellulolytic enzymes readily available for breaking down the main plant polymers found in detritus. It's possible that bacteria with enzyme weaknesses can access simpler carbohydrates as a result of fungi's destruction of plant polymers (Gulis & Suberkropp, 2003; Romani et al., 2006). Enzymes that break down hemicellulose, cellulose, and lignin-like or phenolic plant polymers are secreted by aquatic hyphomycetes. Although it can be restricted, some species of ligninolytic enzymes do degrade lignin more efficiently than others (Gulis et al., 2019). The movement of carbon (C) toward fungal respiration and secondary production, and ultimately to stream detrital food webs, is significantly aided by the decomposition of plant polymers mediated by microbial extracellular enzymes (Gulis & Bärlocher, 2017). A common component of plant cell walls, β-1,4-xylosidase catalyzes the hydrolysis of β-1,4-linkages present in hemicellulose xylooligosaccharides (Sinsabaugh & Shah, 2012). The final steps in the hydrolysis of the cellulose β-1,4-linkages present in the cell wall and fibers of degrading plant litter are catalyzed by β-1,4-glucosidase. In samples of degrading leaf litter, β-1,4-xylosidase and β-1,4-glucosidase activity can be precisely quantified using fluorescent substrate analogs (Hoppe, 1983). These techniques have proven to be trustworthy in assessing the activity of hydrolytic enzymes present in biological materials after being employed in practice for decades. Lignin, one of the most prevalent substances in nature, is a crucial component of vascular plants (Hendel et al., 2005). The connection of lignin with cellulose fibers in plants plays a key role in the movement of carbon in aquatic ecosystems' detrital food webs (Davidson & Jannsens, 2006; Gessner et al., 2007). Fungi and bacteria containing ligninolytic enzymes, such as phenol oxidase (a monooxygenase) and peroxidase, degrade lignin through an oxidative process (Hendel et al., 2005). L-3,4-dihydroxyphenylalanine (L-DOPA) is frequently used as the electron-donating substrate in oxidative enzyme activity assays to measure phenol oxidase activity. L-DOPA is the recommended substrate to be utilized with environmental samples in spectrophotometric evaluation of phenol oxidase activity because of its water solubility and capacity to donate electrons. When DOPA is oxidized, a red tint is produced that can be measured by taking an absorbance reading at 460 nm. L-DOPA can also be supplied along with a very little amount of $H2O2$ to measure the activity of peroxidases (Hendel et al., 2005).

12.5 CONCLUSION

Further research is needed to better understand the underlying mechanisms and processes that influence the composition and function of microbial communities. To get more insight into the morphology, composition, evolution, and roles of microbial communities, in addition to sequencing, morphological and physiological traits (such as the production of extracellular polymeric compounds) should be compiled. Better examination of the processes and underlying mechanisms of microbial community dynamics will be possible when these techniques are combined with theories of community ecology. The large multiscale complexity of microbial systems, however, demonstrates how single-cell metabolism and biological interactions coordinately drive biogeochemical cycles in aquatic ecosystems, which may have an impact on global processes. This review envisages a variety of future research directions that will advance the comprehension of aquatic systems at many geographical and temporal dimensions. In order to relate the composition and function of microbial communities to a wide range of environmental variables at high and broad temporal and spatial resolution, 'omics data, for instance, can be linked to "big data" sources like remote sensing or autonomous profiler units (Buttigieg et al. 2018; Huot et al. 2019). One may be able to reveal hidden patterns in the structure-function connection of extremely complex microbial communities and to clarify the corresponding ecological implications by combining machine learning and artificial intelligence methodologies. To improve our understanding of how aquatic microbial communities work, more environmentally relevant model organisms must be defined. Furthermore, it is yet unknown what many proteins, including sizable protein families, do. Although functional protein characterization is challenging and time-consuming, there is an urgent need for greater efforts in high throughput functional analysis of uncharacterized proteins. A better knowledge of the underlying organismic processes/mechanisms and their environmental regulation will be possible thanks to the intelligent integration of new techniques for investigating cell physiology in combination with "omics" approaches. This information is desperately required for the successful and long-term management of aquatic ecosystems while taking into account the pervasive microbial legacy.

REFERENCES

Alasalvar, C., Taylor, K. D. A., Zubcov, E., Shahidi, F., and Alexis, M. (2002). Differentiation of cultured and wild sea bass (Dicentrarchus labrax): total lipid content, fatty acid and trace mineral composition. *Food Chemistry, 79*(2), 145–150.

Allison, S. D. (2006). Soil minerals and humic acids alter enzyme stability: implications for ecosystem processes. *Biogeochemistry, 81,* 361–373.

Allison, S. D., LeBauer, D. S., Ofrecio, M. R., Reyes, R., Ta, A.-M., & Tran, T. M. (2009). Low levels of nitrogen addition stimulate decomposition by boreal forest fungi. *Soil Biology and Biochemistry, 41*(2), 293–302.

Allison, S. D., Weintraub, M. N., Gartner, T. B., & Waldrop, M. P. (2010). Evolutionary-economic principles as regulators of soil enzyme production and ecosystem function. *Soil Biology, 22,* 229–243.

Allison, S. D., Chao, Y., Farrara, J. D., Hatosy, S., & Martiny, A. C. (2012). Fine-scale temporal variation in marine extracellular enzymes of coastal southern California. *Frontiers in Microbiology, 3,* 301.

Arnosti, C. (2011). Microbial extracellular enzymes and the marine carbon cycle. *Annual Review of Marine Science, 3,* 401–425.

Arnosti, C., Durkin, S., & Jeffrey, W. H. (2005). Patterns of extracellular enzyme activities among pelagic marine microbial communities: Implications for cycling of dissolved organic carbon. *Aquatic Microbial Ecology, 38*(2), 135–145.

Arnosti, C., Steen, A. D., Ziervogel, K., Ghobrial, S., & Jeffrey, W. H. (2011). Latitudinal gradients in degradation of marine dissolved organic carbon. *PLoS One, 6*(12), e28900.

Arnosti, C., Fuchs, B. M., Amann, R., & Passow, U. (2012). Contrasting extracellular enzyme activities of particle-associated bacteria from distinct provinces of the North Atlantic Ocean. *Frontiers in Microbiology, 3,* 425.

Arnosti, C., Bell, C., Moorhead, D. L., Sinsabaugh, R. L., Steen, A. D., Stromberger, M., ... & Weintraub, M. N. (2014). Extracellular enzymes in terrestrial, freshwater, and marine environments: Perspectives on system variability and common research needs. *Biogeochemistry, 117*(1), 5–21.

Azam, F., & Malfatti, F. (2007). Microbial structuring of marine ecosystems. *Nature Reviews Microbiology, 5*(10), 782–791.

Azam, F., Fenchel, T., Field, J. G., Gray, J. S., Meyer-Reil, L. A., & Thingstad, F. (1983). The ecological role of water-column microbes in the sea. *Marine Ecology Progress Series, 10,* 257–263.

Bacosa, H. P., Liu, Z., & Erdner, D. L. (2015). Natural sunlight shapes crude oil-degrading bacterial communities in northern gulf of mexico surface waters. *Frontiers in Microbiology, 6,* 13–25.

Baldrian, P., & Valášková, V. (2008). Degradation of cellulose by basidiomycetous fungi. *FEMS Microbiology Reviews, 32*(3), 501–521.

Baldrian, P., Kolařík, M., Štursová, M., Kopecký, J., Valášková, V., Větrovský, T., ... & Voříšková, J. (2012). Active and total microbial communities in forest soil are largely different and highly stratified during decomposition. *The ISME Journal, 6*(2), 248–258.

Baldwin, A. J., Moss, J. A., Pakulski, J. D., Catala, P., Joux, F., & Jeffrey, W. H. (2005). Microbial diversity in a Pacific Ocean transect from the Arctic to Antarctic circles. *Aquatic Microbial Ecology, 41*(1), 91–102.

Baltar, F. (2018). Watch out for the "living dead": Cell-free enzymes and their fate. *Frontiers in Microbiology, 8*, 2438.

Baltar, F., Arístegui, J., Sintes, E., Van Aken, H. M., Gasol, J. M., & Herndl, G. J. (2009). Prokaryotic extracellular enzymatic activity in relation to biomass production and respiration in the meso-and bathypelagic waters of the (sub) tropical Atlantic. *Environmental Microbiology, 11*(8), 1998–2014.

Baltar, F., Arístegui, J., Gasol, J. M., Sintes, E., Van Aken, H. M., & Herndl, G. J. (2010). High dissolved extracellular enzymatic activity in the deep central Atlantic Ocean. *Aquatic Microbial Ecology, 58*(3), 287–302.

Baltar, F., Arístegui, J., Gasol, J. M., Yokokawa, T., & Herndl, G. J. (2013). Bacterial versus archaeal origin of extracellular enzymatic activity in the Northeast Atlantic deep waters. *Microbial Ecology, 65*(2), 277–288.

Baltar, F., De Corte, D., & Yokokawa, T. (2019). Bacterial stress and mortality may be a source of cell-free enzymatic activity in the marine environment. *Microbes and Environments, 34*(1), 83–88.

Bautz, N. (2019). Effects of temperature on enzyme activity of aquatic litter-associated fungi. Coastal Carolina University/ Honors Theses.

Brown, J. H., Gillooly, J. F., Allen, A. P., Savage, V. M., & West G.B. (2004). Toward a metabolic theory of ecology. *Ecology, 85*, 1771–1789.

Burns, R. G., DeForest, J. L., Marxsen, J., Sinsabaugh, R. L., Stromberger, M. E., Wallenstein, M. D., … & Zoppini, A. (2013). Soil enzymes in a changing environment: Current knowledge and future directions. *Soil Biology and Biochemistry, 58*, 216–234.

Buttigieg, P. L., Fadeev, E., Bienhold, C., Hehemann, L., Offre, P., & Boetius, A. (2018). Marine microbes in 4D—Using time series observation to assess the dynamics of the ocean microbiome and its links to ocean health. *Current Opinion in Microbiology, 43*, 169–185.

Caldwell, B. A. (2005). Enzyme activities as a component of soil biodiversity: A review. *Pedobiologia, 49*(6), 637–644.

Celiker, H., & Gore, J. (2013). Cellular cooperation: Insights from microbes. *Trends in Cell Biology, 23*(1), 9–15.

Chrost, R. J. (1991). Environmental control of the synthesis and activity of aquatic microbial ectoenzymes. In: Chrost, R. J. (ed.) .*Microbial Enzymes in Aquatic Environments*. Springer, New York. 29–59.

Courty, P. E., Hoegger, P. J., Kilaru, S., Kohler, A., Buee, M., Garbaye, J., … & Kües, U. (2009). Phylogenetic analysis, genomic organization, and expression analysis of multi-copper oxidases in the ectomycorrhizal basidiomycete Laccaria bicolor. *New Phytologist, 182*(3), 736–750.

Cycling, B. (2014). Extracellular Enzyme Activity and Microbial.

D'ambrosio, L., Ziervogel, K., MacGregor, B., Teske, A., & Arnosti, C. (2014). Composition and enzymatic function of particle-associated and free-living bacteria: A coastal/offshore comparison. *The ISME Journal, 8*(11), 2167–2179.

Davidson, E. A. & I. A. Jannsens. (2006). Temperature sensitivity of soil carbon decomposition and feedbacks to climate change. *Nature, 440*, 165–173.

Doyle, S. M., Whitaker, E. A., De Pascuale, V., Wade, T. L., Knap, A. H., Santschi, P. H., … & Sylvan, J. B. (2018). Rapid formation of microbe-oil aggregates and changes in community composition in coastal surface water following exposure to oil and the dispersant Corexit. *Frontiers in Microbiology, 9*, 689.

Eusterhues, K., Rumpel, C., Kleber, M., & Kögel-Knabner, I. Durbin, A. M., & Teske, A. (2011). Microbial diversity and stratification of South Pacific abyssal marine sediments. *Environmental Microbiology, 13*(12), 3219–3234.

Eusterhues, K., Rumpel, C., Kleber, M., & Kögel-Knabner, I. (2003). Stabilisation of soil organic matter by interactions with minerals as revealed by mineral dissolution and oxidative degradation. *Organic Geochemistry, 34*(12), 1591–1600.

Fang, C. M., & Moncrieff, J. (2005). The variation of soil microbial respiration with depth in relation to soil carbon composition. *Plant and Soil*, *268*(1), 243–253. https://doi.org/10.1007/s11104-004-0278-4

Ferreira, V. & Chauvet, E. (2011). Future increase in temperature more than decrease in litter quality can affect microbial litter decomposition in streams. *Oecologia*, *167*, 279–291.

Fierer N., Allen A. S., Schimel J. P., & Holden P. A. (2003a). Controls on microbial CO2 production: a comparison of surface and subsurface soil horizons. *Global Change Biology*, *9*, 1322–1332. https://doi.org/10.1046/j.1365-2486.2003.00663.x

Friedline, C. J., Franklin, R. B., McCallister, S. L., & Rivera, M. C. (2012). Bacterial assemblages of the eastern Atlantic Ocean reveal both vertical and latitudinal biogeographic signatures. *Biogeosciences*, *9*(6), 2177–2193.

Frossard, A., Gerull, L., Mutz, M., & Gessner, M. O. (2012). Disconnect of microbial structure and function: Enzyme activities and bacterial communities in nascent stream corridors. *The ISME Journal*, *6*(3), 680–691.

Fuhrman, J. A., Steele, J. A., Hewson, I., Schwalbach, M. S., Brown, M. V., Green, J. L., & Brown, J. H. (2008). A latitudinal diversity gradient in planktonic marine bacteria. *Proceedings of the National Academy of Sciences*, *105*(22), 7774–7778.

Gelsomino, A., & Azzellino, A. (2011). Multivariate analysis of soils: microbial biomass, metabolic activity, and bacterial-community structure and their relationships with soil depth and type. *Journal of Plant Nutrition and Soil Science*, *174*(3), 381–394.

German, D. P., Weintraub, M. N., Grandy, A. S., Lauber, C. L., Rinkes, Z. L., & Allison, S. D. (2011). Optimization of hydrolytic and oxidative enzyme methods for ecosystem studies. *Soil Biology and Biochemistry*, *43*(7), 1387–1397.

Gessner, M.O., Gulis, V., Kuehn, K. A., Chauvet, E. & Suberkropp, K. (2007). Fungal decomposers of plant litter in aquatic ecosystems. In: C. P. Kubicek & I. S. Druzhinina (eds.). *The Mycota: Microbial and Environmental Relationships*. Cambridge University Press, Cambridge. 301–324.

Gillooly, J. F., Brown, J. H., West, G. B., Savage, V. M., & Charnov, E. L. (2001). Effects of size and temperature on metabolic rate. *Science*, *293*, 2248–2251.

Gomez-Pereira, P. R., Fuchs, B. M., Alonso, C., Oliver, M. J., Van Beusekom, J. E., & Amann, R. (2010). Distinct flavobacterial communities in contrasting water masses of the North Atlantic Ocean. *The ISME Journal*, *4*(4), 472–487.

Gomez-Pereira, P. R., Schüler, M., Fuchs, B. M., Bennke, C., Teeling, H., Waldmann, J., … & Amann, R. (2012). Genomic content of uncultured Bacteroidetes from contrasting oceanic provinces in the North Atlantic Ocean. *Environmental Microbiology*, *14*(1), 52–66.

Grossart, H. P., & Simon, M. (1998). Bacterial colonization and microbial decomposition of limnetic organic aggregates (lake snow). *Aquatic Microbial Ecology*, *15*(2), 127–140.

Grossart, H. P., Massana, R., McMahon, K. D., & Walsh, D. A. (2020). Linking metagenomics to aquatic microbial ecology and biogeochemical cycles. *Limnology and Oceanography*, *65*, S2–S20.

Gulis, V. & Bärlocher, F. (2017). Fungi: Biomass, production, and community structure. In: F. R. Hauer & G. A. Lamberti (eds.). *Methods in Stream Ecology*. Academic Press, San Diego, CA. 1:177–192.

Gulis, V. & Suberkropp, K. (2003). Interactions between stream fungi and bacteria associated with decomposing leaf litter at different levels of nutrient availability. *Aquatic Microbial Ecology*, *30*, 149–157.

Gulis, V., Su, R., & Kuehn, K.A. (2019). Fungal decomposers in freshwater environments. In: C. Hurst (ed.). *Advances in Environmental Microbiology*. Springer Nature, New York City. 7:121–155.

Hanif, A., Yasmeen, A., & Rajoka, M. I. (2004). Induction, production, repression, and derepression of exoglucanase synthesis in Aspergillus niger. *Bioresource Technology*, *94*(3), 311–319.

Hedges, J. I., Eglinton, G., Hatcher, P. G., Kirchman, D. L., Arnosti, C., Derenne, S., … & Rullkötter, J. (2000). The molecularly-uncharacterized component of nonliving organic matter in natural environments. *Organic Geochemistry, 31*(10), 945–958.

Hehemann, J. H., Reintjes, G., Klassen, L., Smith, A. D., Ndeh, D., Arnosti, C., … & Abbott, D. W. (2019). Single cell fluorescence imaging of glycan uptake by intestinal bacteria. *The ISME Journal, 13*(7), 1883–1889.

Hendel, B., Sinsabaugh, R. L., & Marxsen, J. (2005). Lignin-degrading enzymes: Phenoloxidase and peroxidase. In: M. A. S. Graça, F. Bärlocher, & M. O. Gessner (eds.). *Methods to Study Litter Decomposition: A Practical Guide*. Springer, Dordrecht. 273–278.

Hoehler, T. M., & Jørgensen, B. B. (2013). Microbial life under extreme energy limitation. *Nature Reviews Microbiology, 11*(2), 83–94.

Hoppe, H. G. (1983). Significance of exoenzymatic activities in the ecology of brackish water: Measurements by means of methylumbelliferyl-substrates. *Marine Ecology*, 11, 299–308.

Hoppe, H. G. (1991). Microbial extracellular enzyme activity: A new key parameter in aquatic ecology. In: Chrost, R. J. (Ed.). *Microbial Enzymes in Aquatic Environments* . Springer, New York. 60–83.

Huot, Y., Brown, C. A., Potvin, G., Antoniades, D., Baulch, H. M., Beisner, B. E., … & Walsh, D. A. (2019). The NSERC Canadian Lake Pulse Network: A national assessment of lake health providing science for water management in a changing climate. *Science of the Total Environment, 695*, 133–668.

IPCC. (2014). *Climate Change 2014: Impacts, Adaptation, and Vulnerability*. Contribution of Working Group II to the Fifth Assessment Report of the Intergovernmental Panel on Climate Change. Cambridge University Press, Cambridge.

Jackson, C. R., Foreman, C. M., & Sinsabaugh, R. L. (1995). Microbial enzyme activities as indicators of organic matter processing rates in a Lake Erie coastal wetland. *Freshwater Biology, 34*(2), 329–342.

Jacobson Meyers, M. E., Sylvan, J. B., & Edwards, K. J. (2014). Extracellular enzyme activity and microbial diversity measured on seafloor exposed basalts from Loihi seamount indicate the importance of basalts to global biogeochemical cycling. *Applied and Environmental Microbiology, 80*(16), 4854–4864.

Jobbágy, E. G., & Jackson, R. B. (2000). The vertical distribution of soil organic carbon and its relation to climate and vegetation. *Ecological Applications, 10*(2), 423–436.

Kalwasińska, A., & Brzezinska, M. S. (2013). Extracellular enzymatic activities in subsurface water of eutrophic Lake Chełmżyńskie, Poland. *Journal of Freshwater Ecology, 28*(4), 517–527.

Kamalanathan, M., Xu, C., Schwehr, K., Bretherton, L., Beaver, M., Doyle, S. M., … & Quigg, A. (2018). Extracellular enzyme activity profile in a chemically enhanced water accommodated fraction of surrogate oil: Toward understanding microbial activities after the Deepwater Horizon oil spill. *Frontiers in Microbiology, 9*, 798.

Kleindienst, S., Paul, J. H., & Joye, S. B. (2015). Using dispersants after oil spills: Impacts on the composition and activity of microbial communities. *Nature Reviews Microbiology, 13*(6), 388–396.

Klonowska, A., Gaudin, C., Fournel, A., Asso, M., Le Petit, J., Giorgi, M., & Tron, T. (2002). Characterization of a low redox potential laccase from the basidiomycete C30. *European Journal of Biochemistry, 269*(24), 6119–6125.

Kramer, C. M., Barkhausen, J., Flamm, S. D., Kim, R. J., & Nagel, E. (2013). "Standardized cardiovascular magnetic resonance (CMR) protocols 2013 update." *Journal of Cardiovascular Magnetic Resonance, 15*(1), 1–10.

Krauss, G. J., Sole, M., Krauss, G., Schlosser, D., Wesenberg, D., & Baerlocher, F. (2011). Fungi in freshwaters: Ecology, physiology and biochemical potential. *FEMS Microbiology Reviews, 35*(4), 620–651.

Lagaert, S., Beliën, T., & Volckaert, G. (2009). Plant cell walls: Protecting the barrier from degradation by microbial enzymes. *Seminars in Cell & Developmental Biology*, 20(9), 1064–1073.

Lakna. (2019, October 1). *What Is the Difference between Intracellular and Extracellular Enzymes - Pediaa.Com*. Pediaa.Com; pediaa.com. https://pediaa.com/what-is-the-difference-between-intracellular-and-extracellular-enzymes/

Lee, C., Wakeham, S., & Arnosti, C. (2004). Particulate organic matter in the sea: The composition conundrum. *AMBIO: A Journal of the Human Environment*, 33(8), 565–575.

Li, Y., Sun, L. L., Sun, Y. Y., Cha, Q. Q., Li, C. Y., Zhao, D. L., ... & Qin, Q. L. (2019). Extracellular enzyme activity and its implications for organic matter cycling in northern Chinese marginal seas. *Frontiers in Microbiology*, 10, 21–37.

Liu, Z., Kobiela, M. E., McKee, G. A., Tang, T., Lee, C., Mulholland, M. R., & Hatcher, P. G. (2010). The effect of chemical structure on the hydrolysis of tetrapeptides along a river-to-ocean transect: AVFA and SWGA. *Marine Chemistry*, 119(1–4), 108–120.

Manivasagan, P., & Kim, S. K. (2014). Extracellular polysaccharides produced by marine bacteria. *Advances in Food and Nutrition Research*, 72, 79–94.

Martinez-Garcia, M., Brazel, D. M., Swan, B. K., Arnosti, C., Chain, P. S., Reitenga, K. G., ... & Stepanauskas, R. (2012). Capturing single cell genomes of active polysaccharide degraders: An unexpected contribution of Verrucomicrobia. *PLoS One*, 7(4), e35314.

Münster, U. (1992). Microbial extracellular enzyme activities in HUMEX Lake Skjervatjern. *Environment International*, 18(6), 637–647.

Ogunlade, T. M., Babaniyi, B. R., Afolabi, F. J., & Babaniyi, G. G. (2021). Physicochemical, heavy metals and microbiological assessment of wastewater in selected abattoirs in Ekiti State, Nigeria. *Journal of Environmental Treatment Techniques*, 9(4), 788–795.

Online-Wikipedia. (2020). Fungal extracellular enzyme activity - Wikipedia; en.wikipedia. org. Retrieved June 21, 2022, from https://en.wikipedia.org/wiki/Fungal_extracellular_enzyme_activity

Orsi, W. D., Richards, T. A., & Francis, W. R. (2018). Predicted microbial secretomes and their target substrates in marine sediment. *Nature Microbiology*, 3(1), 32–37.

Pai, A., Tanouchi, Y., & You, L. (2012). Optimality and robustness in quorum sensing (QS)-mediated regulation of a costly public good enzyme. *Proceedings of the National Academy of Sciences*, 109(48), 19810–19815.

Poll, C., Thiede, A., Wermbter, N., Sessitsch, A., & Kandeler, E. (2003). Micro-scale distribution of microorganisms and microbial enzyme activities in a soil with long-term organic amendment. *European Journal of Soil Science*, 54(4), 715–724.

Pommier, T., Canbäck, B., Riemann, L., Boström, K. H., Simu, K., Lundberg, P., ... & Hagström, Å. (2007). Global patterns of diversity and community structure in marine bacterioplankton. *Molecular Ecology*, 16(4), 867–880.

Priest, F. G. (1984). *Extracellular Enzymes*. Van Nostrand Reinhold, Wokingham. 1–79.

Quigg, A., Passow, U., Chin, W. C., Xu, C., Doyle, S., Bretherton, L., ... & Santschi, P. H. (2016). The role of microbial exopolymers in determining the fate of oil and chemical dispersants in the ocean. *Limnology and Oceanography Letters*, 1(1), 3–26.

Rasse, D. P., Rumpel, C., & Dignac, M.-F. (2005). "Is soil carbon mostly root carbon? Mechanisms for a specific stabilisation." *Plant and Soil*, 269(1–2), 341–356.

Reintjes, G., Arnosti, C., Fuchs, B. M., & Amann, R. (2017). An alternative polysaccharide uptake mechanism of marine bacteria. *The ISME Journal*, 11(7), 1640–1650.

Reintjes, G., Arnosti, C., Fuchs, B., & Amann, R. (2019). Selfish, sharing and scavenging bacteria in the Atlantic Ocean: A biogeographical study of bacterial substrate utilisation. *The ISME Journal*, 13(5), 1119–1132.

Ridley, B. L., O'Neill, M. A., & Mohnen, D. (2001). Pectins: Structure, biosynthesis, and oligogalacturonide-related signaling. *Phytochemistry*, 57(6), 929–967.

Romani, A. M., Fischer, H., Mille-Lindblom, C., & Tranvik, L. J. (2006). Interactions of bacteria and fungi on decomposing litter: Differential extracellular enzyme activities. *Ecology*, 87, 2559–2569.

Schimel, J. (2003). The implications of exoenzyme activity on microbial carbon and nitrogen limitation in soil: A theoretical model. *Soil Biology and Biochemistry*, *35*(4), 549–563.

Sierra, C. A. (2012). Temperature sensitivity of organic matter decomposition in the Arrhenius equation: Some theoretical considerations. *Biogeochemistry*, *108*, 1–15.

Simon, M., Grossart, H. P., Schweitzer, B., & Ploug, H. (2002). Microbial ecology of organic aggregates in aquatic ecosystems. *Aquatic Microbial Ecology*, *28*(2), 175–211.

Sinsabaugh, R. L. (2010). Phenol oxidase, peroxidase and organic matter dynamics of soil. *Soil Biology and Biochemistry*, *42*(3), 391–404.

Sinsabaugh, R. L., & Shah, J. J. F. (2012). Ecoenzymatic stoichiometry and ecological theory. *Annual Review of Ecology, Evolution and Systematics*, *43*(313), 20–12.

Sinsabaugh, R. L., Osgood, M. P., & Findlay, S. (1994). Enzymatic models for estimating decomposition rates of particulate detritus. *Journal of the North American Benthological Society*, *13*(2), 160–169.

Siuda, W., & Chróst, R. J. (2002). Decomposition and utilization of particulate organic matter by bacteria in lakes of different trophic status. *Polish Journal of Environmental Studies*, *11*(1), 53–66.

Smith, D. C., Simon, M., Alldredge, A. L., & Azam, F. (1992). Intense hydrolytic enzyme activity on marine aggregates and implications for rapid particle dissolution. *Nature*, *359*(6391), 139–142.

Stone, M. M., DeForest, J. L., & Plante, A. F. (2014). Changes in extracellular enzyme activity and microbial community structure with soil depth at the luquillo critical zone observatory. *Soil Biology and Biochemistry*, *75*, 237–247.

Teeling, H., Fuchs, B. M., Becher, D., Klockow, C., Gardebrecht, A., Bennke, C. M., ... & Amann, R. (2012). Substrate-controlled succession of marine bacterioplankton populations induced by a phytoplankton bloom. *Science*, *336*(6081), 608–611.

Trasar-Cepeda, C., Carmen Leirós, M., & Gil-Sotres, F. (2008). Hydrolytic enzyme activities in agricultural and forest soils. Some implications for their use as indicators of soil quality. *Soil Biology and Biochemistry*, *40*(9), 2146–2155.

Wallenstein, M. D., & Weintraub, M. N. (2008). Emerging tools for measuring and modeling the in situ activity of soil extracellular enzymes. *Soil Biology and Biochemistry*, *40*(9), 2098–2106.

Wallenstein, M. D., & Richard, G. B. (2011). Ecology of extracellular enzyme activities and organic matter degradation in soil: a complex community-driven process. *Methods of soil enzymology 9*, 35–55.

Wegner, C. E., Richter-Heitmann, T., Klindworth, A., Klockow, C., Richter, M., Achstetter, T., ... & Harder, J. (2013). Expression of sulfatases in Rhodopirellula baltica and the diversity of sulfatases in the genus Rhodopirellula. *Marine Genomics*, *9*, 51–61.

Yang, Y., Liu, W., Zhang, Z., Grossart, H. P., & Gadd, G. M. (2020). Microplastics provide new microbial niches in aquatic environments. *Applied Microbiology and Biotechnology*, *104*(15), 6501–6511.

Yvon-Durocher, G., Caffrey, J. M., Cescatti, A., Dossena, M., del Giorgio, P., Gasol, J. M., & Allen, A.P. (2012). Reconciling the temperature dependence of respiration across timescales and ecosystem types. *Nature,* 487, 472–476.

Ziervogel, K., Steen, A. D., & Arnosti, C. (2010). Changes in the spectrum and rates of extracellular enzyme activities in seawater following aggregate formation. *Biogeosciences*, *7*(3), 1007–1015.

Zinger, L., Amaral-Zettler, L. A., Fuhrman, J. A., Horner-Devine, M. C., Huse, S. M., Welch, D. B. M., ... & Ramette, A. (2011). Global patterns of bacterial beta-diversity in seafloor and seawater ecosystems. *PLoS One*, *6*(9), e24570.

13 Microbial biomass and activity, enzyme activities, and microbial community composition
Long-term effects of aided phytostabilization of trace elements

Gabriel Gbenga Babaniyi, Olaniran Victor Olagoke, and Babafemi Raphael Babaniyi

CONTENTS

13.1 INTRODUCTION

Tropical forests make a substantial contribution to the global carbon (C) cycle as they are highly productive, hold 30% of the Earth's soil C stock and reveal the highest rates of soil respiration of any terrestrial ecosystem (Jobbagy & Jackson 2000; Bond-Lamberty, Wang & Gower, 2004). In the Andes, tropical forests broaden from the lowlands to upwards of 3,000 m asl (above sea level), and they are accepted as one of the most biologically diverse regions of the planet (Myers *et al.*, 2000; Malhi *et al.*, 2010). In the montane cloud forests that take over the middle-upper elevations of the

Andes, wet and relatively cool conditions prevail (Grubb, 1977) which encourage the suppressed rates of decomposition n and accumulation of extensive stocks of soil organic matter (Zimmermann *et al.,* 2009). However, growing concern that atmospheric warming as a result of climate change could lead to significant changes in soil C cycling due to direct effects on microbial breakdown of soil organic matter, soil respiration, and greenhouse gas feedbacks to the atmosphere (Bardgett, Freeman & Ostle, 2008; Craine, Fierer & McLauchlan, 2010; Schindlbacher *et al.*, 2011).

Consequently, changes in plant community composition can lead to shifts in the proportion of recalcitrant (lignins) and accessible (e.g. non-structural carbohydrates) forms of C in plant-derived substrates, altering litter quality (Hattenschwiler & Jorgensen 2010). These indirect, plant-mediated effects of climate change on microbial activity and soil respiration are poorly understood and represent a significant knowledge gap in determining the response of terrestrial C cycling to future climate change (Bragazza *et al.*, 2013; Ward *et al.*, 2013). In addition to direct temperature and precipitation effects climate can indirectly affect soil respiration by influencing plant community productivity and structure, which in turn determines the quantity and quality of C inputs entering the soil (Engelbrecht *et al.*, 2007; Feeley *et al.*, 2011; Garcia-Palacios *et al.*, 2012). Sources of C inputs, including plant litter and rhizodeposition, act as substrates that are mineralized to CO_2 by the soil microbial community. Changes in the quality of these C inputs occur as a result of differences in plant species tissue chemistry that affect the nutrient stoichiometry of their litter and rhizosphere inputs (H€attenschwiler *et al.*, 2008). Consequently, changes in plant community composition can lead to shifts in the proportion of recalcitrant (e.g. lignins) and accessible (non-structural carbohydrates) forms of C in plant-derived substrates, altering litter quality (Hattenschwiler & Jorgensen 2010). These indirect, plant-mediated effects of climate change on microbial activity and soil respiration are poorly understood and represent a significant knowledge gap in determining the response of terrestrial C cycling to future climate change (Bragazza *et al.*, 2013; Ward *et al.*, 2013).

13.2   MICROBIAL BIOMASS AND ACTIVITY

During biological treatment of contaminated soils, the microbial activity measurement is a valuable tool to characterize the physiological state of soil and the biological purification capacity. Qualitative and quantitative determination of soil enzymes may well characterize the microbial activity in soil. "Microbial activity in soil" means all transformations of components of the soil, which are caused by soil microorganisms (Ulrich *et al.*, 2003). Soil microbial biomass is the living component of soil organic matter. It excludes soil animals and plant roots. Although it comprises less than 5% of organic matter in soil, it performs at least three critical functions for plant production in the ecosystem. It is a labile source of carbon (C), nitrogen (N), phosphorus (P), and sulfur (S); it is an immediate sink of C, N, P, and S; and it is an agent of nutrient transformation and pesticide degradation. In addition, microorganisms form symbiotic associations with roots, act as biological agents against plant pathogens, contribute toward soil aggregation (Angers *et al.*, 1992), and participate in soil formation. In the past, interest in microbial biomass assessment in soil has been limited by tedious, time consuming, and unreliable techniques (direct microscopy, culture media) for the enumeration of individual microbial communities.

When soil conditions change, microbial biomass, activity, and diversity adapt quickly. These adaptations can serve as vital indications of how management has affected the soil environment and have a significant impact on ecosystem dynamics (Shen et al., 2016). Estimation of microbial biomass from actual enumeration of the whole suite of microorganisms (bacteria, fungi, actinomycetes, protozoa, nematodes) has been difficult because of incomplete extraction, inappropriate growth media, extremely variable growth habits of microorganisms in soil, and scant information on their biovolumes (Jenkinson, 1988). In the last 20 years, relatively rapid assessment of soil microbial biomass has been possible based on physiological, biochemical, and chemical techniques (Horwath and Paul, 1994). These include chloroform fumigation incubation (Jenkinson and Powlson, 1976), chloroform fumigation extraction (Brookes *et al.*, 1985; Vance *et al.*, 1987), substrate-induced respiration (SIR) (Anderson & Domsch, 1978), and adenosine triphosphate analysis (Jenkinson *et al.*, 1979; Eiland 1983; Webster *et al.*, 1984). Of these, the first two methods have been widely used to estimate microbial biomass in agricultural, pastoral, and forestry systems, rehabilitation of disturbed lands, and pesticide and heavy metals polluted lands and materials. Recently, microbial biomass has even been proposed as a sensitive indicator of soil quality (Karlen *et al.*, 1997) and soil health (Sparling, 1997). But what do the soil microbial biomass numbers really mean? I will examine this question and seek answers in the following sequence:

    i. methods and their limitations to estimate microbial biomass;
   ii. effects of environment, soil, cultural practices and additives on the size of soil microbial biomass;
  iii. source of nutrients and rate of mineralization and its relationship to crop requirements;
  iv. nutrient sink;
   v. pesticide degradation; and
  vi. as an indicator of soil quality.

Probably the most commonly used classical technique for determining the microbial biomass size is that of chloroform fumigation. Fumigation ruptures microbial cells and releases cell walls and cellular contents into the soil. If incubation follows the fumigation, a "flush of decomposition" of the soil organic matter occurs that is due to the decomposition of microorganisms killed during fumigation. When the fumigation and subsequent incubation procedures are conducted under a strict set of conditions, the size of the biomass pool of C can be determined by the size of the carbon dioxide flush using the equation $B = F/kc$ where B is the concentration of biomass C in the soil, F is the amount of carbon dioxide evolved from the fumigated soil minus that from the unfumigated soil, and kc is the fraction of killed biomass C which subsequently is evolved as carbon dioxide. Five assumptions are implicit in the calculation of biomass C from the carbon dioxide flush associated with chloroform fumigation.

1. The C in killed microorganisms is mineralized to carbon dioxide more rapidly than that in living microorganisms.
2. The kill caused by the fumigation is essentially complete.
3. The fraction of microorganisms that die in the unfumigated control is negligible compared to those killed in the fumigated sample.
4. The fraction of the killed biomass that is mineralized (*kc*) is the same in different soils, i.e. a single kc value can be used to estimate biomass C in a wide variety of soils.
5. Fumigation has no effect on the soil other than the killing of the microbial biomass. Both chloroform carbon dioxide and methyl bromide have been used as fumigants. Generally, with chloroform, the kill is 99% or greater but enough microorganisms survive for an inoculum not to be required. However, an inoculum is required when methyl bromide is used. A variation of the fumigation method is to do an extraction of the soil after chloroform fumigation. The C content in the extractant is then compared to the amount of C in a similar soil that has not been fumigated. The difference is due to the C released from the microbial biomass. An extraction efficiency (Kec factor) of 0.45 (or a close similar value) is often used to calculate the microbial biomass C value using the fumigation-extraction method. The chloroform fumigation technique and the application of equations similar to that used for biomass C determinations have also been used to determine the pool sizes of microbial biomass N, P, and S. Microbial S and P measurements have the advantage in that the extraction of the mineral forms of S and P can be carried out immediately after chloroform treatment, eliminating the lengthy delay and unknown mineralization and immobilization rates that occur during incubation. Extraction of C and N constituents within the microbial biomass immediately after chloroform fumigation has not proven as successful. The *k* factors determined for C, N, P, and S are all of the same order of magnitude.

However, exact agreement is not observed and would not be expected because of the different experimental protocols used for each element. Differences in incubation times after fumigation, differences in extractability of the mineral forms of the nutrients, and the relative ratio of the extracted form of the nutrient compared to the total concentration of the nutrient in the biomass fraction all work to cause variation in the k factors. Another method of determining the size of the microbial biomass includes the initial respiratory response method. This method measures the initial maximum respiration rate in a soil upon addition of an easily degraded C source such as glucose. Microcalorimetry determines the microbial biomass by measuring the rate of heat output from soil and relating this heat output to the size of the microbial biomass.

The biomass, activity, and diversity of soil microbial communities could theoretically decrease with a decline in plant diversity due to a fall in the quantity and diversity of litter and rhizodeposits. However, real observations revealed a mixed response of soil microbial biomass, activity, and diversity to rising plant variety, with some showing no reaction at all (Wang et al., 2017). Because most soil microorganisms are heterotrophic, they consume plant exudates or degrade plant material as nourishment. Additionally, plants primarily supply the soil microbial community with carbon resources and other nutrients in the form of plant litter and root exudates. In particular, root exudates provide bacteria with readily accessible sources of carbon and energy. Labile substances, such as enzymes generated by living roots, can also promote microbial growth and activity (Steudel et al., 2012). As a result, the impact of plant diversity on roots may be different from that on soil microbial populations. Additionally, the stress-gradient hypothesis proposes that groups with high levels of physical stress should frequently engage in facilitation relationships, whereas communities with low levels of physical stress should frequently engage in competitive interactions. Additionally, in many plant communities under stressed conditions, facilitation interactions and favorable connections between plant species richness and productivity have been discovered (Wang et al., 2011, 2013). The facilitative effects of plant variety on biomass production in stressful conditions are predicted by the stress-gradient theory, however it is not obvious how plant diversity influences soil microbial populations in stressful environments.

13.3 ENZYMATIC ACTIVITIES

Enzymes are compounds that assist chemical reactions by increasing the rate at which they occur. For example, the food that you eat is broken down by digestive enzymes into tiny pieces that are small enough to travel through your blood stream and enter cells. Enzymes are proteins that are found in all living organisms. Without enzymes, most chemicals reactions within cells would occur so slowly that cells would not be able to work properly. Enzymes function as catalysts. Catalysts accelerate the rate of a chemical reaction without being destroyed or changed. They can be reused for the same chemical reaction over and over, just like a key can be reused to open a door many times. Enzymes are generally named after the substrate affected, and their names usually end in -ase. For example, enzymes that break down proteins are called proteases. While lipases break down lipids, carbohydrases break down carbohydrates. The compounds that enzymes act upon are known as substrates. The substrate can bind to a specific place in the enzyme called the active site. By temporarily binding to the substrate, an enzyme can lower the energy needed for a reaction to occur, thus making this reaction faster. The energy required for a chemical reaction to occur is known as the activation energy. Once the reaction between an enzyme and a substrate is complete, the substrate is changed to a product while the enzyme remains unchanged. The rate of the reaction between an enzyme and a substrate can be affected by different factors. Some of the factors that can affect enzyme activity are temperature, pH, concentration of the enzyme and concentration of the substrate. In living organisms, enzymes work best at certain temperatures and pH values depending on the type of enzyme.

Enzyme activity is measured in vitro beneath circumstances that often do not intimately resemble those in vivo. The objective of measuring enzyme activity is normally to determine the amount of enzyme present under defined conditions, so that activity can be compared between one sample and another, and between one laboratory and another. The conditions chosen are usually at the optimum pH, "saturating" substrate concentrations, and at a temperature that is convenient to control. In many cases the activity is measured in the opposite direction to that of the enzyme's natural function. However, enzymes are catalysts or chemical agents that speed up chemical reactions without being consumed. Most enzymes are proteins that function to reduce energy of activation in chemical reactions. They work on reactants called substrate; the enzyme attaches to the substrate and then the enzyme converts the substrate into a product, while the enzyme remains unaffected. They are protein substances that contain acidic carboxylic groups (COOH–) and basic amino groups (NH$_2$). So, the enzymes are affected by changing the pH value. Each enzyme has a pH value that it works at with maximum efficiency called the optimal pH. If the pH is lower or higher than the optimal pH, the enzyme activity decreases until it stops working

Understanding the function of proteins within the context of their natural cellular environment is perhaps one of the greatest challenges facing disciplines such as biochemistry, cell biology, and animal physiology. Visualizing enzymatic activation and regulation within a cell or an animal in real-time provides a foundation for understanding how a protein functions in a multifaceted milieu of molecules (Weissleder & Ntziachristos, 2003). In contrast to simple gene expression profiling, which can provide information about the regulation of a given process at the level of changes in bulk RNA messages, protein expression and enzyme activation are often difficult to measure and are controlled by a complex set of mechanisms that are independent of transcription. Furthermore, a protein can exert its function in many ways, for example through interactions with other proteins, nucleic acids, and small-molecule-binding partners. This daunting level of complexity in posttranslational regulation has motivated investigators to design methods for monitoring enzyme activity dynamically within the physiologically relevant environment of cells and whole organisms. Most enzymes harbor a set of controls that tightly regulate their activity within the cell. Enzymes can be regulated by multiple mechanisms such as their spatial and temporal expression, binding to small-molecule or protein cofactors and posttranslational modification. Thus, attempts to understand functional regulation of an enzyme using an in vitro approach is often misleading. One reason for the irrelevance of in vitro data is the disruption of organelles and fine compartments, which lead to the release of activators or inhibitors that artificially affect enzymatic activity. Furthermore, the use of fixed in vitro assay conditions might not accurately imitate the conditions found in vivo. Hence, tools for determining enzymatic activity in the context of an intact cell or preferably within a whole organism have great value. Although strategies for monitoring enzyme activity have been developed for a wide range of enzyme families, proteases and kinases have received most attention because of the large size of these families, the understanding of the relevant substrate–enzyme interactions and the potential for identification of new small-molecule drug targets. Attempts to study the cellular and physiological role of these enzymes have focused on the identification of downstream substrates (Ulrich et al., 2003).

However, the process of substrate identification can often be difficult and provides no information about the temporal and spatial regulation of an enzyme. Therefore, new technologies have been developed allowing direct visualization of kinase, phosphatase, and protease activities. These technologies involve both the design of "smart" imaging reagents and the development of techniques and optical instrumentation that allow a sensitive, rapid, and higher solution detection of enzyme activity within cells and whole organisms (Rudin & Weissleder, 2003). Because it is impossible to describe all technologies that can be used to address the question of protein function in a single review, in this article the focus will only be on technologies that provide a direct readout of enzyme activity. In particular, recent advances in the design of novel cleavable substrates, protein reporters, and small-molecule activity-based probes will be discussed. These tools will surely have a dramatic impact on the study of proteins at the level of physiological function. Thus, the biogeochemical process of wetlands is greatly influenced by soil microorganisms, which can also supply nutrients for the growth and operation of plants and soil (Li et al., 2015; Xu et al., 2020). The soil microbial characteristics require in some specific wetlands such as microbial biomass and enzyme activity could indicate the changes in soil characteristics following ecosystem restoration since they are thought to be more sensitive parameters than physico-chemical properties significant indicators. A measure of the active components of soil organic matter is soil microbial biomass (SOM). It is frequently used as a gauge of soil fertility and ecosystem production because of its tight connection to nutrient cycling (Xiao et al., 2015; Singh & Gupta, 2018; Wang et al., 2021). Exudates from plant roots, microbial exudates, and the breakdown products of soil residues are all sources of soil enzymes.

13.4 MICROBIAL COMMUNITY COMPOSITION

Microbial community composition may be one important control on soil processes (Schimel, 1995; Cavigelli & Robertson, 2000; Balser *et al.*, 2002). If the microbial communities residing at depth are simply diluted analogs of the surface microbial communities and exhibit minimal differentiation, the characteristics and properties of microbial processes should be fundamentally similar in the surface and subsurface horizons. However, deeper layers of soil may contain microbial communities that are specialized for their environment and fundamentally distinct from the surface communities (Ghiorse & Wilson, 1988; Zvyagintsev, 1994; Fritze *et al.*, 2000). In this case, the microbial communities in the soil subsurface may function differently from those at the surface and their metabolic properties could not be inferred by studying the microbial communities found in the surface horizons.

The soil microbial biomass acts as the transformation agent of the organic matter in soil. As such, the biomass is both a source and sink of the nutrients C, N, P, and S contained in the organic matter. It is the center of the majority of biological activity in soil. To properly understand biological activity in soil one must therefore, have knowledge of the microbial biomass. Investigating the flow of C and N in the soil, from newly deposited plant or other materials to the mineral forms of carbon dioxide and ammonium or nitrate ions, clearly shows the central role of the microbial biomass. The definition of the soil microbial biomass is the living portion of the soil organic matter, excluding plant roots and soil animals larger than 5×10^{-3} μm^3. The microbial biomass generally comprises approximately 2% of the total organic matter in soil and it may be easily dismissed as of minor importance in the soil. However, this chapter will introduce the biomass as an important agent in controlling the overall biological activity of the soil.

Soil fertility decline is a major biophysical problem confronting crop production in Ghana. Most Ghanaian soils contain low organic matter of less than 1.0% which is inadequate to sustain crop production. Above all, most of the soils are developed on thoroughly weathered parent materials. They are old and have been leached over a long period of time (Benneh et al., 1990) and are, therefore, of low inherent fertility. It is, therefore, obvious that soil fertility decline in Ghana would be on the increase if pragmatic actions are not quickly taken to curtail the situation. One way of doing this is the study of soil microbial biomass dynamics as affected by specific amendments in the various cropping systems. Due to the dominant contribution of microbial biomass in soil metabolism, and its importance as a sink and source of nutrients for plants, microbial biomass is considered to be one of the main determinants of soil fertility (Jenkinson & Ladd, 1981). In Ghana, use of fertilizer and other soil amendments are carried out without taking.

13.5 PHYTOSTABILIZATION OF TRACE ELEMENTS

Phytostabilization is the most feasible technology for the recovery of large areas contaminated by trace elements (Mendez & Maier, 2008). Its advantages are multiple: in situ technique, cost effective, aesthetic value, and restoration of ecosystem functions (Garbisu et al., 2002). We aim to understand the species-specific effects on soil functionality, measured as microbial biomass and enzyme activities related to C, N, and P cycling, as different tree species are expected to affect soil underneath differently due to the contrasted leaf litter quality and root exudates, among others. We selected a trace element contaminated and remediated area (through phytoremediation strategy) in SW Spain where we studied long-term (15 years) effects of afforestation. As soil properties are essential for soil functionality we studied two areas with different soil characteristics (Gil Martínez et al., 2018). Mine spoils of extractory sites represent a permanent threat to surrounding ecosystems and humans, as they are generally contaminated by several trace elements which can be transported through wind erosion and water runoff, or be leached into the groundwater (Vangronsveld & Cunningham, 1998; Eisler, 2004). Mine spoils have generally acidic pH and a low nutrient content. Reduction of the ecological risk associated to mine spoils can be achieved by amendment with alkaline minerals and/or organic matter, making trace elements less mobile by either sorption, complexation or precipitation reactions (Adriano et al., 2004). However, in case of As contamination, amendments should be carefully used with attention paid to the pH value of remediated spoil, to avoid As leaching, which may occur at pH above 7 (Al-Abed et al., 2007). For the Jales mine spoils, pH correlated with As mobility (Bleeker et al., 2003). Neutralization of spoil acidity along with immobilization of excessive trace elements and input of organic matter may create favorable conditions for growth of trace element-tolerant plants. Such plants are excluders, i.e. trace elements retained in roots, and contribute to reduce trace elements leaching by allocating organic matter into the spoils, due to litter fall, deposition of root exudates, and decaying root material. Fresh input of nutrients should also stimulate microbiological and biochemical activity and accelerate the recovery process. Profits of in situ-aided phytostabilization on biochemical parameters of trace element polluted soils have been reported (Kumpiene et al., 2006;

Mench *et al.*, 2006), whereas both short-and long-term effects of phytostabilization on biochemical activity and microbial diversity of trace element-contaminated mine spoils are still poorly understood. Increase of microbial biomass and activity and enzyme activity in spoils amended with organic matter has been reported (Seaker & Sopper, 1988; Perez de Mora *et al.*, 2005). Recovery of microbial activity and diversity in revegetated mine spoils depended on the effectiveness of plant colonization (Machulla *et al.*, 2005).

Phytoremediation can be classified into different applications, such as

1. rhizofiltration;
2. phytoextraction;
3. phytovolatilization;
4. phytodegradation; and
5. phytostabilization.

13.5.1 EFFECT OF AIDED PHYTOSTABILIZATION OF TRACE ELEMENT

It is a biological method included in the Gentle Remediation Options, which, among others, are safer and least interferes with the natural environment (Cundy *et al.*, 2013). This technique is based on the chemical stabilization of heavy metals using various non-organic and/or organic soil additives in connection with adequately chosen plant species (Radziemska *et al.*, 2017). Species which will be resistant to specific conditions present in the soil, such as low pH and high concentrations of heavy metals, ought to be selected. Moreover, they should not accumulate heavy metals in their above-ground parts, thus preventing their further passage to subsequent elements of the food chain, and should be characterized by a fast increase in biomass, ensuring good coverage of the area in a short period of time (Gil-Loaiza *et al.*, 2016) An example of such plants is grasses from the fescue family of grasses, which are commonly used to create a vegetation cover in post-mining areas and slag heaps. Various species of grass, such as red fescue (*Festuca rubra* L.) are the most useful in the process of the aided phytostabilization of heavy metals in soils (Touceda-González *et al.*, 2017). Some literature reports (Gołda & Korzeniowska, 2016) show that *F. rubra* is a suitable species for the revegetation of metal-contaminated soils contaminated by industrial activities such as mining, energy, and fuel production. Furthermore, *F. rubra* has the ability to accumulate Cu, Pb, Mn, and Zn from contaminated soils (Yin *et al.*, 2014). The aim of this technique, besides limiting the bioavailability of heavy metals, is also to restore adequate soil quality (Labidi *et al.*, 2017). Various non-organic materials, such as: CaO, apatite, chalcedonite, septolite, diatomite, dolomite, bentonite, halloysite, hematite or FeO (Kalenik, 2014; Radziemska *et al.*, 2016) and/or organic compounds, e.g., brown coal and wood coal, compost, peat, fly ash, woodchips or wood bark (Gusiatin *et al.*, 2016) are used individually or in combination as soil additives. The search for new sorption materials which can be used as soil additives supporting the soil contaminant immobilizing processes is of key importance in the intensification of heavy metal removal or immobilization processes in soil and the improvement of its quality. The availability, prevalence, price, and effectiveness of removing contaminants from the area with the applied sorption material are undoubtedly of significance; hence, the reason behind the new and intensive search for original and effective additives that may be used in processes of aided phyto-stabilization. Halloysite $[Al_2Si_2O_5(OH)_4 \cdot (H_2O)]$ is an aluminum silicate of volcanic origin, characterized by high porosity (60–70%) and specific surface (65–85 $m^2 \cdot g^{-1}$), high ion-exchange capacity thanks to which it has a high ability to absorb heavy metals, and the ease of both chemical and mechanical treatment. Halloysite from Polish deposits is characterized by a specific platy and tubular structure, with a prevalence of the platy fraction. The inside diameter of tubes is 15 nm, while their length can reach up to 1,000 nm (Jones *et al.*, 2016). Poland is home to one of the largest deposits in the world—the "Dunino" deposit near Legnica (SW Poland), is one of three places in the world, in addition to New Zealand and the USA, with resources estimated at over ten million tons (Sakiewicz *et al.*, 2011). It contains mainly halloysite nanotubes and nanoplates (Sakiewicz *et al.*, 2016).

In addition to air, water, minerals, and live and dead plant and animal stuff, soil also possesses abiotic and biotic qualities. These soil elements can be divided into two groups. Biotic factors, which include all living and extinct soil organisms including plants and insects, fall under the first group. Abiotic variables make up the second group and include all inanimate objects, such as minerals, water, and air. Phosphorus, potassium, nitrogen gas, and these other elements are the most frequently occurring soil minerals that enable plant growth. The less popular minerals calcium, magnesium, and sulfur are also present. The composition of the soil is determined by its biotic and abiotic components (Rabêlo et al., 2021). To ensure sustainable agriculture with good soil quality, it is important to monitor changes in soil properties. Important variables to assess are different physical, chemical, and biological characteristics of the soil, as well as location, vegetation cover, and cultivation practices, including treatments with fertilizers and pesticides, biological additions, or accidental contaminations. When the physical and chemical soil properties are monitored, the conditions for soil storage prior to the analysis are not always critical. In contrast, when monitoring microbial activities, the storage conditions may be decisive for the results. In all microbiological studies freshly collected soil from the field are preferred (Anderson, 1987)

However, for practical reasons this is not always possible. Soil sampling cannot be carried out throughout the year due to climatic conditions, e.g. long wet or dry periods or periods with frozen soil. Thus, there is a need for satisfactory storage methods to preserve the soil. There may also be a need for the storage of soils since sampling and analysis cannot always be carried out simultaneously. The most commonly used method to store soils for microbiological analyses is to place them in cold or frozen storage. Different effects from the storage could be expected depending on whether the soil samples are stored at a few degrees above zero or are frozen. In a refrigerated soil a slow depletion of the available substrate due to ongoing microbial activity (Coxson & Parkinson, 1987) can be expected. Anderson (1987) showed that the storage temperature has a great influence on the survival of microbial biomass in soil as analyzed by SIR. Storage for 70 days at +228C resulted in a 39% loss of biomass, while storage at +28C for 70 days only gave a loss of 18%. Accordingly, studies by Ross (1991) indicated that there was a 41% biomass reduction after 14 months at +48C, as well as a similar reduction of basal respiration rate.

13.5.2 TOLERANCE MECHANISMS OF GRASSES TO TRACE ELEMENT TOXICITY

Trace element tolerance is a syndrome of adaptations at the cellular and biochemical levels rather than a simple physiological characteristic (Baker, 1981). These modifications appear to be primarily concerned with preventing the accumulation of hazardous concentrations at vulnerable cell locations to prevent negative effects (Hall, 2002). The strategies involved are diverse: (i) they involve binding to cell walls, chelation by root exudates, and mycorrhizal connections to immobilize trace elements extracellularly; (ii) tolerance may potentially affect the plasma membrane by inhibiting the absorption of trace elements or by encouraging their outflow pumping from the cytosol; and (iii) there are a number of potential mechanisms for chelating trace elements in the cytosol, including organic acids, amino acids, and peptides, or their compartmentation away from metabolic processes by transport into the vacuole. These mechanisms include the repair of stress-damaged proteins using heat shock proteins or MTs, and the chelation of trace elements by organic acids, amino acids, or peptides (Hall, 2002; Kushwaha et al., 2016). The antioxidative defense machinery is the next crucial "line of defense" against trace element-induced toxicity when the use of these tolerance mechanisms by plants is insufficient to reduce the proportion of free trace elements within the cell (Gratão et al., 2005; Minerals Council of Australia [MCA], 2016; Courtney, 2018). Trace element tolerance is influenced by mechanisms that function at the level of the entire plant, such as root-to-shoot transport. However, since these mechanisms were briefly discussed above for each trace element, they are not included in this section. This section also omitted discussing certain tolerance mechanisms as the release of Hg0 through the leaves to lessen Hg toxicity (Windham et al., 2001).

13.5.3 THE EFFECT OF ROOT EXUDATES ON TRACE ELEMENT AVAILABILITY AND UPTAKE

Numerous metabolites are released by plants from their roots into the rhizosphere to alter the pH or to maintain nutrient bioavailability and respond to environmental trace element stressors (Chen et al., 2017). These exudates are a complex mixture of inorganic ions, gaseous molecules, primarily carbon-based substances (such as amino acids, organic acids, sugars, and phenolics), and high-molecular-weight substances like mucilage and proteins. Gaseous molecules include carbon dioxide (CO_2) and hydrogen gas (H_2) (Kushwaha et al., 2015; Chen et al., 2017). Meanwhile, Malinowski et al. (2004) sought to determine how P nutrition and endophytic bacterial strains affected *F. arundinacea*'s ability to acquire copper. They found that endophyte infection had no effect on the Cu2+-binding activity of this grass's root exudates, but that it was higher (i.e., there was less free Cu2+ present) in the absence of P. When this grass was helped with root exudates from *Belamcanda chinensis* for 4 weeks, the accumulation of Cd, Cu, and Pb by *Echinochloa crus-galli* growing in soils polluted with 600 mg Pb kg−1 soil, 40 mg Cd kg−1 soil, and 100 mg Cu kg−1 soil was 2- to 4-fold higher (Kim et al., 2010). Indicating that root exudates may play a significant role in the effectiveness of phytoextraction, the root exudates also raised the BCF and TF for Cd, Cu, and Pb. The hypothesis that the ability of root exudates to bind trace elements can be crucial for trace element phytostabilization is supported by the fact that Cd bioavailability was reduced as a result of malate exudation by S. bicolor treated to Cd (0, 0.5, and 5 mg L^{-1} in solution). Based on the plant species, physiological conditions, and soil environment, different types and patterns of root exudates are secreted; it appears that depending on these factors, root exudates can support phytoextraction (primarily at sites with low levels of trace element pollution) or phytostabilization (Pinto et al., 2008; Kim et al., 2010). In fact, immobilizing trace elements, which restricts their ability to pass across the plasma membrane, or favoring the production of chelates, which promotes uptake in the cell and the transfer of trace elements from the roots to shoots, can both limit or favor trace element uptake (Kushwaha et al., 2015; Wen et al., 2018; Kumar et al., 2021). However, research on how root exudates affect the tolerance to trace elements in grasses used for phytoextraction or phytostabilization is mostly limited. In terms of plant mineral nutrition, the majority of research to date have concentrated on the function of root exudates in grasses' uptake of Fe and Zn (Chen et al., 2017). For instance, Cakmak et al. (1996) noted that Zn absorption was boosted by the release of phytosiderophores by Zn-deficient A. orientale.

13.6   CONCLUSION

The biomass of soil microorganisms and fungi varied among the different types of eucalyptus. Similar to this, the age and species of the plantation had a big impact on the enzymes involved in carbon cycling. The eucalyptus species had a significant impact on the enzyme involved in sulfur cycling (S), whereas the enzyme engaged in nitrogen cycling (N) increased with plantation aging and was mostly influenced by eucalyptus plantation age. Additionally, the physio-chemical characteristics of the soil, which are influenced by the type of eucalyptus and the age of the plantation, might indirectly alter the soil's microbial biomass and enzyme activity. This brought to light the significance of eucalyptus plantation age and species on soil microbial activities. Grass has unquestionably got a great deal of promise for stabilizing trace elements in soils, sediments, and wastewater. To fully comprehend the mechanisms underlying the absorption, transport, and accumulation of trace elements, however, considerably more research utilizing molecular methods is necessary. Numerous varieties of grass have been studied, but only a small number of As, Cd, Cu, and Zn transporters have been found in a small number of species. Trace element transporter genes express differently depending on the species and tissue. To maximize the phytoremediation of trace elements, it is imperative to comprehend the function of each trace element transporter in grasses. Particularly for Cr, Hg, Ni, and Pb, studies should be done to identify trace element transporters and to clarify the mechanisms behind trace element intake, transport, and accumulation. Acute trace element exposure is the foundation of the majority of current studies that try to better understand the physiological mechanisms that trigger trace element toxicity and how grasses respond to this toxicity. Since a series of signaling events take place during the first stage of trace element exposure to help the plant adjust to the stress, it is challenging to link the results to tolerance mechanisms. However, under field conditions, exposure to a trace element occurs over time, which frequently results in a variety of reactions that are substantially different from those seen in cases of acute exposure. Since most grasses do not endure high dosages of various trace elements for an extended period of time, research must be undertaken under more realistic conditions. Only a few grass species are appropriate for phytofiltration of trace elements, while a variety of grass species may be suitable for phytoextraction of slightly polluted agricultural soils or phytostabilization of soils with low-to-moderate concentrations of trace elements. Utilizing fertilizers and microorganisms that promote plant growth can increase the effectiveness of grasses for phytoextraction or phytostabilization, but the use of such tactics should be carefully considered in each individual situation. Due to the specificity of the trace elements, amendment/microorganism interactions, and plant interactions, increases in the uptake of trace elements by plants can happen instead of greater trace element phytostabilization, and vice versa. Due to its genetic complexity and the concomitant challenges experienced during traditional breeding of grasses, genetic manipulation has been generally disregarded as a means of improving the efficacy of phytoextraction and phytostabilization of trace elements. While still leaving a significant gap that might be investigated during the coming years. Although utilizing solely grasses to restore trace element-contaminated sites has been shown to be an effective technique, growing grasses in conjunction with other plants, such as trees and legumes, helps hasten site restoration and is more suitable from an ecological standpoint.

REFERENCES

Adriano, D. C., Wenzel, W. W., Vangronsveld, J., & Bolan, N. S. (2004). Role of assisted natural remediation in environmental cleanup. *Geoderma* 122(2–4), 121–142.

Al-Abed, S. R., Jegadeesan, G., Purandare, J., & Allen, D. (2007). Arsenic release from iron rich mineral processing waste: Influence of pH and redox potential. *Chemosphere*, 66(4), 775–782.

Anderson, J. C. (1987). An approach for confirmatory measurement and structural equation modeling of organizational properties. *Management Science*, 33(4), 525–541.

Anderson, J. P.E., & Domsch, K. H. (1978). A physiological method for the quantitative measurement of microbial biomass in soils. *Soil Biology and Biochemistry*, 10(3), 215–221.

Angers, D. A. (1992). Changes in soil aggregation and organic carbon under corn and alfalfa. *Soil Science Society of America Journal*, 56(4), 1244–1249.

Baker, A. J. (1981). Accumulators and excluders-strategies in the response of plants to heavy metals. *Journal of Plant Nutrition*, 3(1–4), 643–654.

Balser, T. C., Kinzig, A. P. & Firestone, M. K. (2002). Linking soil microbial communities and ecosystem functioning. In: *The Functional Consequences of Biodiversity: Empirical Progress and Theoretical Extensions* (pp. 265–293).

Bardgett, R. D., Freeman, C., & Ostle, N. J. (2008). Microbial contributions to climate change through carbon cycle feedbacks. *The ISME Journal*, 2(8), 805–814.

Bleeker, W. (2003). The late Archean record: a puzzle in ca. 35 pieces. *Lithos*, 71(2–4), 99–134.

Bond-Lamberty, B. E. N., Wang, C., & Gower, S. T. (2004). Contribution of root respiration to soil surface CO2 flux in a boreal black spruce chronosequence. *Tree Physiology*, 24(12), 1387–1395.

Bragazza, L., Parisod, J., Buttler, A., & Bardgett, R. D. (2013). Biogeochemical plant–soil microbe feedback in response to climate warming in peatlands. *Nature Climate Change*, 3(3), 273–277.

Brookes, P. C., Landman, A., Pruden, G., & Jenkinson, D. S. (1985). Chloroform fumigation and the release of soil nitrogen: a rapid direct extraction method to measure microbial biomass nitrogen in soil. *Soil Biology and Biochemistry* 17(6), 837–842.

Cakmak, I., Öztürk, L., Karanlik, S., Marschner, H., & Ekiz, H. (1996). Zinc-efficient wild grasses enhance release of phytosiderophores under zinc deficiency. *Journal of Plant Nutrition, 19*(3–4), 551–563.

Cavigelli, M. A., & Robertson, G. P. (2000). The functional significance of denitrifier community composition in a terrestrial ecosystem. *Ecology, 81*(5), 1402–1414.

Chen, Y. T., Wang, Y., & Yeh, K. C. (2017). Role of root exudates in metal acquisition and tolerance. *Current Opinion in Plant Biology, 39*, 66–72.

Craine, J. M., Fierer, N., & McLauchlan, K. K. (2010). Widespread coupling between the rate and temperature sensitivity of organic matter decay. *Nature Geoscience, 3*(12), 854–857.

Courtney, R. (2018). Irish mine sites rehabilitation—A case study. In: Prasad, M. N. V., Sajwan, K. S., & Naidu, R. (Eds.). *Bio-Geotechnologies for Mine Site Rehabilitation* (pp. 439–456). Elsevier, Amsterdam.

Coxson, D. S., & Parkinson, D. (1987). Winter respiratory activity in aspen woodland forest floor litter and soils. *Soil Biology and Biochemistry, 19*(1), 49–59.

Cundy, A. B., Bardos, R. P., Church, A., Puschenreiter, M., Friesl-Hanl, W., Müller, I., ... & Vangronsveld, J. (2013). Developing principles of sustainability and stakeholder engagement for "gentle" remediation approaches: The European context. *Journal of Environmental Management, 129*, 283–291.

Eiland, F. (1983). A simple method for quantitative determination of ATP in soil. *Soil Biology and Biochemistry 15*(6), 665–670.

Eisler, R., & Wiemeyer, S. N. (2004). Cyanide hazards to plants and animals from gold mining and related water issues. *Reviews of Environmental Contamination and Toxicology, 183*, 21–54.

Engelbrecht, B. M. J., et al. (2007). Drought sensitivity shapes species distribution patterns in tropical forests. *Nature, 447*(7140), 80–82.

Feeley, K. J., Davies, S. J., Perez, R., Hubbell, S. P., & Foster, R. B. (2011). Directional changes in the species composition of a tropical forest. *Ecology, 92*(4), 871–882.

Fritze, H., Pietikäinen, J., & Pennanen, T. (2000). Distribution of microbial biomass and phospholipid fatty acids in Podzol profiles under coniferous forest. *European Journal of Soil Science, 51*(4), 565–573.

Garbisu, C., Allica, J. H., Barrutia, O., Alkorta, I., & Becerril, J. M. (2002). Phytoremediation: A technology using green plants to remove contaminants from polluted areas. *Reviews on Environmental Health, 17*(3), 173–188.

García-Palacios, P., Maestre, F. T., Bardgett, R. D., & de Kroon, H. (2012). Plant responses to soil heterogeneity and global environmental change. *Journal of Ecology, 100*(6), 1303–1314.

Ghiorse, W. C., & Wilson, J. T. (1988). Microbial ecology of the terrestrial subsurface. *Advances in Applied Microbiology, 33*, 107–172.

Gil-Loaiza, J., White, S. A., Root, R. A., Solís-Dominguez, F. A., Hammond, C. M., Chorover, J., & Maier, R. M. (2016). Phytostabilization of mine tailings using compost-assisted direct planting: Translating greenhouse results to the field. *Science of the Total Environment, 565*, 451–461.

Gil Martínez, M., Domínguez, M. T., Navarro-Fernández, C. M., Crompot, H., Tibbett, M., & Marañón, T. (2018). Long-term effects of trace elements contamination on soil microbial biomass and enzyme activities. *Mine Closure 2018 Proceedings of the 12th International Conference on Mine Closure*: 807, Technische Universität Bergakademie Freiberg.

Gołda, S., & Korzeniowska, J. (2016). Comparison of phytoremediation potential of three grass species in soil contaminated with cadmium. *Environmental Protection and Natural Resources/Ochrona Środowiska i Zasobów Naturalnych, 27*(1), 8–14.

Gratão, P. L., Polle, A., Lea, P. J., & Azevedo, R. A. (2005). Making the life of heavy metal-stressed plants a little easier. *Functional Plant Biology, 32*(6), 481–494.

Grubb, P. J. (1977). Control of forest growth and distribution on wet tropical mountains: with special reference to mineral nutrition. *Annual Review of Ecology and Systematics, 8*(1), 83–107.

Gusiatin, Z. M., & Kulikowska, D. (2016). Behaviors of heavy metals (Cd, Cu, Ni, Pb and Zn) in soil amended with composts. *Environmental Technology*, *37*(18), 2337–2347.

Hall, J. Á. (2002). Cellular mechanisms for heavy metal detoxification and tolerance. *Journal of Experimental Botany*, *53*(366), 1–11.

Hättenschwiler, S., Aeschlimann, B., Coûteaux, M. M., Roy, J., & Bonal, D. (2008). High variation in foliage and leaf litter chemistry among 45 tree species of a neotropical rainforest community. *New Phytologist*, *179*(1), 165–175.

Hättenschwiler, S., & Jørgensen, H. B. (2010). Carbon quality rather than stoichiometry controls litter decomposition in a tropical rain forest. *Journal of Ecology*, *98*(4), 754–763.

Horwath, W. R., & Paul, E. A. (1994). Microbial biomass. In R. W. Weaver, Scott Angle, Peter Bottomley, David Bezdicek, Scott Smith, Ali Tabatabai, Art Wollum (Eds) *Methods of Soil Analysis: Part 2 Microbiological and Biochemical Properties,* 5, 753–773. https://doi.org/10.2136/sssabookser5.2.c36

Jenkinson, D. S., Davidson, S. A., & Powlson, D. S. (1979). Adenosine triphosphate and microbial biomass in soil. *Soil Biology and Biochemistry*, *11*(5), 521–527.

Jenkinson, D. S., & Powlson, D.S. (1976). The effects of biocidal treatments on metabolism in soil—V: A method for measuring soil biomass. *Soil biology and Biochemistry*, *8*(3), 209–213.

Jenkinson, D. S., & Ladd, J. N. (1981). Microbial biomass in soil: measurement and turnover. *Soil Biochemistry*, *5*(1), 415–471.

Jenkinson, D. S. (1988). Soil organic matter and its dynamics. In Wild, A. (Ed) *Russell's Soil Conditions and Plant Growth*. Eleventh edition, Longman Group UK Limited, 564–607.

Jobbágy, E. G., & Jackson, R. B. (2000). The vertical distribution of soil organic carbon and its relation to climate and vegetation. *Ecological Applications*, *10*(2), 423–436.

Jones, S., Bardos, R. P., Kidd, P. S., Mench, M., de Leij, F., Hutchings, T., ... & Menger, P. (2016). Biochar and compost amendments enhance copper immobilisation and support plant growth in contaminated soils. *Journal of Environmental Management*, *171*, 101–112.

Kalenik, M. (2014). Sewage treatment efficacy of sandy soil bed with natural clinoptilolite assist layer. *Ochr. Srodowiska*, *36*, 43–48.

Karlen, D. L., Mausbach, M. J., Doran, J. W., Cline, R. G., Harris, R. F., & Schuman, G. E. (1997). Soil quality: a concept, definition, and framework for evaluation (a guest editorial). *Soil Science Society of America Journal*, *61*(1), 4–10.

Kim, S., Lim, H., & Lee, I. (2010). Enhanced heavy metal phytoextraction by Echinochloa crus-galli using root exudates. *Journal of Bioscience and Bioengineering*, *109*(1), 47–50.

Kumar, V., Pandita, S., Sidhu, G. P. S., Sharma, A., Khanna, K., Kaur, P., ... & Setia, R. (2021). Copper bioavailability, uptake, toxicity and tolerance in plants: A comprehensive review. *Chemosphere*, *262*, 127–810.

Kumpiene, J., Ore, S., Renella, G., Mench, M., Lagerkvist, A., & Maurice, C. (2006). Assessment of zerovalent iron for stabilization of chromium, copper, and arsenic in soil. *Environmental Pollution*, *144*(1), 62–69.

Kushwaha, A., Rani, R., Kumar, S., & Gautam, A. (2015). Heavy metal detoxification and tolerance mechanisms in plants: Implications for phytoremediation. *Environmental Reviews*, *24*(1), 39–51.

Kushwaha, S. K., Pletikosić, I., Liang, T., Gyenis, A., Lapidus, S. H., Tian, Y., ... & Cava, R. J. (2016). Sn-doped Bi1. 1Sb0. 9Te2S bulk crystal topological insulator with excellent properties. *Nature Communications*, *7*(1), 11456.

Labidi, S., Firmin, S., Verdin, A., Bidar, G., Laruelle, F., Douay, F., ... & Sahraoui, A. L. H. (2017). Nature of fly ash amendments differently influences oxidative stress alleviation in four forest tree species and metal trace element phytostabilization in aged contaminated soil: A long-term field experiment. *Ecotoxicology and Environmental Safety*, *138*, 190–198.

Li, J., Zhou, X., Yan, J., Li, H., & He, J. (2015). Effects of regenerating vegetation on soil enzyme activity and microbial structure in reclaimed soils on a surface coal mine site. *Applied Soil Ecology*, *87*, 56–62.

Malhi, K., Mukhopadhyay, S. C., Schnepper, J., Haefke, M., & Ewald, H. (2010). A zigbee-based wearable physiological parameters monitoring system. *IEEE Sensors Journal, 12*(3), 423–430.

Malinowski, D. P., Zuo, H., Belesky, D. P., & Alloush, G. A. (2004). Evidence for copper binding by extracellular root exudates of tall fescue but not perennial ryegrass infected with Neotyphodium spp. endophytes. *Plant and Soil, 267*(1), 1–12.

McCullough, P. R., Stys, J. E., Valenti, J. A., Fleming, S. W., Janes, K. A., & Heasley, J. N. (2005). The XO project: searching for transiting extrasolar planet candidates. *Publications of the Astronomical Society of the Pacific, 117*(834), 783.

Mench, M., Vangronsveld, J., Beckx, C., & Ruttens, A. (2006). Progress in assisted natural remediation of an arsenic contaminated agricultural soil. *Environmental Pollution 144*(1), 51–61.

Mendez, M. O., & Maier, R. M. (2008). Phytostabilization of mine tailings in arid and semi-arid environments—an emerging remediation technology. *Environmental Health Perspectives, 116*(3), 278–283.

Minerals Council of Australia [MCA] (2016). *Mine Rehabilitation in the Australian Minerals Industry. Rehabilitation of Mining and Resources Projects as it Relates to Commonwealth Responsibilities. Canberra, Australia. (Industry Report, Submission 50).* Available online at: https://www.minerals.org.au/sites/default/files/MCA_submission_to_the_Senate_mine_rehabilitation_inquiry_28_Apr_2017.pdf (accessed July 28, 2022).

Myers, C. R., Carstens, B. P., Antholine, W. E., & Myers, J. M. (2000). Chromium (VI) reductase activity is associated with the cytoplasmic membrane of anaerobically grown Shewanella putrefaciens MR-1. *Journal of Applied Microbiology, 88*(1), 98–106.

Pérez-de-Mora, A., Burgos, P., Madejón, E., Cabrera, F., Jaeckel, P., & Schloter, M. (2006). Microbial community structure and function in a soil contaminated by heavy metals: effects of plant growth and different amendments. *Soil Biology and Biochemistry, 38*(2), 327–341.

Pinto, A. P., Sim [Otilde] Es, I., & Mota, A. M. (2008). Cadmium impact on root exudates of sorghum and maize plants: A speciation study. *Journal of Plant Nutrition, 31*(10), 1746–1755.

Rabêlo, F. H. S., Vangronsveld, J., Baker, A. J., Van Der Ent, A., & Alleoni, L. R. F. (2021). Are grasses really useful for the phytoremediation of potentially toxic trace elements? A review. *Frontiers in Plant Science, 12*, 778275. doi: 10.3389/fpls.2021.778275. PMID: 34917111; PMCID: PMC8670575.

Radziemska, M., Mazur, Z., Fronczyk, J., & Matusik, J. (2016). Co-remediation of Ni-contaminated soil by halloysite and Indian mustard (Brassica juncea L.). *Clay Minerals, 51*(3), 489–497.

Radziemska, M., Gusiatin, Z. M., & Bilgin, A. (2017). Potential of using immobilizing agents in aided phytostabilization on simulated contamination of soil with lead. *Ecological Engineering, 102*, 490–500.

Ross, G. M. (1991). Tectonic setting of the Windermere Supergroup revisited. *Geology, 19*(11), 1125–1128.

Rudin, M., & Weissleder, R. (2003). Molecular imaging in drug discovery and development. *Nature Reviews Drug Discovery, 2*(2), 123–131.

Sakiewicz, P., Nowosielski, R., Pilarczyk, W., Gołombek, K., & Lutyński, M. (2011). Selected properties of the halloysite as a component of Geosynthetic Clay Liners (GCL). *Journal of Achievements in Materials and Manufacturing Engineering, 48*(2), 177–191.

Sakiewicz, P., Lutynski, M., Soltys, J., & Pytlinski, A. (2016). Purification of halloysite by magnetic separation. *Physicochemical Problems of Mineral Processing, 52*, 991–1001.

Schimel, D. S. (1995). Terrestrial ecosystems and the carbon cycle. *Global Change Biology, 1*(1), 77–91.

Schindlbacher, A., Rodler, A., Kuffner, M., Kitzler, B., Sessitsch, A., & Zechmeister-Boltenstern, S. (2011). Experimental warming effects on the microbial community of a temperate mountain forest soil. *Soil Biology and Biochemistry, 43*(7), 1417–1425.

Seaker, E. M., & Sopper, W. E. (1988). Municipal sludge for minespoil reclamation: I. Effects on microbial populations and activity (Vol. 17, No. 4, pp. 591-597). American Society of Agronomy, Crop Science Society of America, and Soil Science Society of America.

Shen, Y., Chen, Y., & Li, S. (2016). Microbial functional diversity, biomass and activity as affected by soil surface mulching in a semiarid farmland. *PLoS One*, *11*(7), e0159144.

Singh, J. S., & Gupta, V. K. (2018). Soil microbial biomass: A key soil driver in management of ecosystem functioning. *Science of the Total Environment*, *634*, 497–500.

Sparling, G. P. (1997). Soil microbial biomass, activity and nutrient cycling as indicators of soil health. In Pankhurst, C., Doube, B. M. Gupta, V. V. S. R. (Eds) *Biological Indicators of Soil Health, 101*, 97–119.

Steudel, B., Hector, A., Friedl, T., Löfke, C., Lorenz, M., Wesche, M., & Kessler, M. (2012). Biodiversity effects on ecosystem functioning change along environmental stress gradients. *Ecology Letters*, *15*(12), 1397–1405.

Touceda-González, M., Álvarez-López, V., Prieto-Fernández, Á., Rodríguez-Garrido, B., Trasar-Cepeda, C., Mench, M., … & Kidd, P. S. (2017). Aided phytostabilisation reduces metal toxicity, improves soil fertility and enhances microbial activity in Cu-rich mine tailings. *Journal of Environmental Management*, *186*, 301–313.

Ulrich, S. M., Kenski, D. M., & Shokat, K. M. (2003). Engineering a "methionine clamp" into Src family kinases enhances specificity toward unnatural ATP analogues. *Biochemistry*, *42*(26), 7915–7921.

Vance, E. D., Brookes, P. C. & Jenkinson, D. S. (1987). Microbial biomass measurements in forest soils: determination of kc values and tests of hypotheses to explain the failure of the chloroform fumigation-incubation method in acid soils. *Soil Biology and Biochemistry*, *19*(6), 689–696.

Vangronsveld, J., & Cunningham, S. D. (1998). *Metal-Contaminated Soils: In Situ Inactivation and Phytorestoration*. No. 628.55 V3. Springer, Berlin.

Ward, J., ed. (2013). *The Ecology of Regulated Streams*. Springer Science & Business Media.

Wang, J., Zhang, C. B., Ke, S. S., & Qian, B. Y. (2011). Different spontaneous plant communities in Sanmen Pb/Zn mine tailing and their effects on mine tailing physico-chemical properties. *Environmental Earth Sciences*, *62*(4), 779–786.

Wang, J., Zhang, C. B., Chen, T., & Li, W. H. (2013). From selection to complementarity: The shift along the abiotic stress gradient in a controlled biodiversity experiment. *Oecologia*, *171*(1), 227–235.

Wang, X. Y., Ge, Y., & Wang, J. (2017). Positive effects of plant diversity on soil microbial biomass and activity are associated with more root biomass production. *Journal of Plant Interactions*, *12*(1), 533–541.

Wang, C., Li, H., Sun, X., & Cai, T. (2021). Responses of soil microbial biomass and enzyme activities to natural restoration of reclaimed temperate marshes after abandonment. *Frontiers in Environmental Science*, *9*, 701–610.

Ward, P. J., Jongman, B., Weiland, F. S., Bouwman, A., van Beek, R., Bierkens, M. F., … & Winsemius, H. C. (2013) Assessing flood risk at the global scale: model setup, results, and sensitivity. *Environmental Research Letters*, *8*(4), 044019.

Webster, P. J., Ziebeck, K. R. A., Town, S. L., & Peak, M. S. (1984). Magnetic order and phase transformation in Ni2MnGa. *Philosophical Magazine B*, *49*(3), 295–310.

Weissleder, R., & Ntziachristos, V. (2003). Shedding light onto live molecular targets. *Nature Medicine*, *9*(1), 123–128.

Wen, W., Zhao, H., Ma, J., Li, Z., Li, H., Zhu, X., … & Liu, Y. (2018). Effects of mutual intercropping on Pb and Zn accumulation of accumulator plants Rumex nepalensis, Lolium perenne and Trifolium repens. *Chemistry and Ecology*, *34*(3), 259–271.

Windham, L., Weis, J. S., & Weis, P. (2001). Patterns and processes of mercury release from leaves of two dominant salt marsh macrophytes, Phragmites australis and Spartina alterniflora. *Estuaries*, *24*(6), 787–795.

Xiao, Y., Huang, Z., & Lu, X. (2015). Changes of soil labile organic carbon fractions and their relation to soil microbial characteristics in four typical wetlands of Sanjiang Plain, Northeast China. *Ecological Engineering*, *82*, 381–389.

Xu, J., Liu, B., Qu, Z. L., Ma, Y., & Sun, H. (2020). Age and species of Eucalyptus plantations affect soil microbial biomass and enzymatic activities. *Microorganisms*, *8*(6), 811.

Yin, L., Ren, A., Wei, M., Wu, L., Zhou, Y., Li, X., & Gao, Y. (2014). Neotyphodium coenophialum-infected tall fescue and its potential application in the phytoremediation of saline soils. *International Journal of Phytoremediation*, *16*(3), 235–246.

Zimmermann, N. E., Yoccoz, N. G., Edwards Jr, T. C., Meier, E. S., Thuiller, W., Guisan, A., ... & Pearman, P. B. (2009). Climatic extremes improve predictions of spatial patterns of tree species. *Proceedings of the National Academy of Sciences*, *106*(supplement_2), 19723–19728.

Zvyagintsev, D. G., Dobrovolskaja, T. G., & Lysak, L. V. (2002). Composition of soil bacterial communities: new insight from old and new technologies. *Transactions World Congress of Soil Science*, *9*, 297.

Section III

Microbial remediation

14 Remediation approaches in environmental sustainability

Babafemi Raphael Babaniyi,
Olusola David Ogundele, Ademola Bisi-omotosho,
Ebunoluwa Elizabeth Babaniyi, and
Sesan Abiodun Aransiola

CONTENTS

DOI: 10.1201/9781003394600-17

14.1 INTRODUCTION

The planet was seen in the space first time ever, in mid-20th century in form of a ball which appeared to be fragile with oceans, clouds, greenery, and soils phenomenon. Inadequacies of human to channel its operation along that phenomenon are altering planetary wholeness. Several changes due to planetary system alteration consequently evoke hazards to life. Therefore, as a matter of urgency, this new reality, from which no route of escape, should be recognized with appropriate management strategies (Morelli, 2011). In this context, question regarding sustainability is needless. So, adoption of environmental sustainability approach is the needed tool to keep the planet in its natural state at present and emphases should be strongly laid on issues with polluted and unmitigated environment (Weina and Yanling, 2022). According to agenda 2030 of United Nations with the 17 Sustainable Development Goals at large, provided a new incentive regarding the out-turn of sustainable development, hence, sustainable development is a logical process of employing natural resources with respect to environmental equity concept, together with resolutions of social equity to achieve this goal (Palomares et al., 2021). Significantly, sustainability goal has brought a healthy competition among firms promoting society development (Streimikiene et al., 2021; Zhang et al., 2022). Therefore, adequate knowledge on environmental management is a pivotal in this concept, as it ensures proper restoration of the planet (Fu et al., 2022). The coming together of personnel or experts to rub minds is in the right direction to significantly adduce a new approach toward revitalization of the environment (Olabi et al., 2022).

Climate change is obvious, global greenhouse gases emissions are as result of anthropogenic activities and rapid rise in emerging population along with the demand come with development. Despite stringent efforts to combat generation of greenhouse gases to reduce global emission, developed nations are the leading emitters of greenhouse gases (Gloet and Samson, 2022). Several countries that are most prone to climate change include industrialized zones and under developed country. Many at times, these countries rely on sensitive aspect of the climate, particularly fishing and farming and their adaptive strength diverse because of persevering poverty. The connection between climate change, poverty, political, and security risks together with their relevance should be considered in environmental rehabilitation program (Jewell et al., 2022). There is evidence of continuous declination of biodiversity globally in despite of success recorded in the sustainability scheme. In variably, the rate at which species extinct global is alarming which is approximated to be 1,000 times the number of natural species. Studies showed obvious escalating decrease of species in an ecosystem globally arising from environmental dilapidation (John, 2010; Yang et al., 2022).

Global sea level rose to about 1.7 mm per year in the 20th century, due to increase in the volume of ocean water as a result of sudden rise of temperature, this contributed immensely to the ocean volume with the inflow of water from melting glaciers. For the past few years, there have been observation about accelerating rise of sea-level estimated thus: 3.1 mm/year which correspond with data analysis from satellites and tide gauges, in relation to increasing input from ice sheets of Greenland and Antarctica. Sea level is forecasted to increase appreciably in the coming century if mediation approach is not embarked upon (John, 2010). The IPCC in 2007, forecasted 0.18–0.59 m rise beyond 1990 level towards the tail of the century. Hence, investigation showed that, ocean level increases more than the 2007 projected estimate (John, 2010). Unabated greenhouse gases generation contributed to elevated volume of global sea-level in the range of 1.0–2.0 m, by 2,100. If this occurs, acidification of ocean could be a direct impact of atmospheric release of CO_2. Emergence of industrial revolution had prompted the accumulation of CO_2 in the oceans leading to reduce concentration of CO_2 in the atmosphere consequently, alter the ocean chemistry. This phenomenon could be a litmus revealing possibility of ocean acidification becoming a prime threat to several lower animals, microorganism and can go on to affect food webs and ecological systems, for instance, tropical coral reefs. Normally, concentration of CO_2 in the atmosphere above 450 ppm will cause raise in the temperature along polar area in the oceans which results in the melting of ice, the consequence will be more tensed in Arctic region. However, there have been observation about weight loss in planktonic Antarctic calcifies shell. Basically, the momentum at which chemistry of ocean changes is rapid compared to ocean acidification experienced previously, this is an extinction signal in the history of Earth's vegetations that has to be averted through secondary resuscitation or natural attenuation of environmental phenomenon.

Several factors have been associated with trends of event shaping the nature globally, and some of the global megatrends extend to technological, social, economic, political, and environment concept. Some of the major trend is accelerating urbanization, development of technological innovation, establishment of market, shifting economic power above all, and climate changing. According to united nations population's report of 1960, world statistic was 3 billion, which is about 7.8 billion today and could be above 9 billion by 2,050 (John, 2010) (Figure 14.1).

14.1.1 SOME OF THE FACTOR RESPONSIBLE FOR GLOBAL MEGATRENDS INCLUDE

- Divergence in global population increasing trends: migration, aging population, and development;
- Living in an urban city: expand cities with spiraling pattern of consumption;
- Transforming style of global disease and the possibility of new pandemics;

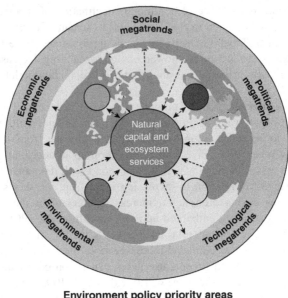

Environment policy priority areas

- Climate change
- Nature and biodiversity
- Natural resources and waste
- Environment, health and quality of life

FIGURE 14.1 A selection of global drivers of environmental change (John, 2010).

- peedy technologies: moving toward unknown;
- Ever growing economy;
- Shifting global power: from a unit-polar to a multi-polar world;
- Strong competitive affinity for resources globally;
- Reduced conservation of natural resources;
- Increase in the severity outcome of climate change;
- unmitigated environmental pollution; and
- Regulatory body and attitude of government: elevated fragmentation, converging result.

Knowledge management plays a vital role in sustainability context, and is central in the viability of the economy (Ikram et al., 2022). Due to negligent compliance of sustainable framework, method of sustainable development should be tailored toward adequate knowledge management and awareness of the consequence of not mediating the impact of change arising from climate (Frolova et al., 2021). Several innovative developments have been jeered toward sustaining the environment. Recently, environmental sustainability guidelines have been a perfect and ethical consideration toward businesses decision-making especially, in terms of siting an industry (Sénéchal and Trentesaux, 2019). Subsequently, green technologies would further strengthen the possibility of achieving environmental sustainability goal (Appolloni et al., 2022). Green technology approach cannot only function toward environmental resuscitation, but a significant means of businesses thriving (Ogbeibu et al., 2021). Currently, many of the studies on green technology shifted attention to direct impact on particular element, instead of accounting for both micro and macro environment (Ahmad et al., 2021).

In an attempt to ascertaining environmental lethal dosage and planetary threshold of the Earth, scientists are making effort to pin point the complex relationship between bio-geophysical processes which determines Earth's ability towards self-regulation. Also, ecologists have indicated threshold in order of importance in the process of ecosystem, when exceeded triggered fundamental change in the proper functioning of the ecosystem. In the recent past, a body of researchers also suggested that some planetary limit humanity should not cross to prevent disaster due to change in environmental contents. According to their suggestion, three major boundaries have been crossed include: loss of biodiversity, climate change together with interference of human with the nitrogen cycle (John, 2010). Therefore, acknowledging the new trend and the uncertainties surrounding the change in planetary system, of course, it is not departing anytime soon, it is here to say. Hence, humanity must work profusely to restore nature.

14.2   ENVIRONMENTAL POLLUTION

The existence of environment is as aged as nature, it is a collective term used to describe the space in which organisms flourish and turn into living avenue of existence to all living and non-living things. Introduction of substances into the environment by humans that could endanger human health, damage ecosystems and living things, destroy buildings or other structures, or interfere with the environment's ability to be used for its intended purposes is regarded as pollution (Ramamohana, 2017) Pollution increases risk of injury, damage to man's health is not limited to physical harm; it also includes offenses to his senses and damage to his property, therefore even smells and sounds that do not physically hurt men can still be considered pollution. Pollution can damage the living things directly by impairment to their health or through the disruption of the ecological systems they are a part of. Pollution could be natural or artificial. Natural occurrences like earthquakes, floods, droughts, cyclones, and many more naturally occurring phenomenon that frequently contaminate the environment are the natural forms of environmental pollution (Ramamohana, 2017). While human activity that contaminate the environment are known as artificial pollution. The various types of environmental pollution include air pollution, water pollution, land contamination, food pollution, radioactive pollution, and more.

Crisis in environment is as a result of ecological and environmental changes brought on by the past and current economic and technological growth. In reality, the 21st century is characterized by socioeconomic, scientific, and technologically advanced, also by grave environmental issues. The ruin of habitats as a result of industrial, urban, and agricultural activities, the decrease couple with the loss of ecological populations as the demerit of excess use of noxious pesticides and herbicides, and the extinction of diverse species as a result of environmental degradation brought on by various forms of pollution, rapid rate of resource exploitation, couple with growing dependence on energy-intensive and ecologically damaging technologies.

There has been a noticeable increased interest in the environment's quality, the disruption of the earth's natural ecosystems, and the depletion of resources during the past ten years due to how quickly environmental deterioration brought on by man is severely affecting the lives of ordinary people (Gu et al., 2020). Also, the process of natural condition modification results in a succession of changes in the biotic and abiotic components of the environment, man's economic activities have a wide range of diverse and extremely complicated effects on the environment. The two major effects that human have on the environment are direct and indirect, because man is aware of the results, both positive and negative, of each program which is initiated to change or modify the natural environment for the economic development of the region concerned, intentional impacts of human actions are preplanned and premeditated. The impacts of anthropogenic changes in the environment become apparent quickly and can be reversed (Khomenko et al., 2021).

On the other hand, human activities that are intended to hasten the rate of economic growth, particularly industrial development, have indirect effects on the environment that are not anticipated and unplanned (Ramamohana, 2017). After a considerable amount of time, when they have accumulated, the indirect effects are felt. The total natural environmental system may change as a result of these indirect human economic activity effects, and the subsequent chain reactions occasionally deteriorate the environment to the point that it becomes life-threatening for people (Manisalidis et al., 2020).

14.2.1   CAUSES OF ENVIRONMENTAL POLLUTION

Today's environmental pollution issue is a complicated result of pressures linked to numerous interrelated elements. There are undoubtedly many different and opposing opinions regarding what might be the fundamental causes of the environmental disaster. The root cause of environmental damage cannot be attributed to a single factor. Simultaneously, activities of these factor could take place and their balance might differ with regards to location over time, the following causes can be identified as the typically underlying factors.

14.2.1.1   Population expansion

Many contemporary scholars believe that population expansion is the primary source of many human issues. Environmental deterioration is also affected by this observation. The multiplier effect of population growth will necessitate a commensurate increase in all requirements for human existence. To meet the daily necessities of life, population increase necessitates abnormal resource exploitation. It leads to population increase in metropolitan areas and migration of people, which opens the door to new issues with human health, ecology, and sustenance (Ramamohana, 2017).

14.2.1.2   General wealth and economic expansion

Affluence with individual consumption rate of products, is a significant factor between poor and the rich, resources, and the environment. Because their wealth is out of proportion to the resources demanded. Therefore, their desires are driven by wealth and not by human needs, the wealthy have an increased propensity to waste resource, which is stifling development in the result of products and services in both developed and developing nations. Amazingly, in spite of notable environmental effects, issues of environmental dilapidation are really attributed to component of affluent. Meanwhile, the lower class (poor) and poverty are usually taken be the culprit of devastated environment (Ramamohana, 2017).

14.2.1.3 Modern technology

Basically, the recent environmental issues could be associated with the type technology adopted in production processes. Development of technologies with significant environmental effects has superseded less harmful ones. This element has played a significant role in the production of synthetic and nonbiodegradable materials including plastics, chemical nitrogen fertilizers, synthetic detergents, synthetic fibers, giant automobiles, petrochemical and other businesses that harm the environment, as well as "disposable culture." As a result of a counter ecological pattern of productive expansion, environmental crises are unavoidable (Slama et al., 2019; Dabrowiecki et al., 2021). Technologies that are eco-friendly still available, unfortunately, are neglected as a result of efficiency and time.

14.2.1.4 Deforestation

Stability of the environment and ecological balance greatly depend on the condition of the forests because, they constitute primary biotic makeup of natural network of the environment. Significantly, current global economies do not favor environment and ecological wellness of natural vegetation. For instance, forests and grasslands have been destroyed so rapidly globally, have reduced noticeably with the birth of many challenges in the environment, include, rapid loss of soil via rain, erosion and wildfires. At both the global and regional levels, shifting cultivation, forests conversion into pastures, overgrazing, lumbering activities, river projects for multipurpose application, and the likes are the main drivers of deforestation. Deforestation causes a number of issues, including environmental degradation through accelerated soil erosion, increased river sediment loads, reservoir siltation, increased frequency and severity of hoods and droughts, altered patterns of precipitation distribution, intensified greenhouse effect intensity, and increased destructive power of atmospheric storms (Ramamohana, 2017). Economic loss from crop damage brought on by more frequent floods and draughts, a decline in agricultural output due to the ruin of fertile soils, a reduction of availability raw materials to the construction industry. As a result, deforestation has a number of negative side effects on the environment (Dabrowiecki et al., 2021).

14.2.1.5 Industrial development

The human society has experienced economic prosperity thanks to rapid industrial development. It has also given the socioeconomic system a new dimension and given the citizens of industrialized nations material comfort, but it has also exacerbated environmental issues on a massive scale. In fact, the glittering results of industrialization have so altered public perception that it is now viewed as the standard of modernity and a fundamental component of a country's socioeconomic development. Quick rates of industrialization led to rapid rates of resource extraction and higher industrial production. Two aspects of industrial development have produced a number of deadly environmental issues as well as widespread environmental issues and ecological imbalance on a local, national, and international scale. Besides useful outcome, industries also generate many unwanted products, like industrial wastes, noxious gases and chemicals precipitates, aerosol and smokes, which harm the environment by contaminating the air, water, land, soils, and other resources. The industrialized nations have raised the number of pollutants that manufacturers emit into component of the environment (water, soil, and air) to an extend of environmental deterioration to the point of no return and have put human society in danger of disintegrating (Nazar et al., 2022). The negative impact of industrialization might alter the totality of the natural world, with sometimes lethal consequences for human civilization. The majority of industrialization's negative effects are connected to pollution and the damage of the environment. Chemical fertilizers, herbicides, and insecticides (outputs of the chemical industries) affect the physical and chemical properties of soils as well as the food chains and food webs by releasing hazardous substances into the environment through application. Similar to this, the discharge of wastes from industries into ponds, tanks, lakes, rivers, and oceans contaminates the water, results to a number of diseases and animal fatalities, and upsets the biological balance of the aquatic ecosystem (Ramamohana, 2017). Large amounts of contaminants, such as ions of chlorine, sulfate, bicarbonate, nitrate, sodium, magnesium, and phosphate, are released into rivers and lakes through sewage effluents as a result of increasing industrial expansion, poisoning the water. The environment is negatively impacted in a number of ways by the emissions of various gases, smokes, ashes, and other particles from factory chimneys. Coal and petroleum combustion have changed the makeup of the atmosphere's natural gases by increasing the amount of carbon dioxide in the atmosphere.

14.2.1.6   Urbanization

People migration from rural to urban areas, as well as the birth and growth of new urban centers brought on by industrial development, is to blame for the rapid exploitation of natural resources and various forms of environmental degradation and pollution in both developed and developing nations. The developed nations of the globe have already reached their maximum level of urbanization. Large slum regions have developed and grown as a result of the concentration of people in crowded metropolitan areas brought on by the accumulation of wealth and the expansion of economic and employment opportunities in urban centers. In actuality, growing urbanization entails a greater concentration of people in a finite number in a place that leads to structural increase, creation of streets, generation of more sewage, storm drains, and automobiles, as well as factories, urban garbage, aerosols, smokes, and dusts, sewage waters, and other environmental issues (Nazar et al., 2022). For instance, the growing urban population utilizes a tremendous amount of water for a variety of applications. Because urban effluents are permitted to be discharged into streams and lakes, utilized waste water, such as sewage water, pollutes them if left untreated. Urban areas grow riskier from the perspective of pollution and environmental issues when integrated with industrial sectors. Huge amounts of gases and aerosols are emitted from industries and chimney of vehicles creating domes of dust over the cities. These dust domes over the cities create "Pollution Domes." Due to high air pollution from gases and aerosols released by vehicles, companies, and home appliances, the quality of the air has been rapidly declining as a result of urban and industrial growth. Large quantities of municipal solid trash contribute to environmental issues in addition to industrial wastes (Ahmad et al., 2021).

14.3 CLASSES OF REMEDIATION TECHNOLOGY

Due to the prevalence of numerous pollutants and the presence of chemicals in quantities that pose health hazards, water, soil, and air contamination is one of the major environmental issues affecting human and animal health (Thompson et al., 2019). Additionally, many places obstruct urban and economic development in nations with large population densities (Hu et al., 2006). The most frequent types of contaminants in water, air, sediments, and soils are metals and organic contaminants; if left untreated, these contaminants could end up becoming serious threat to life (Liu et al., 2018).

Over 120,000 locations in the United States and 3,500,000 hectares of farmland in China are impacted, respectively. More than 350,000 polluted places and an estimated 2.7 million sites that could be contaminated have been found throughout Europe (Song et al., 2017). The anticipated annual cost of decontaminating the regions that have already been identified is 6.5 billion Euros. Over 20 million hectares of land require rehabilitation, with over five million sites being contaminated with heavy metals (Panagos et al., 2013). The actual number of affected locations in Brazil is currently unclear; however, environmental agencies have made efforts to collect this information.

Numerous attempts to mitigate or control contaminants from the application of various methods have been explored to mediate polluted site and water. Following the removal of pollutant, monitoring with assessment procedures could be used in gauging success of the employed technology. Hence, using a remediation technology and then going through the entire management approach after the remediation are both steps in dealing with the issue (Song et al., 2017).

In situ technologies and ex situ technologies are the two primary categories of technology for the remediation of polluted environments. The removal of contaminants from the site of occurrence is known as in situ remediation. Without physically moving the soil or silt, this kind of approach seeks to eliminate toxins from it (Song et al., 2017). Ex situ technologies involve treating or excavating soil or sediment away from the contaminated site. While method of removal is more beneficial in comparison to the process of containment, in situ remediation processes often offer larger economic benefits than ex situ remediation processes (Kuppusamy et al., 2016). Additionally, the in situ methods decrease worker and general public exposure to the contaminated environment. Given the size of the area and the ratio between cost and benefit, in situ remediation techniques are the only choice for treating a sizable polluted site. The in situ technique is used for vast areas of soil and contaminated sediments since it will affect the site less, is easier to operate, and costs less than the ex situ process of treatment (Song et al., 2017).

Many of in situ and ex situ strategies were created recently to clean up the contaminated sites with sediments or mitigating the negative effects. Depending on the method used for remediation, these strategies can be divided into five groups: physical, chemical, biological, thermal, and combined (Huang et al., 2012). Additionally, they can be split into three categories based on how they are carried out: containment, conveyance, and transformation. For the treatment of several inorganic and organic pollutants in the soil, other writers consider the combined approaches to be a superior alternative (Khalid et al., 2017).

14.3.1 PHYSICAL PROCESSES

These procedures involve the physical isolation or removal of pollutants from the environment. The key technologies employed are as follows.

14.3.1.1 Vapor or gaseous extraction

Typically, in unsaturated soils, the method is used in situ to extract volatile and sub-volatile organic pollutants in the soil. The gases are siphoned into the soil pores using vertically or horizontally placed extraction wells in the ground. Before being discharged into the air, these gases need to be treated. For improved control of the volatile production, a contaminated site coverage system can be built. The vapor pressure and solubility of the pollutants affect how effective the technique is (Mulligan et al., 2001). In order to execute the procedure correctly, the humidity, amount of organic matter, porosity, and soil permeability must all be considered. It can also be used to extract gases from the oxygen exchange technique, which involves injecting air or another gas into the subsurface to help with the volatilization or even decomposition of pollutants (Albergaria et al., 2012).

14.3.1.2 Surface capping

With this method, the contaminated area is simply covered in a substance with limited permeability. By preventing the flow of water, this lowers the danger of exposure for people in the contaminated region as well as the mobility of toxins in the soil. Many researchers do not regard it to be a technology, but rather just a method for keeping toxins contained. The soil should instead be used for other civil uses, like carparks, as it fails to perform its environmental tasks (Tomaszewski et al., 2006).

14.3.1.3 Electro-kinetic remediation

Through electrochemical adsorption, this approach can be utilized to remove both organic and inorganic soil, typically in situ. By the use of electrode, the soil is charged with low-density electrical current, which causes the movement of cations to the cathode and anions to the anode via electrical field establishment. Low electrical conductivity and saturated and partially saturated soils make the method more effective (Reddy, 2013).

14.3.2 CHEMICAL PROCESSES

The major techniques for removing pollutants include chemical reactions, whose mechanisms include reduction, oxidation, ion exchange, adsorption, and a combination of these (Song et al., 2017). The following are some of the main chemical technologies.

14.3.2.1 Soil washing

Aqueous solutions are used in this ex situ method to remove pollutants from the surrounding area. The extracted dirt is stirred after being combined with the extraction solution. The cleaned site can be returned to its original location after being washed and should be treated with the media of extraction. The procedure could be carried out in situ (soil flushing), at any rate, the solution for washing is injected via reservoirs and pollutants are solubilized in the extraction solution (Morillo and Villaverde, 2017). This needs to be collected after the polluted area, and the solution needs to have its surface treated. In situ cleaning can transport toxins to the saturated area peradventure, the contaminant gets into groundwater; it becomes necessary to treat groundwater (Morillo and Villaverde, 2017).

14.3.2.2 Stabilization and solidification

By injecting chemicals to transform contaminants to less noxious substance, this approach, based on immobilization of chemical, collects pollutants in situ or ex situ. This method merely stops contaminants from moving in the soil. It is frequently utilized for contaminants that are metallic, radioactive, or extremely poisonous. The potential plant support and microorganisms in situ is minimal because the utilization of the soil is completely altered when the technology is applied (Tajudin et al., 2016). The pollutant is contained in a solid state by the solidifying process. The major materials utilized for solidification are thermoplastics, cement, asphalt, and gray steering wheels. Mud cementing could be introduced in situ with pressure through insertion also could be by mixing with a bulldozer. The resulting hardened block is impermeable and stops pollutants from moving through it. Monitoring is required because the solid matrix may weather over time (Song et al., 2017). While the pollutant is immobilized during the stabilization process, the soil is not solidified. In this procedure, chemicals are added to the soil and pollutants are subjected to physical–chemical processes that result in the formation of precipitated, complex, or absorbed contaminants, decreasing their mobility (Tajudin et al., 2016). There are different materials that could be used, such as lime, phosphates (apatite, ammonium phosphate, bone flour, and hydroxyapatite), alkaline agents (calcium hydroxide and ash flywheels), minerals, or iron oxide (zeolites, goethite, bauxite, and silica gel), and organic matter (peat, bio coal, manure, and chitosan) (Xie et al., 2015).

14.3.2.3 Nanotechnology

For the treatment of polluted sites with inorganic and organic, zero valent metals in nanoscale (palladium, nickel, and iron) have been employed extensively. Particularly, this technology is reliable and can be adopted in situ and ex situ because of its tiny size (usually <100 nm) together with the high surface area coverage potential. Nanomaterials have the ability to dehalogenate organic chemicals and stabilize or decrease inorganic pollutants like arsenic and chromium (Crane and Scott, 2012). Zero valent iron particles, one of the many types of nanoparticles, have been used the most because of their low toxicity and cheap production costs. Regarding this technology's mobility in the water and soil, toxicity, response time for pollutants, time of reactivity, and further research is still needed (Yan et al., 2013).

14.3.3 BIOLOGICAL PROCESS

Through biological processes, living organisms like animals, plants, and microorganisms clean up the environment. These techniques include degrading processes or pollutant transformation. These are the major physiological technologies:

14.3.3.1 Bioaugmentation

Addition of genetically modified microorganism with certain catalytic activity, strain, or consortia enrichment of microbe populations increase their productivity and speeds up the breakdown of pollutants. In order to speed up the deterioration of contaminants on the site, additional organisms which were not in the soil already including sediments, or water are supplied. These bacteria were chosen based on their morphology and ability to break down the pollutant metabolically (Abdulsalam et al., 2011).

14.3.3.2 Bioventilation or bioventing

By introducing oxygen to the soil holes, this approach encourages the development of microorganisms. The availability of oxygen accelerates the metabolic function of microbes in organic pollutants by acting as a media, starting bio decomposition process. Since the metabolism has already begun, oxygen could accept electrons for the generation of power (Lim et al., 2016). The effectiveness of this procedure depends on the soil's ability to allow air to pass through it, therefore its particle size and permeability are crucial factors to take into account (Thomé et al., 2014).

14.3.3.3 Vermiremediation

Worms are used in this method, which has only received a limited amount of research to date, to remove toxins from the soil. Additionally, the worms have the ability to ingest and digest organic pollutants, which changes the soil's structure, biomass, and microbial activity. Additionally, they help co-metabolize soil microbes, enhancing nutrients and minerals that raise the polluted site's organisms live (Rodriguez-Campos et al., 2014).

14.3.3.4 Biostimulation

This entails modifying a number of environmental factors, including nutrients (phosphorus, nitrogen, and potassium), raising dissolution of several pollutants, addition of biopolymers, concentrate acceptors of electron through oxygen addition, or monitoring humidity and temperature to create the ideal requirements for the degradation of microorganisms (Cecchin et al., 2017). Because they might change the surface loads of the soil particles, the application of biostimulation with soil nutrients must be done carefully. This can lessen the interaction between the contamination and the soil, increasing the amount of the contaminant in the free phase of the soil where it may be more easily leached after precipitation (Abed et al., 2015).

14.3.3.5 Phytoremediation

In order to cleanse the soil or sediment, this method uses vital plant's part or plants connected to microbial populations (Lim et al., 2016). It is largely accepted by the general public and is acknowledged as a green technology. Organic and inorganic pollutants can be stabilized, removed, or degraded using it (Song et al., 2017). It is possible to divide this technology into:

14.3.3.6   Phytodegradation

This technique entails the uptake of pollutants by the roots, followed by their breakdown via plant metabolism or the secretion of enzymes to quicken biodegradation within the roots (Germida et al., 2002).

14.3.3.7   Phytoextration

This technique involves the roots removing pollutants, which causes them to accumulate in the tissues of the plants and need the removal of the plant. This method is frequently used to get rid of metallic and radioactive contamination (Song et al., 2017).

14.3.3.8   Phytostabilization

The (often metallic) pollutants in the plant's root system are immobilized by this method. Adsorption, precipitation due to a pH change, the creation of metallic complexes, or a change in the redox state of the pollutants are among ways that contaminants might become immobilized (Lim et al., 2016).

14.3.3.9   Phytovolatization

This process also involves the roots of the plant absorbing pollutants, which are then metabolized, transported, and volatilized by the surfaces. This is only appropriate for soluble and volatile pollutants (Lim et al., 2016).

14.3.3.10   Rhizodegradation

Because of the increase in root oxygenation and a moderate humidity level, this mechanism can be categorized as plant-biodegradation and produces more favorable conditions for bioremediation (Lim et al., 2016). Additionally, the exudates from the roots, which include sugars, aminoacidic chemicals, and other substances, promote microbial proliferation and, as a result, the breakdown of contaminants.

14.3.4   THERMAL PROCESS

Thermal processes entail heating the subsurface, which causes pollutants in the soil or sediment to move around, volatilize, or even be destroyed. Heating by conductive, electric resistance heating, steam heating, and heating through radio frequency are all types of heating techniques (Samaksaman et al., 2016). These procedures make use of the following technologies.

14.3.4.1   Thermal desorption

Increased temperature is used in this method to remove pollutants. Temperature rise causes organic pollutants' vapor pressure to rise, which results in volatilization and, ultimately, contaminants dissolution in the soil. Before being released into the atmosphere, collected gases must be treated. It is critical to assess the contaminant's vapor pressure and concentration, humidity, density, treatment time, and particle distribution for the procedure to work as effectively as possible (Samaksaman et al., 2016). Both in situ and ex situ methods can be used for thermal desorption. In situ heating involves drilling holes in the ground and inserting rods that use electricity to heat the soil. This allows for the heating of greater depths. The method can also be used with heating sheet, like thermal blanket, in the case of more surface-level contamination. Soil sample is used for ex situ treatment, and a desorption device that can be put on the polluted soil site is used to extract contaminants. Pre-treatment of the soil during this procedure to get thinner soil separated out (which needs high temperature) also, get rid of extraneous objects is encouraged. To ensure the approach operates as effectively as possible, it might also be required to lower the soil's humidity (U.S. EPA, 2012).

14.3.4.2 Vitrification

In this method, heat is used to turn the dirt into a glassy matrix. The contaminated soil is heated to a high temperature (about 1500°C), which melts it and creates an inert matrix that traps the toxins. Both in situ and ex situ applications of the method are possible. The heating procedure may result in gas production for organic pollutants. Extremely hazardous and even radioactive pollutants can be removed with this technique. The method is quite costly and is not recommended for soils with significant levels of organic matter or dampness. The soil can only be used for agricultural purposes after then (Meuser, 2012).

14.3.5 COMBINED PROCESSES

Additionally, to these methods, combination procedures could be employed. These require mixing chemicals, physical, and also biological approach with a view to improving remediation effectiveness or address issues that may arise when employing one technology alone. The combined form of the quick but costly physicochemical process and the biological process is a novel approach that has been examined (Cecchin et al., 2017). Nanobioremediaço, a combination of these two processes, uses microorganisms for the last stages of soil degradation while initially treating polluted soil with nanoparticles. As less nanoparticles are supplied and the already-existing microorganisms are utilized, this remediation method ends up being more environmentally friendly (Cecchin et al., 2017).

14.4 AN INSIGHT TO GREEN REMEDIATION TECHNOLOGY IN ENVIRONMENTAL SUSTAINABILITY

The adoption of throughput environmental revitalization implementation with options incorporation to acquire net cleanup actions benefit in the environmental is regarded as green remediation according to U.S. Environmental Protection Agency (U.S. EPA, 2022). The demand mount on the environment during cleanup activities has been reduced with the advent of green remediation strategy, which prevented collateral damage of the environment (U.S. EPA, 2008). The influence of footprint includes the long existing effect on the medial of environmental components such as; pollution of air due to noxious or contaminants like lead and particulate matter;

cycle water variation along local and regional hydrologic network; erosion of soils, depletion of nutrients and changes in geochemical subsurface of the soil; and population reductions with increased ecological diversity and emission of gasses, namely methane (CH_4), carbon dioxide (CO_2), and nitrous oxide (N_2O) among other gasses of greenhouse promoting switch in climate. Environmental sustainability can be achieved via throughput investigation, environmental action policies, and monitoring approaches in site remediation in spite of the chosen strategy for the cleanup (U.S. EPA, 2008). As a result of incentive in cleanup, green remediation advances and to a large gap, it proffers a significant advantage of cleanup activities include: saving of project costs and an increase in the scope of long use or reuse property options without breaching aims of cleanup (U.S. EPA, 2008).

Green remediation strategies depend on sustainable development in which environmental protection does not rule out economic enlargement. The Agency has assembled facts from the scope of EPA programs backing sustainability within the environmental bracket; water, ecosystems and agriculture; environment and energy; toxics and materials (U.S. EPA, 2022). Various schemes, apparatus, and incentives are accessible to assist governments, businesses owners, communities, and each person to be an environmental seneschal, making sustainable decisions, and potentially manage the available resources (U.S. EPA, 2008). Employing green remediation best management practice assist the acceleration of environmental protection pace in lines with strategy of Agency's approach towards enhancing performance of business sectors of the environment. However, green remediation relies on environmentally careful traditions unanimously adopted by individual and business sections, in tune with the US Agency's Sectors guidelines to facility adoption of state-of-the-art procedures for: conservation of water and improvement of water quality, increase energy efficiencies, eradication noxious, management of waste generation, and decrease in greenhouse gases emission and air pollutants (U.S. EPA, National Center for Environmental Innovation, 2007). Rapid climatic change increase concerns which have prompted significant attempt globally to decrease greenhouse gases emissions caused by consumption of fossil fuel and bush burning. Meanwhile, one major aspect of EPA's advancing operation towards green remediation accords considerably emphasis strategies to decrease consumption of energy and emissions of greenhouse gases such as: developing an optimum efficient treatment systems suitable for the technique, usage of renewable energies like wind and solar to tackle power requests of energy-intensive treatment systems or auxiliary appliances, Usage of alternative fuels to power machineries and vehicles, electricity generation from byproducts like methane gas, or collaborating with firm generating power from renewable resources (U.S. EPA, 2008).

Green remediation techniques further emphases the need to protect the natural hydrologic network of the earth. Best remediation management practices involve measures of conserving water, control of stormwater runoff, and process of recycling or treatment of water. Methods for maintaining balance in water level depend on the directives of federal and state ground water protection and management programs and on recent climate-change investigation by government agencies and organizations include U.S. Department of Agriculture, U.S. Geological Survey, and National Ground Water Association (National Ground Water Association online). Green remediation enhances acceptability of sustainable approaches at every site in need of environmental cleanup, the cleanup could be done by federal, state, or local cleanup programs even individually. Past activities like oil spills, leaks, and improper disposal of noxious substances contributed to contamination of landfills, water, and air in many soils across the world.

According to Sustainable Remediation Forum (SURF), sustainable remediation is said to be remediation that preserve wholeness of human existent together with the environment at the same time exploring both natural and manmade products of the environment such as communal and benefits of the economic all-round the cycle of life (U.S. SURF, 2013). In regard to the statement above, sustainable remediation focuses majorly on the environmental protection, economic benefits, with social or communal outcome of remedial processes, this idea goes deeper than stewardship during environmental cleanup, the major priority should not just center on revitalization and restoration of site but the protection of human existence. The approach of Interstate Technology and Regulatory Council (ITRC) differs; they conclude that the integrated idea of green and sustainable remediation (GSR) should identify and deal with the employment of site-specific products, processes required, technologies involvement, coupled with the methods of mitigating the risk of contaminant to receptors when taking cognizant decisions needed to balance communal goals, economic related issues, and environmental impacts (Interstate Technology & Regulatory Committee 2011).

International Organization for Standardization (ISO) has been so active in all of her programs and regulations to put in place guidelines on activities of global professional working to redeem the environment. More so, ISO involvement in GSR is in diverse forms. To start with, International Organization for Standardization Life-cycle assessment (LCA) guidelines, ISO 14040 series, was the bases for numerous instruments and measures utilized for GSR execution (The Horinko Group. 2014). The procedures for environmental analysis together with product or service impact on human wellness rely on LCA. Application of LCA in remediation pave the way for the method regarded as green or sustainable remediation. In addition, ISO 14040 expatiated LCA to be compendium and assessment of the inputs, outputs, and likely effects on the environment within living cycle. Application of LCA could be adopted in remediation processes in various forms for instance, standard to previous methodology, recognizing while placing priority on the possible means to reduce the footprint of cleanups in future, also to consider other available and achievable options of remediation when deciding the techniques to adopt (The Horinko Group, 2014).

The speed at which GSR is growing is on the rise but can be improved in terms of flexibility and affordability of the techniques with the usage of existing approaches that led to its acceptability by the general public. More so, GSR needs to be promoted the more for better usage and attainment of the intended site revitalization. Therefore, more public education, outreach communication, government and individual sensitization, and synergy among those that have awareness of GSR is vital. Also, exchange of ideas and experiences garnered during field work like Best Management Practice, guidelines for implementation and instruments, can be best managed via associations such as international conferences, SURF, ITRC, Association of State and Territorial Solid Waste Management Officials, and many other organizations. In this regard,

SURF had put it upon herself to initiate a workable guideline for studies evaluation which will also, be international database standard of studying cases. Utilization of available knowledge with previous experience is paramount in order to circumvent duplication of efforts or waste of efforts and resources (The Horinko Group, 2014).

In that wise, progress had been made with collective efforts. The development of GSR has been steady though slow but never stopped gaining momentum as the years pass by. Most of the novel technologies and procedures were developed by state-led resourcefulness and individual efforts promoting the establishment of GSR. Green remediation is gaining recognition globally with influence of business owners and partnership of associations such as SURF which provided forum for people to exchange knowledge. Various stakeholders in their capacity are working to advance this field and means of managing the risks. ASTM and ISO guidelines are perfect means to achieve better GSR application in a steady form globally. The framework together with the persistent sharing of experience will surely lead all the facilitator in the same path with better understanding of regulations. At any rate, regulatory bodies in various countries might need to project her workable standard taking clue from international standard, though, they could be flexible, however, a persistent guide pointing towards achieving same goal to safeguard human life and the environment.

14.5 REMEDIATION TECHNOLOGY AN INTERVENTION TO GLOBAL WARMING

Global warming is an observable change of climate, influenced by unusual elevation of normal temperatures of the Earth, that rework balances of weather and ecosystems for a prolong time. This phenomenon is plainly associated to the anthropogenic generation and abundance of greenhouse gases in the space, aggravating effect of the greenhouse gases. To be precise, average temperature of the earth had been on the rise by 0.8°C (33.4°F) if compared toward the tail of 19th century. Apparently, the environmental condition of the plant in the last three decades occurred to be warmer compared to previous decades accounted from the start of the statistical surveys carried out in 1,850. With the current pace of CO_2 emissions in the atmosphere, scientific projection of the CO_2 increase is at 1.5–5.3°C (34.7–41.5°F) of average temperature at the end of 2,100; therefore, action must be taken to avert negative impacts to biosphere and humanity (The Solar Impulse Foundation, 2022). Major activities of human that aggravated emission of greenhouse gases and worming of the glob could be seen in excessive fossil fuel utilization apparently, coal burning, crude oil, gases emitting carbon dioxide, and nitrous oxide are the first sources of global warming. More so, forest exploitation plays an active part regarding climate change. The function of trees is to regulate the climate via absorption of CO_2 in the atmosphere, but when they are cut or burn down, this active impact is lost consequently, accumulated carbon within the trees is emitted into the earth. Intensive farming is another cause of global warming; this can be seen in sudden rise in livestock faming to meet the demand of ever rising population, products protecting plants, and fertilizers. Notably, cattle and sheep manufacture significant volume of methane during food digestion mechanism, but fertilizers release nitrous oxide.

Management of waste such as landfills and incineration produces greenhouse with noxious gases, which are emitted into the soil, water, and atmosphere worsening the effect of greenhouse gases.

Over utilization of natural resources plays an active part in climate change particularly, emissions from international freight transport that result to warming of the globe. Climate change is a prime issue to environmental wellness all over the world, it affects ecosystems, food availability, water resources and stability of the economic in general. Consumption of energy globally is on the rise because of the demand for a better standard of living and ever rising population of the world (Agham, 2017). The good news is on the onset of environmental remediation technology as the solution, if not completely revitalized, it softens the menace caused by global warming.

Decarbonization through renewable energies, and to restore earth to its natural state, renewable form of energy should be the first approach to be considered. The use of fossil fuels should be ignored to pave way for alternatives energies such as solar, wind, geothermal, and biomass (Samer et al., 2020). Among the major energies generation technologies include photovoltaic solar power, solar thermal power for heating and cooling applications, hydropower, biomass power and biofuels (REN21 2019; Østergaard et al., 2020). With regard to transportation, adoption of alternative fuel such as bioethanol, biodiesel, bio-hydrogen, bio-methane, and others is the way forward to decarbonization (Srivastava, 2020; Osman, 2020). Hydrogen fuel generated via electrolysis employing renewable sources is a significant energy technique of environmental decarbonization, an example, electric vehicles using renewable power is another means of decarbonization (Michalski et al., 2019). Decarbonization via renewable energy usage is a key factor in returning earth to her natural form because through this approach, the effect of greenhouse gases with the consequence global warming is minimized.

Capturing and storage of carbon is another efficient remediation technique for decarbonization in industrial zones. This technique involves trapping and separation of CO_2 from processes that uses fossil fuels. Separated CO_2 is kept in geological reservoirs for many years. The capturing technology can be done in three ways: pre-combustion, post-combustion, and oxyfuel combustion. Those technique has unique procedure of trapping and separating CO_2. After capturing of CO_2, it is liquefied and transferred via pipelines to a designated storage materials that are not usable in clean water processes (Vinca et al., 2018). Meanwhile, there are some gaps in this procedure like safety regarding the secured storage and the likelihood of leakage (Ma et al., 2020). Other challenges associated with this technique are public acceptability coupled with the high cost of material and construction (Vinca et al., 2018; Tcvetkov et al., 2019). Also, captured carbon can be used to manufacture chemicals, microalgae, fuels, and concrete building materials, as well as utilization in enhanced oil recovery (Hepburn et al., 2019; Qin et al., 2020). Removal of carbons and other greenhouse gases before discharging industrial effluents to the atmosphere or water is a potential approach of mitigating the adverse impact of greenhouse gases playing an active role in warming of the globe.

Bioenergy carbon capture and storage is part of the major negative emissions strategy. Atmospheric CO_2 captured biologically via photosynthesis processes are used to produce energy through combustion (Pires, 2019). The approach could effectively decrease concentration of greenhouse gas through elimination of CO_2 content of the atmosphere. Apart from combustion path, CO_2 can be trapped during the process of fermentation of ethanol manufacturing or wood pulp gasification emission during production of pulp (Pires, 2019).

Afforestation and reforestation technology, is one of the most prominent remediation approaches globally adopted, which involve planting of trees and some species of plant in a polluted site and new site. Plants make use of CO_2 trapped in the atmosphere for their metabolism and the excess are stored within biomass, organic components that are dead and soils. However, biogenic technique of forestation with its negative discharges, contributed immensely to abatement of climate change. Forestation approach could be adopted through the establishment of new forests, regarded as afforestation and may be re-establishment of previous forest site which have been deforested or degraded that is known as reforestation. In respect to the species of trees, uptake of carbons might span a century in a tree until maturity which then slow down the rate of sequestration significantly (RoyalSociety, 2018). Mitigation of global warming through forestation offers many advantages such as biodiversity, control of food and enhancement of soil, water, and air quality (Harper et al., 2018). Storage of carbon in biomass could be for prolong time but cannot be forever because of natural and anthropogenic interruption. Sometimes, natural disasters like droughts, disease and deforestation due to human action appeared to be impacting negatively on the storage capability of plants. By and large, biogenic storage time is shorter compare to storage in geological materials (Fuss et al., 2018). Also, demand between agricultural land usage and other sectors would be an issue for effective practice of this technology but at the same time, in the case of pollution, the land has to be in good condition for a yieldable agricultural practice. According to limitations of global tropical boundary postulations, an approximated 500 Mha area of land is calculated to be adequate to deploy forestation, which could enhance global removal of carbons from the atmosphere, potential estimate ranging from 0.5 to 3.6 $GtCO_2$ year^{-1} by 2,050 (Fuss et al., 2018). To a large extend, forestation approach for the mitigation of global warming offers direct mechanism and does not necessarily require formal education, experts or sophisticated tools to deploy the technique; this implies that individual can embark on the method at their respective capacity without any itches. More so, this technology does not over looked planting of ornamental plants at homes as they contributed in no small measure in regulating wellness of the environment and capturing of carbons from the surroundings.

Biochar, of recently accepted viable material for capturing carbon with the potential of storing carbon permanently. Biochar is a product of biomass (agricultural and forest residues, specific crops) via pyrolysis conversion process of thermochemical, gasification and hydrothermal carbonization (Oni et al., 2020). Uptake and accumulated carbons in biomass are transformed to char which are buried in the soils for stretched time. Carbon conversion mechanism in the biomass is stable; it resists decomposition (Osman et al., 2020a). Biochar stability in soils makes it a solid technology for removal of carbon (Chen et al., 2019). In regard to the source of feedstock utilized, this technology could effectively eliminate 2.1–4.8 tCO_2/tonne of biochar (RoyalSociety, 2018). Apart from the helpful influence on trapping and storage of carbons from the atmosphere, biochar influence on some greenhouse gas emissions like methane and oxides of nitrogen is noticeable in decrease rate of emissions (Semida et al., 2019). Application of biochar also improves fertility and quality of the soil extensively such as enhanced cycling of nutrient and decreases leaching of nutrient in the soil coupled with high level of nutrient and water holding of the soil and enhancement of microbial activity in the soil which are all attributed to benefit of adopting biochar technology. Moreover, the attributed benefits rely on the physical and chemical makeup of the biochar in line with the nature of feedstock used, pyrolysis conditions, and other processing factors (Oni et al., 2020).

There are many other remediation technologies that can effectively restore earth to natural state from the impact of global warming. Following are the some of the techniques which have significant importance in reducing the environmental pollution. For instance, radiative forcing geoengineering, which is designed to disrupt the radiation of the earth there by reduce earth's temperature; Cirrus cloud thinning technology, is a terrestrial management of radiation technology that increase long-wave radiation emission from the surface of the earth to space to lower temperature of the globe; Surfacebased brightening technology, this increases the albedo of the earth to decrease global temperatures; Spacebased mirrors technology, this is position to reflect some of the incident radiation from the solar to lower temperatures of the globe; Marine sky brightening technology also help to reduce temperature of the earth; Stratospheric aerosol injection technology also, designed to reduce temperature of the globe; Wetland restoration and construction technology, ecosystem of wetlands are highly reach in carbon, wetland enhances sequestration of atmospheric carbon with the aid of photosynthesis also facilitate accumulation of carbon within biomass above and below ground level and within the organic matters in the soil (Villa and Bernal, 2018). Ocean alkalinity enhancement technology, where ocean take up a reasonable amount of carbon yearly through atmospheric diffusion of CO_2 into the water bodies, via photosynthesis of phytoplankton. During the diffusion of CO_2 in water bodies, the reaction of gas with water results in the formation of carbonic acid that later splits to carbonate ions and bicarbonate that accumulated the dissolved inorganic carbon. The reaction further gives off hydrogen ions; this triggered acidity level of the ocean (Renforth and Henderson 2017). pH of the ocean has considerable effect on partial pressure of CO_2 within the content of inorganic carbon, this is the total concentrations of carbon in carbonic acid, carbonate, and bicarbonate ions (Samer et al., 2020). Elevated alkalinity in the ocean reduces partial pressure within the surface of the ocean, leading to more uptake of CO_2 in the oceanic, with a prominent reduction in acidification level of the ocean. High concentration of alkalinity promote conversion of more carbonic acid to bicarbonate and carbonate ions and larger numbers of carbon are accumulated in inorganic form (Renforth and Henderson 2017). The above highlighted technologies and many more available technologies are efficient intervention approach to combat the adverse impact of global warming.

14.6 SUMMARY

Today's environmental pollution issue is a complicated result of pressures linked to numerous interrelated elements. There are undoubtedly many different and opposing opinions regarding what might be the fundamental causes of the environmental disaster. The root cause of environmental damage cannot be attributed to a single factor. Simultaneously, activities of these factor could take place and their balance differs with regard to location over time. Climate change is obvious; several of the greenhouse gases emissions globally are as a result of anthropogenic activities and rapid rise in emerging population along with their demand leading to development. Despite stringent efforts to combat generation of greenhouse gases to reduce global emission, developed nations are the leading emitters of greenhouse gases. Adoption of environmental sustainability approach is the needed tool to keep the planet in its natural state at present and emphases should be strongly laid on issues with polluted and unmitigated environment. The momentum at which chemistry of ocean changes is rapid compared to ocean acidification experienced previously; this is an extinction signal in the history of Earth's vegetations that has to be averted through secondary resuscitation or natural attenuation of environmental phenomenon.

Several factors have been associated with trends of event shaping the nature globally, some of the global megatrends extend to technological, social, economic, political, and environment concept. Some of the major trend is accelerating urbanization, development of technological innovation, establishment of market, and shifting economic power. In an attempt to ascertaining environmental lethal dosage and planetary threshold of the Earth, scientists are making effort to pin point the complex relationship between bio-geophysical processes which determines Earth's ability towards self-regulation. The negative impact of industrialization alters the totality of the natural world, with sometimes lethal consequences to human civilization. Numerous attempts to mitigate or control contaminants from the application of various methods have been explored to mediate polluted site and water. Following the removal of pollutant, monitoring with assessment procedures are used to gauge success of the employed technology. Hence, using a remediation technology and then going through the entire management approach after the remediation are both steps in dealing with the issue in the environment.

REFERENCES

Abdulsalam, S., Bugaje, I. M., Adefila, S. S., and Ibrahim, S. (2011). Comparison of biostimulation and bioaugmentation for remediation of soil contaminated with spent motor oil. *Int. J. Environ. Sci. Technol.*, *8*(1), 187–194.

Abed, R. M., Al-Kharusi, S., and Al-Hinai, M. (2015). Effect of biostimulation, temperature and salinity on respiration activities and bacterial community composition in an oil polluted desert soil. *Int. Biodeterior. Biodegrad.*, *98*, 43–52.

Agham Delphine Tanyi (2017). Environmental remediation and interventions to global warming and climate change, 4th International Conference on Past and Present Research Systems of Green Chemistry, Atlanta, USA, 16–18.

Ahmad, N., Ullah, Z., Arshad, M. Z., Kamran, H., waqas, Scholz, M., and Han, H. (2021). Relationship between corporate social responsibility at the micro-level and environmental performance: The mediating role of employee pro-environmental behavior and the moderating role of gender. *Sustain. Prod. Consum. 27*, 1138–1148. https://doi.org/10.1016/j.spc.2021.02.034

Albergaria, J. T., Maria da Conceição, M., and Delerue-Matos, C. (2012). Remediation of sandy soils contaminated with hydrocarbons and halogenated hydrocarbons by soil vapour extraction. *J. Environ. Manag.*, *104*, 195–201.

Appannagari, R. R. (2017). Environmental pollution causes and consequences: A study. *North Asian Int. Res. J. Soc. Sci. & Human.*, *3*(8), 151–161.

Appolloni, A., Chiappetta Jabbour, C. J., D'Adamo, I., Gastaldi, M., and SettembreBlundo, D. (2022). Green recovery in the mature manufacturing industry: The role of the green-circular premium and sustainability certification in innovative efforts. *Ecol. Econ.*, *193*, 107311. https://doi.org/10.1016/j.ecolecon.2021.107311

Cecchin, I., Reddy, K. R., Thomé, A., Tessaro, E. F., and Schnaid, F. (2017). Nanobioremediation: Integration of nanoparticles and bioremediation for sustainable remediation of chlorinated organic contaminants in soils. *Int. Biodeterior. Biodegrad.*, *119*, 419–428.

Chen H., Ahmed, I., Osman, Chirangano Mangwandi, and David Rooney (2019). Upcycling food waste digestate for energy and heavy metal remediation applications. *Resour. Conserv. Recycl. X*, *3*:100015. https://doi.org/10.1016/j.rcrx.2019.100015

Crane, R. A., and Scott, T. B. (2012). Nanoscale zero-valent iron: Future prospects for an emerging water treatment technology. *J. Hazard. Mater.*, *211*, 112–125.

Dabrowiecki, P., Adamkiewicz, Ł., Mucha, D., Czechowski, P.O., Soliński, M., Chciałowski, A., and Badyda, A. (2021). Impact of air pollution on lung function among preadolescent children in two cities in Poland. *J. Clin. Med*, *10*, 23–75.

Fawzy, S., Osman, A.I., Doran, J., and Rooney, D. W. (2020). Strategies for mitigation of climate change: A review. *Environ. Chem. Lett.*, *18*, 2069–2094. https://doi.org/10.1007/s10311-020-01059-w

Frolova, Y., Alwaely, S. A., and Nikishina, O. (2021). Knowledge management in entrepreneurship education as the basis for creative business development. *Sustainability, 13*, 1167. https://doi.org/10.3390/su13031167

Fu, C., Jiang, H., and Chen, X. (2022). Modeling of an enterprise knowledge management system based on artificial intelligence. *Knowl. Manag. Res. Pract.*, 1–13.https://doi.org/10.1080/14778238.2020.1854632

Fuss, A., Lamb, W. F., Callaghan, M. W., Hilaire, J., Creutzig, F., Amann, T., Beringer, T., de Oliveira Garcia, W., Hartmann, J., and Khanna, T. (2018). Negative emissions—part 2: Costs, potentials and side effects. *Environ. Res. Lett.*, *13*, 063002. https://doi.org/10.1088/1748-9326/aabf9f

Germida, J. J., Frick, C. M., and Farrell, R. E. (2002). Phytoremediation of oil-contaminated soils. In Violante, A., Huang, P.M., Bollag J.-M., and Gianfreda L. (Eds.) *Developments in Soil Science* (Vol. 28, pp. 169–186). Elsevier, Amsterdam.

Gloet, M., and Samson, D. (2022). Knowledge and innovation management to support supply chain innovation and sustainability practices. *Inf. Syst. Manag*, *39*, 3–18. https://doi.org/10.1080/10580530.2020.1818898

Gu, J., Shi, Y., Zhu, Y., Chen, N., Wang, H., Zhang, Z., and Chen, T. (2020). Ambient air pollution and cause-specific risk of hospital admission in China: A nationwide time-series study. *PLoS Med*, *17*, 1003188.

Harper, A. B., Powell, T., and Shu, S. (2018). Land-use emissions play a critical role in land-based mitigation for Paris climate targets. *Nat. Commun.*, *9*, 2938. https://doi.org/10.1038/s41467-018-05340-z

Hepburn, C., Ella, A., John, B., Emily, A., Carter, S. F., Niall, M. D., Jan, C., Minx, P. S., Williams, C. K. (2019). The technological and economic prospects for CO_2 utilization and removal. *Nature*, *575*, 87–97. https://doi.org/10.1038/s41586-019-1681-6

Hu, N., Li, Z., Huang, P., and Tao, C. (2006). Distribution and mobility of metals in agricultural soils near a copper smelter in South China. *Environ. Geochem. Health*, *28*(1), 19–26.

Huang, D., Xu, Q., Cheng, J., Lu, X., and Zhang, H. (2012). Electrokinetic remediation and its combined technologies for removal of organic pollutants from contaminated soils. *Int. J. Electrochem. Sci*, *7*(5), 4528–4544.

Ikram, M., Sroufe, R., Awan, U., and Abid, N. (2022). Enabling progress in developing economies: A novel hybrid decision-making model for green technology planning. *Sustainability*, *14*, 258. https://doi.org/10.3390/su14010258

Interstate Technology & Regulatory Committee (ITRC) (2011). Green and sustainable remediation: State of the science and practice. *GSR-1*, Washington, DC 20001 6. https://higherlogicdownload.s3-external-1.amazonaws.com/ITRC/2c862581-b546-4afc-8bee-ae01f057f5b0_file.pdf?AWSAccessKeyId=AKIAVRDO7IEREB57R7MT&Expires=16 81840866&Signature=7xn5eGVRU%2FjlP1K%2BywUudSmhVMM%3D

Jewell, D. O., Jewell, S. F., and Kaufman, B. E. (2022). Designing and implementing high-performance work systems: Insights from consulting practice for academic researchers. *Hum. Resour. Manag. Rev*, 32, 100–749. https://doi.org/10.1016/j.hrmr.2020.100749

Khalid, S., Shahid, M., Niazi, N. K., Murtaza, B., Bibi, I., and Dumat, C. (2017). A comparison of technologies for remediation of heavy metal contaminated soils. *J. Geochem. Explor.*, *182*, 247–268.

Khomenko, S., Cirach, M., Pereira-Barboza, E., Mueller, N., Barrera-Gómez, J., Rojas-Rueda, D., de Hoogh, K., Hoek, G., and Nieuwenhuijsen, M. (2021). Premature mortality due to air pollution in European cities: A health impact assessment. *Lancet Planet. Health*, *5*, 121–e134.

Kuppusamy, S., Palanisami, T., Megharaj, M., Venkateswarlu, K., and Naidu, R. (2016). In-situ remediation approaches for the management of contaminated sites: A comprehensive overview. *Rev. Environ. Contam. Toxicol.*, *236*, 1–115.

Lim, M. W., Von Lau, E., and Poh, P. E. (2016). A comprehensive guide of remediation technologies for oil contaminated soil—present works and future directions. *Mar. Pollut. Bull.*, *109*(1), 14–45.

Liu, L., Li, W., Song, W., and Guo, M. (2018). Remediation techniques for heavy metal-contaminated soils: Principles and applicability. *Sci. Total Environ.*, *633*, 206–219.

Ma, X., Zhang, X., Tian, D. (2020). Farmland degradation caused by radial diffusion of CO_2 leakage from carbon capture and storage. *J. Clean. Prod.*, 255, 120059. https://doi.org/10.1016/j.jclepro.2020.120059

Manisalidis, I., Stavropoulou, E., Stavropoulos, A., and Bezirtzoglou, E. (2020). Environmental and health impacts of air pollution: A review. *Front. Public Health*, 8, 14. https://doi.org/10.3389/fpubh.2020.00014

McConnico, M. (2010). Environmental challenges in a global context. *The European Environment|State and Outlook, Synthesis,* 128–149. https://www.eea.europa.eu/soer/2010/synthesis/synthesis/environmental-challenges-in-a-global/view

Meuser, H. (2012). *Soil Remediation and Rehabilitation: Treatment of Contaminated and Disturbed Land* (Vol. 23). Springer Science & Business Media, Dordrecht.

Michalski, J., Poltrum, M., and Bunger, U. (2019). The role of renewable fuel supply in the transport sector in a future decarbonized energy system. *Int. J. Hydrog. Energy*, *44*, 12554–12565. https://doi.org/10.1016/j.ijhydene.2018.10.110

Morelli, J. (2011). Environmental sustainability: A definition for environmental professionals. *J. Environ. Sustain.*, *1*(1), 2. https://doi.org/10.14448/jes.01.0002. Available at: http://scholarworks.rit.edu/jes/vol1/iss1/2

Morillo, E., and Villaverde, J. (2017). Advanced technologies for the remediation of pesticide-contaminated soils. *Sci. Total Environ.*, *586*, 576–597.

Mulligan, C. N., Yong, R. N., and Gibbs, B. F. (2001). Remediation technologies for metal-contaminated soils and groundwater: An evaluation. *Eng. Geol.*, *60*(1–4), 193–207.

National Ground Water Association online. Ground Water Protection and Management Critical to the Global Climate Change Discussion. http://www.ngwa.org/PROGRAMS/government/issues/climate.aspx

Nazar, W. and Niedoszytko, M. (2022). Air pollution in Poland: A 2022 narrative review with focus on respiratory diseases. *Int. J. Environ. Res. Public Health*, 19, 895. https://doi. org/10.3390/ ijerph19020895

Ogbeibu, S., Jabbour, C. J. C., Gaskin, J., Senadjki, A., and Hughes, M. (2021). Leveraging STARA competencies and green creativity to boost green organizational innovative evidence: Apraxis for sustainable development. *Bus. Strateg. Environ*, 30, 2421–2440. https://doi.org/10.1002/bse.2754

Olabi, A. G., Obaideen, K., Elsaid, K., Wilberforce, T., Sayed, E. T., and Maghrabie, H. M. (2022). Assessment of the pre-combustion carbon capture contribution into sustainable development goals SDGs using novel indicators. *Renew. Sustain. Energy Rev*, 153, 111–710. https://doi.org/10.1016/j.rser.2021.111710

Oni, B. A., Oziegbe, O., and Oluwole, O. O. (2020). Significance of biochar application to the environment and economy. *Ann. Agric. Sci.* https://doi.org/10.1016/j.aoas.2019. 12.006

Osman, A. I. (2020). Catalytic hydrogen production from methane partial oxidation: Mechanism and kinetic study. *Chem. Eng. Technol.*, 43, 641– 648. https://doi.org/10.1002/ceat. 201900339

Osman, A. I., O'Connor, E., McSpadden, G., Abu-Dahrieh, J., Charles, F., Ala, M., John, H., and Rooney, D. (2020a). Upcycling brewer's spent grain waste into activated carbon and carbon nanotubes for energy and other applications via two-stage activation. *J. Chem. Technol. Biotechnol.*, 95, 183–195. https://doi.org/10.1002/jctb.6220

Østergaard, P. A., Duic, N., Noorollahi, Y., Mikulcic, H., and Kalogirou, S. (2020). Sustainable development using renewable energy technology. *Renew. Energy, 146*, 2430–2437. https://doi.org/10.1016/j.renene.2019.08.094

Palomares, I., Martínez-Cámara, E., Montes, R., García-Moral, P., Chiachio, M., and Chiachio, J. (2021). A panoramic view and swot analysis of artificial intelligence for achieving the sustainable development goals by 2030: Progress and prospects. *Appl. Intell.*, 51, 6497–6527. https://doi.org/10.1007/s10489-021-02264-y

Panagos, P., Van Liedekerke, M., Yigini, Y., and Montanarella, L. (2013). Contaminated sites in Europe: Review of the current situation based on data collected through a European network. *J. Environ. Public Health*, 2013, 20–13.

Pires, J. C. M. (2019). Negative emissions technologies: A complementary solution for climate change mitigation. *Sci. Total Environ.*, 672, 502–514. https://doi.org/10.1016/j. scitotenv.2019.04.004

Qin, Z., Chen, J., Xie, X., Luo, X., Su, T., and Ji, H. (2020). CO2 reforming of CH4 to syngas over nickel-based catalysts. *Environ. Chem. Lett.*, 18, 997–1017. https://doi.org/10.1007/ s10311-020-00996-w

Reddy, K. R. (2013). Electrokinetic remediation of soils at complex contaminated sites: Technology status, challenges, and opportunities. In Manassero, M., et al. (Eds.). *Coupled Phenomena in Environmental Geotechnics* (pp. 131–147). CRC Press, Taylor & Francis Group.

REN21 (2019). Renewables 2019—global status report. https://www. ren21.net/wp-content/ uploads/2019/05/gsr_2019_full_repor t_en.pdf.

Renforth, P., and Henderson, G. (2017) Assessing Ocean alkalinity for carbon sequestration. *Rev. Geophys.*, 55, 636–674. https://doi.org/10.1002/2016RG000533

Rodriguez-Campos, J., Dendooven, L., Alvarez-Bernal, D., and Contreras-Ramos, S. M. (2014). Potential of earthworms to accelerate removal of organic contaminants from soil: A review. *Appl. Soil Ecol.*, 79, 10–25.

RoyalSociety. (2018) Greenhouse gas removal. https://royalsocie ty.org/-/media/policy/projects/greenhouse-gas-removal/royal-society-greenhouse-gas-removal-report-2018.pdf. Accessed 7 June 2022.

Samaksaman, U., Peng, T. H., Kuo, J. H., Lu, C. H., and Wey, M. Y. (2016). Thermal treatment of soil co-contaminated with lube oil and heavy metals in a low-temperature two-stage fluidized bed incinerator. *Appl. Therm. Eng.*, *93*, 131–138.

Semida, W. M., Hamada, R. B., Mamoudou, S., Catherine, R. S., Abd El-Mageed, T. A., Mostafa, M. R., and Nelson, S. D. (2019). Biochar implications for sustainable agriculture and environment: A review. *S. Afr. J. Bot.*, *127*, 333–347. https://doi.org/10.1016/j.sajb.2019.11.015

Sénéchal, O., and Trentesaux, D. (2019). A framework to help decision makers to be environmentally aware during the maintenance of cyber physical systems. *Environ. Impact Assess. Rev*, *77*, 11–22. https://doi.org/10.1016/j.eiar.2019.02.007

Slama, A., Sliwczy´nski, A., Wo´znica, J., Zdrolik, M., Wi´snicki, B., Kubajek, J., Tur´za´nska-Wieczorek, O., Gozdowski, D., Wierzba, W., and Franek, E. (2019). Impact of air pollution on hospital admissions with a focus on respiratory diseases: A time-series multi-city analysis. *Environ. Sci. Pollut. Res. Int*, *26*, 16998–17009.

Song, B., Zeng, G., Gong, J., Liang, J., Xu, P., Liu, Z., …, and Ren, X. (2017). Evaluation methods for assessing effectiveness of in situ remediation of soil and sediment contaminated with organic pollutants and heavy metals. *Environ. Int.*, *105*, 43–55.

Srivastava, R. K. (2020) Biofuels, biodiesel and biohydrogen production using bioprocesses. A review. *Environ. Chem. Lett.*, *18*, 1049– 1072. https://doi.org/10.1007/s10311-020-00999-7

Streimikiene, D., Svagzdiene, B., Jasinskas, E., and Simanavicius, A. (2021). Sustainable tourism development and competitiveness: The systematic literature review. *Sustain. Dev*, *29*, 259–271. https://doi.org/10.1002/sd.2133

Tajudin, S. A., Azmi, M. M., and Nabila, A. T. A. (2016). Stabilization/solidification remediation method for contaminated soil: A review. In *IOP Conference Series: Materials Science and Engineering* (Vol. 136, No. 1, p. 012043). IOP Publishing, Langkawi.

Tcvetkov, P., Cherepovitsyn, A., and Fedoseev, S. (2019). Public perception of carbon capture and storage: A state-of-the-art overview. *Heliyon*, *5*, 02845. https://doi.org/10.1016/j.heliyon.2019.e02845

The Horinko Group (2014). The Rise and Future of Green and Sustainable Remediation. Washington, DC 20037. http://www.thgadvisors.org/wp-content/uploads/2014/03/The-Rise-and-Future-of-Green-and-Sustainable-Remediation.pdf, pp 54

The Solar Impulse Foundation (2022) How to stop global warming? https://solarimpulse.com/. Accessed 18 April 2023.

Thomé, A., Reginatto, C., Cecchin, I., and Colla, L. M. (2014). Bioventing in a residual clayey soil contaminated with a blend of biodiesel and diesel oil. *J. Environ. Eng.*, *140*(11), 06014005.

Thompson, S. O., Ogundele, O. D., Abata, E. O., and Ajayi, O. M. (2019). Heavy metals distribution and pollution indices of scrapyards soils. *Int. J. Curr. Res. Appl. Chem. & Chem. Eng.*, *3*(1), 9–19.

Tomaszewski, J. E., Smithenry, D. W., Cho, Y. M., Luthy, R. G., Lowry, G. V., Reible, D., …, and Sylvestre, M. (2006). Treatment and containment of contaminated sediments. In Danny Reible, D. and Lanczos, T. (Eds.). *Assessment and Remediation of Contaminated Sediments* (pp. 137–178). Springer, Dordrecht.

U.S. Environmental Protection Agency (U.S. EPA) (2012). *Citizen's Guide to Thermal Desorption*. United States Environmental Protection Agency, Office of Solid Waste and Emergency Response. Publication EPA. https://www.epa.gov/sites/default/files/2015-04/documents/a_citizens_guide_to_thermal_desorption.pdf

U.S. Environmental Protection Agency (U.S. EPA), National Center for Environmental Innovation. Cleaner Diesels: Low Cost Ways to Reduce Emissions from Construction Equipment. March 2007. http://www.epa.gov/sectors/construction/

U.S. Environmental Protection Agency (U.S. EPA), Office of Solid Waste and Emergency Response. *Green Remediation: Incorporating Sustainable Environmental Practices into Remediation of Contaminated Sites*. Apr 2008.

U.S. Environmental Protection Agency (U.S. EPA), Office of Solid Waste and Emergency Response. *Green Remediation: Incorporating Sustainable Environmental Practices into Remediation of Contaminate Site*. February 2022.

U.S. Sustainable Remediation Forum, SURF Report—Summer 2013, *SURF Newsletter, 4*(3), 2.

Villa, J. A., and Bernal, B. (2018). Carbon sequestration in wetlands, from science to practice: An overview of the biogeochemical process, measurement methods, and policy framework. *Ecol. Eng., 114*, 115–128. https://doi.org/10.1016/j.ecoleng.2017.06.037

Vinca, A., Rottoli, M., Marangoni, G., and Tavoni, M. (2018). The role of carbon capture and storage electricity in attaining 1.5 and 2 °C. *Int. J. Greenh. Gas Control, 78*, 148–159. https://doi.org/10.1016/j.ijggc.2018.07.020

Weina, A., and Yanling, Y. (2022). Role of knowledge management on the sustainable environment: Assessing the moderating effect of innovative culture. *Front. Psychol, 13*, 861–813. https://doi.org/10.3389/fpsyg.2022.861813

Xie, T., Reddy, K. R., Wang, C., Yargicoglu, E., and Spokas, K. (2015). Characteristics and applications of biochar for environmental remediation: A review. *Crit. Rev. Environ. Sci. Technol., 45*(9), 939–969.

Yan, W., Lien, H. L., Koel, B. E., and Zhang, W. X. (2013). Iron nanoparticles for environmental clean-up: Recent developments and future outlook. *Environ. Sci.: Process. Impacts, 15*(1), 63–77.

Yang, J., Xiu, P., Sun, L., Ying, L., and Muthu, B. (2022). Social media data analytics for business decision making system to competitive analysis. *Inf. Process. Manag., 59*, 102–751. https://doi.org/10.1016/j.ipm.2021. 102751

Zhang L, Xu M, Chen H, Li Y and Chen S. (2022). Globalization, green economy and environmental challenges: State of the art review for practical implications. *Front. Environ. Sci, 10*, 870–271. https://doi.org/10.3389/fenvs.2022.870271

15 Algae for plastic biodegradation
Emerging approach in mitigating marine pollution

Beauty Akter, Mashura Shammi,
and Zobaidul Kabir

CONTENTS

15.1 INTRODUCTION

Globally a large amount of plastic litter enters the oceans from continental sources, industrial and urban activities (effluents runoff), offshore industrial actions (oil and gas explorations), agricultural activities, and aquacultural activities such as fisheries, food and beverage processing, and tourism (Barboza et al., 2018). Plastic pollution is a transfrontier problem that stresses marine ecosystems and is found in all environmental components (La Daana et al., 2022). Microplastics (MPs) are produced as a result of the continual breakdown of plastic goods due to thermal, biogeochemical, and other environmental processes (Barboza *et al.*, 2018). MPs are broken down into nano plastics further. The strong dispersal form of oceanic fluxes transports MPs across the oceans, even to isolated locations such as the polar regions. Consequently, plastic pollution is the most dangerous to marine ecosystems because of its fast disposal rate and limited rate of recovery from the environment (Bansal et al., 2021).

MPs are absorbed/adsorbed/consumed by various primary producers and trophic consumers in the complex food webs of marine ecosystems.

What size are MPs? MPs are little plastic pieces smaller than five millimeters (5 mm or 0.2 inches) in size, with no set lower limit (Deng et al., 2021; Kumar et al., 2021). The common MPs found in the wetlands are polyethylene (PE), polypropylene (PP), polystyrene (PS), polyvinyl chloride, polyethylene terephthalate (PET), polyamide, nylon, and rayon (Dusaucy et al., 2021; Kumar et al., 2021; Miloloža, et al., 2021; Shruti et al., 2021). Metal-oxide-doped microplastic particles (MMPs) are another emerging contaminant recently identified in aquatic systems (Zhang et al., 2020). Synthetic microfibers are another major subcategory of MPs that are derived from textile products during washing and laundry. The microfibers get into surface waters from other sources or become released by wastewater treatment facilities (Peller et al., 2021).

Contamination and accumulation of MPs were low until the mid-1970s in the marine ecosystem. However, since 2012, plastic particles have risen substantially, with the most significant quantities of MPs reported in recent years in marine ecosystems (Dahl et al., 2021). For example, in Almeira, Spain, an intense agricultural industry with plastic-covered greenhouses of 30,000 ha has caused a significant historical accumulation of MPs in the marine ecosystem containing seagrass meadows since the 1970s (Dahl et al., 2021). MPs in different marine ecosystems are increasing alarmingly due to their highly persistent properties (Barboza et al., 2018). Due to weathering processes including photo-oxidation, plastics experience a series of fragmentation and chemical leaching phases in the environment. Recent studies on marine primary producers and consumers have proved MPs to be pervasive toxicants in the marine environment (Bansal et al., 2021). In the world's great lakes, MPs count in water ranges from 10,900 to 2,090,000 MP/km^2 and from 0.27 to 34,000 MP/m^3. In the sediments, MPs ranged from 0.7 to 7,707 MP/kg and from 0.8 to 2,500 MP/m^2. Marine MPs achieve light, buoyant properties through the change in the density of heavier plastic debris while physicochemical and biological processes of the marine system act on it. Water bodies act as temporary reservoirs and provide primary pathways to transport land-sourced MPs before settling down in the sediments (Luo et al., 2018).

MPs are also colonized by an assemblage of bacteria. High temperatures in summer ($31.4 \pm 1.07°C$) promote high bacterial biomasses that have good degrading capacity compared to the winter temperatures ($13.3 \pm 2.49°C$) in the marine environment. Bacterial characterizations by the 16S rRNA method identified a multitude of biofilm-forming microorganisms (Jin et al., 2020). MPs also sorbed pollutants onto their surfaces such as polycyclic aromatic hydrocarbons (PAHs) and polychlorinated biphenyls (PCBs) which considerably decrease in the summer.

Marine MP contamination is a global problem due to the frequency of MP residues in aquatic food webs (zooplanktons, bivalves, crustaceans, fishes, and other marine vertebrates), particularly in species that are eaten by people (Bowley et al., 2020). The exposure route for most marine organisms to MPs is ingestion. After intake, MPs may be absorbed, distributed via the circulatory system, and deposited in various organ sections, which can result in a variety of negative consequences (Barboza et al., 2018).

Enzymatic digestion and density separation are standard approaches to analyzing MPs in soil, water, and other environmental samples. In addition, visual identification in microscopes, and Fourier-transformed infrared, scanning electron microscope photos were used to identify the MP particles (Cao et al., 2022; Dahl et al., 2021; Zhu et al., 2019). The most commonly found MPs in wetlands are fiber, thread, and filament fragments, films, foams, and microbeads (Kumar et al., 2021).

The key to MP removal strategies is the accumulation of algae and MPs. Understanding how algae react to MPs has been facilitated by recent research on the interaction between algae and MPs. In addition, interactions between MPs and biofilm matrices are common in aquatic habitats and crucial to a variety of functions at the ecosystem level. Biofilms may swiftly form on MP surfaces, changing the characteristics of MPs, destiny, and ecotoxicity. Conversely, in aquatic environments, ubiquitous biofilms form on all surfaces and interact with MPs (Kalčíková and Bundschuh 2022; Rummel et al., 2017). When MPs enter aquatic environments, biofilms may develop as a result of the microalgae's secretion of extracellular polymeric substances (EPS). The MPs' density may vary as a result of biofilm buildup, which would help with their transit to the sediments (Gopalakrishnan and Kashian 2022). In addition, studies found that exopolysaccharides or EPS produced by microalgae such as *Chlorella vulgaris* are responsible for cell aggregation when cultured with MP. Additionally, direct contact with MP particles might cause cell aggregation (Demir-Yilmaz et al., 2022). We have largely examined and concentrated on marine MP pollution and its effects in this chapter. Next, we looked at the possibility of (micro)plastic biodegradation by algae and its different enzymes.

15.1.1 A SUMMARY OF MICROPLASTIC CONTAMINATION IN MARINE HABITATS

The operation of the aquatic system is disrupted by the discharge of MPs from both primary and secondary sources, as worldwide demand for plastic materials has grown every day, resulting in severe plastic pollution. The primary producers, including microalgae, can interact with these MPs and components that are leached, as well as marine organisms at all trophic levels (Parsai et al., 2022; Xu et al., 2019). MPs can disrupt the biogeochemical cycles in marine ecologies.

The origin of MPs is generally categorized as primary MPs and secondary MPs. Primary MPs are intentionally synthesized small-size particles for commercial application. For instance, synthetic fabrics, microbeads for toothpaste, cosmetics, and personal care items. Secondary MPs are the product of weathering or disintegration of bigger plastic objects by UV radiation, currents, and waves in the marine environment resulting in secondary MPs from a variety of sources, including plastic bottles, tire wear, agricultural plastic film, artificial grass, paints, and many other sources (Xu et al., 2020).

Irregular-shaped fragments are more abundant than other shaped MPs (Shruti et al., 2021). The specific hydrodynamic circumstances influenced the number and dispersal of MPs. Although seaward flushing and monsoonal flux boosted MP abundance, the NE monsoon reversal of winds and currents impeded MP drift toward the Goa coast (Vanapalli et al., 2021). Because of their minute size, MPs have a tremendous detrimental influence on the ecosystem because of how rapidly they are absorbed into food webs and biogeochemical cycles.

15.1.2 BIOAVAILABILITY AND TOXICITY ON PRIMARY PRODUCERS

The introduction of MPs in the maritime food chain is a primary concern as it may harm marine biota by blocking the digestive tract directly or indirectly. MPs are likely bioaccumulating substances along food chains (Wang et al., 2021a,b). MPs on the ocean surface are consumed by zooplankton, and can transfer across trophic levels (He et al., 2022). The size, as well as the color of MPs, might influence their ingestion in aquatic biota (Vanapalli et al., 2021). Furthermore, when compared to virgin MP, aged MPs such as micro-size polystyrene (mPS) and polyvinyl chloride, displayed altered physicochemical characteristics and environmental behavior (Wang et al., 2021a,b). *Raphidocelis subcapitata*, a type of green algae, was inoculated on the non-weathered and weathered MPs which produced leachates during degradation. Weathered MP leachates were more dangerous than non-weathered material because of their increased toxicity, as shown by lower median effect concentration (EC50) values (Simon et al., 2021).

Nylon MPs inhibited the growth of *M. aeruginosa* in a dose–response study with a maximal inhibition rate of 47.62% at 100 mg/L doses. *M. aeruginosa's* photosynthetic electron transmission and production of phycobiliproteins were impaired. In addition, algal cell membrane was damaged by nylon MPs and consequently, its EPS release was increased leading to oxidative stresses. In addition, transcriptome research revealed that Nylon MPs altered gene expressions in photosynthetic organisms related to the tricarboxylic acid cycle, photosynthesis-antenna proteins, oxidative phosphorylation, carbon fixation, and metabolism of porphyrin and chlorophyll (Zheng et al., 2022).

The type of MP particles, functional groups, sizes, chemical properties, test organisms, and presence of other MPs is a significant factors in determining toxicity (Yi et al., 2019). For instance, the combined action of triphenyltin chloride (TPTCl) and PS MPs on the green alga *Chlorella pyrenoidosa* is distinct from that of either component alone. TPTCl had a 96-hour IC_{50} of 30.64 g/L to the green alga *C. pyrenoidosa*. MPs size had a toxic impact on *C. pyrenoidosa* as well. The 96-hour IC_{50} for 0.55 µm PS was 9.10 mg/L, while there was a non-toxic effect for 5.0 µm PS. Contact of *C. pyrenoidosa* to 0.55 µm PS, resulted in structural impairment. Plasmolysis, vacuolation, and distortion of algal cell membranes are common injuries. The incidence of PS may intensify the toxicity of TPTCl, as indicated by the fact that the 96-hour IC_{50} of TPTCl fell to 9.98 g/L at 5.0 mg/L PS. The reduction in waterborne TPTCl concentration at 0.55 µm PS showed that PS could lessen TPTCl's bioavailability by 15–19% (Yi et al., 2019). At different dosages, the mutual action of mPS and dibutyl phthalate (DBP) produced diverse consequences such as volume change, complexity in cell morphology, and chlorophyll fluorescence intensity (Li et al., 2020). DBP's 96-hour IC_{50} value was 2.41 mg/L. The 96-hour IC_{50} for PS MPs against *C. pyrenoidosa* was 6.90 and 7.19 mg/L for 0.1 and 0.55 µm mPS, respectively. However, slight toxicity was seen with 5 µm mPS. MPS (Fe)-NH2 was more harmful to *C. pyrenoidosa* than MPS (Fe)-COOH and non-iron oxide-doped MP particles in respect of median effective concentration (Zhang et al., 2020).

When treated with PE and PP, *Spirulina* sp. growth dramatically declined (p 0.05) in comparison to the control (0.0312/day) (Hadiyanto et al., 2021a,b,c). In *Skeletonema costatum*, inhibited growth and oxidative stress as well as superoxide dismutase (SOD) reduced more than malondialdehyde, which demonstrated that external damage was more intense than the internal damage (Zhu et al., 2019).

Chlorella pyrenoidosa treated to 1 μm MP also showed significantly lower levels of oxidative stress, photosynthetic pigment concentration, and cell membrane integrity. This concentration also changed the transcript levels of genes associated in photosynthesis and metabolism at the molecular level (Cao et al., 2022).

The generation of EPS which is an integral part of algal biofilm can be impacted by MPs as well. At all temperatures, EPS generation and algal cell density increased in Chlamydomonas as MPs concentration rose from 0 to 0.4 mg/mL. Consequently, the deposition of MPs in EPS also increased (Gopalakrishnan and Kashian 2022).

It is found in a recent study that biodegradable MPs (polylactic acid) and polybutylene succinate have similar toxicity on *C. vulgaris* as polyethylene and polyamide (Su et al., 2022). Marine diatoms (*Phaeodactylum tricornutum*) were exposed to polymethyl methacrylate (PMMA) MPs at varied salinities (25, 35, and 45), which resulted in negative growth with a least EC_{50} rate of 91.75 mg/L. The two variables significantly affected the chlorophyll fluorescence characteristics (Fv/Fm and PSII) and hindered *P. tricornutum*'s ability to photosynthesize.

SOD is an antioxidative substance that emerges in the cell by the stress brought on by MPs. The secretion of EPS is mostly related to the diverse interaction and aggregation of MPs and microalga. MPs cause a remarkable rise in bound EPS of alga *Microcystis aeruginosa*. The cohabitation of MPs and microalgae can accelerate the aging of MPs in terms of apparent surface roughness, lower strength of certain functional groups, charge shifts, and leaking of dangerous chemicals (Su et al., 2022). Furthermore, MPs covered with biofilm may be eaten by benthic planktivores leading to vertical transport to the marine food chain (Gopalakrishnan and Kashian 2022).

The accumulation of MPs and physical harm to algae were seen in SEM photos (Cao et al., 2022; Zhu et al., 2019). Moreover, MPs are anticipated to adhere to the water-biofilm edge or ooze through the biofilm matrix. In such biofilms, when they are susceptible to transformation processes like fragmentation, MPs might gather and absorb or adsorb. Consequently, biofilms could serve as a sink. However, environmental changes may act as stressors to biofilms, causing dieback and thus the release of MPs, which might transform biofilms into a source of MPs (Kalčíková and Bundschuh 2022). It is, therefore, evident, about the damage to the microalga and the MP toxicity may further impact the consumer population of the marine ecosystem. The inhibitory toxic effects of MPs on marine algae are shown in Figure 15.1.

15.1.3 BIOAVAILABILITY AND MICROPLASTIC TOXICITY
ON MARINE CONSUMER POPULATION

Marine consumer organisms mainly uptake MPs through ingestion and the rate of uptake depends on the feeding behavior, size, and concentration of plastic fragments (Beer *et al.,* 2017). It also specifies the degraded nature of MPs. The bioavailability of MPs to zooplanktons, corals, lobsters, sea urchins, worms, fish etc. makes it

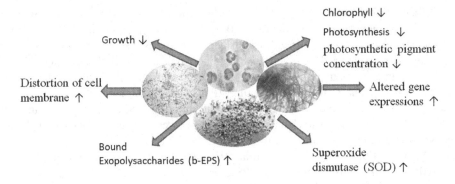

FIGURE 15.1 Microplastics cause harmful inhibitory effects on marine algae. The up arrow indicates an increase while the down arrow indicates a decrease.

persistent in the marine environment. Marine organisms uptaken these nondegradable MPs, and bioaccumulation in the food chain has reached higher tropic levels. Plastic litters are harmful to organisms when they ingest plastics. MPs harm marine organisms in several ways, by clogging the intestinal tract, inhibiting the secretion of the gastric enzyme, misbalancing steroid hormonal levels, and delaying the ovulation process and infertility (Sharma and Chatterjee, 2017) in both marine organisms and human beings (Chia *et al.,* 2020). MPs have been shown to operate as trajectories by absorbing contaminants and are conducive to pollutant bioaccumulation, notably in marine environments, species, and food chains (Amelia et al., 2021). Copepods respond to MPs in species and stage-specific ways. MPs can potentially be bound to organic pollutants (e.g., triclosan, chlorpyrifos, and dibutyl phthalate) and act as vectors for copepods (Bai et al., 2021).

MPs are found in varying amounts in marine mollusks. On average, 15–17 particles per organism were identified most of which were microfibers identified as PET (Polidoro et al., 2022). Furthermore, the marine mollusks also had very high concentrations of pesticides, PCBs and phthalates, and other organic pollutants which was adsorbed/absorbed on the surfaces of MPs.

The buildup of PP MPs is hazardous to marine zooplankton (Jeyavani et al., 2022). MPs are frequently consumed by zooplankton directly or indirectly due to their microsize, color, and buoyancy. Ingestions of MPs damage their intestines, reduce their ability to eat, delay their growth or spawning, shortens their lifetime, and cause aberrant or even lethal gene expression, impact on their behavior, procreation, and offspring (He et al., 2021). Toxicity assessment of PP on marine microcrustacean *Artemia salina* had shown disruption in homeostasis, and the subsequent rise in the rapid release of reactive oxygen that caused death in nauplii (LC_{50} 40.947 g/mL) and meta nauplii (LC_{50} 51.954 g/mL). Swimming behavior in juveniles was altered (Jeyavani et al., 2022). Furthermore, antioxidant biomarker indicators SOD, catalase (CAT), reduced glutathione, glutathione-*S*-Transferase, and acetylcholinesterase enzyme was altered. Additionally, histopathology of juvenile Artemia revealed epithelial cell injury.

The presence of MPs varies among different zooplankton species due to different feeding strategies. For example, in *S. spallanzanii* 42% of MPs were identified with a mean of 1 ± 1.62 MPs/sample, while 93% of *H. carunculata* had 3.35 ± 2.60 MPs/sample. These species are effective bioindicators of the influence of MPs on benthic biota due to their susceptibility to this pollutant (Vecchi et al., 2021). When corals (*Corallium* species) are exposed to MPs, they preferentially absorb PP, ensuing in various biotic effects extending from consumption impairment to mucus production and altered gene expression. MPs can, directly and indirectly, affect the coral microbiome by producing tissue abrasions that allow opportunistic bacteria to proliferate (Corinaldesi et al., 2021). Copepods subjected to nylon fibers exposure had significantly altered prey selectivity (ANOVA, $P < 0.01$), which led to a non-significant 40% drop in algal assimilation rates (ANOVA, $P = 0.07$). On the other hand, copepods subjected to nylon granules experienced non-significant lipid buildup (ANOVA, $P = 0.62$) (Cole et al., 2019).

Water filtering and food consumption both led to an MP buildup in marine crabs, which was distributed throughout organs in the following order: hepatopancreas > intestines > gills > muscles. However, biomagnification did not occur, presumably due to MP egestion (Wang et al., 2021a,b). MPs may also alter the toxicity and bioavailability of contaminants, in addition to having a direct harmful impact (Zocchi et al., 2019). Besides, MP-associated organic chemical pollutants may serve as a different route of co-exposure for these contaminants to marine life (Martín et al., 2022). Types of polymer, sizes, salinity, and temperature are significant parameters for the sorption of hydrophobic organic contaminants from the marine seawater to MPs (Sørensen et al., 2020). Furthermore, polarity, crystallinity, size, the aging process, hydrophobicity, and dissociated forms are the main determinants affecting the sorption process of pollutants on MPs (Martín et al., 2022). Moreover, cumulative exposure may have varied harmful effects depending on whether there are synergistic or antagonistic interactions (Martín et al., 2022). For example, negatively charged MPs decreased TiO_2 NPs toxicity on the *Chlorella* sp. (Thiagarajan et al., 2019).

For example, metal-doped MPs such as MPs(Fe)-NH_2 were strongly adsorbed on the *Daphnia magna* body parts and spread over the antennae, carapace, and apical spine. But, MPs-(Fe)-COOH was primarily deposited inside the daphnia's digestive system, according to optical microscopy (Zhang et al., 2020). When evaluated in the absence of MPs, glyphosate-monoisopropylamine salt instigated the maximum death after one week of exposure (23.3%), whereas glyphosate acid caused the least mortality (12.5%). When combined with MPs, this is inverted, though. The varied sorption qualities of the glyphosate-based chemical formulations can be attributed to the modification in toxicity of the glyphosate formulations brought on by MPs. MP pollution in the oceanic environment and its impacts on the oceanic ecosystem is shown in Figure 15.2.

15.1.4 SEDIMENTS

MP contamination is growing more significant, and new information suggests that the marine environment, particularly sediments, is a crucial sink. Numerous water bodies and sedimentary ecosystems, including beaches, shallow coastal regions, estuaries, fjords, settings on the continental shelf, and deep-water environments, have been shown to contain significant concentrations of MP particles (Yan et al.,

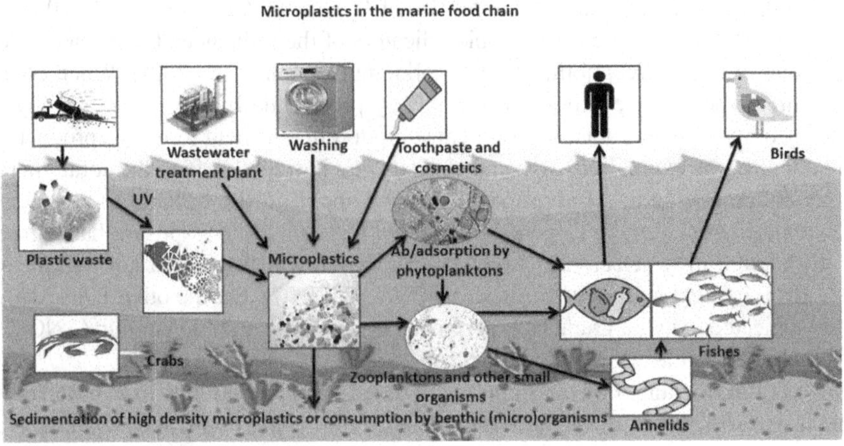

Microplastics in the marine food chain

FIGURE 15.2   Microplastics pollution and its impacts on oceanic ecosystem.

2021). Furthermore, The existence of MPs would change the composition and efficiency of microbial communities in the marine environment, which would alter the carbon/nitrogen cycle, particularly in sediments (Yan et al., 2021). When compared to the low tide line (178 ± 261 mg/m^2), the high tide line had higher MP concentrations in the beach sediments (1323–1228 mg/m^2). Moreover, fresh Cladophora algae easily interacted with MPs through physical entanglement and adsorptive forces. Following algal senescence, these interactions mostly end, and MPs are anticipated to be released in benthic sediments.

15.2   ROLE OF ALGAE AND MICROALGAE IN PLASTIC AND MICROPLASTIC BIODEGRADATION

Because of hydrophobicity, the existence of persistent covalent bonds in the functional groups, plastics, and MPs are challenging to decompose in the natural environment (Liu et al., 2022). Through aerobic and anaerobic mechanisms, plastic polymers can be broken down microbiologically. MPs can be removed using bioremediation, which also promotes environmental sustainability (Manzi et al., 2022). Algae, especially microalgae enhance the degradation process of polymers with the help of synthesized enzymes or toxin systems which are involved to weaken the polymer chemical bonds. In the comparison with other microorganisms, microalgae are considered best for polymer degradation as it does not contain endotoxins and photoautotrophic conditions do not demand organic carbon sources (Chia *et al.*, 2020).

The effect of PS and bisphenol A (BPA) MPs on the toxicity and biodegradation of the microalgae *Chlorella pyrenoidosa* revealed that, over the course of a 16-day culture, PS (5 mg/L) enhanced BPA's growth inhibition of *C. pyrenoidosa* by 1 and 10 mg/L. Similar to this, PS (5 mg/L and 100 mg/L) improved algae's ability to break down BPA (1 mg/L and 10 mg/L) (He et al., 2022).

MP biodegradation is a very effective method to counter the ever-increasing plastic pollution around the globe and to ensure environmentally friendly recycling strategies. Marine single-celled algae (*Phaeodactylum tricornutum*) have a high potential to decompose plastic under marine conditions. Degradation of plastic litter using microalgae is an eco-friendly solution to manage PET waste in salt water environments (Moog *et al.*, 2019).

Degradation of plastic materials using microalgae is a current, attractive, and alternative approach to managing polymer which is usually a cost-effective and efficient alternative that does not produce hazardous materials in comparison with other conventional degradation. The author stated microalgae were found to colonize more on the low-density (LD) polythene sheets. Blue-green filamentous micro algae are much more effective in the biological degradation of polythene sheets (Kumar *et al.*, 2017).

Plastic contents have been used as a carbon and energy source for microbes whereas microalgae can be highly suitable organisms for plastic biodegradation. Unicellular, photosynthetic microalgae *Chlamydomonas reinhardtii* has several advantages as a model organism. It is possible to generate PETase enzyme from *C. reinhardtii* which helps to degrade plastic contents. The demonstration of the performance of PETase was also confirmed qualitatively and quantitatively through scanning electron microscopy and high-performance liquid chromatography (Kim *et al.*, 2020). Some microalgae especially *Scenedesmus dimorphus* (green microalgae), *Anabaena spiroides* (blue-green algae), and *Navicula pupula* (Diatom) degrade plastics very effectively (Falah *et al.*, 2020).

A green microalgae *Uronema africanum* was found to colonize on the surface of LD polyethylene (LDPE) sheets. Green, filamentous algae *Uronema africanum* was found to be grown perpendicular on LDPE sheets and it was seen in the abrasions, erosions, grooves, and ridges of treated LDPE sheets (Sanniyasi *et al.*, 2021). Biodegradation of MP is, therefore, the most convenient approach than any other conventional approach to plastic waste disposal.

15.2.1 FRONTLINE ALGAE AND MICROALGAE AND THEIR MECHANISMS FOR PLASTIC DEGRADATION

The presence of aromatic groups in MP such as PET prevents it from degrading. Biodegradation of MPs is an environmentally friendly approach to balance ever-increasing plastic pollution. Several researchers worked on it and demonstrated the role of algae (especially microalgae) in the degradation of MPs in marine environments. The frontline algae with their enzymatic activity in the degradation of plastic contents are discussed in this section.

Salinity increases the capacity of *Spirulina* species to biodegrade PET (Hadiyanto *et al.*, 2021a). The EPS produced by *Spirulina* stimulates the development of biofilms on the surface of MPs and subsequently cause damage to MP contents. Nutrient-rich biofilm provides a suitable living habitat for other microorganisms like fungi, protozoa, and bacteria. The associations of these microorganisms form a protein structure such as an enzyme which causes a breakdown in the chemical structure of polymeric substances. The *Spirulina platensis* uses Styrofoam as a carbon source, which helps to increase the production of EPS, which in turn helps to form biofilm on the styrofoam surface (Hadiyanto et al., 2021b,c).

Scenedesmus obliquus is a microalga that also produces EPS. Because of induced agglomeration, the hydrodynamic dimension of the EPS-coated PS nanoparticles (PSNPs) enlarged considerably after aging in the EPS-containing solution. When compared to the pure versions, the aged PSNPs showed much lower harmful effects. The surface activation of the PSNPs, as well as the aging lengths, were important factors in the process. The observed improvement in cell viability was well associated with a considerable drop in reactive oxygen species levels, the activity of oxidative stress markers (SOD and CAT), and cell membrane damage. Reduced reactive species levels in cells were substantially linked to higher photosynthetic efficiency measurements (electron transport rate and maximal quantum yield of PS II system).

Degradation of the polymer also depends on the enzyme present and on the types of polymer on which it acts. Interestingly, PETase^{R280A} enzyme secreted by marine single-celled *Phaeodactylum tricornutum* diatom has a high potentiality to degrade PET in the marine environment (Barone *et al.*, 2020). However, some biological degradation of PET requires genetic transformation. For example, using *Phaeodactylum tricornutum* to biodegrade PET needs genetic transformation to secrete PETase enzymes efficiently in the marine environment. The authors demonstrated PETase produced by *P. tricornutum* caused PET degradation under varying conditions (Moog *et al.*, 2019). A green microalgae *Uronema africanum* colonize on dumped LD polythene sheets. The *Uronema africanum* microalgae were used to treat LDPE sheets and it was found LDPE sheets were completely colonized by the algal species which was visible even in the naked eyes. Finally, it was observed there were a lot of abrasions, erosions, and ridges on the treated LDPE sheets (Sanniyasi *et al.*, 2021).

The *Chlamydomonas reinhardtii* is blue-green algae that also produce PETase and degrades PET completely. Two strains of *C. reinhardtii* CC-124 and CC-503 along with PET samples were incubated together at 30 °C for up to 4 weeks. PETase produced by *C. reinhardtii* and found terephthalic acid (TPA), the full degraded form of PET, was distinguished by high-performance liquid chromatography. CC-124 found more potential to secrete PETase. (Kim *et al.*, 2020). Diatoms can also degrade PET by releasing PETase enzyme which is produced by them under mesophilic marine conditions (Falah *et al.*, 2020). *P. tricornutum* has the amazing potential to act as a genetic host and to provide affordable growing conditions for the biodegradation of PET with considerable outcomes. PETase enzyme from *Ideonella sakaiensis* was also found that degrades PET (Amobonye *et al.*, 2020; Moog *et al.*, 2019). Table 15.1 shows the plastic degrading algal species and their enzymes responsible for degrading MPs.

15.2.2 ALGAE FOR BIOPLASTIC PREPARATION

Conventional petro-based plastics consist of synthetic polymers which require a higher amount of money for production and resistance to degradation. Plastic that is derived wholly or partially from bio-based sources and provides a similar function to petroleum-based counter products is called bioplastic (Beckstrom *et al.*, 2020). Green plastics or bioplastics have several benefits compared to petro-based plastics. Green plastics are cellulose-based, require less amount of energy to produce, are non-toxic, reduce fossil fuels and greenhouse gas emissions, and are renewable,

TABLE 15.1

Plastic degrading algae and their enzymes

Algae	Plastic types	Enzymes	References
Spirulina platensis	Styrofoam	Extracellular polymeric substances	Hadiyanto et al. (2021b,c)
Phaeodactylum tricornutum	polyethylene terephthalate (PET)	PETase[R280A]	Barone et al. (2020) Moog et al. (2019) Amobonye et al. (2020)
Uronema africanum	Low-density polyethylene (LDPE)	–	Sanniyasi et al. (2021)
Chlamydomonas reinhardtii	polyethylene terephthalate	PETase	Kim et al. (2020) Amobonye et al. (2020)
Chlorella vulgaris	Polyethylene Terephthalate	PETase	Falah et al. (2020)
Ideonella sakaiensis	Polyethylene Terephthalate	PETase	Moog et al. (2019) Amobonye et al. (2020)

recyclable, decomposable, and environmentally friendly (Thiruchelvi *et al.,* 2020). Researchers are becoming more interested in alga-based biopolymers to achieve a sustainable circular economy. Biopolymers from microalgae provide several benefits including reducing greenhouse gas emissions, being inexpensive, being flexible to grow in diverse environmental conditions, being easily biodegradable, and providing a higher level of biomass yield (Das *et al.,* 2018; Devadas *et al.,* 2021). *Chlorella* from the Chlorellaceae family has cell sizes ranging from 3.2 to 10.2 μm. Chlorella contains 60% protein and can be a valuable source of biopolymer as its protein is characterized as a complex heteropolymer (Devadas *et al.,* 2021).

Biopolymers polyhydroxyalkanoates are synthesized by several microorganisms and thought as a perfect alternative for petro-based polymers as they show similar properties to synthetic polymers and they are fully biodegradable. Microalgae have great potential in this sector as it provides PHAs at a minimal cost. The author reported several microalgae and their PAHs accumulation capacity under different cultural conditions. The *Synechococcus* sp. were grown in Phosphorus deficient conditions and found PAH (55.0%) (w/w dry biomass), *Nostoc muscorum* were cultured in acetate and propionate media and found PAH (31.0%) (w/w dry biomass), *Spirulina platensis* in acetate and CO_2, *Botryococcus braunii* in nitrogen deficient and *Synechococcus elongates* in phosphorus-deficient condition and documented accumulation of PAH (10.0%), (16.4%) and (17.5%) (w/w dry biomass), respectively (Costa *et al.,* 2019).

The presence of Poly Hydroxy Butyrate (PHB) in the algal cells makes algal-based bioplastics biodegradable. This molecule is produced by microorganisms when they are facing stress conditions for nutrient deficiency. The author reported that *Chlorella pyrenoidosa* was cultured in laboratory conditions and the production of bioplastic was successfully achieved. The PHB content was found 27% after 14 days of culture which makes highly biodegradable and environmentally friendly bioplastics (Das *et al.,* 2018).

Microcystis aeruginosa was used to yield polyhydroxybutrate for bioplastic preparation with a concentration of 0.49 ± 0.5 mg/mL where the dry cell weight of biomass was 0.0067 g/L. It was found that the PHB detected in *Spirulina platensis* was 0.006 g/L with a percentage of 5.18% (Abdo *et al.,* 2019). Microalgae can play a very vital role in producing bioplastics as it is easily degradable and encourage an environmentally friendly approach. The algal protein content is also very suitable for bioplastic production. The recyclable properties of bioplastic are enhanced by the PHB content. So, the production of bioplastics using algal biomass will be a great approach to achieving a sustainable circular economy (Thiruchelvi *et al.,* 2020). Using genetic engineering and omics to algae could be a successful method to improve MP biodegradation (Manzi et al., 2022) as well as bioplastic production.

15.3   FUTURE RESEARCH DIRECTION AND CONCLUDING REMARKS

In this chapter, we have reported the potential toxicity of MPs on marine primary producers (micro/macro) algae. Algal growth inhibition, photosynthetic ability, and altered gene expression relevant to photosynthetic pigments are evident in this review. Some of the biofilm-forming algal species show excess exopolysaccharide production while absorbing MPs in their matrices. The biofilm-producing EPS from micro and macro alga requires extensive study to identify potential MP reducers in marine environments. Furthermore, some algal species have shown enzyme PETase-production which can be potentially used for degrading MPs in the marine environment. Some microalgae can be genetically altered to produce PETase to degrade MPs. Other potential enzymes should be recognized from different potential marine algal species to biodegrade MP pollution. In addition, algae-based bioplastic production and its relevant research should be increased for a sustainable circular economy.

REFERENCES

Abdo, S.M., and Ali, G.H. (2019). Analysis of polyhydroxybutrate and bioplastic production from microalgae. *Bulletin of the National Research Centre*, *43*(1), 97. https://doi:10.1186/s42269-019-0135-5

Amelia, T.S.M., Khalik, W.M.A.W.M., Ong, M.C., Shao, Y.T., Pan, H.J., and Bhubalan, K. (2021). Marine microplastics as vectors of major ocean pollutants and its hazards to the marine ecosystem and humans. *Progress in Earth and Planetary Science*, *8*(1), 1–26.

Amobonye, A., Bhagwat, P., and Singh, S. (2020). Plastic biodegradation: Frontline microbes and their enzymes. *Science of the Total Environment.* https://doi.org/10.1016/j.scitotenv.2020.143536

Bai, Z., Wang, N., and Wang, M. (2021). Effects of microplastics on marine copepods. *Ecotoxicology and Environmental Safety*, *217*, 112–243.

Bansal, M., Santhiya, D. and Sharma, J.G. (2021). Behavioural mechanisms of microplastic pollutants in marine ecosystem: Challenges and remediation measurements. *Water, Air, & Soil Pollution*, *232*(9), 1–22.

Barboza, L.G.A., Dick Vethaak, A., Lavorante, B.R.B.O., Lundebye, A.-K., and Guilhermino, L. (2018). Marine microplastic debris: An emerging issue for food security, food safety and human health. *Marine Pollution Bulletin*, *133*, 336–348. https://doi:10.1016/j.marpolbul.2018.05.047

Barone, G.D., Ferizovi, D., Biundo, A., and Lindblad, P. (2020). Hints at the applicability of microalgae and cyanobacteria for the biodegradation of plastics. *Sustainability*, *12*(24), 104–149. https://doi:10.3390/su122410449

Beckstrom, B.D., Wilson, M.H., Crocker, M., and Quinn, J.C. (2020). Bioplastic feedstock production from microalgae with fuel co-products: A techno-economic and life cycle impact assessment. *Algal Research*, *46*, 101–769. https://doi:10.1016/j.algal.2019.101769

Beer, S., Garm, A., Huwer, B., Dierking, J., and Nielsen, T.G. (2017). No increase in marine microplastic concentration over the last three decades – A case study from the Baltic Sea. *Science of the Total Environment*, S0048969717328024–. https://doi:10.1016/j.scitotenv.2017.10.101

Bowley, J., Baker-Austin, C., Porter, A., Hartnell, R., and Lewis, C. (2020). Oceanic hitchhikers – Assessing pathogen risks from marine microplastic. *Trends in Microbiology*. https://doi:10.1016/j.tim.2020.06.011

Cao, Q., Sun, W., Yang, T., Zhu, Z., Jiang, Y., Hu, W., Wei, W., Zhang, Y., and Yang, H. (2022). The toxic effects of polystyrene microplastics on freshwater algae Chlorella pyrenoidosa depends on the different size of polystyrene microplastics. *Chemosphere*, *308*, 136–135.

Chia, W.Y., Tang, D.Y.Y., Khoo, K.S., Kay Lup, A.N., and Chew, K.W. (2020). Nature's fight against plastic pollution: Algae for plastic biodegradation and bioplastics production. *Environmental Science and Ecotechnology*, *4*, 100–065. https://doi:10.1016/j.ese.2020.100065

Cole, M., Coppock, R., Lindeque, P.K., Altin, D., Reed, S., Pond, D.W., Sørensen, L., Galloway, T.S., and Booth, A.M. (2019). Effects of nylon microplastic on feeding, lipid accumulation, and moulting in a coldwater copepod. *Environmental Science & Technology*, *53*(12), 7075–7082.

Corinaldesi, C., Canensi, S., Dell'Anno, A., Tangherlini, M., Di Capua, I., Varrella, S., Willis, T.J., Cerrano, C., and Danovaro, R. (2021). Multiple impacts of microplastics can threaten marine habitat-forming species. *Communications Biology*, *4*(1), 1–13.

Costa, S.S., Miranda, A.L., Morais, M.G.D. (2019). Microalgae as source of polyhydroxyalkanoates (PHAs)—A review, *International Journal of Biological Macromolecules*, *131*, 536–547. https://doi.org/10.1016/j.ijbiomac.2019.03.099

Dahl, M., Bergman, S., Björk, M., Diaz-Almela, E., Granberg, M., Gullström, M., Leiva-Dueñas, C., Magnusson, K., Marco-Méndez, C., Piñeiro-Juncal, N., and Mateo, M.Á. (2021). A temporal record of microplastic pollution in Mediterranean seagrass soils. *Environmental Pollution*, *273*, 116–451.

Das, S.K., Sathish, A., and Stanley, J. (2018). Production of biofuel and bioplastic from Chlorella Pyrenoidosa. *Materials Today: Proceedings*, *5*, 16774–16781.

Demir-Yilmaz, I., Yakovenko, N., Roux, C., Guiraud, P., Collin, F., Coudret, C., Ter Halle, A., and Formosa-Dague, C. (2022). The role of microplastics in microalgae cells aggregation: A study at the molecular scale using atomic force microscopy. *Science of the Total Environment*, *832*, 155–036.

Deng, H., He, J., Feng, D., Zhao, Y., Sun, W., Yu, H., and Ge, C. (2021). Microplastics pollution in mangrove ecosystems: A critical review of current knowledge and future directions. *Science of the Total Environment*, *753*, 142041.

Devadas, V.V., Khoo, K.S., Chia, W.Y., Chew, K.W., Munawaroh, H.S.H., Lam, M.K., Lim, J.W., Ho, Y.C., Lee, K.T., and Show, P.L. (2021). Algae biopolymer towards sustainable circular economy. *Bioresource Technology*. https://doi:10.1016/j.biortech.2021.124702

Dusaucy, J., Gateuille, D., Perrette, Y., and Naffrechoux, E. (2021). Microplastic pollution of worldwide lakes. *Environmental Pollution*, *284*, 117075.

Falah, W., Chen, F., Z.E.B., Hayat, M.T., Mahmood, Q., Ebadi, A., Toughani, M., and LI, E. (2020). Polyethylene terephthalate degradation by microalga *Chlorella vulgaris* along with pretreatment. *Materiale Plastice*, *57* (3), 260–270. https://doi.org/10.37358/MP.20.3.5398

Gopalakrishnan, K., and Kashian, D.R. (2022). Extracellular polymeric substances in green alga facilitate microplastic deposition. *Chemosphere*, *286*, 131–814.

Hadiyanto, H., Khoironi, A., Dianratri, I., Suherman, S., Muhammad, F., and Vaidyanathan, S. (2021a). Interactions between polyethylene and polypropylene microplastics and Spirulina sp. microalgae in aquatic systems. *Heliyon*, *7*(8), e07676.

Hadiyanto, H., Haris, A., Muhammad, F., Afiati, N., and Khoironi, A. (2021b). Interaction between styrofoam and microalgae Spirulina platensis in brackish water system. *Toxics*, *9*(3), 43.

Hadiyanto, H., Muslihuddin, M., Khoironi, A., Pratiwi, W.Z., Fadlilah, M.A.N., Muhammad, F., Afiati, N., and Dianratri, I. (2021c). The effect of salinity on the interaction between microplastic polyethylene terephthalate (PET) and microalgae Spirulina sp. *Environmental Science and Pollution Research*, *29*(5), 7877–7887.

He, M., Yan, M., Chen, X., Wang, X., Gong, H., Wang, W., and Wang, J. (2021). Bioavailability and toxicity of microplastics to zooplankton. *Gondwana Research*, *108*, 120–126.

He, D., Zeng, Y., and Zhou, G. (2022). The influence of microplastics on the toxic effects and biodegradation of bisphenol a in the microalgae Chlorella pyrenoidosa. *Aquatic Ecology*, *56*, 1287–1296.

Jeyavani, J., Sibiya, A., Bhavaniramya, S., Mahboob, S., Al-Ghanim, K.A., Nisa, Z.U., Riaz, M.N., Nicoletti, M., Govindarajan, M., and Vaseeharan, B. (2022). Toxicity evaluation of polypropylene microplastic on marine microcrustacean Artemia salina: An analysis of implications and vulnerability. *Chemosphere*, *296*, 133–990.

Jin, M., Yu, X., Yao, Z., Tao, P., Li, G., Yu, X., Zhao, J.L., and Peng, J. (2020). How biofilms affect the uptake and fate of hydrophobic organic compounds (HOCs) in microplastic: Insights from an In situ study of Xiangshan Bay, China. *Water Research*, *184*, 116–118.

Kalčíková, G., and Bundschuh, M. (2022). Aquatic biofilms—Sink or source of microplastics? a critical reflection on current knowledge. *Environmental Toxicology and Chemistry*, *41*(4), 838–843.

Kim, J.W., Park, S., Tran, Q., Cho, D., Choi, D., Lee, Y.J., and Kim, H. (2020). Functional expression of polyethylene terephthalate-degrading enzyme (PETase) in green microalgae. *Microbial Cell Factories*, *19*, 97. https://doi.org/10.1186/s12934-020-01355-8

Kumar, R., Sharma, P., and Bandyopadhyay, S. 2021. Evidence of microplastics in wetlands: Extraction and quantification in Freshwater and coastal ecosystems. *Journal of Water Process Engineering*, *40*, 101–966.

Kumar, V.R., Kanna, G., and Elumalai, S. (2017). Biodegradation of polyethylene by green photosynthetic microalgae. *Journal of Bioremediation & Biodegradation*, *8*(1). https://doi:10.4172/2155-6199.1000381

La Daana, K.K., Asmath, H., and Gobin, J.F. (2022). The status of marine debris/litter and plastic pollution in the Caribbean Large Marine Ecosystem (CLME): 1980–2020. *Environmental Pollution*, *300*, 118–919.

Li, Z., Yi, X., Zhou, H., Chi, T., Li, W., and Yang, K. (2020). Combined effect of polystyrene microplastics and dibutyl phthalate on the microalgae Chlorella pyrenoidosa. *Environmental Pollution*, *257*, 113604.

Liu, L., Xu, M., Ye, Y., and Zhang, B. (2022). On the degradation of (micro) plastics: Degradation methods, influencing factors, environmental impacts. *Science of the Total Environment*, *806*, 151–312.

Luo, W., Su, L., Craig, N.J., Du, F., Wu, C., and Shi, H. (2018). Comparison of microplastic pollution in different water bodies from urban creeks to coastal waters, *Environmental Pollution*. https://doi.org/10.1016/j.envpol.2018.11.081.

Manzi, H.P., Abou-Shanab, R.A., Jeon, B.H., Wang, J., and Salama, E.S. (2022). Algae: A frontline photosynthetic organism in the microplastic catastrophe. *Trends in Plant Science*, *27*, 1159–1172.

Martín, J., Santos, J.L., Aparicio, I., and Alonso, E. (2022). Microplastics and associated emerging contaminants in the environment: Analysis, sorption mechanisms and effects of co-exposure. *Trends in Environmental Analytical Chemistry*, e00170. https://doi. org/10.1016/j.teac.2022.e00170

Miloloža, M., Bule, K., Ukić, Š., Cvetnić, M., Bolanča, T., Kušić, H., Bulatović, V.O., and Grgić, D.K. (2021). Ecotoxicological determination of microplastic toxicity on algae Chlorella sp.: Response surface modeling approach. *Water, Air, & Soil Pollution, 232*(8), 1–16.

Moog, D., Schmitt, J., Senger, J., Zarzycki, J., Rexer, K., Linne, U., Erb, T., and Maier, U.G. (2019). Using a marine microalga as a chassis for polyethylene terephthalate (PET) degradation. *Microbial Cell Factories, 18*(1), 171. https://doi:10.1186/s12934-019-1220-z

Parsai, T., Figueiredo, N., Dalvi, V., Martins, M., Malik, A., and Kumar, A. (2022). Implication of microplastic toxicity on functioning of microalgae in aquatic system. *Environmental Pollution, 308*, 119–626.

Peller, J., Nevers, M.B., Byappanahalli, M., Nelson, C., Babu, B.G., Evans, M.A., Kostelnik, E., Keller, M., Johnston, J., and Shidler, S. (2021). Sequestration of microfibers and other microplastics by green algae, Cladophora, in the US Great Lakes. *Environmental Pollution, 276*, 116695.

Polidoro, B., Lewis, T., and Clement, C. (2022). A screening-level human health risk assessment for microplastics and organic contaminants in near-shore marine environments in American Samoa. *Heliyon, 8*(3), e09101.

Rummel, C.D., Schäfer, H., Jahnke, A., Arp, H.P.H., and Schmitt-Jansen, M. (2022). Effects of leachates from UV-weathered microplastic on the microalgae Scenedesmus vacuolatus. *Analytical and Bioanalytical Chemistry, 414*(4), 1469–1479.

Sanniyasi, E., Gopal, R.K., Gunasekar, D.K., and Raj, P.P. (2021). Biodegradation of low-density polyethylene (LDPE) sheet by microalga, Uronema africanum Borge. *Scientific Reports, 11*, 172–133. https://doi.org/10.1038/s41598-021-96315-6

Sharma, S., and Chatterjee, S. (2017). Microplastic pollution, a threat to marine ecosystem and human health: A short review. *Environmental Science and Pollution Research*. https:// doi:10.1007/s11356-017-9910-8

Shruti, V.C., Pérez-Guevara, F., and Kutralam-Muniasamy, G. (2021). The current state of microplastic pollution in the world's largest gulf and its future directions. *Environmental Pollution, 291*, 118–142.

Simon, M., Hartmann, N.B., and Vollertsen, J. (2021). Accelerated weathering increases the release of toxic leachates from microplastic particles as demonstrated through altered toxicity to the green algae raphidocelis subcapitata. *Toxics, 9*(8), 185.

Sørensen, L., Rogers, E., Altin, D., Salaberria, I., and Booth, A.M. (2020). Sorption of PAHs to microplastic and their bioavailability and toxicity to marine copepods under co-exposure conditions. *Environmental Pollution, 258*, 113–844.

Su, Y., Cheng, Z., Hou, Y., Lin, S., Gao, L., Wang, Z., Bao, R., and Peng, L. (2022). Biodegradable and conventional microplastics posed similar toxicity to marine algae Chlorella vulgaris. *Aquatic Toxicology, 244*, 106–097.

Thiagarajan, V., Iswarya, V.P.A.J., Seenivasan, R., Chandrasekaran, N., and Mukherjee, A. (2019). Influence of differently functionalized polystyrene microplastics on the toxic effects of P25 TiO2 NPs towards marine algae Chlorella sp. *Aquatic Toxicology, 207*, 208–216.

Thiruchelvi, R., Das, A., and Sikdar, E. (2020). Bioplastics as better alternative to petro plastic. *Materials Today: Proceedings*, S221478532035269X–. https://doi:10.1016/j. matpr.2020.07.176

Vanapalli, K.R., Dubey, B.K., Sarmah, A.K., and Bhattacharya, J. (2021). Assessment of microplastic pollution in the aquatic ecosystems–An indian perspective. *Case Studies in Chemical and Environmental Engineering, 3*, 100–071.

Vecchi, S., Bianchi, J., Scalici, M., Fabroni, F., and Tomassetti, P. (2021). Field evidence for microplastic interactions in marine benthic invertebrates. *Scientific Reports*, *11*(1), 1–12.

Wang, T., Hu, M., Xu, G., Shi, H., Leung, J.Y., and Wang, Y. (2021a). Microplastic accumulation via trophic transfer: Can a predatory crab counter the adverse effects of microplastics by body defence? *Science of the Total Environment*, *754*, 142099.

Wang, Z., Fu, D., Gao, L., Qi, H., Su, Y., and Peng, L. (2021b). Aged microplastics decrease the bioavailability of coexisting heavy metals to microalga Chlorella vulgaris. *Ecotoxicology and Environmental Safety*, *217*, 112–199.

Xu, C., Zhang, B., Gu, C., Shen, C., Yin, S., Aamir, M., and Li, F. (2020). Are we underestimating the sources of microplastic pollution in terrestrial environment? *Journal of Hazardous Materials*. https://doi.org/10.1016/j.jhazmat.2020.123228

Xu, X., Wang, S., Gao, F., Li, J., Zheng, L., Sun, C., He, C., Wang, Z., and Qu, L. (2019). Marine microplastic-associated bacterial community succession in response to geography, exposure time, and plastic type in China's coastal seawaters. *Marine Pollution Bulletin*, *145*, 278–286. https://doi.org/10.1016/j.marpolbul.2019.05.036

Yan, B., Liu, Q., Li, J., Wang, C., Li, Y., and Zhang, C. (2021). Microplastic pollution in marine environment: Occurrence, fate, and effects (with a specific focus on Biogeochemical Carbon and Nitrogen Cycles). In Muthu, S.S. (Ed.). *Microplastic Pollution* (pp. 105–126). Springer, Singapore.

Yi, X., Chi, T., Li, Z., Wang, J., Yu, M., Wu, M., and Zhou, H. (2019). Combined effect of polystyrene plastics and triphenyltin chloride on the green algae Chlorella pyrenoidosa. *Environmental Science and Pollution Research*, *26*(15), 15011–15018.

Zhang, F., Wang, Z., Song, L., Fang, H., and Wang, D.G. (2020). Aquatic toxicity of iron-oxide-doped microplastics to Chlorella pyrenoidosa and Daphnia magna. *Environmental Pollution*, *257*, 113–451.

Zheng, X., Liu, X., Zhang, L., Wang, Z., Yuan, Y., Li, J., Li, Y., Huang, H., Cao, X., and Fan, Z. (2022). Toxicity mechanism of Nylon microplastics on Microcystis aeruginosa through three pathways: Photosynthesis, oxidative stress and energy metabolism. *Journal of Hazardous Materials*, *426*, 128094.

Zhu, Z.L., Wang, S.C., Zhao, F.F., Wang, S.G., Liu, F.F., and Liu, G.Z. (2019). Joint toxicity of microplastics with triclosan to marine microalgae Skeletonema costatum. *Environmental Pollution*, *246*, 509–517.

Zocchi, M., and Sommaruga, R. (2019). Microplastics modify the toxicity of glyphosate on Daphnia magna. *Science of the Total Environment*, *697*, 134–194.

16 Bioremediation of dye

Oluwafemi A. Oyewole, Daniel Gana,
Binta Buba Adamu, Abraham O. Ayanwale,
and Evans C. Egwim

CONTENTS

DOI: 10.1201/9781003394600-19

16.1   INTRODUCTION

Dyes are organic substances that permanently and irreversibly alter the natural color of whatever they are applied to. Dyes achieve this color change by adhering to compatible surfaces in solutions or through the formation of covalent complexes, physical adsorption or mechanical retention (Guha, 2019; Benkhaya *et al.*, 2020a,b). Dyes exhibit a high level of solubility in water due to the functional groups present in their structure. This characteristic makes it hard to remove them using conventional methods. They also possess chromophore within their structures that allow them to impact color on the right substrates (Drumond *et al.*, 2013). The color fixing property is related to auxotrophic groups that are polar and able to bind to polar groups of substrates.

16.2   CLASSIFICATION OF DYES

16.2.1   Classification based on source

Dyes are classified as synthetic and natural dyes. Natural dyes are the colorants that are mostly obtained from vegetable or animal (mainly insects) sources and do not undergo any chemical processing. Many natural dyes are extracted from flowers, roots, and fruits of certain dye producing plants. Natural dyes are eco-friendly, least polluting, and least toxic thus safe for use in the textile industry. The use and acceptance of synthetic dyes became widespread in the 19th century when it replaced natural dyes. Synthetic dyes became widely accepted due to their cost-effectiveness and consistency in the production of colors (Yusuf *et al.*, 2017). Synthetic dyes are able to produce different color shades and offered a wider range of application. However, the only setback associated with synthetic dyes is their toxicity, which negatively impact the environment. Also, synthetic dyes are known to contain non-biodegradable components, which imbued them with their recalcitrant properties (Benkhaya *et al.*, 2020a,b). The synthetic dyes are aromatic compounds that have aromatic rings within their structures. The rings of synthetic dyes contain delocalized electrons and varied functional groups. Colors in the dyes arise from the chromogene-chromophore, an electron acceptor within the dye molecule. Its dyeing capacity relies on the presence of auxochrome groups that donate valance electrons.

The chromogene is a complex aromatic structure containing benzene, anthracene, naphthalene, rings that carry binding chromophores that have conjugated links and several pairs of delocalized electrons, which form conjugated systems. Azo groups ($-N=N-$), ethylene group ($=C=C=$), carbonyl group ($=C=O$), nitro ($-NO_2$), carbon-sulfur ($=C=S$), nitrozo ($-N=O$), chinoid, or carbon–nitrogen groups ($=C=NH$) mainly act as chromophores.

The auxochrome groups can be easily ionized and are responsible for the binding capacity of dye molecules to textiles. Commonly known auxochrome groups are sulfonate (SO_3H), amino ($-NH_2$), carboxyl ($-COOH$), and the hydroxyl ($-OH$) (Saini, 2017).

16.2.2   Classification based on chemical structures/applications

Classifying dyes based on chemical structure is the most recommended means of classification. Under this criterion, dyes are classified as azo and anthraquinone dyes. The other chemical classes of dyes employed in industrial scale include phthalocyanine, arylmethane, xanthene, and indigoid derivatives. While Azo dyes were observed to offer cost-effectiveness, the anthraquinone dyes were expensive and weaker in producing the desired results. Based on their usage and applications, dyes are classified by the two main cloth fibers, which are cotton and polyester. Other classifications include direct, basic, acidic, vat, sulfur, disperse, azoic, and solvent dyes (Saini, 2017; Benkhaya et al., 2020a,b). The use of dyes is not restricted to the textile industry only; it extends to painting, printing, food, and cosmetics industries and a few others (Hunger, 2002).

Over the past decades, there is a continuous upward trend in the use of synthetic dyes globally, it is estimated that over 7×10^7 tons of synthetic dyes are manufactured yearly across the globe, with over 10,000 of such dyes utilized in industrial activities (Benkhaya et al., 2020a,b). These dyes are utilized in batch, continuous, or semi-continuous processes (Drumond et al., 2013).

16.3 CHEMICAL STRUCTURE OF AZO AND ANTHRAQUINONE DYES

16.3.1 AZO DYES

Azo dye refers to a group of organic textile colorants that came into limelight in the 1880s. It accounts for over 70% of all dyes produced annually (Concerns, 2013; Sarkar et al., 2017). The term "Azo" came from the presence of an azo group (R-N=N-R') where R and R' are usually aryl. Azo dyes are important members of the azo family (i.e. compounds containing R-N=N-R'). Dyes containing an azo functional group make up a large family of organic dyes. Azo dyes are classified as azo, diazo, trisazo, or poliazo, based on the number of azo groups within their structure (Pereira & Alves, 2012).

The most commonly encountered types of azo dyes are the acid dyes for polyamide and protein substrate like silk, wool, and nylon, disperse dyes for hydrophobic substrates like acetate and polyester and the reactive and direct dyes for cellulose based substrates like paper, linen, rayon, and cotton. Azo dyes are formed via two important steps: (i) the conversion of an aromatic amine (represented as Am) to a diazo compound via a process known as diazotization, (ii) the reaction of the formed diazo compound with a phenol, naphthol aromatic amine or compounds that possess an active methylene group to produce the corresponding azo dye via a process known as diazo coupling.

The processes involved in the formation of azo dyes are shown in the following equations:

STEP I: Diazotization

$$\text{Am-N}_2 \rightarrow \text{Am}^+ \qquad\qquad \text{Equation 1}$$

STEP II: Diazo coupling

$$\text{Am-N}_2^+ + \text{Am'-OH} \rightarrow \text{Am-N=N-Am'-OH} \qquad\qquad \text{Equation 2}$$

Azo dyes have found use in treating leather articles, textiles, cosmetics, waxes, mineral oils, papers, fibers, and some foodstuff. Although azo pigments are closely related to azo dyes, they tend to exhibit high insolubility in water and other solvents. Azo dyes are effective at 60°C hence their popularity for dyeing cloths. An Azo-free-dyes will require 100°C for effectiveness thus leading to waste of valuable resources and time. Azo dyes are available in different colors with distinctive color fastness, which allows them to adhere to surfaces at a rate four times higher than other form of dyes. These characteristics makes azo dyes an invaluable addition to the textile industry (Benkhaya *et al.*, 2020a,b).

Even with the level of effectiveness and benefits displayed by azo dyes, some of them like acid violet 7, Congo red, Disperse Red 1 and 3, reactive black, malachite green, and reactive brilliant red, may pose a serious health challenge to animals and humans especially when used on certain textiles or present in water supplies (Sarkar *et al.*, 2017). When azo dyes from dyeing factories gets discharged into water systems, they can damage ecosystem. This has been a major challenge particularly in developing countries where there is little or no restrictions guiding the use of dye. Sometimes, the functional group (azo group) in azo dyes can be cleaved to produce probable hazardous substances known as aromatic amines. About 24 carcinogenic aromatic amines have been identified from the cleavage of azo groups in dyes and about 5% of azo dyes are capable of cleaving to form these dangerous human carcinogens (Concerns, 2013).

16.3.2 ANTHRAQUINONE DYES

The anthraquinone dyes are large group of dyes with the anthraquinone as their functional group. Anthraquinone is the second largest categories of industrial dyes after the azo dyes. They feature a vast collection of colors appearing in the completely visible spectrum (Islam & Mostafa, 2019). The anthraquinone ring in itself, is colorless but can be manipulated to form red to blue dyes by introducing an electron donor such as amino or hydroxyl group in the 1–4-, 5-, or 8-positions. The point of substitution and the functional group being added determines the color of the anthraquinone dye. Anthraquinone dyes are often classified as carbonyl dyes together with indigo dyestuff.

Members of the anthraquinone family can be found in natural and synthetic dyes. They are often represented in vat and mordant even though they are reactive and disperse dyes. Anthraquinone dyes are characterized by an excellent level of light fastness. Most red dyes contain the anthraquinone structure. These dyes come from both plant and animal origins and can combine with metal salts to form complexes. The resulting metal complex from this combination exhibit good wash fastness (Deitersen *et al.*, 2019). In as much as anthraquinone is a parent chain for anthraquinone dyes, not every colored derivative of anthraquinone is a dye. The three major types of anthraquinone dyes include neutral, anionic, and cationic (Deitersen *et al.*, 2019).

The formation of anthraquinone dyes proceeds *via* the synthesis of the nucleus or the electrophilic substitution of the unsubstituted anthraquinone. The most common methods of producing anthraquinone dyes is by nitration and sulfonation in the α- and β-positions. About 80% of all the important anthraquinone dyes were based on anthraquinone sulfonic acids. This method has been replaced by the nitration of anthraquinone, which is triggered by environmental considerations thus gaining more traction. It has been observed that the sulfonation of anthraquinones generates large volumes of waste dilute acids that may lead to environmental pollution (Hunger, 2002).

16.4 INDUSTRIAL DISCHARGE OF DYE TO THE ENVIRONMENT

Textile industries are the major consumers of dyes; they make use of large quantities of water and give off 90% of this as waste. Wastewater from textile industry serves as a nuisance to environmental health as the constituents being majorly dyes are recalcitrant and toxic. Textile industries serve as potent threat to the environment if effluents are not managed properly (Mondal *et al.*, 2017). The industrial discharge of dyes negatively affects the economic development of any nation. The impact of this activity has been felt in major exporters of dyes like China, the Europe, India, and the USA (Yaseen & Scholz, 2019).

Although industries accounts for the bulk of dye discharge, it is pertinent to note that both natural and synthetic dyes are also present in household wastes. They can be discharged from homes, wastewater treatment plants and several types of industries with emphasis on those that produce, extract, and/or apply dyes. In the textile industry, dyes are mostly washed off in the course of the dyeing process and huge amounts of residual dyes end up in the effluent. The higher concentrations of dyes in effluent are as a result of the dyes escaping conventional effluent water treatment processes and are thus discharged into the environment (dos Santos *et al.*, 2007). The unspent dyes from industrial processes, normally gives coloration to the effluent, and the color concentrations vary between 10 and 200 ppm.

Dye bearing waste water from the industries possess some effects including carcinogenicity and aesthetic pollution of the environment. This effect is brought about by the degrading nature of their products (Sivakumar & Palanisamy, 2008). Effluents from dye industry come with a high concentration of colors, suspended solids, pH, biochemical oxygen demand, chemical oxygen demand, salts, temperature, and metals (Yaseen & Scholz, 2019).

Several factors can be responsible for the high toxicity of industrial waste water contaminated by dyes. The factors may include the presence of acetic acids, nitrates, vat dyes, naphthol, Sulfur, mercury, chromium, nickel, arsenic, cadmium, cobalt, and cobalt. The other possible chemicals that may be present in textile effluent are formaldehyde-based dye fixing agents, auxiliary chemicals, hydrocarbon-based softeners, chlorinated stain removers, and non-biodegradable dyeing chemicals.

16.5 ENVIRONMENTAL IMPACT OF DYES

The liquid and solid effluents from the dyeing industry serves as major contributing factor to the destruction of the environment, aquaculture, agriculture, ecology, and public health. In ancient times, people have relied solely on the use of natural dyes for adding color to textile materials. Due to insufficiency, and increased demand for dye, the dyeing industry had to diversify into the production of synthetic dyes to meet the market demand at the expense of environmental and human health. The synthetic dyes used in the manufacturing industry contain wide range of chemicals napthol, sulfur, nitrates, enzymes, and soaps. These chemicals are not retained in the finished product and end up being discharged into the environment where they retain their toxicity and carcinogenicity and are less eco-friendly (Mia *et al.*, 2019). It was reported by Kjellstrom *et al.* (2006) that water, soil, and air contaminated by effluents are associated with heavy disease burdens. When these effluents (solid or liquid) are released into the environment, they constitute a nuisance and predispose the environment to the various negative environmental impacts.

The dyeing industries require large tons of water for their manufacturing processes. This water is mixed with hazardous chemicals that are later released into the environment as waste water. Waste water from the dyeing industry is one of the most heavily polluted water due to its composition (Mia *et al.*, 2019). Every year, the textile industry converts about 2,000,000 tons of dye to effluent during dyeing, finishing, and printing operations. This huge amount of waste water is associated with inefficient dyeing processes employed by many small to medium scale companies. Majority of these companies cannot afford waste water treatment before its discharge. The concentration of dyes converted into effluent differ from one country to the other and is determined by the need for dyes (Mia *et al.*, 2019). Effluents from textile dyeing industry can reduce soil productivity and crop yield (Islam & Mostafa, 2019).

The increase in demand for synthetic dyes tends to produce a corresponding increase in the production of waste water after the dyeing process. The waste water (the major source of water pollution) may be heavily contaminated with dyes, which are non-biodegradable and possess a high thermal and photo stability.

Of the 800,000 tons of dye produced in the world, up to 15% of the synthetic dyes are lost during their industrial applications. Due to their invaluable role in the cosmetic, leather, pharmaceutical, food, paper printing, and textile industries, this concentration tends to increase every year (Hassaan & Nemr, 2017).

Azo dyes are highly electron deficient since they possess a strong nitrogen–nitrogen bond at the center of the ring. The nitrogen bond also contributes to their carcinogenicity and toxicity especially, during a reductive cleavage of the bonds. Azo dyes can alter the chemical composition and the physical features of the soil, deteriorate the overall health of water bodies and threatens the autochthonous micro-fauna and flora existing in such environments. The toxic nature of dyes can kill soil microorganisms, which in turn may affect the agricultural productivity (Hassaan & Nemr, 2017).

In waterbodies, dyes may persist as pollutants and provides biomagnification as they cross the entire food chains. This effect makes the organisms at higher trophic levels to display higher contamination levels as compared to the preys. The recalcitrant nature of the dyes is harmful to the environment because it influences the nature and functioning of ecosystems (Mondal *et al.*, 2017). Long term exposure to the dyes has detrimental impacts on human health and aquatic life. This scenario is common in complexed metal dyes. The complexed metal dyes are commonly used in the textile industry because of their longer half-lives and resistance. Their detrimental impacts arise from the fact that they possess heavy metals such as nickel, cobalt, copper, and chromium (Elango *et al.*, 2017). The heavy metals cause harm in the aquatic environment because they become absorbed into the gills of fish and eventually accumulate in their tissues. They get into the human body through the food chain and cause various pathologies. The recalcitrant character also causes other challenges in terms of oxidative stress, which originates from chromium textile dyes. It eventually damages the growth and development of plants during assimilation of CO_2 and photosynthesis.

Synthetic dyes cause cancer especially bladder cancer arising from the presence of 2-naphathalamine and benzidine (Tüfekci et al., 2007). Other diseases caused by dyes include disorders of the central nervous system and dermatitis. Dyes also cause irritations to the skin and eyes owing to the acute toxicity that often caused by inhalation and oral ingestion. Some people may develop contact dermatitis, allergic reactions and occupational asthma (Drumond et al., 2013). They arise from the formation of a conjugate between human serum albumin and reactive dyes (Elango et al., 2017). The dye acts as an antigen and produces immunoglobin, which combines with histamine to induce the allergic reaction. Textile Dyes have long term hazardous impacts on human health because of their genetoxity. Their mutagenic potential arises from the ability to intercalate with the helical structures within the DNA (Lellis et al., 2019).

It is easy to spot azo dyes in water even at a concentration <1 ppm. This tends to affect the water–gas solubility, transparency, and aesthetic merit. The reducing light penetration caused by azo dyes can decrease photosynthesis, deplete the dissolved oxygen concentrations and deregulate the natural cycles in aquatic communities. Exposure to azo dyes can have an acute or chronic effect on organisms, in most cases, its severity is based on the concentration of the dyes and the exposure duration. For example, aromatic amine like 1,4-diamino benzene can cause permanent chemosis, exopthamlmose, lacrimation, blindness, contact dermatitis, skin irritations, vertigo, hypertension, gastritis, vomiting, acute tubular necrosis supervene, and rhabdomyolysis. After ingestion, aromatic amine containing azo parent chain can lead to edema of the face, pharynx, larynx, neck, tongue, and the irritation of the respiratory tract. Since aromatic amines are mobilized by sweat or water, they exhibit a high level of absorption through expose body surfaces such as the skin and mouth. The metabolism of soluble azo dyes by the liver enzymes has a severe effect on the liver and other digestive tract (Hassaan & Nemr, 2017).

During wet processing, a lot of gaseous emissions are released into air thus tampering with its purity. It is pertinent to note that air pollution occurs via the release of different levels of gases like SO_2, NO_2, CO_2 and others. Researchers have identified these gases as the second greatest pollution challenge for the printing and dyeing industries. Dye may occur as particles in the air, which can be poisonous when inhaled. After inhalation, these dye particles in the air can affect a person's immune system and in extreme cases, may lead to respiratory sensitization during the next encounter. The common symptoms linked with respiratory sensitization includes, wheezing, coughing, sneezing, watery eyes, and itching. They may also produce some symptoms similar to asthma (Mia et al., 2019).

The most commonly encountered health challenges associated with the presence of dye particles in the air arise from the exposure of chemicals acting as irritants. They are capable of causing sore eyes, sneezing, blocked or itchy noses, and skin irritations. These dye particles and irritants include ammonia, formaldehyde-based resins, acetic acid, optical whiteners, bleach, caustic soda, soda ash, and some shrink-resist chemicals. Some disperse, vat, and reactive dyes have also been recognized to be skin sensitive (Mia *et al.*, 2019). Thus, the textile dyes have numerous detrimental impacts on the environment owing to their unique characteristics.

16.6   REGULATIONS GOVERNING DYE DISCHARGE TO THE ENVIRONMENT

Every product must be carefully analyzed before being released into the market and environment. It is the analyses of these products that will determine their level of safety. Apart from identifying their technical and application properties, there is need to establish the toxicological and ecological safety. Dyes are no exception to this rule as all manufacturers, customers and processors, must to a large extent, find out if the product is safe for use or not. Before any dye is accepted into the market, it must have met the requirements stated by the legal provisions in that region (Hunger, 2002).

The textile industries have been identified as a key contributor of heavy pollutants found in the air, water bodies, and land. Their ability to generate lots of noise, waste water, and solids has been a major source of concern around the world. The indiscriminate discharge of untreated effluents from industrial activities into the environments, necessitate the need for effective regulation for the discharge of dyes (Concerns, 2013). As harmful as dyes can be to the environment, there is no any universal law regulating the discharge of waste waters from the textile industries into the environment, including those containing azo dyes. Also, the permissible level of dyes in effluent discharge varies from countries to countries, with little or no information about these values in many countries (Drumond *et al.*, 2013).

Developed countries such as the USA, Canada, European nations, and Australia have enforced an environmental legalization that establishes dye effluent limits. In most cases, the legislation enforced by the USA and the EU has served as the reference point for many countries who wish to enforce legislations on the release of dyes into the environment. Countries like Thailand and Morocco have successfully adopted the US and EU systems respectively. In countries like Malaysia, Pakistan, and India, the use of effluent contamination limits is recommended but not compulsory. Countries like China where strict environmental management strategies have been enforced, there is a great deal of attention to pollutants from all sources. The Chinese authorities ensure that over 90% of effluents from government-owned dyeing industries are adequately treated with about 70% of such wastewater attaining the standard required by the national regulating bodies. Furthermore, privately controlled enterprises are compelled to adhere to the national standards, majority of the wastewater from the private sector are treated, with only a small fraction of such wastewater discharged as untreated effluents (Drumond *et al.*, 2013).

Due to level of toxicity and harm they can inflict on the environment, the EU, in 2002, banned the use of azo dyes since they could break down into any of the 24 possible carcinogenic products (Brüschweiler & Merlot, 2017). In 2009, Nigeria identified dye and paints as hazardous wastes as part of its National implementation plan for Stockholm Convention on Persistent Organic Pollutants (Federal Ministry of Environment, FMENV, 2009). The azo group can transform into aromatic amines by cleavage of the bonds. Aromatic amines are potentially dangerous and may be responsible for the carcinogenic and mutagenic activities of azo dyes. Aromatic amines are produced from the incomplete synthesis of azo dyes or its degradation, as shown in the Equation 3. The digestive tract of animals plays an integral role in meeting the requirement for this reaction. Although the regulations guiding the discharge of dyes may differ from one country to the other, it must be according to the following principles as stated by Hunger (2002).

$$\underset{\text{Azo dye}}{\text{A-N} = \text{N-B}} \xrightarrow[\text{pH 6}]{\text{Na}_2\text{S}_2\text{O}_4} \underset{\text{amines}}{\text{A-NH}_2 + \text{B-NH}_2} \qquad \text{Equation 3}$$

The EU and other industrialized nations require that all chemicals (including dyes) be registered with the right body. Registering the product will require important data such as physiochemical, toxicological, and ecotoxicological studies. The physiochemical status such as solubility, melting/boiling point, and vapor pressure must be well established while the toxicological parameters include mutagenicity and acute toxicity tests. The ecotoxicological studies evaluate the impact of such product when they get into the environment, including its toxicity to animals, biodegradability, and biological oxygen demand characteristics. Some of the important chemical and environmental laws employed by different countries concerning the use and release of dye into the environment was reported by Hunger (2002) (Table 16.1).

Conventional physio-chemical methods of treating dye contamination, which include ion exchange, membrane processes, and activated carbon absorption are in use to treat dye-containing wastewaters from textile industries before their discharge. A very effective physio-chemical remediation for dyes in wastewater is the Advanced Oxidation Process (AOP) (Garrido-Cardenas et al., 2019).

AOP is based on the principle that free metallic radicals when generated can be oxidized by complex organic compounds. AOP results in complete or partial remediation of dyes and can be used as a pretreatment before bioremediation. In general, the physio-chemical method has a limitation since the toxic organic compounds are not totally eradicated but transformed to less toxic compounds. From a bioremediation approach, activated sludge may be used for the treatment of dye contamination (Mondal et al., 2017).

The establishment of the Ecological and Toxicological Association of Dyestuffs Manufacturing Industry in 1974, brought about an increased need to decontaminate dyes before releasing them into the environment. This legislation was not only focused on reducing harmful effect of dye in the environment but to also protect final consumers and dye users while cooperating with public and government concerns over the toxicological impacts of dyes (Robinson et al., 2001; Pereira & Alves, 2012). Some of the commonly utilized methods of dye remediation were discussed by Robinson et al. (2001) and Pereira and Alves (2012) to include:

TABLE 16.1

Overview of remediation strategies of dye contamination

S/no	Country/region	Legislation
1	European Union	• Directive for the Classification, Packaging, and Labelling of Dangerous Substances (67/548/EEC), and several amendments and adaptions, as last amended by Dir. 2001/59/EC • Directive for the Classification, Packaging, and Labelling of Dangerous Preparations (88/379/EEC), and amendments. • Restrictions on the Marketing and Use of Certain Dangerous Substances and Preparations 76/769/EEC, and amendments, as last amended by Dir. 2002/61/EC • Directive on the Control of Major Accident Hazards involving Dangerous Substances (96/82/EC) • Protection of Workers from the Risk to Exposure to Carcinogens at Work (89/391/EEC, 90/394/EEC) • Directive concerning Integrated Pollution Prevention (96/61/EC) • MSDS Directive 91/155/EEC
2	Germany	List of water-endangering substances under the Federal Water Act (2001)
3.	Switzerland	• Poison Law (1969) • Environmental Protection Act (1985)
4.	USA	• Toxic Substance Control Act (1976) • OSHA Hazard Communication Standard (1985)
6	Canada	• Canadian Environmental Protection Act (1994)
7	Japan	• Chemical Substance Control Law (1973) • Industrial Safety and Health Law (1972)
8	Australia	• National Industrial Chemicals Notification and Assessment Scheme (1990)

16.6.1 CHEMICAL METHODS OF DYE CONTAMINATION REMEDIATION

This involves the use of oxidizing agents like hydrogen peroxide (H_2O_2) for the remediation of dyes. Its simplicity makes it one of the most commonly utilized methods of treating effluent from dye industries. The oxidizing agent can cleave the aromatic ring of dyes thus removing them from dye-containing effluent. Some of the commonly utilized oxidative processes used for removing dye from dye containing effluent include but are not limited to ozonation, photochemical, sodium hypochloride (NaOCl), cucurbituril, electrochemical destruction, and Fentons reagent ($H_2O_2 \pm$ Fe(II) salts).

In photochemical reduction, the degradation of dye molecules into H_2O and CO_2 occurred *via* treatment with ultraviolet (UV) light in the presence of hydrogen peroxide (H_2O_2). The UV light activates the cleaving of the hydrogen peroxide (H_2O_2) into two hydroxyl radicals (HO–), leading to the oxidation of organic material thus removing them from effluent (Gomes da Silva & Faria, 2003).

16.6.2 Physical Method for Dye Contamination Remediation

Researchers like Kandisa and Saibaba (Kandisa and Saibaba, 2016; Mokif, 2019), explained how adsorption techniques are efficient at producing excellent results during the bioremediation of dye. The adsorption agents are highly efficient in removing pollutant that are too stable for conventional methods. Adsorption techniques are based on two mechanisms: ion exchange and adsorption. The method is economically feasible. The use of physical methods/treatments is influenced by several physio-chemical factors such as particle size, contact time, sorbent surface area, pH, and dye/sorbent interactions. The commonly utilized mechanism used for this process include the use of wood, peat, activated carbon, mixture of coal and fly ash, silica gel, irradiation, ion exchange, membrane filtration, electro-kinetic coagulation, and other materials like clay, corn cob, and hulls.

16.6.3 Biological Techniques of Dye Contamination Remediation

This process involves the use of living cells and their products for breaking down the complex dye structures (Bhatia *et al.*, 2017). It places so much attention on the ability of microorganisms to produce different enzymes and metabolites that can cleave the complex dye structures and remove them from effluents thus keeping the environment safe. Some of the commonly utilized methods include the decolorization of dyes by white rot fungi, microbial cultures, and adsorption by living/dead microbial biomass.

16.7 THE CONCEPT OF BIOREMEDIATION OF DYE CONTAMINATED ENVIRONMENTS

Microorganisms are found everywhere from heavily contaminated soils and water bodies to production plants and assembly location. These microorganisms are equipped with the right resources for survival in all environments (including regions where extremes of temperature are prevalent). As they interact with the biotic and abiotic factors in their environment, they are forced to produce enzymes and metabolites that assist them in survival and food production. These abilities are very useful in degrading complex, harmful molecules to their smaller units, which are less toxic to the environment.

The use of microorganisms for the bioremediation of dye contaminated soils have been reported by several researchers and their mechanism of action proceeds *via* two important processes: aerobic and anaerobic. There are also reports on how these processes can be combined for better results (Erkurt *et al.*, 2010; Dave *et al.*, 2015; Jamee and Siddique, 2019).

16.8 MICROORGANISMS INVOLVED IN BIOREMEDIATION OF DYE

16.8.1 Bacteria

Several reports have shown the role of aerobic, facultative aerobic, and anaerobic bacterial species in the degradation of dyes. The first attempt to isolate a pure bacterial culture that can degrade azo dyes was dated to as far back as the 1970s when pure cultures of *Bacillus cereus, Aeromonas hydrophila*, and *Bacillus subtilis* were utilized (Dave *et al.*, 2015). The use of bacterial monoculture such as the *Pseudomonas* sp., *Pseudomonas luteola,* and *Proteus mirabilis* has demonstrated a great promise in the breakdown of azo dyes under anaerobic conditions (Saratale *et al.*, 2011). In addition to this, cultures of *Comamonas* sp., *Rhizobium radiobacter, Sphingomonas* sp., *Exiguobacterium* sp., and *Desulfovibrio desulfuricans,* were reported to decolorize various types of commercial dyes. Under aerobic conditions, azo dyes cannot undergo bacterial degradation, since the azo reductase is inhibited by the presence of O_2. However, some species of *A. hydrophila* and *Micrococcus* sp., have been shown to demonstrate significant degradation of azo dyes under aerobic conditions (Dave *et al.*, 2015).

Decolorization of polymeric dyes by live bacteria is related to capabilities of the isolate to form ligands. Pasti and Crowford (Paszczynski *et al.*, 1992) proposed that peroxidases within the bacteria convert the azo dye to cation radicals that can be easily attacked by the water or hydrogen peroxide nucleophile. This process results in simultaneous splitting of the azo linkage to generate intermediates (Srinivasan & Viraraghavan, 2010). These intermediates further undergo redox reactions to produce stable intermediates.

Bacterial species are also capable of degrading dyes in consortium. Mixed cultures containing three isolates of *P. polymyxa, M. luteus,* and *Micrococcus* sp., were found to decolorize nine dyes even though they were incapable of achieving such a feat while acting singly. A similar result was obtained with a consortium of four bacterial isolates containing, *Stenotrophomonas acidaminiphila, P. fluorescence, B. cereus,* and *P. putida* (Moosvi *et al.*, 2005). Table 16.2 gives a summary of different bacterial species capable of degrading dye and the type of dye they degrade.

Fungal cells are able to degrade dyes because they are capable of producing laccase, lignin modifying enzymes, lignin peroxidase, and manganese peroxidase. These compounds have unspecific activity and this attribute is utilized in the initial process of mineralization of dyes (Vrsanska *et al.*, 2016). The extents of contribution of laccase, MnP, and LiP in mineralization of dyes vary depending on the genera of fungi. Biosorption mechanisms also play important roles in the process of decolorizing dyes using living fungi (Srinivasan & Viraraghavan, 2010). The mechanism for decolorization of dyes

TABLE 16.2
Bacterial species and the type of dye degraded

Bacterial species	Dye degraded	Reference
Enterobacter sp. CV–S1	Crystal violet	Roy *et al.* (2018)
Bacillus megaterium KY848339	Azo dye acid red	Ewida *et al.* (2019)
Bacillus cereus	Dye contaminated effluent	Ayisa *et al.* (2018)
Bacillus amyloliquefaciens	alizarin red S	Ito *et al.* (2018)
Micrococcus luteus	Congo red	Ito *et al.* (2018)
Pleurotus sp.	Methylene blue	Van Der Maas *et al.* (2018)
Alcaligenes aquatilis	Synazol red 6HBN	Ajaz *et al.* (2019)

in wastewater using dead fungi is biosoprtion and it entails various physicochemical interactions such as deposition, adsorption, and ion exchange.

Fu and Viraraghavan (Fu & Viraraghavan, 2003) investigated investigated the role of functional groups present in fungal biomass of *Asperigillus niger* such as amino, lipid fractions, carboxyl, and phospahate in the biosorption of dyes. They discovered that amino and carboxyl groups were the main binding sites during biosorption of Acid Blue 9 while amino group was the main binding site in the biosorption of Acid Blue 29 (Fu & Viraraghavan, 2003). The primary mechanism for this process is belived to be electrostatic attraction.

Although very little information is available on the use of yeast for the degradation and decoloration of dyes, some researchers have reported the activity of yeast strains like *Issatchenkia occidental,* and *Candida zeylanoides* (Pereira & Alves, 2012). Crystal violet dyes can also be degraded using oxidative yeasts within four days (Srinivasan & Viraraghavan, 2010). Some yeast stratins such as *Candida Zeylanoides* can degrade several azo dyes and the reduction products from this process incldue sulfalinic acids, metalinilic acid. Aksu (2003) also discovered that *Saccharomyces cerevisiae* is efective in removing dyes from mollasses media.

16.8.2 Algae and cyanobacteria

Algal and cyanobacterial species can utilize aniline, a degradation product from the breakdown of azo dyes. There are reports on the ability of *Spirogyra* species and blue-green algae to treat azo dye effluents. Blue-green algal species can remove pyrene from solutions *via* biotransformation or bioaccumulation. Cyanobacteria and algal species like *C. vulgaris, Volvox aureus, Elkatothrix viridis, Oscillatoria rubescens, Nostoc linckia*, and *Lyngbya lagerlerimi* were isolated from sites heavily contaminated with dyes and are proposed to possess the right enzymes for the degradation of dyes (El-Sheekh et al., 2009).

16.9 MECHANISM OF DYE BIOREMEDIATION

Bioremediation is an important method of recovering polluted soils and water from heavy dye contamination. It serves as an attractive alternative to the physical and chemical methods of recovery since it uses natural products like microbial enzymes and metabolites. Its cost-efficiency and eco-friendliness makes it one of the best methods with a wide range of public acceptance. Several microorganisms have been isolated and shown to possess the right apparatus for breaking down dyes in contaminated soils and water bodies. These microorganisms utilize several mechanisms, which can be aerobic, anaerobic or a combination of both depending on their biochemistry and nature of the contaminated location (Keharia & Madamwar, 2003).

16.9.1 Aerobic mechanism of bioremediation

This method of treating dye is inefficient for two reasons; dyes are highly stable to biological oxidation and they exhibit a poor absorption to activated sewage sludge. It is pertinent to note that dyes are designed to be oxidatively stable. Most highly soluble anionic dyes can pass through aerobic biological treatments undegraded. Highly soluble anionic dyes like Reactive Black, Reactive Blue 19, C1 reactive violet 15, were discovered to remain active even after 20 days of incubation with aerobic cultures (Stolz, 2001). Many fungal species especially those belonging to the white rot family have been shown to degrade dyes aerobically. This activity has been reported to be caused by the ability of these fungal species to produce enzymes like manganese independent peroxidases, manganese peroxidase, lignin peroxidase, and laccase. Aerobic bacteria like *Flavobacterium* ATCC 39723 and *Streptomyces* sp. can produce extracellular peroxidases that can degrade xenobiotic compounds including dyestuff (Keharia & Madamwar, 2003).

16.9.2 ANAEROBIC MECHANISM OF BIOREMEDIATION

Anaerobic bacteria have the capacity to degrade azo dyes in the absence of air. The reduction of azo dyes is an ubiquitous capacity of most anaerobic microorganisms (Stolz, 2001). The major interest in the use of anaerobic bacteria in the degradation of azo dyes emanated from the metabolism of azo dyes used as food additives. In an anoxic environment, the uncharged azo dyes are susceptible to biological oxidation thus reducing them to their corresponding amines which are less toxic. The use of bacterial anaerobic techniques have been identified as one of the most effective methods of dye bioremediation (Keharia & Madamwar, 2003).

Under anoxic conditions, bacterial species can mediate the degradation of azo dyes *via* a process azo linkage reductive cleavage. The cleavage of azo bonds is characterized by a group of soluble nonspecific cytoplasmic enzymes known as azo reductases. The azo reductase facilitates electron transfer via soluble flavins or other redox mediators to the azo dyes, reducing them to amines in the process (Russ *et al.*, 2000).

The anaerobic process offers a wide range of merits in terms of aeration requirements, low sludge degeneration, and the production of methane gas. The only limitation to this mechanism of dye bioremediation is the production of aromatic amines, which may be resistant to further enzymatic degradation.

16.9.3 CONSORTIA OF AEROBIC AND ANAEROBIC MECHANISMS

Although the anaerobic degradation of dyes occurs at a very fast rate, there is an almost impossible complete mineralization of molecules, thus generating a high accumulation of aromatic amines.

Amines are toxic, carcinogenic, and mutagenic, they can persist in the environment for a long time due to their recalcitrant nature, thus, endangering the health of man and animal. The aromatic amines that are recalcitrant to anaerobic cleavage are easily degraded by aerobic bacteria thus making treatment with both aerobic and anaerobic cultures an excellent idea (Keharia & Madamwar, 2003). The success of this method was first demonstrated for Yellow 3, a sulfonated azo dye mordant reported by Stolz (2001).

Both processes (anaerobic/aerobic) treatment is carried sequentially or simultaneously as the case may be. The sequential process may combine the anaerobic and aerobic step either alternatively in the same vessel or in a continuous system in separate vessels. The sequential and simultaneous treatments require the auxiliary substrates, which serve as a source of energy for bacteria in the anaerobic zones and as a source of reduction equivalents for the azo bond cleavage.

16.10   ADVANTAGES AND LIMITATIONS OF BIOREMEDIATION OF DYE

Bioremediation of dye is an excellent process aimed at recovering polluted sites without causing more harm. Since their activity are enzyme based, they are effective at removing recalcitrant compound. The bioremediation of dyes renders them harmless by breaking their complex bonds and reducing them to their lowest units (which, are often non-toxic). The overall process is harmless and may not pose a threat to other living organisms. The bioremediation of dyes terminates the transportation of pollutants from contaminated soil to water bodies and the air. It is cost-effective and easy to set up.

Even with the level of success achieved with the bioremediation of dyes, it is a very slow process, which may take years to complete. The entire process is only limited to compounds that are biodegradable and may even produce by-products that can constitute a nuisance in the environment.

16.11   FACTORS INFLUENCING DYE BIOREMEDIATION

16.11.1   THE NATURE OF THE DYE

Azo dyes are stable to aerobic reduction thus posing a serious challenge during bioremediation with obligate aerobes. This reduces the activity of organisms and their enzymes in breaking down the dye molecule into its corresponding smaller units. Some dye molecules are highly toxic and may even trigger a mutation in microorganisms used for bioremediation. Some dye molecules may be difficult to absorb. The by-products of dye biodegradation can be toxic or clogged reaction vessels thereby reducing efficiency of bioremediation (Pereira & Alves, 2012).

Azo dyes containing methyl, methoxy, sulfo or nitro groups have low biodegrability and may persist longer in the environment than those containing the hydroxyl or amino group. There are also reports on how azo dyes with a limited membrane permeability (e.g. sulfonated azo dyes) are impossible to be reduced intracellularly (Erkurt et al., 2010).

16.11.2   NATURE OF THE ENVIRONMENT

The microbial degradation of dye is very much dependent on the environment. As stated earlier, bioremediation of dye is faster and more effective in anoxic environments. This also plays a pivotal role in monitoring the system and determining the choice of organism for the activity. The environment may also reduce access to dye molecules for effective biodegradation (Vikrant et al., 2018). Environmental parameters like temperature, and pH are important players in dyes biodegradation. For instance, Ayisa et al. (2019) reported how temperature and pH affect the biodegradation of dye by Pseudomonas aeruginosa.

16.11.3 Type of organism involved

Although fungi, bacteria, and algae have been reported to take part in the bioremediation of dyes, it is pertinent to note that bacteria shows a better decolorization and remediation than other group of microorganisms (Jadhav *et al.*, 2016). Azo dyes exhibit a high resistance to aerobic bacterial degradation with exception to few species that produce azo reductase, an enzyme which reduces azo dyes. Strains of bacteria that reduces azo dyes aerobically may require organic carbon source to enhance the process.

16.11.4 Availability of nutrient

Nutrients have an effect on bioremediation of dyes using microorganisms. The process of dye biodegradation can be enhanced by adjusting the initial growth conditions (Srinivasan & Viraraghavan, 2010). Starch, celloboise, glucose, and maltose are good sources of carbon that is utilized in the decolorization of cotton bleaching effluents using white rot fungus. Lower concentrations of ammonium chloride and high concentration of glucose enhance the performance of *P. chrysosportium* when used to decolorize methyl violet (Sathian *et al.*, 2013). Higher concentrations of nitrogen inhibit bioremediation of Congo Red dye but it has no effect on bioremediation of dyes by *Cyathus bulleri*. Limiting the concentration of certain nutrients favors bioremediation of certain dyes using specific microorganism.

16.12   FUTURE ADVANCES IN DYE BIOREMEDIATION

An increased attention will be channeled toward exploring the efficacy of biological systems toward the degradation of recalcitrant dyes. Information from recent findings have shown that utilizing a microbial cocktail consisting of algae, fungi, and bacteria will produce a better resistance and adaptability to the environmental conditions surrounding the biodegradation site. Microbial genomes can be engineered for the development of microbial strains with better adaptation and degradation of extremely recalcitrant dyes. Reports on the direct usage of microbe-based dye degradation mechanisms *via* direct biocatalysts (like enzymes) may play an important role in reducing the time involved in preparing microbial cultures for bioremediation. Since biocatalysts are industrially lucrative, they can be reused after minimal processing thus reducing operational cost while boosting productivity. The only challenge associated with this method of treating dye polluted sites will be the cost-intensive nature of biocatalysts (Danish Khan *et al.*, 2015; Das & Mishra, 2019).

Scientists are currently looking into the use of Microbial fuel cells (MFCs) as profitable alternatives for the generation of electricity and degradation of dyes. This technology will require a combined effort from biologists, chemists, chemical engineers, and genetic engineers to be achieved. It will also open new levels of research in dye bioremediation in terms of operational capabilities, stability, and kinetics (Mishra *et al.*, 2020).

REFERENCES

Ajaz, M., Rehman, A., Khan, Z., Nisar, M. A., & Hussain, S. (2019). Degradation of azo dyes by Alcaligenes aquatilis 3c and its potential use in the wastewater treatment. *AMB Express*, *9*(1), 64. https://doi.org/10.1186/s13568-019-0788-3

Aksu, Z. (2003). Reactive dye bioaccumulation by Saccharomyces cerevisiae. *Process Biochemistry*, *38*(10), 1437–1444. https://doi.org/10.1016/S0032-9592(03)00034-7

Ayisa, T. T., Oyeleke, S. B., Oyewole, O. A., Adamu, B. B., Umar, Z., & John, W. C. (2018). Biodecolourization of dye-contaminated textile effluents using *Bacillus cereus* N27. *Nigerian Journal of Biotechnology*, *34*(1), 133. https://doi.org/10.4314/njb.v34i1.17

Ayisa T. T., Oyewole O. A., Oyeleke, S. B., Adamu B. B, & Ahmadu, H. (2019). The effects of temperature and pH on decolourization of dye contaminated soil by *Pseudomonas aeruginosa* DM1. *Journal of Laboratory Science 6*(1), 34–40.

Benkhaya, S., M'rabet, S., & El Harfi, A. (2020a). A review on classifications, recent synthesis and applications of textile dyes. In *Inorganic Chemistry Communications* (Vol. 115, p. 107891). Elsevier B.V. https://doi.org/10.1016/j.inoche.2020.107891

Benkhaya, S., M'rabet, S., & El Harfi, A. (2020b). Classifications, properties, recent synthesis and applications of azo dyes. In *Heliyon* (Vol. 6, Issue 1). Elsevier Ltd. https://doi.org/10.1016/j.heliyon.2020.e03271

Bhatia, D., Sharma, N. R., Singh, J., & Kanwar, R. S. (2017). Biological methods for textile dye removal from wastewater: A review. *Critical Reviews in Environmental Science and Technology*, *47*(19), 1836–1876. https://doi.org/10.1080/10643389.2017.1393263

Brüschweiler, B. J., & Merlot, C. (2017). Azo dyes in clothing textiles can be cleaved into a series of mutagenic aromatic amines which are not regulated yet. *Regulatory Toxicology and Pharmacology*, *88*, 214–226. https://doi.org/10.1016/j.yrtph.2017.06.012

Concerns, E. (2013). *The Environmental, Health and Economic Impacts of Textile Azo Dyes.* House of Parliament-Parliament of Science and Technology, 1–5. https://static.igem.org/mediawiki/2014/2/29/Goodbye_Azo_Dye_POSTnote.pdf

Danish Khan, M., Abdulateif, H., Ismail, I. M., Sabir, S., & Zain Khan, M. (2015). Bioelectricity generation and bioremediation of an azo-dye in a microbial fuel cell coupled activated sludge process. *PLoS One*, *10*(10), e0138448. https://doi.org/10.1371/journal.pone.0138448

Das, A., & Mishra, S. (2019). Complete biodegradation of azo dye in an integrated microbial fuel cell-aerobic system using novel bacterial consortium. *International Journal of Environmental Science and Technology*, *16*(2), 1069–1078. https://doi.org/10.1007/s13762-018-1703-1

Dave, S. R., Patel, T. L., & Tipre, D. R. (2015). Bacterial degradation of azo dye containing wastes. *Environmental Science and Engineering (Subseries: Environmental Science)*, *9783319109411*, 57–83. https://doi.org/10.1007/978-3-319-10942-8_3

Deitersen, J., El-Kashef, D. H., Proksch, P., & Stork, B. (2019). Anthraquinones and autophagy – Three rings to rule them all? In *Bioorganic and Medicinal Chemistry* (Vol. 27, Issue 20, p. 115042). Elsevier Ltd. https://doi.org/10.1016/j.bmc.2019.115042

dos Santos, A. B., Cervantes, F. J., & van Lier, J. B. (2007). Review paper on current technologies for decolourisation of textile wastewaters: Perspectives for anaerobic biotechnology. In *Bioresource Technology* (Vol. 98, Issue 12, pp. 2369–2385). https://doi.org/10.1016/j.biortech.2006.11.013

Drumond Chequer, F. M., de Oliveira, G. A. R., Anastacio Ferraz, E. R., Carvalho, J., Boldrin Zanoni, M. V., & de Oliveir, D. P. (2013). Textile dyes: Dyeing process and environmental impact. *Eco-Friendly Textile Dyeing and Finishing*. https://doi.org/10.5772/53659

Elango, G., Rathika, G., & Elango, S. (2017). Physico-chemical parameters of textile dyeing effluent and its impacts with casestudy. *International Journal of Research in Chemistry and Environment*, *7*(1), 17–24. www.ijrce.org

El-Sheekh, M. M., Gharieb, M. M., & Abou-El-Souod, G. W. (2009). Biodegradation of dyes by some green algae and cyanobacteria. *International Biodeterioration and Biodegradation, 63*(6), 699–704. https://doi.org/10.1016/j.ibiod.2009.04.010

Erkurt, E. A., Erkurt, H. A., & Unyayar, A. (2010). *Decolorization of Azo Dyes by White Rot Fungi* (pp. 157–167). https://doi.org/10.1007/698_2009_48

Ewida, A. Y. I., El-Sesy, M. E., & Abou Zeid, A. (2019). Complete degradation of azo dyeacid red 337 by Bacillus megaterium KY848339.1 isolated from textile wastewater. *Water Science, 33*(1), 154–161. https://doi.org/10.1080/11104929.2019.1688996

Federal Ministry of Environment (FMENV) (2009). *National Implementation Plan for the Stockholm Convention Federal Republic of Nigeria National Implementation Plan for the Stockholm Convention on Persistent Organic Pollutants (POPS)*. Final Report Federal Ministry of Environment, Abuja.

Fu, Y., & Viraraghavan, T. (2003). Column studies for biosorption of dyes from aqueous solutions on immobilised Aspergillus niger fungal biomass. *Water SA, 29*(4), 465–472. https://doi.org/10.4314/wsa.v29i4.5054

Garrido-Cardenas, J. A., Esteban-García, B., Agüera, A., Sánchez-Pérez, J. A., & Manzano-Agugliaro, F. (2019). Wastewater treatment by advanced oxidation process and their worldwide research trends. *International Journal of Environmental Research and Public Health, 17*(1), 170. https://doi.org/10.3390/ijerph17010170

Gomes da Silva, C., & Faria, J. L. (2003). Photochemical and photocatalytic degradation of an azo dye in aqueous solution by UV irradiation. *Journal of Photochemistry and Photobiology A: Chemistry, 155*(1–3), 133–143. https://doi.org/10.1016/s1010-6030(02)00374-x

Guha, A. K. (2019). A review on sources and application of natural dyes in textiles. *International Journal of Textile Science, 8*(2), 38–40. https://doi.org/10.5923/j.textile.20190802.02

Hassaan, M. A., & Nemr, A. El. (2017). Health and environmental impacts of dyes: Mini review. *American Journal of Environmental Science and Engineering, 1*(3), 64–67. https://doi.org/10.11648/j.ajese.20170103.11

Hunger, K. (2002). *Industrial Dyes: Chemistry, Properties, Applications*. Wiley-VCH Verlag GmbH & Co. KGaA, Germany. https://doi.org/10.1002/3527602011

Islam, M., & Mostafa, M. (2019). Textile dyeing effluents and environment concerns - A review. *Journal of Environmental Science and Natural Resources, 11*(1–2), 131–144. https://doi.org/10.3329/jesnr.v11i1-2.43380

Ito, T., Shimada, Y., & Suto, T. (2018). Potential use of bacteria collected from human hands for textile dye decolorization. *Water Resources and Industry, 20*, 46–53. https://doi.org/10.1016/j.wri.2018.09.001

Jadhav, I., Vasniwal, R., Shrivastava, D., & Jadhav, K. (2016). Microorganism-based treatment of azo dyes. *Journal of Environmental Science and Technology, 9*(2), 188–197). Asian Network for Scientific Information. https://doi.org/10.3923/jest.2016.188.197

Jamee, R., & Siddique, R. (2019). Biodegradation of synthetic dyes of textile effluent by microorganisms: An environmentally and economically sustainable approach. *European Journal of Microbiology and Immunology, 9*(4), 114–118. https://doi.org/10.1556/1886.2019.00018

Kandisa, R. V., & Saibaba KV, N. (2016). Dye removal by adsorption: A review. *Journal of Bioremediation & Biodegradation, 07*(06), 1–4. https://doi.org/10.4172/2155-6199.1000371

Keharia, H., & Madamwar, D. (2003). Bioremediation concepts for treatment of dye containing wastewater: A review. *Indian Journal of Experimental Biology* 41(9), 1068–1075.

Kjellstrom, T., Lodh, M., McMichael, T., Ranmuthugala, G., Shrestha, R., & Kingsland, S. (2006). Air and water pollution: Burden and strategies for control. In: Disease Control Priorities in Developing Countries, Jamison, D. T., Breman, J. G., Measham, A. R., et al., editors. Oxford University Press, New York, 817–832.

Lellis, B., Fávaro-Polonio, C. Z., Pamphile, J. A., & Polonio, J. C. (2019). Effects of textile dyes on health and the environment and bioremediation potential of living organisms. *Biotechnology Research and Innovation*, *3*(2), 275–290. https://doi.org/10.1016/j.biori.2019.09.001

Mia, R., Selim, M., Shamim, A. M., Chowdhury, M., Sultana, S., Armin, M., Hossain, M., Akter, R., Dey, S., & Naznin, H. (2019). Review on various types of pollution problem in textile dyeing & printing industries of Bangladesh and recommandation for mitigation. *Journal of Textile Engineering & Fashion Technology*, *5*(4), 220–226. https://doi.org/10.15406/jteft.2019.05.00205

Mishra, S., Nayak, J. K., & Maiti, A. (2020). Bacteria-mediated bio-degradation of reactive azo dyes coupled with bio-energy generation from model wastewater. *Clean Technologies and Environmental Policy*, *22*(3), 651–667. https://doi.org/10.1007/s10098-020-01809-y

Mokif, L. A. (2019). Removal methods of synthetic dyes from industrial wastewater: A review. *Mesopotamia Environmental Journal*, *5*(1), 23–40. https://doi.org/10.31759/mej.2019.5.1.0040

Mondal, P., Baksi, S., & Bose, D. (2017). Study of environmental issues in textile industries and recent wastewater treatment technology. *World Scientific News*, *61*(2), 98–109. www.worldscientificnews.com

Moosvi, S., Keharia, H., & Madamwar, D. (2005). Decolourization of textile dye Reactive Violet 5 by a newly isolated bacterial consortium RVM 11.1. *World Journal of Microbiology and Biotechnology*, *21*(5), 667–672. https://doi.org/10.1007/s11274-004-3612-3

Paszczynski, A., Pasti-Grigsby, M. B., Goszczynski, S., Crawford, R. L., & Crawford, D. L. (1992). Mineralization of sulfonated azo dyes and sulfanilic acid by Phanerochaete chrysosporium and Streptomyces chromofuscus. *Applied and Environmental Microbiology*, *58*(11), 3598–3604. https://doi.org/10.1128/aem.58.11.3598-3604.1992

Pereira, L., & Alves, M. (2012). Dyes-environmental impact and remediation. *Environmental Protection Strategies for Sustainable Development*, 111–162. https://doi.org/10.1007/978-94-007-1591-2_4

Robinson, T., McMullan, G., Marchant, R., & Nigam, P. (2001). Remediation of dyes in textile effluent: A critical review on current treatment technologies with a proposed alternative. *Bioresource Technology*, *77*(3), 247–255. https://doi.org/10.1016/S0960-8524(00)00080-8

Roy, D. C., Biswas, S. K., Saha, A. K., Sikdar, B., Rahman, M., Roy, A. K., Prodhan, Z. H., & Tang, S. S. (2018). Biodegradation of Crystal Violet dye by bacteria isolated from textile industry effluents. *PeerJ*, *2018*(6). https://doi.org/10.7717/peerj.5015

Russ, R., Rau, J., & Stolz, A. (2000). The function of cytoplasmic flavin reductases in the reduction of azo dyes by bacteria. *Applied and Environmental Microbiology*, *66*(4), 1429–1434. https://doi.org/10.1128/AEM.66.4.1429-1434.2000

Saini, R. D. (2017). Textile organic dyes: Polluting effects and elimination methods from textile waste water. *International Journal of Chemical Engineering Research*, *9*(1), 975–6442. http://www.ripublication.com

Saratale, R. G., Saratale, G. D., Chang, J. S., & Govindwar, S. P. (2011). Bacterial decolorization and degradation of azo dyes: A review. *Journal of the Taiwan Institute of Chemical Engineers*, *42*(1), 138–157. https://doi.org/10.1016/j.jtice.2010.06.006

Sarkar, S., Banerjee, A., Halder, U., Biswas, R., & Bandopadhyay, R. (2017). Degradation of synthetic azo dyes of textile industry: a sustainable approach using microbial enzymes. *Water Conservation Science and Engineering*, *2*(4), 121–131. https://doi.org/10.1007/s41101-017-0031-5

Sathian, S., Radha, G., Shanmugapriya, V., Rajasimman, M., & Karthikeyan, C. (2013). Optimization and kinetic studies on treatment of textile dye wastewater using Pleurotus floridanus. *Applied Water Science*, *3*(1), 41–48. https://doi.org/10.1007/s13201-012-0055-0

Sivakumar, P., & Palanisamy, P. N. (2008). Low-cost non-conventional activated carbon for the removal of reactive red 4: Kinetic and isotherm studies. *Rasayan Journal of Chemistry, 1*(4), 871–883.

Srinivasan, A., & Viraraghavan, T. (2010). Decolorization of dye wastewaters by biosorbents: A review. *Journal of Environmental Management,* 91(10), 1915–1929. https://doi.org/10.1016/j.jenvman.2010.05.003

Stolz, A. (2001). Basic and applied aspects in the microbial degradation of azo dyes. *Applied Microbiology and Biotechnology,* 56(1–2), 69–80. https://doi.org/10.1007/s002530100686

Tüfekci, N., Sivri, N., & Toroz, İ. (2007). Pollutants of textile industry wastewater and assessment of its discharge limits by water quality standards. *Turkish Journal of Fisheries and Aquatic Sciences,* 7(2). http://www.trjfas.org/abstract.php?lang=en&id=319

Van Der Maas, A. S., Da Silva, N. J. R., Da Costa, A. S. V., Barros, A. R., & Bomfeti, C. A. (2018). The degradation of methylene blue dye by the strains of pleurotus sp. With potential applications in bioremediation processes. *Revista Ambiente e Agua, 13*(4), 1. https://doi.org/10.4136/ambi-agua.2247

Vikrant, K., Giri, B. S., Raza, N., Roy, K., Kim, K. H., Rai, B. N., & Singh, R. S. (2018). Recent advancements in bioremediation of dye: Current status and challenges. *Bioresource Technology,* 253, 355–367. https://doi.org/10.1016/j.biortech.2018.01.029

Vrsanska, M., Voberkova, S., Langer, V., Palovcikova, D., Moulick, A., Adam, V., & Kopel, P. (2016). Induction of laccase, lignin peroxidase and manganese peroxidase activities in white-rot fungi using copper complexes. *Molecules, 21*(11), 1553. https://doi.org/10.3390/molecules21111553

Yaseen, D. A., & Scholz, M. (2019). Textile dye wastewater characteristics and constituents of synthetic effluents: A critical review. In *International Journal of Environmental Science and Technology,* 16(2), 1193–1226. Center for Environmental and Energy Research and Studies. https://doi.org/10.1007/s13762-018-2130-z

Yusuf, M., Shabbir, M., & Mohammad, F. (2017). Natural colorants: Historical, processing and sustainable prospects. *Natural Products and Bioprospecting,* 7(1), 123–145). https://doi.org/10.1007/s13659-017-0119-9

17 Recent advancements in the bioremediation of heavy metals from the polluted environment by novel microorganisms

G. Mary Sandeepa, B. Lakshmanna, and M. Madakka

CONTENTS

17.1 INTRODUCTION

The environmental pollution caused by the handling and exploration of heavy metals and crude oil is greatly threatening the health of human and ecosystem. These threats have deleterious effects on life forms which play an essential role

DOI: 10.1201/9781003394600-20

in food chain maintenance and destroy the energy level in the environment (Chen and Zhong, 2019). Because of various man-made activities, the imbalance of ecosystem has been increasing (Agarwal et al., 2019). The industrial use of heavy metals has economic significance and becoming important pollutants in that ecosystem. Bioaccumulation and non-biodegradability in environment is leading to metal toxicity and it is becoming significant environmental treat (Gautam et al., 2014). Many number of inorganic metals which are used in small quantities for redox and metabolic functions are nickel, magnesium, chromium, calcium, copper, manganese, sodium and zinc. Heavy metals like lead, aluminum, gold, cadmium, silver, and mercury don't have any role biologically but are toxic in nature.

17.2   ENVIRONMENTAL OCCURRENCE OF HEAVY METALS

17.2.1   ARSENIC (AS)

This element is present everywhere in the environment and can be present in low quantities virtually in all environmental matrices (ATSDR, 2000). Methylated metabolites, monomethyl arsenic acid, dimethyl arsenic, and trimethyl arsenic oxide and the important inorganic forms are trivalent and pentavalent arsenite. The natural phenomena like erosion of soil, volcanic eruptions, and anthropogenic activities are major environmental pollution by arsenic (ATSDR, 2000).

Large compounds containing arsenic are producing industrially and also using in manufacturing herbicides, insecticides, algicides, fungicides, and sheep dips. The arsenic-containing compounds also have been utilized in treatment of yaws, amoebic dysentery, syphilis, and trypanosomiasis (Centeno et al., 2005). They also applied in veterinary medicines for the extermination of tape warms in cattle and sheep (Tchounwou et al., 1999).

17.2.2   CADMIUM (CD)

Cd is widely present in earth's crust about 0.1 mg/kg concentration. Sedimentary rocks have highest level of cadmium compounds and marine phosphates have 15 mg cadmium/kg. The important industrial application of cadmium is the production of pigments, batteries, and alloys (Wilson, 1988). In recent years, cadmium is using in batteries which has been shown significant growth but commercially decreased in big countries because of ecological concern

17.2.3   CHROMIUM (CR)

This occurred naturally in earth's crust in oxidative states such as chromium (II) and chromium (VI). The metal processing industries, stainless steel welding, tannery facilities, chromate product ion, ferrochrome, and chrome pigment are the largest contribution of chromium. The main chromium pollution is due to [Cr (VI)] a hexavalent form (ATSDR).

17.2.4 Lead (Pb)

This is in a bluish – grey color and naturally occurs in earth's crust in small quantities. Anthropogenic activities like mining, fuels burning, and manufacturing industries are contributing the release of high amounts of lead in to the environment. Lead has various domestic, industrial, and agricultural applications. Lead is widely utilized in the of lead-acid batteries production, ammunitions devices to shield X-rays, and metal products like solder and pipes.

17.2.5 Mercury (Hg)

This metal is a wide spread pollutant and environmental toxicant, which makes an extreme damage in tissue and leads to adverse health effects (Sarkar, 2005). Animals and humans are exposed to different types of mercury forms in the environment like organic mercury compounds, inorganic mercury, mercury vapor, and mercuric compounds. This metal is used in the dentistry (dental amalgams), batteries, switches and thermostats making industries, caustic soda production, and as fungicides in wood processing and pharmaceutical products preservatives (Tchounwou et al., 2003).

17.2.6 Nickel (Ni)

Weathering of rocks, volcanic emissions, nickel compounds' solubilization from soil, and many anthropogenic activities like battery, electroplating catalyst electronic equipment industries are the major natural sources of Nickel through which it enters in to air, water, and soil (Duda-Chodak and Blaszczyk, 2008).

17.2.7 Zinc (Zn)

It is the major metal found in industrial effluents (Plum et al., 2010). Electroplating, galvanization, manufacturer of batteries, and metallurgical industries are the major sources if Zn (Plum et al., 2010).

17.2.8 Copper (Cu)

It is recognized as one of the heavy metal pollutants. This metal is occurred inside the earth. It is found rarely in nature in ores (chalcopyrite) as well as in combined state (Igiri et al., 2018).

17.3 HEAVY METAL TOXICITY TOWARD MICROBES

The metal ability to cause harmful effects on microbes is called metal toxicity and it depends on the absorbed dose and bioavailability of that heavy metal (Rasmussen et al., 2000). Many mechanisms are involved in heavy metal toxicity, such as inactivating the vital enzymes, acting as redox catalysts in ROS production, ion regulation destruction and directly affects the protein and DNA synthesis (Gauthier et al., 2014).

The heavy metals change the biochemical and physiological properties of microorganisms. Cadmium (Cd) and Chromium (Cr) can induce denaturation and oxidative damage as well as lower the bioremediation potential of microorganisms. Cr can react with carboxyl and thiol groups of enzymes and can alter the enzyme's structure and activity (Cervantes et al., 2001). Cationic Cr (III) complexes react with phosphate (negative charge) of DNA electro statically and would affect replication, transcription and mutagenesis (Cervantes et al., 2001). The ROS produced by Cu (I) and Cu (II) will act as electron carriers and may cause severe damage to proteins, DNA, and lipids (Giner-Lamia et al., 2014). Stabilization of superoxide radicals will occur by aluminum which can damage DNA (Booth et al., 2015). Heavy metals can adhere to cell surface and can enter through trans membrane carriers or ion channels (Chen et al., 2014) they also can bind to enzymes by competitively and uncompetitively with substrates which can cause configurational change in structure (Gauthier et al., 2014). Lead (Pb) and cadmium (Cd) have deleterious effects on microorganisms like cell membrane damage and destroying the structure of DNA. This is because of ligand binding or metal binding to native binding sites (Olaniran et al., 2013).

The fungi cell wall mainly composed of peptides and polysaccharide which have good capacity for heavy metal binding. Because of their role as enzyme inhibitors and protein denaturing agents heavy metals are found to be toxic to fungi. Heavy metals like mercury can bind to SH groups which involved in the regulation of enzyme and can cause irreversible inactivation. But cadmium can bind to aromatic amino acids residue in enzyme formation and cause oxidation damage. Heavy metals show deleterious effects on reproduction of many fungi because spore formation and germination is very sensitive when compared with mycelia (Pawlowska and Charvat, 2004). Fungi are considered good alternative of the biodegradation for toxic heavy metals in the environment; they are considered more resistant to heavy metals than bacteria (Rajapasksha et al., 2004).

17.4   MICROBIAL RESISTANCE MECHANISMS AGAINST HEAVY METALS

Several types of microbes are involved in bioremediation of heavy metals. Microbes exhibit different mechanisms to resist against metals or die in heavy metal stress conditions. Different resistance mechanisms of microbes are (I) metal ion active transport across the membrane, (II) extracellular barriers, (III) intracellular sequestration, (IV) extracellular sequestration, and (V) heavy metal reduction.

Microbial metabolism produce products like oxygen, hydrogen, and H_2O_2. These are utilized for metal's oxidation and reduction. Microbial metabolites may also mediate solubilization or precipitation. Dissolved metal chelates will form during aerobic oxidation of organic acids and in fermentation of inorganic acids. During anaerobic fermentation, organic acids will be produced from cellulose; it may be used as a reduced carbon source for precipitation of metals and sulfate reduction (EPA, 2006). Phosphate, CO_2, H_2S are produced by microbes which stimulates the precipitation of phosphates, non-dissolved carbonates and sulfides of metals like cadmium, copper, chromium, nickel, lead, and mercury. Sulfate-reducing bacteria is used in the removal of heavy metals and radionuclides from liquid waste of nuclear facilities, sulfate containing mining drainage waters, saw dust, or wood straw.

P. putida strain have the ability to accumulate cadmium, copper, zinc intra cellularly with the help of proteins, which are rich in cysteine having low molecular weight (Higham et al., 1986). *Rhizobium leguminosarum* cells also exhibit the cadmium ions' intracellular sequestration by glutathione (Lima et al., 2006). The decontamination of metal ions is occurred by intracellular and extracellular precipitation, energetic uptake of metals and conversion of valence with metals accumulation to their spores and mycelium. The fungal outer cell wall behaves like a metal ligand for metal ion labeling and makes removal of inorganic metals (Xie et al., 2016). *Geobacter* spp. and *Desulfuromonas* spp. can decrease harmful metals to non-toxic or less toxic metals. A strict anaerobic organism *G. metallireduceris* can reduce manganese (Mn) from Mn (IV) (lethal) to Mn (II) and poisonous U (VI) to U (IV). The high toxic Cr (VI) will be converted to less toxic Cr (III) by *G. sulfurreducens* and *G. metallireducen.* Large amount of H_2S is generated by sulfate reducing bacteria which causes metal cation precipitation (Luptakova and Kusnierova, 2005). The strain *Vibrioharveyi* can precipitate divalent lead which is soluble to complex lead phosphate salt (Mire et al., 2004). Methylation by microorganisms plays an essential role in metal biodegradation, bacteria such as *Bacillus* spp., *E. coli* spp., *Pseudomonas* spp. can methylates Hg (II) to gaseous form of methyl mercury. Arsenic (As) biomethylation converts arsenic (As) to gaseous arsenic, Selenium biomethylation forms volatile dimethyl selenide and lead (Pb) biomethylation forms dimethyl lead in contaminated top soil. Microorganism can change metal ions to reducing state to oxidation state, so they can decrease harmfulness (Jyoti and Harsh, 2014). Enzymatic activity of microbes reduce the heavy metal ions which results very less toxic form of chromium and mercury formation (Barkay et al., 2003).

17.5   FUNGAL BIOREMEDIATION OF HEAVY METAL

Fungal cell walls are rigid and formed with lipids, chitin, mineral ions, proteins, polyphosphates and polysaccharides.

Nowadays, fungi are using as biosorbents for the remediation of hazardous metals because of their excellent metal up taking and recovery capacities (Akar et al., 2005; Fu et al., 2012). According to many studies lifeless and active fungal cells play an essential role in the inorganic chemicals adhesion (Karakagh et al., 2012). According to Srivastava and Thakur (2006). *Aspergillus* spp. is using for degradation of chromium in tannery waste water. At pH 6, in a bioreactor 85% of chromium was removed from synthetic medium, but 65% was removed from the tannery effluents. The organic pollutants hinder the growth of the organism in tannery effluents. Park et al. (2005) studied that dead fungal biomass of *Saccharomyces cerevisiae*, *Aspergillus niger*, *Penicillium chrysogenum*, and *Rhizopus oryzae* can be converted hazardous Cr (VI) to less or non-hazardous Cr (III). *Candida sphaerica* produces biosurfactants which removes 95% of Fe, 90% of Zn, and 79% of Pb. The metal ion surfactant complexes can interact with metals directly before detachment of soil. Yeast spp. also accumulates copper (52–68%) and nickel Ni (57–71%). The pH (optimum 3–5) and metal ion concentration affects the process. Luna et al. (2016) studied that *Candida sphaerica* produces biosurfactants which can remove 95% of Fe, 90% of Zn, and 79% of Pb. These surfactants interact directly with heavy metals and can form complexes before soil detachment. *Candida* spp. Can accumulate 57–71% of Ni and 52–68% of Cu.

17.6 CONSORTIA OF MICROBES IN REMEDIATION OF HEAVY METALS

Different species which can work in synergy with each other have been implicated to clean up the heavy metal polluted site and other organic contaminants. Microorganism exhibits positive interactions like synergism, commensalism, and mutualism which results in growth, survival, and metabolism which are important for bioremediation. In commensalism microbes transform the pollutants without being able to use the energy derived from the metabolism to support the growth. Microbes which are cometabolizing can interact with other community members through commensalism and produce cometabolites which can serve as substrates for other microbes. Cometabolism is the important one which can serve as substrates for other microbes. The advantage of Cometabolism is to reduce the time length for the degradation of contaminant. Secondly, there is a synergism in consortium in which microbes interact each other. So they benefit from one other by supplying nutrients for their needs. Scientists studied the efficiency of bacterial consortia in removal of chromium, cadmium, lead, zinc, cobalt, and copper, at approximately 75–85% within 2 hours of contact duration.

17.7 PHYCOREMEDIATION

It is the using of different cyanobacteria and algae and for the heavy metals bioremediation (Chabukdhara et al., 2017). Algae produce more biomass when compared to other microbial biosorbents. These biosorbents have biosorption capacity and these are used for heavy metal bioremediation (Abbas et al., 2014). Hazardous metals are adsorbed or integrated into the algal cells. Algal cells have different chemical groups on their surfaces like carboxyl, hydroxyl, amide and phosphate which can act as metal attachment sites (He and Chen, 2014). Goher et al. (2016) worked on *Chlorella vulgaris* dead cells to remove cadmium (Cd^{2+}), lead (Pb^{2+}) and copper (Cu^{2+}) from aqueous solution under various conditions of pH, contact time, and biosorbent dosage. According to their results, *C. vulgaris* biomass is highly efficient for the removal of 97.7% of copper (CU^{2+}), 95.5% of cadmium, and 99.4% of lead from the mixed solution of 50 mg/dm^3 of each metal ion.

17.8 MICROBE-MEDIATED NANOBIOREMEDIATION OF HEAVY METALS

Palladium nanoparticles mediated by *Clostridium pasteruriaanum* from sandy aquifer material were tested for nanoparticles efficiency in bioremediation. This study gives positive results in remediation of hexavalent chromium Cr (VI) which leads hydrogen gas production (Chidambaram et al., 2010). This research finds that the Cr (VI) removal rate is 7.2 g and demonstrated that nanocatalysts significantly increase chromium removal rates than traditional in situ biostimulation strategies. The same study was done on Cr (VI) reduction by using polyvinyl alcohol, carbon nanotubes and sodium alginate matrix immobilized on *Pseudomonas aeruginosa* (Pang et al., 2011). The immobilized cells reduced 80 mg/L Cr (VI) up to 84% and further complete reduction occur within 24 hours. Immobilized calcium alginate beads with *Shewanella oneidensis* and which is stabilized with CNT also showed that a best biofunctionalized mechanism for toxic Cr (VI) reduction to Cr (III) in sewage (Yan et al., 2013). *Lysinibacillus sphaericus* is used in magnetic oxide nanoparticles synthesis and was used for removing of hexavalent chromium in the polluted environment (Kumar et al., 2019). According to their study the biofilm of exopolysaccharides from *Lysinibacillus sphaericus* acts as a good stabilizing, reducing and capping agent, having many binding sites for different metal ions. The functionalized exopolysaccharide magnetic oxide nanoparticles exhibited the capacity to absorb Cr (VI). Algae also showed efficiency in nanobioremediation. *Chlorococcum* sp. can produce iron nanoparticles which increased the 4 mg/L Cr (VI) reduction to Cr (III) up to 92% (Subramaniyam et al., 2015). Phyco-synthesized iron nanoparticles mediated by algal biomolecules showed high reactivity, efficient reduction, and greater stability of toxic pollutants in the environment.

Phyco-synthesized nanoparticles (iron) mediated by algal biomolecules showed higher efficiency, reactivity, and stability in the reduction of toxic contaminants in the environment.

As a functionalized agent the *Chlorella vulgaris* is incorporated in ultrafine bimetallic (TiO$_2$/Ag) chitosan nanofiber mats gives the significance of algae in the photoremoval method of Cr (VI) under UV treatment (Wang et al., 2017). Organic substances such as carboxylate acids and chlorophylls are released by *C. vulgaris* was showed that increased photocatalytic reduction of Cr (VI) on TIO$_2$/Ag hybrid nanomaterial. The same study was done on effluents from pharmaceutical industries having hazardous metals such as lead and chromium (Kumari et al., 2020). Phyco-synthesized silver nanoparticles functionalized by *Bacillus cereus* with aluminium for an efficient biodegradation. The bacterial cell mediated nanoadsorbent method removes the 98.13% of Cr and 98.76% of Pb in the pharmaceutical industries waste effluents. The lead sulfide (PbS) nanoparticles produced from *Rhodosporidium diobovatum* break down pb (II) nanoparticle easily and make them less toxic and useful forms by fungi (Torimiro et al. 2021). In one study using a bioactive polymer chitosan nanoparticles (CS-NP) synthesized from *Cunninghamells elegans* (Alsharari et al., 2018). These NCt particles have a greater bioremediation and biosorption capacity on Pb (II) and Cu (II) ions in contaminated area than chitosan. The *Pseudomonas aeruginosa* successfully remove cadmium in cadmium contaminated water and increased the reduction of cadmium which in turn accelerated the cadmium sulfide nana particles biosynthesis (Raj et al., 2016). Similarly, a study on bioremediation of cadmium contaminated soil treatment with both nanohydroxyapatite and *Bacillus subtills* effectively degraded cadmium (Liu et al., 2018) them promotes the growth of rhizosphere community and bacterial diversity in that site. According to recent research bioremediation of both lead and cadmium was possible (Zhu et al., 2020). In this method *E. coli* and nanoparticles of metals were used in the bioremediation of heavy metals. Selenium nanoparticles which are produced with the presence of *Citrobacter freudii* were used to effectively to remediate mercury polluted soil (Wang et al., 2017). Here, elemental mercury (HgO) was converted to mercuric selenide (HgSe) under aerobic and anaerobic conditions, 39.1–48.6% and 45.8–57.1% bioremediation value was reported.

17.8.1 MOLECULAR AND GENETIC BASIS OF METAL TOLERANCE IN MICROORGANISMS

Directly or indirectly microorganisms play an important role to maintain the bio-geochemical cycles in ecosystem which help in mineral nutrient recycling like phosphorous, nitrogen, sulfur and many different metallic ions like Fe, Al and Cu which contributes the sustainability of life. In bioremediation, the metal microbe interaction occur which involves complexation, adsorption, precipitation, reduction, oxidation, and methylation. In addition, the metal ions and microbes reactions have been broadly classified into six different types namely accumulation intracellularly, cell wall association of metals, metal-siderophore interaction, extracellular polymer metal interaction, bacterial metabolite's immobilization and volatilization, and transformation of the metals. Bacteria can grow everywhere because of their iniquitousness they have an opportunity to expose a wide range of metal toxicity (Nanda et al., 2019). During the evolution they developed resistance mechanisms to tolerate heavy metals like copper, arsenic, chromium, nickel, mercury, lead, etc. Metal resistance in them is because of metal resistance genes present on the chromosomes and plasmids. These genetic determinants were first discovered in plasmids. Operon system in plasmids can serve as genetic determinants for heavy metals. For example, czc operon in *R. metallidurans* CH34 plasmid PMOL30 acts as a heavy metal-resistance genetic determinant (e.g. zinc, cadmium and cobalt). PMOL28 is an another plasmid found in these bacterial systems, which has the *cnr* operon that act as resistance genetic determinant for nickel, cobalt, and chromium. According to Nies (1999) and Cooksey bacterial plasmids confer resistance against titanium and mercury. The copper resistance genetic determinant is cop operon was discovered in pseudomonas sp. on Plasmid as well as chromosome. When copper is present this operon is induced and it encodes proteins like CopA, sopA, CopC, and Cop D. The copper ions are accumulated by Cop proteins in compartments in the cell periplasm as well as in outer membrane. In *E. coli*, *P. aeruginosa*, and *B. subtilis* the *ars* operons are having s similar genetic determinant's structure. To understand the characterization of these genetic determinants become important for the characterization and understanding of tolerance, which will be an important to devise an efficient biodegradation strategy. In *P. aeruginosa*, Pd resistance N6P6 is mediated by *bmtA* genes expression. Metallothionein is encoded by these genes which will maintain intracellular homeostasis. When Pb (II) ion concentration increases the *bmtA* genes expression. bmtA genes also induced by other heavy metals like copper, mercury, arsenic and cadmium. Cadmium resistance is mediated by RND-driven zinc exporter in Gram-negative bacteria. The Czc system. Nickel exporter, Ncc in *R. metallidurans* plasmid has three structural genes *czcA*, *czcB*, and *czcC*. For the above three mentioned heavy metals *czcA* is essential one and encodes for a cation carrier antiporter which is present in the cell membrane. CzB encodes for CzB protein and plays a role in cation transportation. This gene if deleted it will result in complete loosing of Cs (II) and Zn (II) resistance. *Czc* gene encodes the outer membrane protein CzcC protein. These proteins form a complicated membrane cation efflux pump (CzcABC complex) for Cd (II) detoxification. Czc operon expression is also controlled by czcR and czcD genes. Cd resistance is mediated by glutathione in Yeast *S. cerevisiae*. Glutathione bound to Cd (II) ions to form cadmium-bisglutathionato complex, is then transported into the vacuole viaYCF1p transporter, which is ABC transporter. In cyanobacteria, the operon has two genes *smtA* and *smtB*, which control the cadmium resistance. *smtA* gene encodes Mts protein which is important for Cd resistance. smtB gene regulates smtA expression and the promoter and operator regions are present between these two genes. Arsenic resistance is present in *ars* operon, which is present both in plasmids as well as chromosomes. For example, in *S. aureus* it is found only in the plasmids but in *E. coli* it is present on plasmids as well as in chromosomes. In *E. coli* R773, plasmid has five genes, viz. *arsR, arsD, arsA, arsB,* and *arsC* so it is known as arsRDA. In periplasm of Gram-negative bacteria, Hg (II) combines with Hg (II)-binding protein (MerP), which helps in preventing the mercury toxicity in periplasmic protein.

17.8.2 Genetic engineering of microorganisms

With the advancement of genetic engineering technology, microbes are genetically engineered with desired characteristics like metal tolerance, stress tolerance, metal accumulation ability, metal-chelating peptides and proteins over expression. Scientists engineered microbes to produce trehalose and made it reduces 1 mM Cr (VI) to Cr (III). The genetically modified *Chlamydomonas reinhardtii* showed significant improvement in Cd toxicity tolerance and its accumulation (Ibuot et al. 2017). The examples of genetically engineered microorganisms for heavy metal bioremediation are *E. coli* ArsR (ELP153AR) to degrade As (III) (Kostal et al. 2004) and *Saccharomyces cerevisiae* (CP2 HP3) to degrade Cd^{2+} and Zn^{2+} (Vinopal et al. 2007).

REFERENCES

Abbas, H.S., Ismail, M.I., Mostafa, M.T., and Sulaymon, M.T. "Biosorption of heavy metals: A review," *J Chem Sci Technol*, vol. 3, pp. 74–102, 2014.

Akar, T., Tunali, S., and Kiran, I. "*Botrytis cinerea* as a new fungal biosorbent for removal of Pb(II) from aqueous solutions," *Biochem Eng J*, vol. 25, no. 3, pp. 227–235, 2005.

Alsharari, S.F., Tayel, A.A., and Moussa, S.H., "Soil emendation with nano-fungal chitosan for heavy metals biosorption," *Int J Biol Macromol,* vol. 118, pp. 2265– 2268, 2018.

Agarwal, P., Gupta, R., and Agarwal, N., "Advances in synthesis and applications of microalgal nanoparticles for wastewater treatment," *J Nanotechnol,* 2019: Article ID 7392713, https://doi.org/10.1155/2019/7392713

Agency for Toxic Substances and Disease Registry (ATSDR), *Toxicological Profile for Arsenic TP-92/09.* Georgia: Center for Disease Control, Atlanta; 2000.

Barkay, T., Miller, S.M., and Summers, A.O. "Bacterialmercur resistance from atoms to ecosystems," *FEMS Microbiol Rev*, vol. 27, no. 2–3, pp. 355–384, 2003.

Booth, S.C., Weljie, A.M., and Turner, R.J. "Metabolomics reveals differences ofmetal toxicity in cultures of Pseudomonas pseudoalcaligenes KF707 grown on different carbon sources," *Front Microbiol*, vol. 6, p. 827, 2015.

Centeno, J.A, Tchounwou, P.B., Patlolla, A.K., Mullick, F.G., Murakat, L., Meza, E., Gibb, H., Longfellow, D., and Yedjou, C.G. "Environmental pathology and health effects of arsenic poisoning: A critical review," in Naidu R, Smith E, Smith J, Bhattacharya P, editors. *Managing Arsenic in the Environment: From Soil to Human Health.* Adelaide: CSIRO Publishing Corp., pp. 311–327, 2005.

Cervantes, C., Campos-Garc´ıa, J., Devars, S., et al. "Interactions of chromium with microorganisms and plants," *FEMS Microbiol Rev*, vol. 25, no. 3, pp. 335–347, 2001.

Chabukdhara, M., Gupta, S.K., and Gogoi, M. "Phycoremediation of heavy metals coupled with generation of bioenergy," in Gupta, S, Malik, A, Bux, F, editors. *Algal Biofuels.* Cham: Springer, pp. 163–188, 2017.

Chen, S., Yin, H., and Ye, J. et al. "Influence of co-existed benzo[a]pyrene and copper on the cellular characteristics of Stenotrophomonas maltophilia during biodegradation and transformation," *Bioresour Technol*, vol. 158, pp. 181–187, 2014.

Chen, S., and Zhong, M. "Bioremediation of petroleum contaminated soil," *Intechopen,* 2019. https://doi.org/10.5772/intechopen.90289.

Chidambaram, D., Hennebel, T., Taghavi, S., Mast, J., Boon, N., Verstraete, W., van der Lelie, D., and Fitts, J.P. "Concomitant microbial generation of palladium nanoparticles and hydrogen to immobilize chromate," *Environ Sci Technol,* vol. 44, no. 19, pp. 7635–7640, 2010.

Duda-Chodak, A., and Blaszczyk, U. "The impact of nickel on human health," *J. Elementol,* vol. 13, pp. 685–693, 2008.

Fu, Q.Y., Li, S., and Zhu, Y.H. "Biosorption of copper (II) from aqueous solution by mycelial pellets of Rhizopus oryzae," *Afr J Biotechnol*, vol. 11, no. 6, pp. 1403–1411, 2012.

Gautam, R.K., Soni, S., and Chattopadhyaya, M.C. "Functionalized magnetic nanoparticles for environmental remediation," in Chaudhery, MS and Ketaki, KP, editors. *Handbook of Research on Diverse Applications of Nanotechnology in Biomedicine, Chemistry, and Engineering.* Duxford: Woodhead Publishing, pp. 518–551, 2014.

Gauthier, P.T., Norwood, W.P., Prepas, E.E., and Pyle, G.G. "Metal-PAH mixtures in the aquatic environment: A review of co-toxicmechanisms leading tomore-than-additive outcomes," *Aquat Toxicol*, vol. 154, pp. 253–269, 2014.

Giner-Lamia, J., L´opez-Maury, L., Florencio, F.J., and Janssen, P.J. "Global transcriptional profiles of the copper responses in the cyanobacterium synechocystis sp. PCC 6803," *PLoS One*, vol. 9, no. 9, e108912, 2014.

Goher, M.E., El-Monem, A.M.A., Abdel-Satar, A.M., Ali, M.H., Hussian, A.-E.M., and Napi´orkowska-Krzebietke, A. "Biosorption of some toxic metals from aqueous solution using nonliving algal cells of Chlorella vulgaris," *J Elementol.*, vol. 21, no. 3, pp. 703–714, 2016.

He, J., and Chen, J.P. "A comprehensive reviewon biosorption of heavy metals by algal biomass: Materials, performances, chemistry, and modeling simulation tools," *Bioresour Technol*, vol. 160, pp. 67–78, 2014.

Higham, D.P., Sadler, P.J., and Scawen, M.D. "Cadmium-binding proteins in pseudomonas putida: Pseudothioneins," *Environ Health Perspect*, vol. 65, pp. 5–11, 1986.

Ibuot, A., Dean, A. P., McIntosh, O. A., & Pittman, J. K. (2017). Metal bioremediation by CrMTP4 over-expressing Chlamydomonas reinhardtii in comparison to natural wastewater-tolerant microalgae strains. *Algal Research,* 24, 89–96.

Igiri, B.E., Okoduwa, S.I., Idoko, G.O., Akabuogu, E.P., Adeyi, A.O., and Ejiogu, I.K. "Toxicity and bioremediation of heavy metals contaminated ecosystem from tannery wastewater: A review," *J Toxicol*, vol. 2018, pp. 1–16, 2018. https://doi.org/10.1155/2018/ 2568038.

Jyoti, B., and Harsh, K.S.N. "Utilizing Aspergillus niger for bioremediation of tannery effluent," *Octa J Environ Res*, vol. 2, no. 1, pp. 77–81, 2014.

Karakagh, R.M., Chorom, M., Motamedi, H., Kalkhajeh, Y.K., and Oustan, S. "Biosorption of Cd and Ni by inactivated bacteria isolated from agricultural soil treated with sewage sludge," *Ecohydrol Hydrobiol*, vol. 12, no. 3, pp. 191–198, 2012.

Kostal, J., Yang, R., Wu, C. H., Mulchandani, A., & Chen, W. (2004). Enhanced arsenic accumulation in engineered bacterial cells expressing ArsR. *Applied and Environmental Microbiology,* 70(8), 4582–4587.

Kumar, H., Sinha, S.K., Goud, V.V., and Das, S. "Removal of Cr(VI) by magnetic iron oxide nanoparticles synthesized from extracellular polymeric substances of chromium resistant acid-tolerant bacterium *Lysinibacillus sphaericus,*" *J Environ Health Sci Eng,* vol. 17, pp. 1001–1016, 2019.

Kumari, V., and Tripathi, A. "Remediation of heavy metals in pharmaceutical effluent with the help of *Bacillus cereus*-based green-synthesized silver nanoparticles supported on alumina," *Appl Nanosci,* vol. 10, pp. 1709–1719, 2020.

Lima, A.I.G., Corticeiro, S.C., and de Almeida Paula Figueira, E.M. "Glutathione-mediated cadmium sequestration in Rhizobium leguminosarum," *Enzyme Microb Technol,* vol. 39, no. 4, pp. 763–769, 2006.

Liu, W., Zuo, Q., Zhao, C., Wang, S., Shi, Y., Liang, S., Zhao, C., and Shen, S. "Effects of *Bacillus subtilis* and nanohydroxyapatite on the metal accumulation and microbial diversity of rapeseed (*Brassica campestris L.*) for the remediation of cadmium-contaminated soil," *Environ Sci Pollut Res,* vol. 25, no. 25, pp. 25217–25226, 2018.

Luna, J.M., Rufino, R.D., and Sarubbo, L. A. "Biosurfactant from Candida sphaerica UCP0995 exhibiting heavy metal remediation properties," *Process Saf Environ Prot,* vol. 102, pp. 558–566, 2016.

Luptakova, A., and Kusnierova, M. "Bioremediation of acidmine drainage contaminated by SRB," *Hydrometallurgy,* vol. 77, no. 1–2, pp. 97–102, 2005.

Mire, E., Tourjee, J.A., O'Brien, W.F., Ramanujachary, K.V., and Hecht, G.B. "Lead precipitation by *Vibrio harveyi*: Evidence for novel quorum-sensing interactions," *Appl Environ Microbiol,* vol. 70, no. 2, pp. 855–864, 2004.

Nanda, M., Kumar, V., and Singh, D. K. "Multimetal tolerance mechanisms in bacteria: The resistance strategies acquired by bacteria that can be exploited to 'clean-up' heavy metal contaminants from water," *Aquat Toxicol,* vol. 212, pp. 1–10, 2019.

Nies, D.H. "Microbial heavy-metal resistance," *Appl Microbiol Biotechnol,* vol. 51, pp. 730–750, 1999.

Olaniran, A.O., Balgobind, A., and Pillay, B. "Bioavailability of heavy metals in soil: Impact on microbial biodegradation of organic compounds and possible improvement strategies," *Int J Mol Sci,* vol. 14, no. 5, pp. 10197–10228, 2013.

Pang, Y., Zeng, G.M., Tang, L., Zhang, Y., Liu, Y.Y., Lei, X.X., Wu, M.S., Li, Z., and Liu, C. "Cr (VI) reduction by *Pseudomonas aeruginosa* immobilized in a polyvinyl alcohol/ sodium alginate matrix containing multi-walled carbon nanotubes," *Bioresour Technol,* vol. 102, no. 22, pp. 10733–10736, 2011.

Park, D., Yun, Y.-S., Jo, J.H., and Park, J.M. "Mechanismof hexavalent chromium removal by dead fungal biomass of Aspergillus niger," *Water Res,* vol. 39, no. 4, pp. 533–540, 2005.

Pawlowska T.E., and Charvat I. "Heavy-metal stress and developmental pattern of arbuscular mycorrizal fungi," *Appl Environ Microbiol,* vol. 70, pp. 6643–6649, 2004.

Plum, L.M., Rink, L., and Haase, H. "The essential toxin: Impact of zinc on human health," *Int J Environ Res Public Health,* vol. 7, no. 4, pp. 1342–1365. https://doi.org/10.3390/ijerph7041342

Raj, R., Dalei, K., Chakraborty, J., and Das, S. "Extracellular polymeric substances of a marine bacterium mediated synthesis of CdS nanoparticles for removal of cadmium from aqueous solution," *J Colloid Interface Sci,* vol. 462, pp. 166–175, 2016.

Rajapasksha R.M.C.P., Tobor-Kaplan M.A., and Baath F. "Metal toxicity affects fungal and bacterial activities in soil differently," *Appl Environ Microbiol,* vol. 70, pp. 2966–2973 2004.

Rasmussen, L.D. Sørensen, S.J., Turner, R.R., and Barkay, T. "Application of a mer-lux biosensor for estimating bioavailable mercury in soil," *Soil Biol Biochem,* vol. 32, no. 5, pp. 639–646, 2000.

Sarkar, B.A. "Mercury in the environment: Effects on health and reproduction," *Rev Environ Health,* vol. 20, pp. 39–56, 2005.

Srivastava, S., and Thakur, I.S. "Isolation and process parameter optimization of Aspergillus sp. for removal of chromium from tannery effluent," *Bioresour Technol,* vol. 97, no. 10, pp. 1167–1173, 2006.

Subramaniyam, V., Subashchandrabose, S.R., Thavamani, P., Megharaj, M., Chen, Z., Naidu, R., "Chlorococcum sp. MM11—a novel phyco-nanofactory for the synthesis of iron nanoparticles", *J. Appl. Phycol.,* 27(5), 1861–1869, 2015.

Tchounwou, P.B., Ayensu, W.K., Ninashvilli, N., Sutton, D. "Environmental exposures to mercury and its toxicopathologic implications for public health," *Environ Toxicol,* vol. 18, pp. 149–175, 2003.

Tchounwou, P.B., Wilson, B., and Ishaque, A. "Important considerations in the development of public health advisories for arsenic and arsenic-containing compounds in drinking water," *Rev Environ Health,* vol. 14, no. 4 pp. 211–229, 1999.

Torimiro, N., Daramola, O.B., Oshibanjo, O.D., Otuyelu, F.O., Akinsanola, B.A., Yusuf, O.O., ... & Omole, R.K. (2021). Ecorestoration of heavy metals and toxic chemicals in polluted environment using microbe-mediated nanomaterials. *International Journal of Environmental Bioremediation & Biodegradation,* 9, 8–21.

Vinopal, S., Ruml, T., & Kotrba, P. (2007). Biosorption of Cd2+ and Zn2+ by cell surface-engineered Saccharomyces cerevisiae. *International Biodeterioration & Biodegradation,* 60(2), 96–102.

Wang, L., Zhang, C., Gilles, F.G., and Pan, M.G. "Algae decorated TiO2/Ag hybrid nanofiber membrane with enhanced photocatalytic activity for Cr(VI) removal under visible light," *Chem Eng J,* vol. 314, pp. 622–630, 2017.

Wang, X., Zhang, D., Pan, X., Lee, D.J., Al-Misned, F.A., Mortuza, M.G., and Gadd, G.M. "Aerobic and anaerobic biosynthesis of nano-selenium for remediation of mercury contaminated soil," *Chemosphere,* vol. 170, pp. 266–273, 2017.

Wilson, D.N. Association Cadmium. Cadmium – market trends and influences, Cadmium 87 Proceedings of the 6th International Cadmium Conference, London: pp. 9–16, 1988.

Xie, Y., Fan, J., Zhu, W., et al. "Effect of heavy metals pollution on soilmicrobial diversity and bermudagrass genetic variation," *Front Plant Sci,* vol. 7, p. 775, 2016.

Yan, F.F., Wu, C., Cheng, Y.Y., He, Y.R., Li, W.W., and Yu, H.Q. "Carbon nanotubes promote Cr (VI) reduction by alginate-immobilized *Shewanella oneidensis* MR-1," *Biochem Eng,* vol. 77, pp. 183–189, 2013.

Zhu, N., Zhang, B., and Yu, Q. "Genetic engineering-facilitated co-assembly of synthetic bacterial cells and magnetic nanoparticles for efficient heavy metal removal," *ACS Appl Mater Interfaces,* vol. 12, no. 20, pp. 22948–22957, 2020.

18 Bioremediation approaches for treatment of heavy metals, pesticides and antibiotics from the environment

M. Srinivasulu, M. Subhosh Chandra, D. Mallaiah,
G. Jaffer Mohiddin, and G. Narasimha

CONTENTS

18.1 INTRODUCTION

Heavy metal contamination is one of the burning areas of environmental research. Despite natural existence, various anthropomorphic sources have contributed to an abnormally highest concentration of heavy metals in the soil and water environment. They are categorized by their long persistence in natural environment leading to serious health consequences in humans, animals and plants even at lowest concentrations (1 or 2 µg in few cases) (Adikesavan et al. 2019). The heavy metals are naturally occurring elements which are found in the earth's shell. The sources heavy metal pollution are the anthropomorphic activities like mining, smelting, industrial production, using of metals and metal containing compounds for domestic and the agricultural applications. However, heavy metal pollution also occur naturally. These sources have been reported to contribute to human exposure and contamination of environment by various researchers (Zouboulis et al. 2004; He et al. 2005; Rahman and Bastola 2014).

DOI: 10.1201/9781003394600-21

Due to increasing the population, urbanization and rapid industrialization are documented as significant challenges to the groundwater resources management in the developing countries. Many research reports have revealed that heavy metals pollution existence in many countries, thus representing it as a global problem. Considerable concentrations of toxic heavy metals (As, Cd, Cr, Zn, Fe, Pb, Mn, Cu, Ni, etc.) in soil, surface and ground water have been reported in different countries like China, Hong Kong, Germany, Turkey, India, Italy, Greece, Iran, Bangladesh, etc. (Wuana and Okieimen 2011; Kaonga et al. 2017). Above all these, due to lack of awareness on the suitable effluent disposal and failure to imply stringent regulatory standards have added to the source of environmental deterioration (Khalid et al. 2017). Thus, these factors have ended up in the formation of enormous amounts of solid waste in different toxic forms which eventually pollute the whole ecosystem. The disposed wastewaters also affect the quality of surface water and soil, which on constant proceeding without suitable care may cross acceptable limits prescribed by international regulatory agencies (EPA 1992, 2002). The majority of industrial scale remediation, which involve chemical, physical and biological methods, are employed as single remediation strategies. Although the accomplishment of these processes, they also face certain drawbacks like low efficiency, toxic sludge generation and high cost. On the other hand, this can be overcome by upgrading them as integrated processes, which shown more efficacy for heavy metal remediation as reported by several researchers (Mao et al. 2016; Selvi and Aruliah 2018).

Pyrethroid insecticides have been using continuously to control the insect pests in the agriculture, horticulture and forestry. Because pyrethroids are considered safer alternative to organophosphate pesticides (OPs), their applications considerably increased while the use of OPs was banned/restricted. Though pyrethroids insecticides have benefits in the agriculture, their widespread and continuous use is a major problem as they pollute the aquatic and terrestrial ecosystems and affect non-target organisms. Pyrethroids were not easily degraded immediately after application, because their residues were detected in soils, therefore, there is an urgent need to decontaminate pyrethroid polluted environments (Mariusz and Zofia Piotrowska-Seget 2016). Pyrethroids are broad-spectrum insecticides and due to the presence of chiral carbon differentiate among different forms of pyrethroid pesticides. Microbial approaches have been emerged as a popular way to reduce pyrethroid toxicity to marine life and mammals. Among microorganisms, particularly bacterial and fungal strains can be effectively degrades pyrethroids into nontoxic compounds. Various strains of bacteria and fungi such as *Bacillus* sp., *Raoultella ornithinolytica*, *Psudomonas flourescens*, *Brevibacterium* sp., *Aspergillus* sp., *Candida* sp., *Acinetobactor* sp., *Candia* spp. and *Trichoderma* sp., were used for the degradation/remediation of pyrethroid insecticides (Pankaj et al. 2019).

Pyrethroids are the most commonly used global pesticides. The natural sources of pyrethroids are the flowers of *Chrysanthemum cinerariaefolium* and, in 1949, the allethrin has been developed as the first synthetic pyrethroid insecticide (Ensley 2018; Gammon et al. 2019; Xu et al. 2019). Chrustek et al. (2018) reported that pyrethroids to be 2,250 times more toxic to insects than higher animals, and the toxicity of pyrethroids is attributed to the disruption of sodium and chloride channels. At elevated concentrations pyrethroids inhibit the function of gamma amino butyric acid (GABA) gated chloride ion channel (Bradberry et al. 2005; Gammon et al. 2019). Pyrethroids are largely used to control insect pests in agriculture, horticulture, forestry and household. Pyrethroid insecticides are considered comparatively safe but their widespread use makes them harmful for humans and animals (Burns and Pastoor 2018; Bordoni et al. 2019).

The pesticides are biocidal substances used to protect plants from weeds, fungal infection and insects (FAO-WHO Joint Report 2012). Based on the chemical composition these pesticides are divided into different types: organophosphates, organochlorides, pyrethroids, sulfenylurea, carbamates, biopesticides, etc. Though, based on their inhibitory actions, these pesticides are mainly divided into three types: insecticides, fungicides and herbicides (Dam 1974). The fungicides, like carbendazim, are widely used in agriculture and forestry to protect arable crops (rice, maize, cotton cereals, oil, etc.), fruits, vegetables, ornamentals and medicinal herbs from fungal diseases, like sheath blight, brown spot, scab, leaf spot, etc., as they suppress the metabolism of the target phytopathogenic fungi (WHO Report 1993).

Antibiotic is one of the major discoveries and have brought a revolution in the field of medicine for the human therapy. Furthermore, in addition to medicinal uses, antibiotics also have wide applications in the agriculture and animal husbandry. In developing nations, antibiotics use have helped to increase the life expectancy by lowering the deaths due to bacterial infections, but the risks associated with antibiotics pollution is largely affecting people. The antibiotics are released into environment partially degraded/undegraded and creating antibiotic pollution; therefore, its bioremediation is a major challenging task (Mohit Kumar et al. 2019).

The antibiotics are most commonly approved medicaments in modern medicine (Hernandez et al. 2012). Between 2005 and 2009, the units of antibiotics sold in India particularly increased about 40%. The sales of cephalosporins were particularly striking, with sales increasing by 60%, though some enhance was seen in most classes of antibiotics, which makes India as the world's largest user of antibiotics (Shah et al. 2015).

The antibiotics play a significant role in the management of the infectious diseases in humans, animals, livestock and aquacultures throughout world. The discharge of huge amount of antibiotics into the water and soil make a great threat to the microorganisms in the environment (Mariusz Cycon et al. 2019).

The antibiotics are natural, synthetic and semisynthetic chemical compounds, which exhibits the antimicrobial activities (Catteau et al. 2018). The antibiotics are maybe most successful family of drugs used to treat the microbial infections in the humans and animals with definite action on the target organisms. Penicillin was discovered accidentally by Alexander Fleming in 1928, since after that many antibiotics were produced for the treatment of human, plants and animal health. The antibiotics have been used to enhance the growth in animal farming such as cattle, hogs and poultry and to improve the efficiency feeding (Cowieson and Kluenter 2018), even though they are banned in EU in 2006, however still using China and India (Ronquillo and

Hernandez 2017) particularly in the livestock and agriculture industry. The increased use of the antimicrobial compounds for the benefit of humans, animals and agriculture indicated their regular and continuous release into the environment (Nielsen et al. 2018). Along with target population antibiotics also influence the non-target population with highest toxicity impact (Grenni et al. 2018). The major failure of the antibiotics and other pharmaceutical industries is mainly due to the development of antibiotic resistance in the microorganisms (Tacconelli et al. 2018). HGT, resistance in the bacterial population indicate the severe health risks in the humans and animals (Kivits et al. 2018).

18.2   REMEDIATION OF HEAVY METALS BY BACTERIA

Uptake of the heavy metals by microorganisms takes place by the bioaccumulation, which is an active process or by adsorption, which is known as passive method. Numerous microorganisms such as bacteria, fungi and algae were used to clean the heavy metal polluted environments (Kim et al. 2015). Use of metal-resistant strains in single or as consortium and immobilized form for the decontamination/remediation of heavy metals yielded effectual results because immobilized form have more chemosorption sites in order to biosorb heavy metals.

The microbial biomass has diverse biosorptive capabilities, which also differs very much among microbes. On the other hand, the biosorption capability of each microbial cell mainly depends on its pretreatment and the experimental conditions. Microbial cell must adjust to change in physical, chemical and configuration of bioreactor to increase biosorption (Fominaand and Gadd 2014). The bacteria are most important biosorbents because of their ubiquity, size and capability to grow under controlled conditions and resilience to the environmental situations (Srivastava et al. 2015). The *Bacillus pumilus* and *Brevibacterium iodinium* removes more than 87% and 88% of lead (Pb) with a decrease of 1000 mg/L to 1.8 mg/L in 96 hours. In other study Singh et al. (2013) used indigenous facultative anaerobic *Bacillus cereus* to detoxify hexavalent chromium. The *B. cereus* has a tremendous capability of 72% Cr(VI) removal at 1,000 μg/mL chromate concentration. The bacteria have capable of reducing Cr(VI) under a wide array of temperatures (25–40°C) and pH (6–10) with optimum at 37°C and initial pH 8.0. Numerous heavy metals have been tested using the bacterial species such as *Pseudomonas*, *Flavobacterium*, *Enterobacter*, *Micrococcus* sp. and *Bacillus* sp. Their immense biosorption capability is due to highest surface to volume ratios and the potential active chemosorption sites (teichoic acid) on the cell wall (Mosa et al. 2016). The bacteria were more stable and survived better when they were in mixed cultures (Sannasi et al. 2006). Therefore, consortia of cultures were metabolically superior for biosorption of metals and were more appropriate for field application (Kader et al. 2007). The *Micrococcus luteus* was used to remove a vast quantity of Pb from a synthetic medium. Under the ideal environments, the elimination ability was 1965 mg/g (Puyen et al. 2012).

Abioye et al. (2018) investigated the biosorption of chromium (Cr), lead (Pb) and cadmium (Cd) in the tannery effluent by *B. megaterium*, *Bacillus subtilis*, *Aspergillus niger* and *Penicillium* sp. *B. megaterium* showed the highest lead (Pb) decrease (2.13–0.03 mg/L), followed by *B. subtilis* (2.13–0.04 mg/L). *A. niger* showed the highest capability to reduce the chromium (Cr) concentration (1.38–0.08 mg/L) followed by *Penicillium* sp. (1.38–0.13 mg/L) whereas *B. subtilis* showed the maximum ability to decrease the concentration of cadmium (Cd) (0.4–0.03 mg/L) followed by *B. megaterium* (0.04–0.06 mg/L) at 20 days. Kim et al. (2015) designed a batch system using zeolite immobilized *Desulfovibrio desulfuricans* for the elimination of chromium (Cr^{6+}), copper (Cu) and nickel (Ni) with removal efficiency of 99.8%, 98.2% and 90.1%, respectively. Ashruta et al. (2014), reported that the efficient removal of chromium, zinc, cadmium, lead, copper and cobalt by bacterial consortia at around 75–85% in less than 2 hours duration.

18.3 REMEDIATION OF HEAVY METALS BY FUNGI

The fungi are extensively used as biosorbents for the elimination of toxic metals with outstanding capacities for metal uptake and recovery (Fu et al. 2012). Most studies showed that active and lifeless fungal cells play a significant role in the adhesion of inorganic chemicals (Magyarosy et al. 2002; Tiwari et al. 2013). The *Aspergillus* sp. was used for the elimination of chromium from tannery waste water, 85% of chromium was removed at pH 6 in a bioreactor system from the synthetic medium, when compared to a 65% removal from the tannery effluent Srivastava and Thakur (2006). This might be the presence of organic pollutants that hamper the growth of organism. *Coprinopsis atramentariais* studied for its ability to bioaccumulate 76% of Cd^{2+}, at a concentration of 1 mg/L of Cd^{2+} and 94.7% of Pb^{2+}, at a concentration of 800 mg/L f Pb^{2+}. Therefore, it has been documented as an effective accumulator of heavy metal ions for mycoremediation (Lakkireddy and Kues 2017). Park et al. (2005) reported that dead fungal biomass of *Aspergillus niger, Rhizopus oryzae, Saccharomyces cerevisiae* and *Penicillium chrysogenum* could be used to convert toxic Cr(VI) to less toxic or nontoxic Cr(III). Luna et al. (2016) also observed that *Candida sphaerica* produces biosurfactants with a removal efficiency of 95%, 90% and 79% for Fe (iron), zinc (Zn) and lead (Pb), respectively. These surfactants could form complexes with metal ions and interact directly with heavy metals before detachment from the soil. *Candida* spp. accumulate substantial quantity of nickel Ni (57–71%) and copper Cu (52–68%), but the process was affected by initial metal ion concentration and pH (optimum 3–5) (Donmez and Aksu 2001). In recent years, biosurfactants have gained importance due to their little toxicity, biodegradable nature and diversity. Mulligan et al. (2001) assessed the viability of using surfactin, rhamnolipid and sophorolipid for the removal of heavy metals (Cu and Zn). Single wash by 0.5% rhamnolipid removed 65% of copper (Cu) and 18% of the zinc (Zn), whereas 4% sophorolipid removed 25% of the copper (Cu) and 60% of zinc (Zn). The numerous strains of yeast like *Hansenula polymorpha, S. cerevisiae, Rhodotorula pilimanae, Pichia guilliermondii, Yarrowia lipolytica* and *Rhodotorula mucilage* were used to bio-convert Cr(VI) to Cr(III) (Ksheminska et al. 2008; Chatterjee et al. 2012). Various types of bacteria and fungi used for bioremediation of heavy metals are presented in Table 18.1.

TABLE 18.1
Different types of bacteria/fungi used for the bioremediation of heavy metals

S. no.	Name of the bacteria used for bioremediation of heavy metals	Name of the heavy metal	Reference
1.	*Bacillus pumilus, Brevibacterium iodinium*	Lead (Pb)	Srivastava et al. (2015)
2.	*Bacillus cereus*	Hexavalent chromium	Singh et al. (2013)
3.	*Micrococcus luteus*	Pb	Puyen et al. (2012)
4.	*B. megaterium, Bacillus subtilis, B. megaterium*	Chromium (Cr), lead (Pb), and cadmium (Cd)	Abioye et al. (2018)
5.	*Desulfovibrio desulfuricans*	Chromium (Cr^{6+}), copper (Cu), and nickel (Ni)	Kim et al. (2015)
	Name of the fungi used for bioremediation of heavy metals		
7.	*Aspergillus* sp.	Chromium	Srivastava and Thakur (2006)
8.	*Coprinopsis atramentariais*	$Cd^{2+} Pb^{2+}$	Lakkireddy and Kues (2017)
9.	*Aspergillus niger, Rhizopusoryzae, Saccharomyces cerevisiae*, and *Penicillium chrysogenum*	Cr (VI)	Park et al. (2005)
10.	*Candida sphaerica*	Fe (iron), zinc (Zn), and lead (Pb)	Luna et al. (2016)
11.	*Hansenula polymorpha, S. cerevisiae, Rhodotorula pilimanae, Pichia guilliermondii, Yarrowia lipolytica* and *Rhodotorula mucilage*	Cr (VI)	Chatterjee et al. (2012) and Ksheminska et al. (2008)

18.4 REMEDIATION OF PYRETHROIDS BY BACTERIA

Extensive use and the frequent applications of synthetic pyrethroids have led to their permanent accumulation in the environment, particularly in soils they are toxic to target and also non-target organisms (Meyer et al. 2013; Xu et al. 2015). Owing to their high hydrophobic properties, pyrethroids bind strongly to the soil particles and organic matter, which permits them to leach into the ground water and to form residues of these compounds, thereby adversely affecting the ecosystem (Gu et al. 2008; Xu et al. 2015). The reports of several studies have showed that pyrethroids have various harmful effects on soil biology that involve different quantitative and qualitative changes in the soil microflora, alterations in the activities of enzymes and changes in the nitrogen balance of the soil (inhibition of N_2 fixing and nitrifying microorganisms as well as interference with ammonification). The direct and indirect effect of pyrethroids on the soil microbiological aspects in turn affects the soil fertility plant growth (Tejada et al. 2015; Das et al. 2016).

The pyrethroid insecticides undergo several pathways once they reached to the soil environment, including transformation or degradation, volatilization, sorption, desorption uptake by plants, overflow into surface water and transfer into ground water. Alteration or degradation is one of the key processes that directs the environmental fate and transportation of pyrethroids, which also consist of different processes including abiotic degradation (e.g., oxidation, hydrolysis and photolysis) and biodegradation. In these processes, pyrethroids were transformed into a degradation product or totally mineralized. But, the structure of a pyrethroid determines its innate biodegradation (Zhao et al. 2013; Xu et al. 2015; Zhang et al. 2016). During the degradation of pyrethroid insecticides in the environment ester cleavage is a main process that resulted in the release of 3-phenoxybenzyl alcohol, cyclopropane acid, 3-phenoxybenzoic acid or 3-phenoxybenzaldehyde (3-PBA). 3-Phenoxybenzyl alcohol is also an intermediate in the photocatabolism of pyrethroid insecticides, which undergo an oxidation and resulted in the formation of corresponding carboxylic acid (Wang et al. 2011; Xiao et al. 2015). Among the mentioned metabolites, 3-PBA frequently accumulates in the soil and further degradation of pyrethroids and 3-PBA may be partial or inhibited due to its strong antimicrobial activity (Xia et al. 2008; Chen et al. 2011b).

The microorganisms show a potential of pyrethroid degradation in the liquid media and can also able to degrade them in the soil environment. On the other hand, the potential of these microbes to utilize the pyrethroids in the soils and their use in remediation of pyrethroid-contaminated soils has only confirmed for a few bacteria of the genera *Acinetobacter* (Akbar et al. 2015a), *Brevibacillus* (Akbar et al. 2015a), *Bacillus* (Zhang et al. 2016), *Ochrobactrum* (Akbar et al. 2015b), *Rhodococcus* (Akbar et al. 2015b), *Serratia* (Cycoń et al. 2014), *Sphingomonas* (Akbar et al. 2015a), *Stenotrophomonas* (Chen et al. 2011a), *Catellibacterium* (Zhao et al. 2013), *Streptomyces* (Chen et al. 2012a) and fungi, which includes *Candida* (Chen et al. 2012b).

Among the biological approaches, which include biostimulation, attenuation and bioaugmentation, the last one appear the most promising for the elimination of pyrethroids and their residues from the soil (Akbar et al. 2015a). The importance of bioaugmentation with microorganisms in the clean-up of polluted soil also has been demonstrated a few years back in relative to the other pesticides which includes organochlorinated pesticides (Sáez et al. 2014), triazines (Silva et al. 2015), organophosphorus pesticides (Aceves-Diez et al. 2015), chloroacetamide (Zheng et al. 2012), carbamate (Pimmata et al. 2013) and derivatives of phenoxyacetic acid (Önneby et al. 2014). The results of these studies caused an increasing interest in the screening of new pyrethroid-degrading strains and search for more efficient bioremediation approaches (Zhang et al. 2016).

While it has shown in the studies on bioremediation of pyrethroid polluted soils, the inoculated bacterial or fungal strains were able to degrade different pyrethroids with high effectiveness. But, majority of these studies associated with degradation of cypermethrin and which were carry out in the laboratory under controlled conditions. The dose used in these studies covers a nastiest case scenario of pesticide concentrations in the soils and shows extremely highest potential of pyrethroid-degrading microbial strains to remove pesticides from the polluted soils (Mariusz et al. 2016). Akbar et al. (2015a) shows that cypermethrin at the dosage of 200 mg/kg soil has been almost completely removed (90–100%) from the soil inoculated (1 × 10⁷ cells/g soil) with *Acinetobacter calcoaceticus* MCm5, *Brevibacillus parabrevis* FCm9 or *Sphingomonas* sp. RCm6 in 42 days. At the same time, the initial dosage of cypermethrin in the non-sterilized control soil has decreased about 44%. The capability of *Bacillus megaterium* JCm2, *O. anthropi* JCm1 or *Rhodococcus* sp. JCm5 to degrade cypermethrin in soil was reported by Akbar et al. (2015b); in these studies, the rate of degradation of pyrethroids in the case of bioaugmented soil has been reached 88–100% at 42 days of the experiment.

18.5   REMEDIATION OF PYRETHROIDS BY FUNGI

The metabolic and ecological prospective of fungi makes them appropriate for bioremediation and treatment of waste (Harms et al. 2011). The cell-free fungal extracts were known to efficiently degrade chlorpyrifos and pyrethroids (Yu et al. 2006). A β-cyhalothrin degradation by various fungi have been reported which includes *Trichoderma viridae* strain 2211, *Trichoderma viridae* strain 5-2, *Phanerochaete chryosogenum*, *Aspergillus terricola* and *A. niger*. The study followed by the extraction and identification of major degradation metabolites (Birolli et al. 2019).

A radiolabeled (¹⁴C) permethrin had used to know the mechanism of the degradation of pyrethroid in the soil and sediment. It has been observed the *R*-enantiomer of both *trans* and *cis* permethrin were mineralized quickly when compared to *S*-enantiomer and degradation product of *cis* permethrin was more persistent in soil (Qin and Gan 2006). The enantioselective degradation of the pyrethroids have been performed at southern California under the field conditions (soil and sediment) and the enantioselective degradation of *cis*-bifenthrin, cypermethrin and permethrin take place at half-life of 270–277 days, 52–135 days and 99–141 days, respectively. In the absence of enantioselectivity in biodegradation represents the preferential condition for the transformation (Qin et al. 2006).

The fungal enzymes have been reported for the biodegradation of pesticide. A few fungal enzymes catalyze the esterification, hydroxylation, dehydrogenation and deoxygenation during the process of degradation. *A. niger* YAT carries the etherification reaction during the biodegradation of cypermethrin. Similar to the bacterial biodegradation of pyrethroid metabolites, the degrading enzymes of fungal strains have also been confirmed (Maqbool et al. 2016).

Each living cell that can survive in the different environmental conditions must have metabolic pathways, which helps to obtain required food (nutrition) from the surroundings (soil, water). The oxygenases (monooxygenases and dioxygensases) play an important role in the pesticide biodegradation by common pathways (Bhatt et al. 2019).

18.6   REMEDIATION OF FUNGICIDES BY BACTERIA

Fungicide, carbendazim is primarily hazardous to the living organisms, including humans. According to NPIRSD Reports (2016), 99.83% of carbendazim may cause genetic defects, 99.65% of it can harm fertility and 93.92–98.96% of it is extremely toxic to aquatic life with long lasting effects. Its mutagenic, carcinogenic and teratogenic properties were reported as carbendazim may cause harm to the immune, nervous or endocrine systems (Xu et al. 2007). Hence, in order to protect our living organisms from the harmful effects of carbendazim, this toxicant is required to be eliminated from the environment. The conventional methods such as physicochemical methods, chemical degradation in aqueous solution by UV-TiO_2 photocatalysis, hydrolysis, oxidation, evaporation, volatilization, etc., are time consuming and costly. Conversely, climate-dependent process like phytoremediation has portability problems (Mazellier 2003). The method of bioremediation or biodegradation has different advantages over conventional techniques; complete breakdown of organic contaminants into other nontoxic chemicals, minimal requirement of equipment/chemicals and therefore, low cost of treatment per unit volume of soil/groundwater when compared to the other remediation technologies; also, implemented as an in situ or ex situ method depending on the conditions (Agnieszka 2010). The degradation rates of carbendazim by physical and abiotic chemical methods were reported as slow, by the microbial metabolism thought to be the principal degradative process in natural soils. A limited number of carbendazim-degrading bacterial strains were reported. The potential and ecologically competitive microorganisms are required to remediate a range of carbendazim polluted environments (Xinjian et al. 2013). Therefore, bacterial biodegradation proved to be an ecofriendly, efficient and cost-effective alternative process. So degradation by microorganisms has been evolved as a unique area of interest to researchers throughout the world. Few aerobic bacterial strains (*Pseudomonas* sp., *Nocardioides* sp., *Rhodococcus jialingiae*, etc.) were reported to show the capability to degrade carbendazim to 2-aminobenzimidazole, benzimidazole and 2-hydroxybenzimidazole, by specific pathways and can be use them as carbon and energy sources; however, their degradation rates were very slow in the synthetic media (up to 99% in 72 hours). Moreover, they have low activity or survival in the real environment as they are subjected to abiotic (such as pH, temperature, inorganic nutrients, etc.) and biotic (predation, competition, etc.) stress; however this is not happened in the laboratory environment (Pandey et al. 2010). Only few reports are available on the biodegradation of cabendazim in India (Sharma and Arya 2014). Therefore, a novel bacterial isolate *Bacillus licheniformis* JTC-3 has been reported, for fast and efficient biodegradation of carbendazim from the agrowastes.

A triazole foliar fungicide, propiconazole (PCZ), has been using in the agriculture. The intake of this fungicide yearly is about 7373 g a.i./ha. The degradation of pesticides by microbes gained the extensive interest in the agricultural field and environmental microbiology. Only few reports are available on the fungicide degradation by bacteria and in the previous years, many reports have revealed that bacterial community were effective degraders of many classes of fungicides: vinclozolin (Lee et al. 2008), captan (Megadi et al. 2010), tubeconazole (Nicole et al. 2009), benzimidazole (Cycon et al. 2011) and thiram (Sherif et al. 2011). Furthermore, previously available reports were revealed that PCZ utilization by microbes is not possible because of its ability to strongly adsorb with the soil's organic matter (Woo et al. 2010). Likewise, the laboratory scale degradation was conducted in the different soils by the commercial grade of PCZ alone or in the combination with benazone, dichlorprop and 2-methyl-4-chlorophenoxyacetic acid. It was reported that the maximum persistence of PCZ was found in the agricultural soil since its traces were observed after 84 days period of the experiment (Thorstensen and Lode 2001). A fungicide benomyl has an extraordinary stimulating effect on the growth and multiplication of different micro-organisms. The highest percent of increase in microbial counts have been always recorded in *Pseudomonas* sp. followed by organic nitrogen users and fungi. The addition of NPK has improved microbial counts significantly which in turn reflected positively on degradation rates of benomyl. The degradation of more than 90% of benomyl at 8.0 mg/g soil was reached in the control and in the NPK amended soil after 360 and 60 days, respectively (Randa et al. 2015).

18.7   REMEDIATION OF ANTIBIOTICS BY BACTERIA

The recent research confirms that the total antibiotics usage reached 92,700 tons in China, of which veterinary antibiotics accounted for up to 52% (Zhang et al. 2015). The antibiotics and its metabolites are released into water, sediments and soils from the feces or urine (excretion rates range from 40 to 90%) (Wang et al. 2018). Actually, about 84% of the total excretion of antibiotics (54,000 tons) stemmed from the live-stock industries in China (Zhang et al. 2015). Unluckily, the residual antibiotics are released into the agricultural soils as soon as animal wastes are used as fertil-izer (Wang et al. 2015). For instance, sulfonamides and tetracyclines are generally detected in the agricultural soils at concentrations of 6–33 and 5–25 mg/kg, respec-tively (Ji et al. 2012). In the presence of antibiotics, some microbial genes may mutate or the gene expression change, which lead to the development of antibiotic resistance (Zhu et al. 2013). The horizontal gene transfer of antibiotic resistance genes (ARGs) via mobile genetic elements is a great hazard to humans (Qiao et al. 2018). Hence, developing a potential removal method for antibiotics and ARGs from soil is utmost demanding.

The process of biodegradation is a low-cost, ecofriendly method for the removal of organic pollutants from soils, but electron acceptor deficiency restricts the sus-tainability of this technology (Li et al. 2017). Interestingly, soil microbial fuel cells (MFCs) represent a promising remediation technology that provides an inexhaust-ible electron acceptor and simultaneously employs microorganisms as catalysts to directly convert chemical energy into electricity (Li et al. 2016). The soil MFC tech-nology showed an excellent effectiveness in the degradation of polycyclic aromatic hydrocarbons (PAHs) (Yu et al. 2017), petroleum hydrocarbon (Li et al. 2016a), phenol (Huang et al. 2011) and pesticides (Cao et al. 2015), in soils. The biocurrent generated from soil MFCs be able to stimulate the growth of functional micro-bial populations and thus improves their biodegradation efficacy (Li et al. 2016a). Nevertheless, the remarkable internal resistance of soil limits electron transfer and the further improvement of soil MFC performance (Li et al. 2017). Amendment for soil MFCs, mixing the conductive carbon fiber into the soils reduced internal resistance by 58% and increased the degradation rates of petroleum hydrocarbons by 100–329% (Li et al. 2016b). In recent times, MFC technology has been carried out well in degrading antibiotics and inhibiting ARGs in wastewater (Catal et al. 2018; Wang et al. 2018), However little has been reported regarding its effect on polluted soils.

During the biodegradation process, functional microorganisms (e.g. exoelectrogens and degraders) cannot operate alone and complicated community network architecture is present in their microworld. For instance, Anaerolineaceae fermentative bacteria break down small molecule saccharides into short chain fatty acids and H_2, thus providing the electron donors for electrochemically active bacteria such as Geobacteraceae and denitrifying bacteria such as Rhodocyclaceae and Comamonadaceae (Lu et al. 2015). Furthermore, homoacetogens were found to enhance the conversion rate of acetate in MFCs by providing substances for exoelectrogens and methanogens, even these synergistic interactions occurred among fermentative bacteria, homoacetogens, exoelectrogens and methanogens (Hari et al. 2016).

After the management to farm animals, parent compounds as well as metabolites of the veterinary antibiotics were excreted and reached to the agricultural soils either directly by grazing animals or by the application of manure to land after a storage period. Haller et al. (2002) quantified Sulfamethazine (SMZ) in the manure of the treated pigs and calves of the six farms mainly in the range of mg/kg. By the application of liquid manure to fields, the risk of contaminating other environmental compartments is thus heightened (Winckler and Grafe 2001). Christian et al. (2003) found that SMZ residual concentration both in surface water (7 ng/L) and soil samples (15 µg/kg, dry weight) 7 months after a liquid manure application, indicating a high stability of SMZ in soil. These non-lethal concentrations be able to select resistant microorganisms (Andersson and Hughes 2012), which then potentially transfer resistance to other soil bacteria, including human pathogens, by gene transfer (Wellington et al. 2013).

In order to decrease these risks for the environment and human health, an effectual long term approach is strived, to clean up soils from SMZ, without destroying the soils, instead preserving them for further agricultural usage. For the decontamination of soils from the antibiotics, which are frequently applied to the fields, the bioremediation approach should not only be efficient, but also sustainable. In this case, sustainability means that the microorganisms should be applied to the soil only once and they should sustain their degradation capacity eventually and degrade the contaminant again, whenever next it applied (Natalie et al. 2016).

Till now, studies on SMZ removal have been focusing on the decontamination of waste water reactors with activated sludge (García-Galán et al. 2011; Oliveira et al. 2016), electrochemical SMZ removal from the aqueous solutions (Saidi et al. 2013), SMZ removal from water and soil using biochar (Rajapaksha et al. 2015), gamma irradiation in sewage and aqueous solution (Liu et al. 2014) and other adsorption removal techniques (Qiang et al. 2013). All these techniques were not developed for the large scale soil remediation of entire agricultural areas.

Oliveira et al. (2016) observed that SMZ degradation has mostly studied in the activated sludge systems and anaerobic waste water treatment; however, information about successful SMZ degradation is limited and where a high success was reported, the study has been conducted with disproportional higher concentrations of SMZ (90 mg/L). In their study, SMZ was biodegraded at the environmental concentration of 100 µg/L in anaerobic conditions. Because it was dependent on the availability of the easily degradable organic matter, a cometabolic degradation of SMZ was suggested.

The only study showing a metabolic SMZ degradation conducted by indigenous soil microbes was done by Topp et al. (2013). The higher SMZ degradation by native soil microorganisms have been observed in the laboratory experiments after long-term application of SMZ to the field (once per year for 10 years). Topp et al. reported that for the accomplishment of this study, it is necessary to use higher SMZ concentrations than the ones reported in soil (10 mg/kg).

Since the decontamination studies, mentioned before, focused on SMZ degradation in aqueous and sometimes anaerobic systems, they observed the need to provide an approach that can effectively accelerates the SMZ mineralization in soil. This approach should be victorious at a concentration of 1 mg/kg which is close to the environmental conditions (Haller et al. 2002) and should be directly applied to soil and avoid the destruction of the soil structure and relocation as carried out by the chemical soil extraction, soil combustion, or other harsh ex situ soil remediation techniques. In this soil inoculation approach, not only a single strain capable of degrading a soil contaminant is applied, but a microbial community. The microbial community has been enriched from an aliquot of this Canadian soil, from which Topp et al. (2013) also isolated the single SMZ degrading strain earlier. Moreover, this microbial community was attached to a protective material (defined clay particles) to improve the survival of the microorganisms in the new and foreign soil environment. The efficient approach, for the enhanced biodegradation of other organic chemicals in soils has been presented, which make sure that the introduced function of accelerated mineralization survives and establishes in the new soil environment (Wang et al. 2013).

The microbial degradation may contribute to disappearance of the antibiotics in soil. Few bacteria that degrade antibiotics were isolated from antibiotics contaminated soils. For instance, strains belonging to the genera *Microbacterium* (Topp et al. 2013), *Burkholderia* (Zhang and Dick 2014), *Stenotrophomonas* (Leng et al. 2016), *Labrys* (Mulla et al. 2018), *Ochrobactrum* (Mulla et al. 2018) and *Escherichia* (Wen et al. 2018) were capable of degrading sulfamethazine, penicillin G, tetracycline, erythromycin and doxycycline in liquid cultures, respectively. The other bacteria belonging to the genera *Klebsiella* (Xin et al. 2012), *Acinetobacter*, *Escherichia* (Zhang et al. 2012), *Microbacterium* (Kim et al. 2011), *Labrys* (Amorim et al. 2014) and *Bacillus* (Erickson et al. 2014) that were capable of degrading chloramphenicol, sulfapyridine, sulfamethazine, ciprofloxacin, norfloxacin and ceftiofur have been isolated from patients, sediments, sludge, animal feces and seawater. Particularly, a *Microbacterium* sp. showed degradation of sulfamethazine in soil and, when introduced into agricultural soil, increased the mineralization of that antibiotic about 44–57% (Hirth et al. 2016). The main role of microorganisms in the antibiotic degradation or transformation in soil was confirmed by results of many studies performed in the sterile and non-sterile soils. The half-life or DT50 values of antibiotics were much lower in soils with autochthonous microorganisms when compared to those obtained from the sterilized soils. For instance, Pan and Chu (2016) shows that when applied to soil at a dosage of 0.1 mg/kg soil, erythromycin disappeared faster in non-sterile soil when compared to sterile soil, with DT50 of 6.4 and 40.8 days, respectively. Sulfachloropyridazine applied at 10 mg/kg soil has been degraded almost three times faster in the soils with autochthonous microorganisms (half-life 20–26

TABLE 18.2

Different types of bacteria used for the bioremediation of antibiotics

S.no.	Name of the bacteria used for bioremediation of antibiotics	Name of the antibiotic	Reference
1.	*Microbacterium*	Sulfamethazine	Topp et al. (2013)
	Burkholderia	Penicillin G	Zhang and Dick (2014)
	Stenotrophomonas	Tetracycline	Leng et al. (2016)
	Labrys	Erythromycin	Mulla et al. (2018)
	Ochrobactrum	Doxycycline	Mulla et al. (2018)
	Escherichia		Wen et al. (2018)
2.	*Klebsiella*	Chloramphenicol	Xin et al. (2012)
	Acinetobacter	Sulfapyridine	Zhang et al. (2012)
	Escherichia	Sulfamethazine	Kim et al. (2011)
	Microbacterium	Ciprofloxacin	Amorim et al. (2014)
	Labrys	Norfloxacin	Erickson et al. (2014)
	Bacillus	Ceftiofur	

days) compared to sterile soils (half-life 68–71 days) (Accinelli et al. 2007). Zhang et al. (2017) also shows that microbial activity plays a main role in the biotransformation of sulfadiazine in soils, with a DT50 of 8.48, 8.97 and 10.22 days (non-sterile soil) and 30.09, 26.55 and 21.21 days (sterile soil) for concentrations of 4, 10 and 20 mg/kg soil, respectively. A similar trend has also been observed for other antibiotics, such as norfloxacin (Pan and Chu 2016), sulfamethazine (Accinelli et al. 2007; Pan and Chu 2016), sulfamethoxazole (Zhang et al. 2017) and tetracycline (Lin and Gan 2011; Pan and Chu 2016). Different types of bacteria used for bioremediation of antibiotics are presented in Table 18.2.

18.8 REMEDIATION OF ANTIBIOTICS BY FUNGI

Unlimited and irresponsible use of antibiotics has resulted in their accumulation in the environment. This has led to the emergence of multiple drug resistant microorganisms. The effect of ciprofloxacin (CIP) was determined on radial growth and biomass of *P. ostreatus*. The titrimetric and spectrophotometric analyses were performed to assess the degradation potential of *P. ostreatus* toward CIP. It has been observed that CIP has a stimulatory effect on growth and enzyme activity of *P. ostreatus*. The maximum enzyme (glucanase, ligninases, laccase) production was found at the highest concentration of CIP (500 ppm). The antibiotic degradation has been estimated about 68.8, 94.25 and 91.34% after 14 days of incubation at 500 ppm CIP using Titrimetric, Indigo carmine and Methyl orange assay, respectively (Singh et al. 2017).

Because of nondegradable nature, most of the antibiotics persist in the environment for longer periods. The antibiotics such as CIP degrade very slowly and may persist in soil in its original form for up to 1–4 months, hence creating a microenvironment for the development of antibiotic-resistant strains (Laxminarayan et al.

2013). CIP is one of the most commonly used second generation broad spectrum quinolone, it has been detected in the domestic waste water in the concentrations up to 1000–6000 ng/l, that causes the probable occurrence of selective pressures and the consequent selection of resistant bacteria (Batt et al. 2007).

Due to the extensive and unhindered use of antibiotics has aggravated to such magnanimous proportions that it is becoming increasingly complicated to treat diseases caused by such resistant bacterial strains. Therefore, it is necessary to search for new methods for effectual degradation of antibiotics persisting in the environment (Singh et al. 2017).

The fungi such as *Gloeophyllum striatum* (Wetzstein et al. 1999), *Phanerochaete chrysosporium* (Martens et al. 1996) and trametes versicolor (Rodríguez-Rodríguez et al. 2012) have been reported for their use in the bioremediation of antibiotics. *Pleurotus ostreatus* (*P. ostreatus*), is a temperate edible mushroom which forms oyster shaped fruiting bodies that be able to grown on different agricultural wastes in the temperature range of 25–28 °C (Ahmed et al. 2009). The capability of *P. ostreatus* since the bioremediation agent which has been attributed to the production of various enzymes such as laccase, lignin peroxidases, manganese peroxide and xylanases these are very important for the various metabolic reactions such as substrate utilization and degradation of pollutants (Jegatheesan et al. 2012; Singh et al. 2012).

18.9   CONCLUSIONS

Increasing in the concentration of heavy metals released into the environment by anthropogenic activities and naturally occurring metals cause great threat to the aquatic animals and humans. The uptake of heavy metals by microbes occurs via bioaccumulation. Application of metal-resistant strains as single, consortium and immobilized form for the bioremediation of heavy metals yielded effective results as the immobilized form might have more chemosorption sites to biosorb heavy metals. The bacteria are more stable and can survive better when they are in mixed culture. Most of the studies shows that active and dead fungal cells play an important role in the adhesion of inorganic chemicals, the efficacy of *Aspergillus* sp. has used for the removal of chromium tannery waste water 85% of chromium has removed at pH 6 in the bioreactor system from a synthetic medium, when compared to a 65% removal from the tannery effluent. The dead fungal biomass of *Aspergillus niger*, *Rhizopus oryzae*, *Saccharomyces cerevisiae* and *Penicillium chrysogenum* can be used to change toxic Cr(VI) to less toxic or nontoxic Cr(III).

Results obtained from many studies showed that pyrethroids may have many harmful effects on the soil biology which involve quantitative and qualitative changes in soil microflora, the changes in the activity of enzymes and inhibition of N_2 fixing and nitrifying microorganisms as well as interference by ammonification. Among the biological approaches, that include attenuation, biostimulation and bioaugmentation, the last one appeared the most promising for the elimination of pyrethroids and their residues from soil. The metabolic and ecological potential of fungi makes them appropriate for the bioremediation and waste treatment. The cell-free extracts of fungi were known to effectively degrade chlorpyrifos and pyrethroids. The fungal enzymes have been reported for pesticide biodegradation. The oxygenases (monooxygenases and dioxygensases) play an important role in the biodegradation of pesticides via the common pathways. The degradation rates of carbendazim by the physical and abiotic chemical methods were reported very slow, by the microbial metabolism thought to be the major degradative process in the natural soils. Therefore, bacterial biodegradation proved to be an ecofriendly, cost-effective and efficient.

The antibiotics and their metabolites are excreted into water, sediments and soils in the urine or feces. Due to the presence of antibiotics, some microbial genes may mutate (or the gene expression modify), which lead to the development of resistance to antibiotics. The horizontal gene transfer of ARGs by mobile genetic elements is an immeasurable hazardous to humans. Hence, developing a removal method for antibiotics and ARGs in soil is increasingly desirable. The microbial degradation could contribute to the disappearance of antibiotics in soil. The bacteria which degrade the antibiotics were isolated from the antibiotics contaminated soils. The bacterial strains belongs to the genera *Microbacterium, Burkholderia, Stenotrophomonas, Labrys, Ochrobactrum* and *Escherichia* were able to degrade sulfamethazine, penicillin G, tetracycline, erythromycin and doxycycline in liquid cultures, respectively.

REFERENCES

Abioye OP, Oyewole OA, Oyeleke SB et al. (2018) Biosorption of lead, chromium and cadmiumin tannery effluent using indigenous microorganisms. *Brazilian J Biol Sci* 59:25–32. https://doi.org/10.21472/bjbs.050903

Accinelli C, Koskinen WC, Becker JM, Sadowsky MJ (2007) Environmental fate of two sulfonamide antimicrobial agents in soil. *J Agric Food Chem* 55:2677–2682. https://doi.org/10.1021/jf063709j

Aceves-Diez AE, Estrada-Castañeda KJ, Castañeda-Sandoval LM (2015) Use of Bacillus thuringiensis supernatant from a fermentation process to improve bioremediation of chlorpyrifos in contaminated soils. *J Environ Manage* 157:213–219. https://doi.org/10.1016/j.jenvman.2015.04.026

Adikesavan S, Rajasekar A, Jayaraman T et al. (2019) Integrated remediation processes toward heavy metal removal/recovery from various environments-a review. *Front Environ Sci.* https://doi.org/10.3389/fenvs.2019.00066

Agnieszka K, Jacek K, Wojeciech D (2010) Biodegradation of Carbendazim by epiphytic and neustonic bacteria of eutrophic Chełmżyńskie Lake. *Polish J Microbiol* 57:221–230.

Ahmed SA, Kadam JA, Mane VP et al. (2009) Biological efficiency and nutritional contents of *Pleurotus florida* (Mont.) Singer cultivated on different agro-wastes. Nat Sci 7(1):44–48.

Akbar S, Sultan S, Kertesz M (2015a) Determination of cypermethrin degradation potential of soil bacteria along with plant growth-promoting characteristics. *Curr Microbiol* 70: 75–84. https://doi.org/10.1007/s00284-014-0684-7

Akbar S, Sultan S, Kertesz M (2015b) Bacterial community analysis of cypermethrin enrichment cultures and bioremediation of cypermethrin contaminated soils. *J Basic Microbiol* 55:819–829. https://doi.org/10.1002/jobm.201400805

Amorim CL, Moreira IS, Maia AS et al. (2014) Biodegradation of ofloxacin, norfloxacin, and ciprofloxacin as single and mixed substrates by *Labrys portucalensis* F11. *Appl Microbiol Biotechnol* 98:3181–3190. https://doi.org/10.1007/s00253-013-5333-8

Andersson DI, Hughes D (2012) Evolution of antibiotic resistance at non-lethal drug concentrations. *Drug Resist Update* 15(3):162–172. https://doi.org/10.1016/j.drup.2012.03.005

Ashruta GA, Nanoty V, Bhalekar U (2014) Biosorption of heavy metals from aqueous solution using bacterial EPS. *Int J Life Sci* 2(3): 373–377.

Batt AL, Kim S, Aga DS (2007) Comparison of the occurrence of antibiotics in four fullscale wastewater treatment plants with varying designs and operations. *Chemosphere* 68(3):428–435. https://doi.org/10.1016/j.chemosphere.2007.01.008

Bhatt P, Pathak VM, Joshi S et al. (2019) Major metabolites after degradation of xenobiotics and enzymes involved in these pathways, in *Smart Bioremediation Technologies* (Cambridge, MA: Academic Press) 205–215. https://doi.org/10.1016/B978-0-12-818307-6.00012-3

Birolli WG, Arai MS, Nitschke M, Porto ALM (2019) The pyrethroid (±)-lambda-cyhalothrin enantioselective biodegradation by a bacterial consortium. *Pestic Biochem Physiol* 156:129–137. https://doi.org/10.1016/j.pestbp.2019.02.014

Bordoni L, Nasuti C, Fedeli D et al. (2019) Early impairment of epigenetic pattern in neurodegeneration: additional mechanisms behind pyrethroid toxicity. *Exp Gerontol* 124:110629. https://doi.org/10.1016/j.exger.2019.06.002.

Bradberry SM, Cage SA, Proud foot AT, Vale JA (2005) Poisoning due to pyrethroids. *Toxicol Rev* 24:93–106. https://doi.org/10.2165/00139709-200524020-00003

Burns CJ, Pastoor TP (2018) Pyrethroid epidemiology: a quality-based review. *Crit Rev Toxicol* 48:297–311. https://doi.org/10.1080/10408444.2017.1423463

Cao X, Song HL, Yu CY, Li XN (2015) Simultaneous degradation of toxic refractory organic pesticide and bioelectricity generation using a soil microbial fuel cell. *Bioresour Technol* 189:87–93. https://doi.org/10.1016/j.biortech.2015.03.148

Catal T, Yavaser S, Enisoglu-Atalay V (2018) Monitoring of neomycin sulfate antibiotic in microbial fuel cells. *Bioresour Technol* 268:116–120. https://doi.org/10.1016/j. biortech.2018.07.122

Catteau L, Zhu F, Van Bambeke J, Quetin L (2018) Natural and hemi-synthetic pentacyclic triterpenes as antimicrobials and resistance modifying agents against *Staphylococcus aureus*: a review. *Phytochem Rev* 1–35. https://doi.org/10.1007/s11101-018-9564-2

Chatterjee S, Chatterjee CN, Dutta S (2012) Bioreductionof chromium (VI) to chromium (III) by a novel yeast strain *Rhodotorula mucilaginosa* (MTCC9315). *Afr J Biotechnol* 1:14920–14929.

Chen S, Lai K, Li Y et al. (2011a) Biodegradation of deltamethrin and its hydrolysis product 3-phenoxybenzaldehyde by a newly isolated Streptomyces aureus strain HP-S-01. *Appl Microbiol Biotechnol* 90:1471–1483. https://doi.org/10.1007/s00253-011-3136-3

Chen S, Yang L, Hu M, Liu J (2011b) Biodegradation of fenvalerate and 3-phenoxybenzoic acid by a novel Stenotrophomonas sp. strain ZS-S-01 and its use in bioremediation of contaminated soils. *Appl Microbiol Biotechnol* 90:755–767. https://doi.org/10.1007/ s00253-010-3035-z

Chen S, Geng P, Xiao Y, Hu M (2012a) Bioremediation of β-cypermethrin and 3-phenoxy-benzaldehyde contaminated soils using Streptomyces aureus HP-S-01. *Appl Microbiol Biotechnol* 94:505–515. https://doi.org/10.1007/s00253-011-3640-5

Chen S, Luo J, Hu M, Geng P et al. (2012b) Microbial detoxification of bifenthrin by a novel yeast and its potential for contaminated soils treatment. *PLoS One* 7:e30862. https://doi. org/10.1371/journal.pone.0030862

Christian T, Schneider RJ, Färber HA et al. (2003) Determination of antibiotic residues in manure, soil, and surface waters. *Acta Hydrochim Hydrobiol* 31(1):36–44. https://doi. org/10.1002/aheh.200390014

Chrustek A, Hołyńska-Iwan I, Dziembowska I (2018) Current research on the safety of pyre-throids used as insecticides. *Medicina* 54:61. https://doi.org/10.3390/medicina54040061

Cowieson AJ, Kluenter AM (2018) Contribution of exogenous enzymes to potentiate the removal of antibiotic growth promoters in poultry production. *Anim Feed Sci Technol.* https://doi.org/10.1016/j.anifeedsci.2018.04.026

Cycon M, Wojcik M, Piotrowska-Seget Z (2011) Biodegradation kinetics of the benzimidazole fungicide thiophanate-methyl by bacteria isolated from loamy sand soil. *Biodegradation* 22:573–583. https://doi.org/10.1007/s10532-010-9430-4

Cycoń M, Zmijowska A, Piotrowska-Seget Z (2014) Enhancement of deltamethrin degradation by soil bioaugmentation with two different strains of Serratia marcescens. *Int J Environ Sci Tech* 11:1305–1316. https://doi.org/10.1007/s13762-013-0322-0

Dam W (1974) Photolysis of methyl benzimidazol-2-yl carbamate. *Chemosphere* 5:239–240.

Das R, Das SJ, Das AC (2016) Effect of synthetic pyrethroid insecticides on N_2-fixation and its mineralization in tea soil. *Eur J Soil Biol* 74:9–15. https://doi.org/10.1016/j.ejsobi.2016.02.005

Donmez G, Aksu Z (2001) Bioaccumulation of copper (ii) and nickel (ii) by the non-adapted and adapted growing Candida SP. *Water Res* 35(6):1425–1434. https://doi.org/10.1016/S0043-1354(00)00394-8

Ensley SM (2018) Pyrethrins and pyrethroids: veterinary toxicology. *Basic Clin Princ* 39:515–520. https://doi.org/10.1016/B978-0-12-811410-0.00039-8

EPA (1992) *Framework for Ecological Risk Assessment. EPA/630/R-92/001*. Risk Assessment Forum, U.S. Environmental Protection Agency, Washington, DC.

EPA (2002) *Environmental Protection Agency, Office of Groundwater and Drinking Water. Implementation Guidance for the Arsenic Rule*. EPA Report 816-D-02-005, Cincinnati, OH.

Erickson BD, Elkins CA, Mullis L et al. (2014) A metallo-β-lactamase is responsible for the degradation of ceftiofur by the bovine intestinal bacterium *Bacillus cereus* P41. *Vet Microbiol* 172:499–504. https://doi.org/10.1016/j.vetmic.2014.05.032

FAO-WHO Joint Report (2012) Food and Agriculture Organization USA Pesticide residues in food: REPORT. Joint Food and Agriculture Organization/World Health Organization Meeting on Pesticide Residues. http://www.fao.org/agriculture/crops/thematic-sitemap/theme/pests/jmpr/en/ (accessed 15.11.16)

Fominaand M, Gadd GM (2014). Biosorption: current perspectiveson concept, definition and application. *Bioresource Technol* 160:3–14. https://doi.org/10.1016/j.biortech.2013.12.102

Fu QY, Li S, Zhu YH (2012) Biosorption of copper (II) from aqueous solution by mycelial pellets of *Rhizopus oryzae. Afr J Biotechnol* 11(6):1403–1411. https://doi.org/10.5897/AJB11.2809

Gammon DW, Liu Z, Chandrasekaran A et al. (2019) Pyrethroid neurotoxicity studies with bifenthrin indicate a mixed Type I/II mode of action. *Pest Manag Sci* 75:1190–1197. https://doi.org/10.1002/ps.5300

García-Galán MJ, Rodríguez-Rodríguez CE, Vicent T et al. (2011) Biodegradation of sulfamethazine by trametes versicolor: removal from sewage sludge and identification of intermediate products by UPLC–QqTOF-MS. *Sci Total Environ* 409(24):5505–5512. https://doi.org/10.1016/j.scitotenv.2011.08.022

Grenni P, Ancona V, Caracciolo AB (2018) Ecological effects of antibiotics on natural ecosystems: a review. *Microchem J* 136:25–39. https://doi.org/10.1016/j.microc.2017.02.006

Gu XZ, Zhang GY, Chen L, Dai RL, Yu YC 2008. Persistence and dissipation of synthetic pyrethroid pesticides in red soils from the Yangtze river delta area. *Environ Geochem Health* 30:67–77. https://doi.org/10.1007/s10653-007-9108-y

Haller MY, Müller SR, McArdell CS (2002) Quantification of veterinary antibiotics (sulfonamides and trimethoprim) in animal manure by liquid chromatography-mass spectrometry. *J Chromatogr* 952(1–2):111–120. https://doi.org/10.1016/s0021-9673(02)00083-3

Hari AR, Katuri KP, Logan BE, Saikaly PE (2016) Set anode potentials affect the electron fluxes and microbial community structure in propionate-fed microbial electrolysis cells. Sci Rep 6:38690.

Harms H, Schlosser D, Wick LY (2011) Untapped potential: exploiting fungi in bioremediation of hazardous chemicals. *Nat Rev Microbiol* 9(3):177-192. https://doi.org/ 10.1038/nrmicro2519

He ZL, Yang XE, Stoffella PJ (2005) Trace elements in agroecosystems and impacts on the environment. *J Trace Elem Med Biol* 19:125–140. https://doi.org/10.1016/j.jtemb.2005.02.010

Hernandez F, Rivera A, Ojeda A et al. (2012) Photochemical degradation of the CIP antibiotic and its micro-biological validation. *J Environ Sci Eng A* 1:448–453.

Hirth N, Topp E, Dörfler U et al. (2016) An effective bioremediation approach for enhanced microbial degradation of the veterinary antibiotic sulfamethazine in an agricultural soil. *Chem Biol Technol Agric* 3:29. https://doi.org/10.1186/s40538-016-0080-6

Huang DY, Zhou SG, Chen Q et al. (2011) Enhanced anaero-bic degradation of organic pol-lutants in a soil microbial fuel cell. *Chem Eng J* 172:647–653. https://doi.org/10.1016/j. cej.2011.06.024

Jegatheesan M, Kumaran MS, Eyini M (2012) Enhanced production of laccase enzyme by the white-rot mushroom fungus Pleurotus florida using response surface methodology. *Int J Curr Res* 4(10):025–031.

Ji XL, Shen QH, Liu F et al. (2012) Antibiotic resistance gene abundances associated with antibiotics and heavy metals in animal manures and agricultural soils adjacent to feed lots in Shanghai, China. *J Hazard Mater* 235–236:178–185. https://doi.org/10.1016/j. jhazmat.2012.07.040

Kader J, Sannasi P, Othman O et al. (2007) Removal of Cr (VI) from aqueous solutions by growing andnon-growing populations of environmental bacterial consortia. *Global J Environ Resource* 1:12–17.

Kaonga CC, Kosamu IB, Lakudzala DD et al. (2017). A review of heavy metals in soil and aquatic systems of urban and semi-urban areas in Malawi with comparisons to other selected countries. *Afr J Environ Sci Technol* 11:448–460. https://doi.org/10.5897/ AJEST2017.2367

Khalid S, Shahid M, Khan N et al. (2017) A comparison of technologies for remedia-tion of heavy metal contaminated soils. *J Geochem Explor* 182:247–268. https://doi. org/10.1016/j.gexplo.2016.11.021

Kim DW, Heinze TM, Kim BS et al. (2011) Modification of norfloxacin by a *Microbacterium* sp. strain isolated from a wastewater treatment plant. *Appl Environ Microbiol* 77:6100– 6108. https://doi.org/10.1128/AEM.00545-11

Kim IH, Choi JH, Joo JO et al. (2015) Development of a microbe-zeolite carrier for the effective elimination of heavy metals from seawater. *J Microbiol Biotechnol* 25(9):1542–1546.

Kivits T, Broers HP, Beeltje H et al. (2018) Presence and fate of veterinary antibiotics in age-dated ground water in areas with intensive livestock farming. *Environ Pollut* 241:988–998. https://doi.org/10.1016/j.envpol.2018.05.085

Ksheminska H, Fedorovych D, Honchar T et al. (2008) Yeast tolerance to chromium depends on extracellular chromate reduction and Cr (III) chelation. *Food Technol Biotechnol* 46(4):419–426.

Lakkireddy K, Kues U (2017) Bulk isolation of basidiosporesfrom wild mushrooms by elec-trostatic attraction with low risk of microbial contaminations. *AMB Express* 7(1):28. https://doi.org/10.1186/s13568-017-0326-0

Laxminarayan R, Duse A, Wattal C et al. (2013) Antibiotic resistance-the need for global solu-tions. *Lancet Infect Dis* 13(12):1057–1098.

Lee JB, Sohn HY, Shin KS et al. (2008) Microbial biodegradation and toxicity of vinclozolin and its toxic metabolite 3,5-dichloroaniline. *J Microbiol Biotechnol* 18:343–349.

Leng Y, Bao J, Chang G et al. (2016) Biotransformation of tetracycline by a novel bacterial strain *Stenotrophomonas maltophilia* DT1. *J Hazard Mater* 318:125–133. https://doi. org/10.1016/j.jhazmat.2016.06.053

Li AXJ, Wang X, Zhang YY et al. (2016a) Salinity and conductivity amendment of soil enhanced the bioelectrochemical degradation of petroleum hydrocarbons. *Sci Rep* 6:32861.

Li XJ, Wang X, Wan LL et al. (2016b) Enhanced biodeg-radation of aged petroleum hydro-carbons in soils by glucose addition in microbial fuel cells. *J Chem Technol Biotechnol* 91:267–75. https://doi.org/10.1002/jctb.4660

Li XJ, Wang X, Zhao Q et al. (2016c) Carbon fiber enhanced bioelectricity generation in soil microbial fuel cells. *Biosens Bioelectron* 85:135–141. https://doi.org/10.1016/j.bios.2016.05.001

Li XJ, Wang X, Weng LP et al. (2017) Microbial fuel cells for organic-contaminated soil remedial applications: a review. *Energy Technol* 5:1156–1164. https://doi.org/10.1002/ente.201600674

Lin K, Gan J (2011) Sorption and degradation of wastewater-associated non-steroidal anti-inflammatory drugs and antibiotics in soils. *Chemosphere* 83:240–246. https://doi.org/10.1016/j.chemosphere.2010.12.083

Liu Y, Hu J, Wang J (2014) Fe^{2+} enhancing sulfamethazine degradation in aqueous solution by gamma irradiation. *Radiat Phys Chem* 96:81–87. https://doi.org/10.1016/j.radphyschem.2013.08.018

Lu L, Xing DF, Ren ZY (2015) Microbial community structure accompanied with electricity production in a constructed wetland plant microbial fuel cell. *Bioresour Technol* 195:115–121. https://doi.org/10.1016/j.biortech.2015.05.098

Luna JM, Rufino RD, Sarubbo LA (2016) Biosurfactant from *Candida sphaerica* UCP0995 exhibiting heavy metal remedi-ation properties. *Proc Safe Environ Protec* 102:558–566. https://doi.org/10.1016/j.psep.2016.05.010

Magyarosy A, Laidlaw R, Kilaas R (2002) Nickel accumulation and nickel oxalate precipitationby *Aspergillus niger*. *Appl Microbiol Biotechnol* 59(2–3):382–388. https://doi.org/10.1007/s00253-002-1020-x

Mao X, Han FX, Shao X et al. (2016) Electro-kinetic remediation coupled with phytoremediation to remove lead, arsenic and cesium from contaminated paddy soil. *Ecotoxicol Environ Saf* 125:16–24 https://doi.org/10.1016/j.ecoenv.2015.11.021

Maqbool Z, Hussain S, Imran M et al. (2016) Perspectives of using fungi as bioresource for bioremediation of pesticides in the environment: a critical review. *Environ Sci Pollut Res* 23:16904–16925. https://doi.org/10.1007/s11356-016-7003-8

Mariusz C, Zofia Piotrowska-Seget Z (2016) Pyrethroid-degrading microorganisms and their potential for the bioremediation of contaminated soils: a review. *Front Microbiol*. https://doi.org/10.3389/fmicb.2016.01463

Mariusz C, Agnieszka M, Zofia PS (2019) Antibiotics in the soil environment-degradation and their impact on microbial activity and diversity. *Front Microbiol*. https://doi.org/10.3389/fmicb.2019.00338

Martens R, Wetzstein HG, Zadrazil F et al. (1996) Degradation of the fluoroquinolone enrofloxacin by wood-rotting fungi. *Appl Environ Microbiol* 62(11):4206–4209.

Mazellier PE, Leroy J, Laat D, Legube B (2003) Degradation of carbendazim by UV/H_2O_2 investigated by kinetic modeling. *Environ Chem Lett* 1:68–72. https://doi.org/10.1007/s10311-002-0010-7

Megadi BV, Tallur NP, Mulla IS, Ninnekar HZ (2010) Bacterial degradation of fungicide captan. *J Agric Food Chem* 58:12863–12868. https://doi.org/10.1021/jf1030339

Meyer BN, Lam C, Moore S, Jones RL (2013) Laboratory degradation rates of 11 pyrethroids under aerobic and anaerobic conditions. *J Agric Food Chem* 61:4702–4708. https://doi.org/10.1021/jf400382u

Mohit K, Shweta J, Kushneet KS et al. (2019) Antibiotics bioremediation: perspectives on its ecotoxicity and resistance. *Environ Int* 124:448–461. https://doi.org/10.1016/j.envint.2018.12.065

Mosa KA, Saadoun I, Kumar K et al. (2016) Potential biotechnological strategies for the cleanup of heavy metals and metalloids. *Front Plant Sci* 7:303. https://doi.org/10.3389/fpls.2016.00303

Mulla SI, Hu A, Sun Q et al. (2018) Biodegradation of sulfamethoxazole in bacteria from three different origins. *J Environ Manage* 206:93–102. doi: 10.1016/1.jenvman.2017.10.029

Mulligan CN, Yong RN, Gibbs BF (2001) Remediation technologies for metal-contaminated soils and groundwater: an evaluation. *Eng Geol* 60(1–4):193–207. https://doi.org/10.1016/S0013-7952(00)00101-0

Natalie H, Edward T, Ulrike D et al. (2016) An effective bioremediation approach for enhanced microbial degradation of the veterinary antibiotic sulfamethazine in an agricultural soil. *Chem Biol Technol Agric* 3:29. https://doi.org/10.1186/s40538-016-0080-6

Nicole TS, Priscila SC, Maria DCR, Peralbaand Marco AZA (2009) Biodegradation of tebuconazole by bacteria isolated from contaminated soils. *J Environ Sci Health B Pestic Food Contam Agric Wastes* 45:67–72. https://doi.org/10.1080/03601230903404499

Nielsen KM, Gjoen T, Asare NYO et al. (2018) Tronsmo antimicrobial resistance in wildlife potential for dissemination opinion of the panel on Microbial Ecology of the Norwegian Scientific Committee for Food and Environment. VKM Report.

NPIRSD Report (2016) National Pesticide Information Retrieval System's Database on Carbendazim. http://npirspublic.ceris.purdue.edu/ppis/ (accessed 11.08.17).

Oliveira GHD, Santos-Neto AJ, Zaiat M (2016) Evaluation of sulfamethazine sorption and biodegradation by anaerobic granular sludge using batch experiments. *Bioprocess Biosyst Eng* 39(1):115–124. https://doi.org/10.1007/s00449-015-1495-3

Önneby K, Håkansson S, Pizzul L, Stenström J (2014) Reduced leaching of the herbicide MCPA after bioaugmentation with a formulated and stored Sphingobium sp. *Biodegradation* 25:291–300. https://doi.org/10.1007/s10532-013-9660-3

Pan M, Chu LM (2016) Adsorption and degradation of five selected antibiotics in agricultural soil. *Sci Total Environ* 545–546:48–56. https://doi.org/10.1016/j.scitotenv.2015.12.040

Pandey G, Dorrian SJ, Russell RJ (2010) Cloning and biochemical characterization of a novel carbendazim (methyl-1H-benzimidazol-2-ylcarbamate)-hydrolyzing esterase from the newly isolated Nocardioides sp. strain SG-4G and its potential for use in enzymatic bioremediation. *Appl Environ Microbiol* 76:2940–2945. https://doi.org/10.1128/AEM.02990-09

Pankaj B, Yaohua H, Hui Z, Shaohua C (2019) Insight into microbial applications for the biodegradation of pyrethroid insecticides. *Front Microbiol*. https://doi.org/10.3389/fmicb.2019.01778.

Park D, Yun YS, Jo JH, Park JM (2005) Mechanismofhex-avalent chromium removal by dead fungal biomass of *Aspergillus niger*. *Water Res* 39(4):533–540. https://doi.org/10.1016/j.watres.2004.11.002.

Pimmata P, Reungsang A, Plangklang P (2013) Comparative bioremediation of carbofuran contaminated soil by natural attenuation, bioaugmentation and biostimulation. *Int Biodeter Biodegr* 85:196–204. https://doi.org/10.1016/j.ibiod.2013.07.009

Puyen ZM, Villagrasa E, Maldonado J et al. (2012) Biosorption of lead and copper by heavy-metal tolerant *Micrococcus luteus* DE2008. *Biores Technol* 126:233–237. https://doi.org/ 10.1016/j.biortech.2012.09.036

Qiang Z, Bao X, Ben W (2013) MCM-48 modified magnetic mesoporous nano-composite as an attractive adsorbent for the removal of sulfamethazine from water. *Water Res* 47(12):4107–4114. https://doi.org/10.1016/j.watres.2012.10.039

Qiao M, Ying GG, Singer AC, Zhu YG (2018) Review of antibiotic resistance in China and its environment. *Environ Int* 110:160–172. https://doi.org/10.1016/j.envint.2017.10.016

Qin S, Budd R, Bondarenko S et al. (2006) Enantioselective degradation and chiral stability of pyrethroids in soil and sediment. *J Agric Food Chem* 54:5040–5045. https://doi.org/10.1021/jf060329p

Qin S, Gan J (2006) Enantiomeric differences in permethrin degradation pathways in soil and sediment. *J Agric Food Chem* 54:9145–9151. https://doi.org/10.1021/jf0614261

Rahman PKSM, Bastola S (2014) Biological reduction of iron to the elemental state from ochre deposits of Skelton Beck in Northeast England. *Front Environ Sci* 2:22–25. https://doi.org/10.3389/fenvs.2014.00022

Rajapaksha AU, Vithanage M, Ahmad M et al. (2015) Enhanced sulfamethazine removal by steam-activated invasive plant-derived biochar. *J Hazard Mater* 290:43–50. https://doi.org/10.1016/j.jhazmat.2015.02.046

Randa HE, Adil AE, Awad GO, Ashraf MS (2015) Microbial degradation of the fungicide Benomyl in soil as influenced by addition of NPK. *Int J Curr Microbiol App Sci* 4(5):756–771.

Rodríguez-Rodríguez CE, García-Galán MA, Blánquez P et al. (2012) Continuous degradation of a mixture of sulfonamides by *Trametes versicolor* and identification of metabolites from sulfapyridine and sulfathiazole. *J Hazard Mater* 213–214:347–354. https://doi.org/10.1016/j.jhazmat.2012.02.008

Ronquillo MG, Hernandez JCA (2017) Antibiotic and synthetic growth promoters in animal diets: review of impact and analytical methods. *Food Control* 72:255–267. https://doi.org/10.1016/j.foodcont.2016.03.001

Sáez JM, Alvarez A, Benimelli CS, Amorosso MJ (2014) Enhanced lindane removal from soil slurry by immobilized Streptomyces consortium. *Int Biodeterior Biodegr* 93:63–69. https://doi.org/10.1016/j.ibiod.2014.05.013

Saidi I, Soutrel I, Fourcade F et al. (2013) Flow electrolysis on high surface electrode for biodegradability enhancement of sulfamethazine solutions. *J Electroanal Chem* 707:122–8. https://doi.org/10.1016/j.jelechem.2013.09.006

Sannasi P, Kader J, Ismail BS, Salmijah S (2006) Sorption of Cr (VI), Cu(II) and Pb(II) by growing and non-growing cells of a bacterial consortium. *Biores Technol* 97(5):740–747. https://doi.org /10.1016/j.biortech.2005.04.007

Sarmah AK, Meyer MT, Boxall ABA (2006) A global perspective on the use, sales, exposure pathways, occurrence, fate and effects of veterinary antibiotics (VAs) in the environment. *Chemosphere* 65(5):725–759. https://doi.org/10.1016/j.chemosphere.2006.03.026

Selvi A, Aruliah R (2018) A statistical approach of zinc remediation using acidophilic bacterium via an integrated approach of bioleaching enhanced electrokinetic remediation (BEER) technology. *Chemosphere* 207:753–763. https://doi.org/10.1016/j.chemosphere.2018.05.144

Shah Z, Dighe A, Londhe V (2015) Pharmacoeconomic study of various brands of antibiotic medications in India. *World J Parma Res* 4(3):1600–1606.

Sharma A, Arya A (2014) Screening, isolation and characterization of Brevibacillus borstelensis for the bioremediation of carbendazim. *J Environ Sci Sustain* 2:12–14.

Sherif AM, Elhussein AA, Osman AG (2011) Biodegradation of fungicide thiram (TMTD) in soil under laboratory conditions. *Am J Biotechnol Mol Sci* 1:57–68. https://doi.org/10.5251/ajbms.2011.1.2.57.68

Silva VP, Moreira-Santos M, Mateus C, et al. (2015) Evaluation of Arthrobacter aurescens strain TC1 as bioaugmentation bacterium in soils contaminated with the herbicidal substance terbuthylazine. *PLoS One* 10:e144978. https://doi.org/10.1371/journal.pone.0144978

Singh KS, Robinka K, Loveleen K (2017) Biodegradation of ciprofloxacin by white rot fungus *Pleurotus ostreatus*. *3 Biotech* 7(1):69. https://doi.org/10.1007/s13205-017-0684-y

Singh MP, Pandey AK, Vishwakarma SK et al. (2012) Extracellular xylanase production by *Pleurotus* species on lignocellulosic wastes under in vivo condition using novel pretreatment. *Cell Mol Biol (Noisy-le-grand)* 58(1):170–173. https://doi.org/10.1170/T937

Singh N, Tuhina V, Rajeeva G (2013) Detoxification of hexavalent chromium by an indigenous facultative anaerobic *Bacillus cereus* strain isolated from tannery effluent. *Afr J Biotechnol* 12(10):1091–1103. https://doi.org/10.4314/ajb.v12i10

Srivastava S, Agrawal SB, Mondal MK (2015) A review on progress of heavy metal removal using adsorbents of microbial and plant origin. *Environ Sci Pollution Res* 22:15386–15415. https://doi.org /10.1007/s11356-015-5278-9

Srivastava S, Thakur IS (2006) Isolation and process parameter optimization of *Aspergillus* sp. for removal of chromium from tannery effluent. *Biores Technol* 97(10):1167–1173. https://doi.org/10.1016/j.biortech.2005.05.012

Tacconelli E, Carrara E, Savoldi A et al. (2018) Discovery, research, and development of new antibiotics: the WHO priority list of antibiotic-resistant bacteria and tuberculosis. *Lancet Infect Dis* 18(3):318–327. https://doi.org/10.1016/S1473-3099(17)30753-3

Tejada M, García C, Hernández T, Gómez I (2015) Response of soil microbial activity and biodiversity in soils polluted with different concentrations of cypermethrin insecticide. *Arch Environ Contam Toxicol* 69:8–19. https://doi.org/10.1007/s00244-014-0124-5

Thorstensen CW, Lode O (2001) Laboratory degradation studies of bentazone, dichlorprop, MCPA, and propiconazole in Norwegian soils. *J Environ Qual* 30:947–953. https://doi.org/10.2134/jeq2001.303947x

Tiwari S, Singh SN, Garg SK (2013) Microbially enhanced phytoextraction of heavy-metal fly-ash amended soil. *Commun Soil Sci Plant Anal* 44(21):3161–3176. https://doi.org/1 0.1080/00103624.2013.832287

Topp E, Chapman R, Devers-Lamrani M et al. (2013) Accelerated biodegradation of veterinary antibiotics in agri-cultural soil following long-term exposure, and isolation of a sulfamethazine-degrading Microbacterium sp. *J Environ Qual* 42(1):173–178. https://doi.org/10.2134/jeq2012.0162

Wang B, Ma Y, Zhou W et al. (2011) Biodegradation of synthetic pyrethroids by Ochrobactrum tritici strain pyd-1. *World J Microbiol Biotechnol* 27:2315–2324. https://doi.org/10.1007/s11274-011-0698-2

Wang C, Qu GZ, Wang TC et al. (2018) Removal of tetracycline antibiotics from wastewater by pulsed corona discharge plasma coupled with natural soil particles. *Chem Eng J* 346:159–170. https://doi.org/10.1016/j.cej.2018.03.149

Wang F, Fekete A, Harir M et al. (2013) Soil remediation with a microbial community established on a carrier: strong hints for microbial communication during 1,2,4-trichlorobenzene degradation. *Chemosphere* 92(11):1403–1409.

Wang FH, Qiao M, Chen Z (2015) Antibiotic resistance genes in manure-amended soil and vegetables at harvest. *J Hazard Mater* 299:215–221. https://doi.org/10.1016/j.jhazmat.2015.05.028

Wang L, You LX, Zhang JM et al. (2018) Biodegradation of sulfadiazine in microbial fuel cells: reaction mechanism, biotoxicity removal and the correlation with reactor microbes. *J Hazard Mater* 360:402–411. https://doi.org/10.1016/j.jhazmat.2018.08.021

Wellington EMH, Boxall ABA, Cross P et al. (2013) The role of the natural environment in the emergence of antibiotic resistance in Gram-negative bacteria. *Lancet Infect Dis* 13(2):155–165. https://doi.org/10.1016/S1473-3099(12)70317-1

Wen X, Wang Y, Zou Y et al. (2018) No evidential correlation between veterinary antibiotic degradation ability and resistance genes in microorganisms during the biodegradation of doxycycline. *Ecotoxicol Environ Saf* 147:759–766. https://doi.org/10.1016/j.ecoenv.2017.09.025

Wetzstein HG, Stadler M, Tichy HV et al. (1999) Degradation of CIP by basidiomycetes and identification of metabolites generated by the brown rot fungus *Gloeophyllum striatum*. *Appl Environ Microbiol* 65(5):1556–1563.

WHO (1993) *World Health Organization* International Programme on Chemical Safety: Geneva. Environ Heal Criter, Carbendazim. http://www.who.int/ipcs/publications/ehc/ehc_numerical/en/

Winckler C, Grafe A (2001) Use of veterinary drugs in intensive animal production. *J Soils Sediments* 1(2):66–70. https://doi.org/10.1007.BF0298771

Woo C, Daniels B, Stirling R, Morris P (2010) Tebuconazole and propiconazoletolerance and possible degradation by Basidiomycetes: a wood-based bioassay. *Int Biodeterior Biodegrad* 64:403–408. https://doi.org/10.1016/j.ibiod.2010.01.009

Wuana RA, Okieimen FE (2011) Heavy metals in contaminated soils: a review of sources, chemistry, risks and best available strategies for remediation. ISRN 20. https://doi.org/10.5402/2011/402647

Xia WJ, Zhou JM, Wang HY, Chen XQ (2008) Effect of nitrogen on the degradation of cypermethrin and its metabolite 3-phenoxybenzoic acid in soil. *Pedosphere* 18:638–644. https://doi.org/10.1016/S1002-0160(08)60058-2

Xiao Y, Chen S, Gao Y et al. (2015) Isolation of a novel beta-cypermethrin degrading strain Bacillus subtilis BSF01 and its biodegradation pathway. *Appl Microbiol Biotechnol* 99:2849–2859. https://doi.org/10.1007/s00253-014-6164-y

Xin Z, Fengwei T, Gang W et al. (2012) Isolation, identification and characterization of human intestinal bacteria with the ability to utilize chloramphenicol as the sole source of carbon and energy. *FEMS Microbiol Ecol* 82:703–712. https://doi.org/10.1111/j.1574-6941.2012.01440.x

Xinjian Z, Yujie H, Harvey PR et al. (2013) Isolation and characterization of Carbendazim-degrading Rhodococcus erythropolis djl-11. *PLoS One* 8(10):e74810. https://doi.org/10.1371/journal.pone.0074810

Xu H, Li W, Schilmiller AL et al. (2019) Pyrethric acid of natural pyrethrin insecticide: complete pathway elucidation and reconstitution in *Nicotiana benthamiana*. *New Phytol* 223:751–765. https://doi.org/10.1111/nph.15821

Xu JL, He J, Wang ZC et al. (2007) Rhodococcus qingshengii, a carbendazim-degrading bacterium. *Int J System Evolut Microbiol* 57:2754–2757. https://doi.org/10.1099/ijs.0.65095-0

Xu Z, Shen X, Zhang XC, Liu W, Yang F (2015) Microbial degradation of alpha-cypermethrin in soil by compound-specific stable isotope analysis. *J Hazard Mater* 295:37–42. https://doi.org/10.1016/j.jhazmat.2015.03.062

Yu B, Tian J, Feng L (2017) Remediation of PAH polluted soils using a soil microbial fuel cell: influence of electrode interval and role of microbial community. *J Hazard Mater* 336:110–118. https://doi.org/10.1016/j.jhazmat.2017.04.066

Yu YL, Fang H, Wang X et al. (2006) Characterization of a fungal strain capable of degrading chlorpyrifos and its use in detoxification of the insecticide on vegetables. *Biodegradation* 17:487–494. https://doi.org/10.1007/s10532-005-9020-z

Zhang H, Zhang Y, Hou Z et al. (2016) Biodegradation potential of deltamethrin by the Bacillus cereus strain Y1 in both culture and contaminated soil. *Int Biodeter Biodegrad* 106:53–59. https://doi.org/10.1016/j.ibiod.2015.10.005

Zhang Q, Dick WA (2014) Growth of soil bacteria, on penicillin and neomycin, not previously exposed to these antibiotics. *Sci Total Environ* 493:445–453. https://doi.org/10.1016/j.scitotenv.2014.05.114 et al. Zhang QQ, Ying GG, Pan CG et al. (2015) Comprehensive evaluation of antibiotics emission and fate in the river basins of China: source analysis, multimedia modeling, and linkage to bacterial resistance. *Environ Sci Technol* 49:6772–6782. https://doi.org/10.1021/acs.est.5b00729

Zhang WW, Wen YY, Niu ZL et al. (2012) Isolation and characterization of sulfonamide-degrading bacteria Escherichia sp. HS21 and Acinetobacter sp. HS51. *World J Microbiol Biotechnol* 28:447–452. https://doi.org/10.1007/s11274-011-0834-zZhang Y, Hu S, Zhang H et al. (2017) Degradation kinetics and mechanism of sulfadiazine and sulfamethoxazole in an agricultural soil system with manure application. *Sci Total Environ* 607–608:1348–1356. https://doi.org/10.1016/j.scitotenv.2017.07.083

Zhao H, Geng Y, Chen L et al. (2013) Biodegradation of cypermethrin by a novel Catellibacterium sp. strain CC-5 isolated from contaminated soil. *Can J Microbiol* 59:311–317. https://doi.org/10.1139/cjm-2012-0580

Zheng J, Li R, Zhu J et al. (2012) Degradation of the chloroacetamide herbicide butachlor by Catellibacterium caeni sp. nov DCA-1^T. *Int Biodeter Biodegrad* 73:16–22. https://doi.org/10.1016/j.ibiod.2012.06.003

Zhu YG, Johnson TA, Su JQ et al. (2013) Diverse and abundant antibiotic resistance genes in Chinese swine farms. *Proc Natl Acad Sci USA* 110:3435–3440. https://doi.org/10.1073/pnas.1222743110

Zouboulis AI, Loukidou MX, Matis KA (2004) Biosorption of toxic metals from aqueous solutions by bacteria strains isolated from metalpolluted soils. *Process Biochem* 39:909–916. https://doi.org/10.1016/S0032-9592 (03)00200-0

19 Current advanced technological tools for the bioremediation of pesticides

Srinivasan Kameswaran, Bellamkonda Ramesh, Gopi Krishna Pitchika, M. Srinivasulu, and M. Subhosh Chandra

CONTENTS

DOI: 10.1201/9781003394600-22

19.1 INTRODUCTION

Pollution is an issue that affects the entire world. Pollutants are released into the environment in large quantities every year. A total of 6×10^6 chemical compounds have been created, with thousands of new chemicals created each year. More than 450 million kg of pollutants are discharged globally in air and water, according to Third World Network data. Environmentalists all over the world are working to solve this problem in a variety of ways, yet many compounds are still utilised without regard for their negative consequences (Shukla et al. 2010). Traditional methods for removing these chemicals, such as landfilling, high-temperature incineration, and chemical decomposition (e.g. UV oxidation, base-catalysed dechlorination), can be effective, but they come with a number of drawbacks, including complexity, cost, and lack of public acceptance, particularly incineration, which can expose workers and nearby residents to contaminants by causing air pollution. Hazardous materials excavation, handling, and transportation all pose considerable risks.

The cost of removing 1 m³ of soil from a one-acre polluted site is estimated to be between \$0.6 and 2.5 million dollars (McIntyre 2003). Furthermore, these traditional methods are not always sufficient (Dixon 1996). There needs to be a new alternative method that either entirely destroys the contaminants or turns them into harmless compounds.

Bioremediation, or the use of living organisms to clean up polluted environments, is a new technology (Samanta et al. 2002; Singh 2008). It is a cost-effective and environmentally beneficial technology that uses natural biological activity to eliminate or render harmless certain pollutants. It has a high level of public approval and can frequently be performed on-site (Vandervoort 1997). Bioremediation has various advantages, according to Hussain et al. (2009) that have made it a preferred methodology for remediating contaminated sites over alternative physicochemical processes. It's getting more and more appealing as a method of cleanup. Bacteria, yeast, and fungi are the most common biological agents utilised (Strong and Burgess 2008).

The type of the contaminants, moisture content, pH, nutritional condition, temperature, microbial diversity of the site, and oxidation–reduction potential (redox potential) are the most significant criteria for bioremediation (Dua et al. 2002). Microorganisms utilise pollutants as a nutrition or energy source in this process (Tang et al. 2007). Phytoremediation (plants) and rhizoremediation are also included (plants and microbes interaction). These are the bioremediation processes that have progressed the greatest. Bioremediation, when utilised in conjunction with phytoremediation and rhizoremediation, has a substantial impact on the fate of hazardous wastes and can be used to eliminate undesired substances from the biosphere (Ma and Burken 2003). Bioremediation does have some drawbacks. Only biodegradable substances are allowed. Biodegradation products can be more persistent or poisonous than the parent substance, for example, TCP; the biodegradation product of chlorpyrifos has been found to be more persistent and dangerous than chlorpyrifos itself. Biological processes are frequently very particular as well.

Petroleum hydrocarbons, polycyclic aromatic hydrocarbons, halogenated hydrocarbons, herbicides, solvents, and metals are all common pollutants that can be bioremediated. Pesticides are particularly important among these toxins because they are commonly employed to reduce insect infestations of crops and so safeguard crops from potential yield losses and product quality degradation (Damalas 2009). Pesticides are an important part of modern agriculture since they are absolutely vital for cost-effective pest management (Gouma 2009). Excessive and repeated use of pesticides, on the other hand, causes environmental degradation. Pesticides have been found in rising amounts in both soils and streams during the last decade (European Environment Agency 1999; Heath et al. 2010; Surekha et al. 2008).

Approximately 90% of agricultural pesticides are spread through the air, soil, and water instead of reaching their target species. As a result, they can be found in the air, surface and ground water, sediment, soil, vegetables, and to some extent, meals. Furthermore, many soil-applied pesticides are purposely introduced into the soil environment for the control of soil-borne pests and pathogens, resulting in unacceptably high levels of residues and metabolites in soil (Gamon et al. 2003; Shalaby and Abdou 2010). Pesticides are also responsible for 6.3% of total volatile emissions in the environment (Yates et al. 2011). Pesticides, in particular, can easily penetrate into the tissues of living species, resulting in bioaccumulation. Because of their eco-friendliness, bioremediation approaches for treating pesticides in soil have gotten a lot of interest and have been used successfully in a lot of nations (Enrica 1994; Mohammed 2009; Ritmann et al. 1988). It has been shown to be an effective method for decontaminating pesticide-polluted locations in the current environment (Mervat 2009).

19.2   BIOREMEDIATION AFFECTING FACTORS

Soil is the ultimate "sink" for pesticides used in agricultural and public health care. Soil serves as a repository for a variety of bacteria. It takes in a variety of compounds and works as a scavenger for hazardous pollutants. The binding of individual and mixture pesticides in soil is affected by a variety of parameters such as soil type, moisture, pH, organic matter, and temperature. This, in turn, has an impact on the efficacy of bioremediation procedures (Alexander 1999; Magan et al. 2010; Tao and Yang 2011; Vidali 2001).

19.2.1 MOISTURE LEVEL

Microorganisms require water for development and for the passage of nutrients and products across the cell wall during the biodegradation process (Aislabie and Lloyd-Jones 1995; Diez et al. 1999). A suitable amount of moisture should be present for growth. Low moisture levels can stifle bacterial activity, while too much moisture can fill the tiny pores between particles, limiting oxygen transmission. The ideal moisture level for soil bioremediation is 25–85% of the water holding capacity (Vidali 2001).

Moisture and temperature are the two key environmental elements that determine pesticide behaviour in soil, according to Taylor and Spencer (1990), with moisture having a greater relative weight than soil temperature. Chlorpyrifos, an organophosphorus insecticide, was studied in soils under various moisture regimes by Awasthi and Prakash (1997). Chlorpyrifos was rapidly destroyed in all airdry soils, but slightly more slowly in soils with higher field capacity and/or under submerged conditions. Low moisture and nutrient content may cause triazines and other xenobiotic chemicals to survive longer in the soil (Yadav and Loper 2000). Tao and Yang (2011) analysed fluroxypyr degradation and discovered that 20% moisture had the longest half time for fluroxypyr decay, indicating that fluroxypyr degradation is limited by a low water holding capacity.

19.2.2 OXYGEN CONCENTRATION AND NUTRIENT AVAILABILITY

The ability of microbial degraders to remain active in the natural environment determines their degradation efficiency. As a result, bioaugmentation or biostimulation could improve bioremediation by increasing the ability of inoculated microorganisms or promoting the activity of indigenous microbial degraders (Alexander 1999). Briceño et al. (2007) discovered that soil bacteria and fungi have the ability to breakdown or mineralise a variety of pesticide groups. According to the research, adding organic amendments and nutrients can impact pesticide adsorption, transport, and biodegradation. For degradation, the most commonly proposed carbon: nitrogen: phosphorous (C:N:P) ratio is 100:10:1 (Oudot and Dutrieux 1989). Mohamed et al. (2011) discovered that adding NPK fertilisers to soil boosted oxyfluorfen (herbicide) biodegradation. Fluroxypyr decomposition is also aided by organic matter in the soil (Jung et al. 2008; Tao and Yang 2011). For aerobic degradation, the optimal oxygen content for bioremediation is >0.2 mg/L and >10% airfilled pore space.

19.2.3 pH

The pH of the soil is significant because it influences the availability of nutrients. Most microbial species can only thrive in a specific pH range. Because neutral and ionic forms have extremely distinct adsorption capabilities, several researchers have noticed that pH plays a major role in the adsorption of chemicals containing acidic functional groups on activated carbon and soil (Cea et al. 2005, 2007; Diez et al. 1999; Kookana and Rogers 1995).

19.2.4 TEMPERATURE

The pace of biodegradation is influenced by temperature, which controls the rate of enzymatic reactions within microorganisms. "For every 10°C increase in temperature, the pace of enzyme reactions in the cell roughly doubles" (Nester et al. 2001). The temperature to which bacteria can withstand has a limit. Atrazine and lindane, according to Paraiba and Spadotto (2002), act as dangerous chemicals in areas where soil temperatures are low or very low, with top soil temperatures less than 20°C. Paraiba et al. (2003) also discovered that soil temperature influences the leaching capability of thirty pesticides into groundwater, implying that contamination by these substances varies depending on climatic conditions. Tao and Yang (2011) discovered that the best temperature for fluroxypyr breakdown is 25–35°C, as well as the optimum temperature for microbial activity. According to Mohamed et al. (2011), increasing the temperature has a substantial impact on oxyfluorfen biodegradation in soil. At 40°C, degradation was better (55.2–78.3%) than at 28°C (17.5–36.6%).

19.3 CONCERNS ABOUT PESTICIDES

Pesticides include insecticides, fungicides, herbicides, and nematicides, which are used to prevent or inhibit plant diseases and insect pests. In Asia, India is the major producer of pesticides. The Indian Pesticide Industry, with 82,000 MT of production in 2005–2006, is rated second in Asia (after China) and twelfth in the world for pesticide use, with 90,000 tonnes produced annually (Boricha and Fulekar 2009). Only about 2%–3% of pesticides applied are really used, and the rest remains in the soil and water, polluting the ecosystem (World Health Organization 1990).

These toxins can also enter the food chain, posing a threat to ecosystems and humans (Liu and Xiong 2001). Some soil bacteria can breakdown xenobiotic chemicals (such as pesticides) entirely or degrade them to an intermediate form that can be used by other microbes (Oshiro et al. 1996; Shabir et al. 2008). However, pesticides can sometimes be transformed into a form that is considerably more hazardous than the original. These are a source of substantial public, scientific, and regulatory concern because to their toxicity, mutagenicity, carcinogenicity, and genotoxicity.

19.3.1 PESTICIDES HAVE A LONG-TERM EFFECT

Pesticide persistence in soil can range from a week to several years, depending on the pesticide's structure and the availability of soil components. Phosphates, for example, are highly poisonous and only last three months, whereas chlorinated hydrocarbon insecticides like chlordane are known to last at least 4–5 years, and in some cases more than 15 years. Pesticides persist, posing harm to animal and human health. Longer persistence results in the buildup of residues in soil, which may lead to higher bioaccumulation by plants, to the point where human and livestock ingestion of plant products may be harmful. Pesticide residues have been thoroughly documented in a variety of environmental matrices (soil and water) all over the world. To detoxify the environment, it is critical to remove pesticide residues from soil and water.

19.3.2 PESTICIDES AND THEIR CONSEQUENCES

Pesticide buildup in the soil environment has a negative impact on soil health and productivity. Pesticides can alter the soil microflora permanently, for example, by inhibiting N_2 fixing soil bacteria like *Rhizobium, Azotobacter*, and *Azospirillum*, as well as cellulolytic and phosphate-solubilising microbes.

Pesticide residues in animal products and other foods eventually accumulate in humans, particularly in the adipose tissue, blood, and lymphoid organs, causing immunodeficiency, autoimmunity, and hypersensitivity reactions such as eczema, dermatitis, allergic respiratory diseases, and other conditions. Many pesticides are known to cause chromosome alterations in humans and animals, resulting in liver and lung cancer.

19.4 PESTICIDE BIODEGRADATION IN SOIL

Pesticides can be degraded and converted into simpler, non-toxic chemicals by soil microbes. "Biodegradation" is the term for this process. The metabolic activities of bacteria, fungi, actinomycetes, and plants play an important part in the degradation process (Nawaz et al. 2011). The majority of pesticides that reach the soil are bio-degradable, although certain pesticides are completely resistant to biodegradation. These are referred to be "recalcitrant" (Aislabie and Lloyd-Jones 1995; Mulchandani et al. 1999; Richins et al. 1997).

19.5 BIOREMEDIATION TECHNIQUES

Before implementing any remediation technique in a given contaminated site to meet multiple requirements, it must be evaluated for safety and efficacy. Microorganisms should be able to breakdown pollutants at a fair rate and to a given level of regulatory compliance. Toxic byproducts must not be formed, inhibiting compounds must not be present, pollutants must be bioavailable, and circumstances must be tuned and maintained to enable microbial growth and activity, according to Andreoni and Gianfreda (2007). Different bioremediation approaches are based on three essential principles: the pollutant's potential to undergo biological alteration (biochemistry), its accessibility to microorganisms (bioavailability), and the possibility of optimising biological activity (bioactivity) (Dua et al. 2002).

19.6 IN SITU BIOREMEDIATION

These procedures allow for in-place treatment, minimising the need for excavation and the transportation of toxins. These are the least expensive and, in most cases, the most preferred solutions because they cause the least amount of disruption at the application site. Because microbial species with chemotactic abilities can travel into an area containing pollutants, chemotaxis is crucial for the study of in situ bioremediation. The efficiency of in situ bioremediation can be improved by enhancing bacteria's chemotactic behaviour, according to Strobel et al. (2011). In the case of pesticides, in situ treatment can be carried out using a variety of methods. The use of fungi as a pesticide degrader is one of the best examples. The use of white rot fungi in pesticide bioremediation has received a lot of attention in recent years. White rot fungus feed on woody tissues that contain lignin, a refractory chemical, in their native habitat (Fragoeiro 2005). The enzymes needed to decompose lignin, as well as other poisonous and resistant substances, are found in white rot fungi. In laboratory tests, it was discovered that white rot fungi may degrade pesticides 45–75% more effectively than control samples (Fragoeiro 2005).

In situ bioremediation of organochlorine pesticides using solid organic carbon and zero-valent iron in soil and groundwater was proven to be a highly successful approach for the removal of persistent organochlorine pesticides by Seech et al. (2008). Qureshi et al. (2009) used combined biostimulation and bioaugmentation procedures to perform in situ bioremediation of organochlorine pesticide contaminated sites and found that they are extremely efficient, however initial acclimatisation for 2–3 months was necessary in open field settings.

19.6.1 IN SITU TREATMENTS

19.6.1.1 Bioventing

It's a way to keep bioremediation going by infusing oxygen and/or nutrients into the soil (Shanahan 2004). Nitrogen and phosphorus are two nutrients that are commonly given (Rockne and Reddy 2003). These are then used to disseminate the tainted soil uniformly. In actuality, the soil texture determines the dispersal. Biovented oxygen and nutrients cannot diffuse throughout fine-textured soils, such as clays, due to low permeability. Fine-textured soils are extremely challenging to maintain moisture content because their smaller pores and large surface area allow them to retain water. Fine-textured soils take a long time to drain from waterlogged conditions, preventing oxygen from reaching soil bacteria all throughout the contaminated region (United States Environmental Protection Agency 2006). Bioventing works best in well-drained, medium-textured, and coarse-textured soils. A basic bioventing system consists of a well and a blower that pumps air into the soil through the well (Lee et al. 2006). Frutos et al. (2010) used a bioventing device to achieve a 93% reduction in phenanthrene levels after 7 months of therapy.

19.6.1.2 Biosparging

It entails injecting pressurised air beneath the water table to raise groundwater oxygen levels and speed up the biological breakdown of pollutants by naturally present microbes. It increases the contact between soil and groundwater by increasing mixing in the saturated zone. The simplicity and low cost of placing small-diameter air injection sites allows for a great deal of flexibility in the system's design and construction. To determine the application of these processes and the optimum placement of air, the extent and kind of contamination, as well as soil properties, must be determined before to implementing bioventing or biosparging (Baker 1999).Neither of these methods is acceptable for chemicals that are likely to volatilise quickly. With pesticide-contaminated locations, these approaches haven't been applied very often. At an averaged groundwater temperature of 18°C, Kao et al. (2008) found that the biosparging system eliminated more than 70% of BTEX within a ten-month corrective period. As a result, biosparging appears to be a potential approach for treating BTEX-contaminated groundwater.

Lambert et al. (2009) claim that biosparging increases both aerobic biodegradation and volatilisation, and that it is commonly utilised in residual hydrocarbon source zone treatment. In order to measure the level of remediation achieved in terms of both mass removed and mass discharged into groundwater, this method was used in pulsed mode to a known source of gasoline contamination. Around 80% of the pentane and 50% of the hexane in place were volatilised and eliminated, but only around 4% of the aromatic hydrocarbons were. Limited biodegradation of hydrocarbons was confirmed by CO_2 and O_2 monitoring in the off-gas.

19.6.1.3   Bioaugmentation

Bioaugmentation is the process of bringing microorganisms to a contaminated place to help in deterioration. However, there are certain disadvantages: (a) introduced microorganisms seldom compete effectively enough with native populations to develop and maintain useful population levels and (b) most soils with long-term exposure to biodegradable waste include indigenous microbes that are also excellent degraders. Bioaugmentation is frequently combined with biostimulation, in which sufficient amounts of water, nutrients, and oxygen are supplied into the contaminated site to boost the activity of introduced microbial degraders (Couto et al. 2010) or stimulate cometabolism (Lorenzo 2008). The idea behind biostimulation is to increase the degradation potential of a polluted matrix by accumulating amendments, nutrients, or other limiting variables, and it's been used to a wide range of xenobiotics (Kadian et al. 2008). In situ bioremediation, on the other hand, has certain drawbacks: (1) it is not appropriate for all soils, (2) total degradation is difficult to achieve, and (3) natural circumstances (e.g. temperature) are difficult to control for effective biodegradation.

19.7   EX SITU BIOREMEDIATION

Ex situ bioremediation involves excavating polluted soil and treating it elsewhere. It includes the following treatments:

19.7.1   LANDFARMING

It is a method of excavating dirt and mechanically separating it via sifting. After that, the polluted soil is layered over clean soil, allowing natural processes to cleanse, breakdown, and immobilise toxins (United States Environmental Protection Agency 2006). The polluted soil layer is subsequently covered with a synthetic, concrete, or clay membrane. Mixing with oxygen is done by ploughing, harrowing, or milling. To enhance the restoration process, nutrients and moisture may be provided. Crushed limestone or agricultural lime is also used to manage the pH of the soil (keeping it near 7.0) (Van Deuren et al. 2002).

Pesticides contaminated soils lend themselves well to land farming (Felsot et al. 1992). It is a full-scale bioremediation system in which the size and location of the spreading operation are determined by the pesticide application rate in order to avoid dangerous pesticide concentrations in soil, groundwater, or crops. In general, the rate is similar to the pesticide label rate, which is the recommended rate of pesticide application per unit of land or soil.

Large quantities of pesticides that were delivered to Africa for locust control in the 1950s did not reach at the correct time or place, causing them to become obsolete. The Africa Stockpiles Programme, launched by the FAO, is aimed to clear Africa of stockpiles and dispose of them in an environmentally appropriate manner. Even in the most industrialised countries, decontaminating polluted land is a difficult and expensive task. It was practically inconceivable to deal with these issues in Mali and Mauritania.

FAO produced a site-specific remediation plan for each place based on the analytical results and environmental surveys. To improve local pesticide biodegradation, remediation was typically centred on the addition of organic matter and land-farming. The goal of this treatment was to facilitate and expedite natural pesticide breakdown in the soil as much as feasible. This has been demonstrated to be particularly effective in the case of organophosphate and carbamate insecticides. Organochlorine pesticides are long-lasting and take a long time to breakdown. Landfarming in Mali began in July 2008 in Molodo. Four months later, lab analysis of soil samples from the same places revealed a two-order-of-magnitude decrease in organophosphate content. Dieldrin concentrations at the same location remained unaltered, as expected.

19.7.2 Biopiling

A treatment bed, or mound of contaminated soil, an aeration system, an irrigation or nutrient system, and a leachate collection system are all part of the basic biopile system. Biodegradation is aided by controlling factors such as moisture, heat, nutrients, oxygen, and pH. The irrigation or nutrient system is buried beneath the soil and uses vacuum or positive pressure to transport air and nutrients. To minimise runoff, evaporation, and volatilisation, as well as increase solar heating, soil mounds can be up to 20 feet high and covered with plastic. If volatile organic chemicals in the soil volatilise into the air stream, the air leaving the soil could be treated to remove or destroy the volatile organic compounds before they reach the atmosphere. Treatment usually lasts 3–6 months (Wu and Crapper 2009).

19.7.3 Composting

Organic wastes are decomposed by bacteria at elevated temperatures, usually between 55°C and 65°C. Heat generated during the breakdown process raises the temperature, resulting in enhanced pollutant solubility and increased metabolic activity in composts. Composts have a high level of substrate, which can lead to organic pollutants co-metabolising. Composts have a larger and more diversified microbial community than soils. The following are the basic steps of windrow composting. First, large boulders and debris are removed from contaminated soils by excavating and screening them (Blanca et al. 2007; Bouwer and Zehnder 1993). The soil is moved to a composting pad with a makeshift building to offer confinement and weather protection. Straw, alfalfa, manure, agricultural wastes, and wood chips are utilised as bulking agents and as a carbon source addition. Windrows are made up of layers of soil and additives stacked in long stacks. By turning the windrow with a commercially available windrow turning machine, the windrow is thoroughly blended. The concentrations of moisture, pH, temperature, and explosives are all monitored. The windrows are dismantled at the end of the composting period, and the compost is transported to the ultimate disposal spot. Composting is a method for removing pesticides from contaminated locations, and numerous significant corporations, including W.R. Grace and Astra Zeneca, have developed and patented composting systems.

Many researches on the fate of pesticides during composting have been undertaken, according to winter. Pesticides will most likely behave differently in compost than they do in soil. Only 12% of the initial aldrin, 3% of the dieldrin, and less than 15% of the monolinuron and imugan added to municipal solid waste and biosolids feedstock were destroyed after composting, according to Muller and Korte (1975) and Muller et al. (1976). The herbicides buturon and heptachlor, on the other hand, were destroyed in 55% of cases. Composting may not be ideal for treating feedstock polluted with these persistent pesticides, according to the authors, who found no indication of pesticide mineralisation. The three weeks period used in this study, according to critics, is insufficient to assess the possibilities for removing persistent chemicals.

The usefulness of volatilisation in the rehabilitation of pesticide contaminated compost was demonstrated by Petruska (1985). After three weeks of composting cow manure and sawdust, losses owing to volatilisation reached 22% for diazinon and 50% for chlordane. Chlordane was not much mineralised, but diazinon was significantly altered albeit at a moderate rate of mineralisation. Racke and Frink (1989) also presented information on a pesticide's limited mineralisation during composting. During the composting of municipal biosolids, approximately 97% of the insecticide carbaryl was converted, however only 5% of this was due to mineralisation. They halted their trial after only 20 days, which drew criticism as well. However, both investigations revealed that significant pesticide mineralisation during composting may necessitate extra processing time.

Lemmon and Pylypiw (1992) evaluated the persistence of diazinon, chlorpyrifos, pendimethalin, and isofenphos after composting with grass clippings, finding that pesticides were undetectable shortly after application and dissipated quickly after composting. Rao et al. (1995) discovered that the pesticide atrazine mineralised minimally during composting with a variety of wood-derived substrates. There was no identifiable atrazine after 160 days of composting, however a maximum of 7% had been mineralised. Unmineralised atrazine appears to have leached or formed a combination with the humic components, inhibiting further transformation.

After composting grass clippings, Vandervoort (1997) noticed a similar type of outcome, namely lower quantities of chlorpyrifos, 2,4-D, isoxaben, triclopyr, clopyralid, and fluprimidol. After 128 days, pesticide levels were below the standard detection limit, and samples from the inside of compost piles that had been turned were generally lower than samples from the outside and static piles. Malathion, chlorpyrifos-methyl, and lindane residues were completely reduced following composting in just 8 days, according to Frenich et al. (2005), but endosulfan residues were only partially decomposed. Mohamed et al. (2011) discovered that composting improved the breakdown of the herbicide oxyfluorfen.

19.7.4 BIOREACTORS

To encourage the action of microorganisms, polluted soil is mixed with water and nutrients, then stirred by a mechanical bioreactor. This procedure is more suited to clay soils than other ways, and it is also a rather quick process (United States Environmental Protection Agency 2006). Bioslurry reactors, fermenters, prepared bed reactors, and a variety of enclosed systems may be used if the potential risks from discharges and emissions are particularly substantial (Riser-Roberts 1998). Fulekar (2009) used *Pseudomonas aeruginosa* to pesticide fenvalerate in a scale-up bioreactor and found that this method would be advantageous to the pesticide industry for fenvalerate bioremediation.

19.7.5 PRECIPITATION OR FLOCCULATION

The major method of treating metal-laden industrial wastewaters has long been precipitation. Because of the efficacy of metals precipitation in these applications, the method is being examined and selected for use in remediating ground water containing heavy metals and associated radioactive isotopes. The conversion of soluble heavy metal salts to insoluble salts that will precipitate is the goal of this approach. Physical methods such as clarifying (settling) and/or filtering can subsequently be used to remove the precipitate from the treated water. The pH modification, the addition of a chemical precipitant, and flocculation are commonly used in the process. Metals usually precipitate as hydroxides, sulphides, or carbonates from the solution. Electrostatic surface charges can hold very small particles in suspension during the precipitation process. These charges induce clouds of counterions to develop surrounding the particles, creating repulsive forces that prevent aggregation and limit the efficiency of subsequent solid liquid separation processes.

Chemical coagulants are frequently used to overcome the particles' repulsive forces. Inorganic electrolytes (such as alum, lime, ferric chloride, and ferrous sulphate), organic polymers, and synthetic polyelectrolytes containing anionic or cationic functional groups are the three main forms of coagulants. Following the addition of coagulants, low sheer mixing in a flocculator is used to increase particle interaction, allowing particle development through the sedimentation phenomena known as flocculant settling.

19.7.6 MICROFILTRATION

Micro filtration is performed by membranes with pore sizes of 0.1–10 m. To remove dissolved particles from waste water, microfiltration membranes are utilised at a constant pressure. More than 90% of the original waste water is recoverable and reuseable.

19.7.7 ELECTRO DIALYSIS

Electro dialysis uses electrical current and specific membranes that are semi-permeable to ions based on their charge. Membranes that permeate cations and anions are alternatively positioned with flow channels in between and electrodes on each side of the membranes. The electrodes pull their opposing ions through the membranes, removing them from the water. Banasiak et al. (2011) recently employed this technology to remove the insecticide endosulfan from polluted water.

19.8 PESTICIDE DEGRADATION BY BACTERIA AND FUNGI

The primary agents used in bioremediation are natural organisms, either indigenous or foreign (Prescott et al. 2002). The microorganisms used vary depending on the chemical composition of the polluting agents and must be carefully chosen because they can only survive in a narrow spectrum of chemical pollutants (Dubey 2004; Prescott et al. 2002). Because a contaminated site is likely to contain a wide range of pollutants, a wide range of microorganisms is likely to be necessary for efficient cleanup (Wanatabe et al. 2001).

Bacteria from the genera *Alcaligenes, Flavobacterium, Pseudomonas*, and *Rhodococcus* show outstanding pesticide breakdown abilities (Aislabie and Lloyd-Jones 1995; Boricha and Fulekar 2009; Mulchandani et al. 1999; Richins et al. 1997). Actinomycetes have a lot of potential when it comes to pesticide biotransformation and biodegradation. Actinomycetes belonging to the genera *Arthrobacter, Clavibacter, Nocardia, Rhodococcus, Nocardioides,* and *Streptomyces* have the ability to degrade pesticides, according to DeSchrijver and DeMot (1999).Although the metabolic process for pesticide breakdown by actinomycetes has not been widely researched, these bacteria are known to produce extracellular enzymes that degrade a wide spectrum of complex chemical molecules. The presence of several monooxygenases and dioxygenases is a frequent feature of aerobic actinomycetes (Larkin et al. 2005).

Actinomycetes capable of digesting pesticides have been discovered and characterised in recent investigations, and various strains of this group have been proposed for soil decontamination (Benimeli et al. 2003). Pesticides such as atrazine, lindane, diuron, metalaxyl, terbuthylazine, DDT, dieldrin, gamma-hexachlorocyclo-hexane (g-HCH), aldrin, heptachlor, lindane, mirex, chlordane, and others have been degraded by white-rot fungi (Bending et al. 2002; Hickey et al. 1994; Kennedy et al. 1990; Magan et al. 2010; Mougin et al. 1994; Quintero et al. 2007; Singh et al. 2004). *Phanerochaete chrysosporium* and *Trametes versicolor* are two white-rot fungi with a lot of biodegradation potential (Bastos and Magan 2009; Bending et al. 2002; Fragoeiro and Magan 2008; Pointing 2001; Rubilar et al. 2007; Tortella et al. 2005).

Pesticides, such as organochlorines, striazines, triazinones, carbamates, organophosphates, organophosphonates, acetanilides, and sulphonylureas, have a wide range of chemical structures. Mineralisation of these pesticides is challenging with a single isolate, and total breakdown frequently necessitates the use of consortia of bacteria.

19.9 PHYTOREMEDIATION

Phytoremediation, also known as plant-assisted bioremediation, is an approach for cleaning and restoring soil and wastewater that has been used for over 300 years and has now emerged as a viable strategy for removing a variety of soil contaminants. It is a low-cost, ecologically friendly, and efficient approach that may be used in the field (Brooks 1998; Chaney et al. 1997; Trapp and Karlson 2001; Zavoda et al. 2001). Above all, phytoremediation monitoring is a simple process. There's also the option of recovering and repurposing precious items. The use of plant root systems' unique and selective capacities, as well as uptake, transformation, volatilisation, and rhizodegradation, are the key processes involved in phytoremediation. Plants facilitate pollutant breakdown by creating a suitable microenvironment surrounding their roots during this process. Endophytic bacteria, as well as rhizospheric bacteria, are engaged in the breakdown of hazardous pollutants in the soil environment.

Endophytic bacteria are non-pathogenic bacteria that live in the internal tissues of plants and can help plants flourish by manufacturing a variety of natural products and contributing to the biodegradation of pollutants in the soil (Sessitsch et al. 2002). Germaine et al. (2006) successfully demonstrated the use of bacterial endophytes to reduce levels of harmful pesticide residues in agricultural plants. They infected pea plants (*Pisum sativum*) with a poplar endophyte that could digest 2,4 dichlorophenoxyacetic acid (2,4-D), which resulted in a greater ability to remove 2,4-D from the soil and no accumulation of 2,4-D in the tissues.

Phytoremediation has been used to remove pesticides from the environment, and there has been a lot of research done on it. Plant roots absorb pesticides onto their surfaces, while dead roots add organic matter to the soil, which can improve pesticide sorption on soil organic matter, where microbial transformations can take place (Karthikeyan et al. 2004). In contaminated soils when *Kochia* sp. plants were planted, increased degradation of atrazine, metolachlor, and trifluralin was detected, according to preliminary research (Coats and Anderson 1997). The cleanup of atrazine-contaminated soil and groundwater has also been found to be successful using deep-rooted poplar trees (Nair et al. 1993). Dosnon-Olette et al. (2011) investigated the ability of the aquatic plant *Lemna minor* to remove the herbicides isoproturon and glyphosate from contaminated water using phytoremediation.

Plants were exposed to five different pesticide concentrations (0–20 g l^{-1} for isoproturon and 0–120 g l^{-1} for glyphosate) in culture conditions for 4 days, with growth rate and chlorophyll a fluorescence as endpoints. Because impacts on growth rate and chlorophyll fluorescence were low (25%) at exposure doses of 10 g l^{-1} for isoproturon and 80 g l^{-1} for glyphosate, this concentration was chosen to evaluate herbicide removal. After 4 days of incubation, removal yields for isoproturon and glyphosate were 25% and 8%, respectively.

Khan et al. (2011) investigated organophosphate pesticide phytoremediation. Experiments were conducted to examine pesticide uptake by plants and residue in the soil under controlled and experimental conditions for this aim. Both organic materials and the antibiotic streptomycin bind to organophosphate pesticides. The enzyme p-nitrophenol 4-hydroxylase isolated from root and shoot is inhibited by streptomycin. HPLC analysis verified the results obtained from the UV visible spectrophotometer. In unsterilised soil, wheat plants enhanced the uptake/ degradation of p-nitrophenol, methyl parathion, and hydroquinone by 94.7%, 64.85%, and 55.8%, respectively. Hydrolysis of methyl parathion produces p-nitrophenol, which is then converted to hydroquinone, releasing nitrite. The release of nitrite by leaf and root extracts, as well as the presence of hydroquinone in the reaction mixture, indicate that the enzyme p-nitrophenol 4-hydroxylase is active.

Using biotechnological technologies, phytoremediation can be made more effective. Bacteria (rhizospheric or endophytic) can be genetically modified to breakdown harmful chemicals in the environment by gene transfer. However, genetic engineering of endophytic and rhizospheric bacteria, as well as transgenic plants, appears to be a potential technique for environmental cleanup (McGuinness and Dowling 2009). Herbicide phytoremediation has been extensively researched using ordinary plants. For phytoremediation of foreign chemicals in contaminated soil and water, transgenic plants designed to metabolise herbicides and long-lasting pollutants can be utilised (Kawahigashi 2009).

19.9.1 Phytoextraction

Plants or algae are used in phytoextraction or phytoaccumulation to convert pollutants from soils, sediments, or water into harvestable plant biomass. Contaminants are absorbed by the root system and stored in the root biomass, or they are transported up into the stems or leaves. Until it is harvested, a living plant may continue to absorb toxins. Because a lesser level of pollutant remains in the soil after harvest, the growth/ harvest cycle must normally be repeated over several crops in order to achieve a meaningful cleanup. The cleaned soil can now support different vegetation after the process. Phytoextraction can be done either naturally (with hyperaccumulators) or artificially (with chelates to increase bioavailability) (Utmazian and Wenzel 2006).

Continuous phytoextraction relies on the plants' ability to collect pollutants in their shoots over time. To accomplish this, plants must have effective detoxifying systems for accumulated toxins. When it comes to heavy metals like nickel, resistance in *Thlaspi goesingense* is a key factor in nickel hyperaccumulation in hydroponically grown plants. As a result, manipulating metal tolerance in plants will be critical to the creation of effective phytoremediation crops. It will be critical to understand the present molecular and biochemical methods plants use to avoid metal toxicity in order to build hypertolerant plants capable of accumulating large metal concentrations. Mercury-resistant *Arabidopsis thaliana* was recently shown to efficiently remove mercury from solution by overexpressing bacterial mercury reductase. Synthetic metal chelates, such as ethylene diaminetetraacetic acid, are added to soils to enhance metal accumulation by plants in induced or chelate-assisted phytoextraction. Metal-enriched plant wastes can be disposed of as hazardous waste or used for metal recovery if it is economically possible.

Mukherjee and Kumar (2011) studied the ability of mustard (*Brassica campestris*) and maize (*Zea maize*) to remove the organochlorine pesticide endosulfan from soil, finding that plant uptake and phytoextraction with maize and mustard contributed 47.2% and 34.5%, respectively, while other degradation processes accounted for 38.7% and 35.9%. After growing the crop plants in soil, the accumulated endosulfan in the soil decreased by 55%–91%, indicating that plant uptake and phytoextraction may be the primary processes for endosulfan removal by the plant.

19.9.2 RHIZOFILTRATION

Rhizofiltration is more concerned with the treatment of contaminated groundwater than it is with the treatment of dirty soils. The pollutants are absorbed by the plant roots or adsorbed onto the root surface. Plants used for rhizofiltration are first acclimated to the pollutant before being planted in situ. Plants are cultivated hydroponically in clean water rather than soil until they have formed a substantial root system. To acclimate the plant, the water supply is replaced with a polluted water supply once a large root system has been established. After the plants have been accustomed to their surroundings, they are placed in a polluted area, where the roots absorb both the polluted water and the poisons. The roots are gently removed and disposed of once they have become saturated.

Rhizofiltration is assumed to be dependent on physicochemical processes (such ion exchange and chelation), and it can even happen on dead roots (Anonymous 2009). Aside from its reliance on surface absorption as the major method for removing contaminants from waste streams, it is also linked to the biosorption process, which involves the use of living or dead microbial, fungal, or other biomass to absorb significant amounts of materials. Several plants, including mustard, rye, corn, and sunflower, have the natural ability to absorb and precipitate pollutants from solutions. When there are low concentrations of containment and a significant volume of waste, it is especially effective and cost-efficient. As a result, rhizofiltration could be particularly useful for radionuclide-contaminated water. Pletsch et al. (1999) discovered that periodically treating a location can reduce pollution to acceptable levels in radioactively contaminated pools when sunflowers were cultivated.

19.9.3 PHYTOSTABILISATION

Phytostabilisation is the process of vegetation immobilising hazardous pollutants in soils and retaining polluted soils and sediments in place. Windblown dust, a major source of human exposure at hazardous waste sites, is avoided by establishing rooted plants. Because of the huge volume of water transpired by plants, hydraulic management is achievable in some circumstances, preventing leachate migration to groundwater or receiving waters. Another way to keep toxins in the subsurface stable is to keep water from reacting with the waste and causing it to migrate. This is a standard strategy for landfill covers, but it can also be used to reduce groundwater plume surface water recharge.

Plants having fibrous root systems, such as grasses, herbaceous species, and wetland species, are commonly utilised for phytostabilisation. These species' typical rooting depths are 30–60 cm for upland species and 30 cm for wetland species (Anonymous 2009). Phytostabilisation covers are simple soils or sediments that have been planted with vegetation that has been chosen to control bulk soil mobility and/or inhibit contaminant migration by phytosequestration. Other plants, such as halophytes and hyperaccumulators, can be chosen for their ability to phytoextract and accumulate pollutants into aboveground tissues in addition to phytosequestering contaminants in the rhizosphere. Moving toxins into the plant obviously carries additional hazards; nevertheless, depending on the total human health and ecological dangers connected with the location, this component of a phytostabilisation cover application for soils/sediments may still be appropriate.

19.9.4 PHYTODEGRADATION (PHYTOTRANSFORMATION)

Phytodegradation is the breakdown of contaminants into smaller units by plant enzymes (Newman and Reynolds 2004) or the release of exudates that aid in the degradation of pollutants in the rhizosphere via co-metabolism. Plant molecules do not completely break down complex and resistant substances to basic molecules (water, carbon dioxide, etc.), hence phytotransformation refers to a change in chemical structure rather than a complete breakdown of the compound.

Plants behave similarly to the human liver when dealing with these xenobiotic chemicals (foreign compound/pollutant), so the name "Green Liver Model" is used to describe phytotransformation. Plant enzymes increase the polarity of xenobiotics after they have been taken up by adding functional groups such as hydroxyl groups (-OH). Phase I metabolism is the process through which the human liver raises the polarity of medicines and foreign substances.

While enzymes such as Cytochrome P450s are responsible for the early reactions in the human liver (Yoon et al. 2008), nitroreductases play the same role in plants. Plant biomolecules such as glucose and amino acids are added to the polarised xenobiotics in the second step of phytotransformation, known as Phase II metabolism, to raise the polarity even more (known as conjugation). This is comparable to the mechanisms that occur on the reactive centres of xenobiotics in the human pancreas, where glucuronidation [addition of glucose molecules by the UGT (e.g. UGT1A1) class of enzymes (Mendez and Maier 2008)] and glutathione addition events occur. In phytotransformation, the plants minimise toxicity (with few exceptions) and sequester the xenobiotics.

Hussain et al. (2009) found several of the enzymes produced by plants that can degrade pesticides. The enzymes in concern were aromatic dehalogenase for DDT, nitrilase for herbicides, O-demethylase for alachlor, and metolachlor and phosphatase for organophosphates.

19.9.5 PHYTOVOLATILISATION

The volatilisation of pollutants from the plant, either from the leaf stomata or from the plant stems, is known as phytovolatilisation (Anonymous 2009). Radial diffusion has also been documented in stem tissues (Davis et al. 1999; Narayanan et al. 1999; Zhang et al. 2001). For example, methyl-*tert*-butyl ether (MTBE) can escape into the atmosphere via leaves, stems, and bark (Hong et al. 2001). Radial diffusion from the stem, rather than transpiration from the leaves (Ma and Burken 2003; Newman and Reynolds 2004), was found to be the predominant dissipation mechanism in tree core samples of hybrid poplars subjected to trichloroethylene (TCE) (Ma and Burken 2003). Phytovolatilisation can occur when a breakdown product formed from the parent contaminant's rhizodegradation or phytodegradation occurs. This effect was investigated in poplars for trichloroethene or its breakdown products absorption and phytovolatilisation (Anonymous 2009).

Ethylene dibromide, TCE, MTBE, and carbon tetrachloride are among the pollutants that benefit from phytovolatilisation (CTC). Insect activity in grain bins was controlled with these. As the water passes through the plant's vascular system from the roots to the leaves, the contaminant may be adjusted. With increasing distance from the roots, the concentration of volatile chemicals in the xylem generally declines (Ma and Burken 2003). Compounds containing double bonds, such as TCE, could be rapidly oxidised in the environment by hydroxyl radicals once released. Phytovolatilisation, on the other hand, may not be a viable solution in some situations (e.g. limited air circulation). MTBE, for example, has a lengthy half-life in the atmosphere and might end up in shallow groundwater after rainstorms (Pankow et al. 1997). Simple mass balance models can be used to estimate whether phytovolatilisation provides a major risk to humans and/or the environment in such instances (Ma and Burken 2003; Narayanan et al. 1999). Nonetheless, in comparison to other emissions, the rate of release of volatile chemicals from plant tissues is often low. Other phytoremediation applications that can be used on specific sites include:

19.9.5.1   Riparian buffer strips

These are linear bands of permanent vegetation adjacent to an aquatic ecosystem that are designed to trap and remove various nonpoint source pollutants, such as herbicide and pesticide contaminants, nutrients from fertilisers, and sediments from upland soils from both overland and shallow subsurface flow (via phytodegradation, phytovolatilisation, and rhizodegradation) (Lowrance et al. 1984, 1986; Peterjohn and Correll 1984; Pinay and Decamps 1988). In the absence of suitable buffer strips, water treatment plants and other costly restoration procedures will be required. Riparian buffers, on the other hand, perform a variety of roles in addition to improving water quality (Budd et al. 1987; O'Laughlin and Belt 1995). Buffers can help remove pesticides that are strongly linked to the soil, although their efficiency varies. Because each pesticide has different mobility and soil binding qualities, buffers' ability to retain pesticides varies. Grass buffers have been shown to reduce herbicide losses by more than 50% (http://www.fwi.co.uk/academy/article/116941/grass-buffer-strips-around-osr.html).

19.9.5.2   Plants cap

As a cap or cover, plants grow over landfill sites. Plastic or clay caps are commonly used on such sites. They aid with erosion control, contaminant leaching, and the degradation of the underlying landfill.

19.10   RHIZOREMEDIATION

Because phytoremediation has limitations, a new concept of rhizoremediation has been developed. In this method, naturally occurring rhizospheric microorganisms or/and selected bacteria obtained by enrichment process are employed to degrade selected contaminants after ensuring that the best food source, namely root exudates, is present in the soil. Actually, the rhizosphere has a lot of remedial potential (Anderson et al. 1993; Cunningham et al. 1996; Davis et al. 1998, 2003). The stimulation of microbial populations caused by root exudates and enhanced soil moisture, oxygen, and nutritional conditions could be the reason for this. Many factors influence the success of rhizoremediation, including soil conditions, climate, compatible plant species, and associated rhizosphere bacteria (Anonymous 2009; Prasad et al. 2010).

Rhizoremediation can have an instantaneous influence on pollutant concentrations after planting in some circumstances. In other circumstances, the plant may need multiple seasons to interact with a contaminated zone at depth. It may also depend on whether the plant is actively or indirectly involved in the contaminant's remediation. Several laboratory experiments have been conducted to determine the rhizosphere's impact on pesticide breakdown (Anderson and Coats 1995; Anderson et al. 1994; Hoagland et al. 1994; Kruger et al. 1997a,b; Nair et al. 1993; Perkovich et al. 1995; Shann and Boyle 1994; Zablotowicz et al. 1994). Sugars, alcohols, and organic acids found in plant root exudates provide a glucose source for the soil microflora, promoting microbial development and activity. Microbes may respond to some of these chemicals as chemotactic signals. Plant roots also loosen the soil and carry water to the rhizosphere, boosting microbial activity further (Dzantor 2007). The potential of ryegrass for rhizosphere bioremediation of chlorpyrifos in mycorrhizal soil was explored by Korade and Fulekar (2009).

Plant growth-promoting rhizobacteria (PGPR) are rhizosphere microorganisms that are tightly connected with roots. Furthermore, rhizosphere bacteria play an important role in plant nutrient recycling, soil structure maintenance, noxious chemical detoxification, and insect control (Rajkumar et al. 2009, 2010). Plant root exudates, on the other hand, supply sustenance to rhizosphere bacteria, raising rhizosphere microbiological activity, which stimulates plant development and reduces metal toxicity in plants. PGPR and Arbuscular Mycorrhizal Fungi are two rhizosphere microorganisms engaged in plant interactions with the soil milieu that have gained significance all over the world to treat soil (Ma et al. 2011). When a contaminant is present in a soil, bacteria, yeast, and fungi naturally choose organisms that prefer that chemical as a source of food and energy. Specific species selected by utilising the pollutant as a primary food source can have microbial populations that are several orders of magnitude higher than organisms that do not metabolise the contaminant. The rate of contaminant breakdown, metabolisation, or mineralisation in the soil is determined by the bioactivity of the soil, which is mostly generated from the proteins and enzymes of soil organisms. However, the presence of electron acceptors or donors, cometabolites, inorganic nutrients, plant vitamins and hormones, pH, and/or water can all limit contaminant degradation.

Plants and soil bacteria in the rhizosphere develop a symbiotic interaction in general. Plants give nutrients that microorganisms require to thrive, and microbes create a healthier soil environment for plant roots to grow in. Plants loosen the soil and, in particular, transport oxygen and water into the rhizosphere.

Plants also emit certain phytochemicals (sugars, alcohols, carbs, and so on) that serve as a primary source of food (carbon) for soil organisms that help to maintain a healthier soil environment. Alternatively, the phytochemical that is secreted could be an allelopathic agent that prevents other plants from growing in the same soil. Plants are protected against competition, soil diseases, poisons, and other substances that are naturally present or might otherwise grow in the soil environment in exchange for exporting these phytochemicals.

19.11 PESTICIDE DEGRADATION THROUGH GENETICS

Several studies have been published in order to examine the genetic basis of pesticide biodegradation, with a focus on the role of catabolic genes and the use of recombinant DNA technology. Only a few microbes have pesticide degrading genes that have been identified. The majority of catabolic degradation genes are found on chromosomes, although in a few situations, these genes can also be found in plasmids or transposons. Recent breakthroughs in metagenomics and whole genome sequencing have opened up new possibilities for discovering novel pollutant degradative genes and their regulatory elements in both culturable and nonculturable microbes found in the environment. Enzymes responsible for the breakdown of numerous pesticides have been found in mobile genetic elements such as plasmids and transposons. Isolation and characterisation of pesticide degrading microorganisms, as well as novel techniques for isolating and analysing nucleic acids from soil microorganisms, will provide unique insights into the molecular mechanisms that lead to the development of increased pesticide degradation.

Understanding the genetic basis of the methods by which microorganisms biodegrade contaminants and interact with the environment is critical for the successful deployment of in situ remediation technology (Husain et al. 2009). Yan et al. (2007) identified a number of microbial enzymes that may hydrolyse pesticides, including organophosphorus hydrolase (OPH; encoded by the *opd* gene). This gene has been discovered in bacterial populations that can utilise organophosphate insecticides as a carbon source, and these strains have been isolated from all around the world. Although these plasmids have a lot of genetic variability, the area carrying the *opd* gene is much conserved. *Pseudaminobacter, Achrobacter, Brucella,* and *Ochrobactrum* genes for methyl-parathion hydrolase (encoded by the *mpd* gene) were identified by comparison with the gene *mpd* from *Pleisomonas* sp. M6 strain (Zhongli et al. 2001), the gene for the OPH has 996 nucleotides, a typical promoter sequence of the promoter TT (Zhang et al. 2005).

19.12 BIOREMEDIATION OF PESTICIDES THROUGH GENETIC ENGINEERING

Microorganisms respond to various types of stress in different ways and increase fitness in a contaminated environment. Using genetic engineering approaches, this process can be hastened. I the amplification, disruption, and/or modification of targeted genes that encode enzymes in metabolic pathways, (ii) the reduction of pathway bottlenecks, (iii) the enhancement of redox and energy generation, and (iv) the recruitment of heterologous genes to give new characteristics to metabolic pathways have all been made possible by recombinant DNA and other molecular biological techniques (Megharaj et al. 2011; Shimizu 2002). Various genetic techniques have been created and applied to optimise biodegradation enzymes, metabolic pathways, and organisms (Pieper and Reineke 2000). New information on metabolic pathways and degrading bottlenecks continues to emerge, necessitating the need to expand the molecular toolset already accessible. Even in a single changed organism, the added genes or enzymes must be integrated into the regulatory and metabolic network for appropriate expression (Cases and de Lorenzo 2005; Pieper and Reineke 2000; Shimizu 2002).

Genetically engineered microbes were the first to exhibit detoxification of organophosphate pesticides, and the genes encoding these hydrolases have been cloned and expressed in *E. coli, Streptomyces lividans, P. pseudoalcaligenes, Pichia pastoris,* and *Yarrowia lipolytica* (Fu et al. 2004; Shen et al. 2010; Wang et al. 2012; Wu et al. 2004; Yu et al. 2009).

Phytoremediation, or the use of plants to clean up polluted soil and water resources, is another approach that has been utilised. It is recognised as a cost-effective, aesthetically beautiful, and ecologically beneficial "Green technology" (Eapen et al. 2007; Singh et al. 2011a,b). Plants, on the other hand, lack the catabolic pathways necessary for full degradation/mineralisation of externally supplied organic molecules. Plants' ability to breakdown organic pollutants can be further boosted by introducing effective heterologous genes involved in the degradation of organic pollutants into plants (Singh et al. 2011a,b).

Unfortunately, the rates of hydrolysis of various enzymes for members of the OP compound family vary substantially, ranging from diffusion-controlled hydrolysis for paraoxon to several orders of magnitude slower hydrolysis for malathion, chlorpyrifos, and other pesticides (Casey and Grunden 2011). Despite the use of site-directed mutagenesis to improve OPH's substrate specificity and stereoselectivity (Casey and Grunden 2011; Van Dyk and Brett 2011), the ability to deduce substrate-specific changes is now limited to the active-site residues.

Two noteworthy publications have suggested that directing evolution could be a biological approach for efficient decontamination. Directed evolution was recently used to generate OPH variants with up to 25 fold improvements in hydrolysis of methyl parathion (Cho et al. 2002), a substrate that hydrolyses 30 fold less efficiently than paraoxon, and others report directed evolution of OPH to improve hydrolysis of a poorly hydrolysable substrate, chlorpyrifos (1,200 fold less efficient than paraoxon). The best variation hydrolyses chlorpyrifos at a rate equivalent to wildtype OPH's hydrolysis of paraoxon, resulting in a 700 fold improvement (Cho et al. 2004).

19.13   GENOMIC AND FUNCTIONAL GENOMICS APPLICATIONS

19.13.1   Metagenomics Applications in Pesticides Bioremediation

Multiple interacting elements, such as pH, water content, soil structure, climatic fluctuations, and biotic activity, contribute to the complexity of microbial diversity. According to current estimates, more than 99% of microorganisms found in natural habitats are not easily culturable, making them unavailable for biotechnology or basic study (Zhou et al. 2010). The discovery of technologies to isolate nucleic acids from environmental sources during the last two decades has opened a door to a hitherto unknown diversity of microbes. Researchers can investigate natural microbial communities without cultivating them by analysing nucleic acids isolated directly from ambient samples (Maphosa et al. 2012; Zhou et al. 2010).

Each organism in an ecosystem has its own set of genes, and the "metagenome" is made up of all of the community members' genomes. Metagenome technology (metagenomics) has accumulated DNA sequences that are currently being utilised in novel biotechnological applications (Rayu et al. 2012; Rajendhran and Gunasekaran 2008). Metagenome searches will always result in the identification of previously undiscovered genes and proteins due to the overwhelming majority of non-culturable microorganisms in any environment (Rayu et al. 2012; Rajendhran and Gunasekaran 2008).

19.13.2 Functional Genomics Applications in Pesticide Bioremediation

Functional genomics incorporates several traditional molecular genetics and biological techniques, such as the examination of phenotypic changes caused by mutagenesis and gene disruption, in its broadest sense (Zhao and Poh 2008). Functional genomics is a relatively new subject that combines important novel technologies with bioinformatics to do genome-wide research. Proteomics for protein identification, characterisation, expression, interactions, and transcriptome profiling using microarrays are among the emerging techniques, as is metabolic engineering (Zhao and Poh 2008). Proteomics' application in environmental bioremediation research provides a global perspective of the protein compositions of microbial cells and is a viable way to address bioremediation's molecular mechanisms. Functional genomics, when combined with proteomics, provides insight into global metabolic and regulatory networks, which can help researchers better understand gene functions.

In a functional genomics method, the basic idea is to broaden the area of biological research from examining a single gene or protein to studying all genes or proteins in a systematic manner. The traditional method for evaluating gene function is to use gene disruption or complementation to determine which gene is required for a specific biological function under a given circumstance. The precise study of gene activities becomes possible with the addition of technologies such as transcriptomics and proteomics to classic genetic techniques (Rajendhran and Gunasekaran 2008; Rayu et al. 2012; Zhao and Poh 2008).

Metabolic engineering combines a systematic investigation of metabolic and other processes with molecular biological tools to design and apply reasonable genetic alterations to improve cellular attributes (Koffas et al. 1999). Understanding microbial physiology will allow them to adapt to their host cells and become more efficient bioremediation processes, which would be impossible to achieve through evolution (Rayu et al. 2012).

Scientists will be better able to answer questions like how oxygen stress, nutrient availability, or high contaminant concentrations along different geochemical gradients or at transitional interfaces affect the structure and function of the organohalide respiring community with these new genomics tools. Finally, by tracking the general structure and function of the microbial community, as well as key functional players, educated decisions may be made about how to best adjust field conditions to accomplish effective pesticide bioremediation.

19.14   IMMOBILISATION OF CASE CELLS AS A STRATEGY FOR IMPROVING PESTICIDE BREAKDOWN EFFICIENCY

Because it allows for the maintenance of enzymatic activity over lengthy periods of time, cell immobilisation has been used for biological pesticide elimination (Chen and Georgiou 2002; Martin et al. 2000; Richins et al. 2000). The ability to use a high cell density, avoid cell washout even at high dilution rates, easy separation of cells from the reaction system, repeated use of cells, and better protection of cells from harsh environments have all been demonstrated to be significant advantages of whole cell immobilisation over conventional biological systems using free cells. Immobilised cells have been shown to be far more tolerant of perturbations in the reaction environment and less vulnerable to hazardous compounds, making them particularly appealing for the treatment of toxic compounds like pesticides (Ha et al. 2008). Furthermore, the increased degradation capability of immobilised cells is related to the cells' protection from inhibitory compounds in the environment. The rates of deterioration for repeated operations increased with each batch, demonstrating that cells grew more acclimated to the reaction conditions over time (Ha et al. 2009).

There are two kinds of cell immobilisation processes: those that rely on physical retention (entrapment and inclusion membrane) and those that rely on chemical bonding (biofilm production) (Kennedy and Cabral 1983). Various inorganic (clays, silicates, glass, and ceramics) and organic (cellulose, starch, dextran, agarose, alginate, chitin, collagen, keratin, and others) materials, or substrates can be employed in cell immobilisation procedures (Arroyo 1998). Although entrapment in polymeric gels has become the primary method for immobilising cells, immobilised cells on supports have been used more frequently in the biodegradation of xenobiotics like insecticides (Lusta et al. 1990).

In order to degrade pesticides, it is necessary to look for materials that have favourable properties for cell immobilisation, such as physical structure, ease of sterilisation, and the ability to use it repeatedly, but most importantly, the support must be inexpensive enough to be used for pesticide degradation in the future. Thus, the approaches can be divided into two categories: active methods that cause microorganisms to be captured in a matrix, and passive methods that rely on microorganisms' natural or synthetic surfaces to create biofilms.

A biofilm, on the other hand, is a cohesive complex structure of microorganisms organised in colonies and cell products like extracellular polymers (exopolymer) that grows attached to a suspension bracket or in solid surface static (static biofilm) (Davey and O'Toole 2000; Nicolella et al. 2000). The biofilm creation process consists of numerous phases, beginning with surface attack or recognition, followed by growth and use of diverse carbon and nitrogen sources for the synthesis of sticky compounds. In the meantime, a stratified organisation is taking place, which is influenced by oxygen gradients and other abiotic factors. Colonisation is the term for this process. After then, the biofilm goes through an intermediate period of development, which varies depending on the presence of nutrients from the medium or friction with the surrounding water flow. Finally, cells may split and colonise various surfaces after a period of biofilm ageing (Yañez-Ocampo et al. 2009).

Because biofilms form at a solid-liquid interface, the flow rate flowing through them effects the physical detachment of microorganisms, the hydrodynamic plays a significant part in their formation. They have a system of channels that allow nutrients and trash to be transported; this is important when the environment is changed and microbes are deprived of molecules necessary for their development. Resistance to host defences and antimicrobial treatments is another feature of biofilms. While different control elements impact different microorganisms, the colonies arranged and incorporated in an exopolymer form an impervious barrier that only affects the most surface bacteria. When biofilm cells are liberated, they can migrate and be deposited in new niches while still keeping the features of a biofilm adhering to a surface. Microorganisms exchange information with one another. This is known as quorum sensing, and it involves signalling molecules that facilitate intercellular communication to regulate and express certain genes (Betancourth et al. 2004; Singh and Walker 2006).This property is influenced by cell density; for example, a biofilm with a high cell density increases the production of resistance genes that aid in survival and protection (Dorigo et al. 2007). Bacteria can also create chemicals that promote colony growth while inhibiting the growth of other germs, putting disease microorganisms in a better position within the biofilm (Villena and Gutiérrez-Correa 2003). Supports can be made of synthetic or natural materials.

The tezontle (in Nahuatl, tezt means rock and zontli means hair), a native volcanic rock of Morelos state, has shown promising results in the breakdown of pesticide combinations (central Mexico). This rock is extremely porous, has a huge contact surface area, and can be disinfected and reused. The existence of micropores helps bacteria to colonise and form microcolonies. This material's immobilisation approach is based on the colonisation of tezontle micropores via the creation of a biofilm. Following that, a current containing pesticide wastes is carried through to allow contact with the immobilised microorganisms, allowing biodegradation to take place. This method has proven to be quite effective, and it can be utilised to degrade pesticide wastes (Yaez-Ocampo 2009). A bacterial community was immobilised in a biofilm on tezontle in our lab, and it shown a significant capability for the removal of a mixture of organophosphate pesticides, which are commonly used in agricultural and stockbreeding in Mexico. Furthermore, this material with immobilised cells was packaged in an up-flow reactor, which resulted in increased bacterium survival as well as more efficient pesticide removal (Yañez-Ocampo 2009).

Furthermore, various investigations suggest that a variety of materials possess the properties required to immobilise microbes. The use of diverse plant fibres as a substrate for immobilising bacterial consortiums to digest xenobiotics, for example, offers significant benefits. The utilisation of natural structural materials for cell entrapment, such as petiolar felt-sheath of palm, has brought a new dimension to a range of immobilisation matrices. Reusability, freedom from toxicity issues, mechanical strength for necessary support, and open spaces inside the matrix for growing cells, avoiding rupture and diffusion problems, are all advantages of such biostructures. These findings suggest that additional forms of biomaterials derived from various plant sources could be employed for cell entrapment.

The loofa sponge (*Luffa cylindrica*) was employed as a carrier material for immobilising various microorganisms for xenobiotic adsorption or degradation. This sponge has been used to immobilise organisms such as *Porphyridium cruentrum, Chlorella sorokiniana, Funalia trogii,* and *Penicillium cyrlopium,* as well as dyes and chlorinated compounds, for nickel and cadmium II treatment. The sponges are produced in huge quantities in Mexico, where they are currently utilised for bathing and dishwashing. Loofa grows well in both tropical and subtropical regions. They're made up of an interconnected vacuum within an open network of matrix support materials and are light and cylindrical in shape. Their potential as carriers for cell immobilisation is quite strong due to their random lattice of small cross sections combined with very high porosity. The sponges are made up of interconnecting spaces inside an open network of fibres and are robust and chemically stable. The sponges are favourable for cell attachment due to their random lattices with small cross sections combined with high porosity (Akhtar et al. 2004; Iqbal and Edyvean 2004; Mazmanci and Unyayara 2005). Our study group used this sponge, and we discovered methyl parathion removal efficiency of 75%.

19.15   ADVANTAGES OF PESTICIDES BIOREMEDIATION

Because bioremediation is a natural process, the public views it as an acceptable waste treatment method for contaminated materials like soil. Bioremediation can theoretically be used to completely destroy a wide range of pollutants. Many chemicals that are considered dangerous by law can be converted into non-hazardous products. This removes the risk of future responsibility related with contaminated material treatment and disposal. Instead of moving toxins from one environmental medium to another, such as from land to water or air, target pollutants can be completely destroyed. It also has a lower environmental impact. Bioremediation can frequently be done on-site, with little or no disturbance to routine operations. This also avoids the need to transfer large amounts of waste off-site, as well as the potential health and environmental risks that might occur during transit. To clear up contamination, bioremediation makes use of natural resources. Bioremediation has the potential to be less expensive than other hazardous waste cleaning solutions.

19.16   DISADVANTAGES OF PESTICIDES BIOREMEDIATION

Bioremediation can also be troublesome since it can result in partial pollutant breakdown, which can result in hazardous and potentially volatile compounds (Rockne and Reddy 2003). Throughout the process, extensive monitoring is also required (Federal Remediation Technology Roundtable 2006). The cleanup process is often quite expensive, labour intensive, and might take many months to reach acceptable levels. Pesticide bioremediation, in particular, necessitates a longer treatment period than other alternative detoxification methods (Rockne and Reddy 2003). Extrapolating from laboratory scale studies to field operations is similarly problematic. Even if genetically engineered bacteria are discharged into the environment, it becomes impossible to remove them after a certain period of time. Another issue with both in situ and *ex situ* methods is that they have the potential to cause significantly more harm than the pollutant itself. Biological processes are frequently much specialised. Important site concerns include the availability of metabolically competent microbial populations, adequate environmental growth conditions, and optimal concentrations of nutrients and contaminants.

19.17 FINALLY, SOME THOUGHTS

Understanding the molecular mechanisms of pesticide degradation is critical for developing novel alternatives and/or efficient technologies for the treatment of pesticide residues or bioremediation of contaminated locations. This information could be used to address pesticide residues in the field (such as waste from cleaning pesticide containers) or obsolete pesticides in the future. Furthermore, by using ways to improve the efficiency of degradation, such as cell immobilisation (bacteria or fungus), we may be able to eliminate the existence of obsolete pesticides and waste generated, lowering pesticide dangers to the environment and human health.

REFERENCES

Aislabie J, Lloyd-Jones G (1995) A review of bacterial degradation of pesticides. *Aust J Soil Res* 33:925–942.

Akhtar N, Iqbal J, Iqbal M (2004) Removal and recovery of nickel (II) from aqueous solution by Loofa sponge immobilized biomass of *Chlorella sorokiniana* characterization studies. *J Hazard Mater* 108(1–2):85–94.

Alexander M (1999) *Biodegradation and Bioremediation*. Academic Press, San Diego, CA.

Anderson TA, Coats JR (1995) Screening rhizosphere soil samples for the ability to mineralize elevated concentrations of atrazine and metolachlor. *J Environ Sci Health (Part B)* 30:473–484.

Anderson TA, Guthrie EA, Walton BT (1993) Bioremediation in the rhizosphere. *Environ Sci Tech* 27(13):2630–2636.

Anderson TA, Kruger EL, Coats JR (1994) Biological degradation of pesticide wastes in the root zone of soils collected at an agrochemical dealership. In *Bioremediation Through Rhizosphere Technology*, Anderson TA, Coats JR, eds., ACS Symposium Series 563, American Chemical Society, Washington, DC, 199–209.

Andreoni V, Gianfreda L (2007) Bioremediation and monitoring of aromatic-polluted habitats. *App Microbiol Biotechnol* 76:287–308.

Anonymous (2009) *ITRC (Interstate Technology and Regulatory Council), Phytotechnology Technical and Regulatory Guidance and Decision Trees, Revised*. PHYTO-3, Washington, DC.

Arroyo M (1998) Inmovilización de enzimas. Fundamentos, métodos y aplicaciones. *Ars Pharmaceut* 39(2): 23–39.

Awasthi M, Prakash NB (1997) Persistence of chlorpyrifos in soils under different moisture regimes. *Pest Sci* 50(1):1–4.

Baker RS (1999) Bioventing systems: a critical review. In *Bioremediation of Contaminated Soils*, Adriano DC, Bollag JM, Frankenberger Jr. WT, Sims RC, eds. American Society of Agronomy, Crop Science Society of America, Soil Science Society of America, Madison, WI, 595–630.

Banasiak L, Schafer A, Van der Bruggen B (2011) Sorption of pesticide endosulfan by electrodialysis membranes. *Cheml Eng J* 166(1):233–239.

Bastos AC, Magan N (2009) *Trametes versicolor*: potential for atrazine bioremediation in calcareous clay soil, under low water availability conditions. *Int Biodeterior Biodegrad* 63:389–394.

Bending GD, Friloux M, Walker A (2002) Degradation of contrasting pesticides by white rot fungi and its relationship with ligninolytic potential. *FEMS Microbiol Lett* 212:59–63.

Benimeli CS, Amoroso MJ, Chaile AP, Castro GR (2003) Isolation of four aquatic streptomycetes strains capable of growth on organochlorine pesticides. *Biores Technol* 89(2):133–138.

Betancourth M, Botero JE, Rivera SP (2004) Biopelículas: una comunidad microscópica en desarrollo. *Colombia Médica* 35:3–1.

Blanca AL, Angus JB, Spanova K, Lopez-Real J, Russell NJ (2007) The influence of different temperature programmes on the bioremediation of polycyclic aromatic hydrocarbons (PAHs) in a coal-tar contaminated soil by in-vessel composting. *J Hazard Mater* 14:340–347.

Boricha H, Fulekar MH (2009) *Pseudomonas plecoglossicida* as a novel organism for the bioremediation of cypermethrin. *Bio Med* 1(4):1–10.

Bouwer EJ, Zehnder AJB (1993) Bioremediation of organic compounds putting microbial metabolism to work. *Trends Biotech* 11:287–318.

Briceño G, Palma G, Durán N (2007) Influence of organic amendment on the biodegradation and movement of pesticides. *Crit Rev Environ Sci Tech* 37:233–271.

Brooks RR (1998) Geobotany and hyperaccumulators. In *Plants That Hyperaccumulate Heavy Metals*, Brook RR, ed. CAB International, Wallingford, Vol. 8, 55–94.

Budd WW, Cohen PL, Saunders PR, Steiner FR (1987) Stream corridor management in the Pacific Northwest; I: Determination of stream corridor widths. *Environ Manag* 11:587–597.

Cases I, de Lorenzo V (2005) Promoters in the environment: transcriptional regulation in its natural context. *Nat Rev Microbiol* 3:105–118.

Casey MT, Grunden AM (2011) Hydrolysis of organophosphorus compounds by microbial enzymes. *Appl Microbiol Biotechnol* 89:35–43.

Cea M, Seaman JC, Jara A, Fuentes B, Mora ML, Diez MC (2007) Adsorption behavior of 2,4-dichlorophenol and pentachlorophenol in an allophanic soil. *Chemosphere* 67:1354–1360.

Cea M, Seaman JC, Jara A, Mora ML, Diez MC (2005) Describing chlorophenol sorption on variable-charge soil using the triple-layer model. *J Colloid Interf Sci* 292(1):171–178.

Chaney RL, Malik M, Li YM, Brown SL, Brewer EP, Angel JS, Baker AJ (1997) Phytoremediation of soil metals. *Curr Opin Biotech* 8:279–283.

Chen W, Georgiou G (2002) Cell-surface display of heterologous proteins: from high throughput screening to environmental applications. *Biotechnol Bioeng* 5:496–503.

Cho MH, Mulchandani CA, Chen W (2002) Bacterial cell surface display of organophosphorus hydrolase for selective screening of improved hydrolysis of organophosphatenerve agents. *Appl Environ Microbiol* 68:2026–2030.

Cho MH, Mulchandani CA, Chen W (2004) Altering the substrate specificity of organophosphorus hydrolase for enhanced hydrolysis of chlorpyrifos. *Appl Environ Microbiol* 70:4681–4685.

Coats JR, Anderson TA (1997) *The Use of Vegetation to Enhance Bioremediation of Surface Soils Contaminated with Pesticide Wastes*, USEPA, Office of Research and Development, Washington, DC.

Couto MNPFS, Monteiro E, Vasconcelos MTSD (2010) Mesocosm trials of bioremediation of contaminated soil of a petroleum refinery: comparison of natural attenuation, biostimulation and bioaugmentation. *Environ Sci Poll Res* 17(7):1339–1346.

Cunningham SD, Anderson TA, Schwab AP, Hsu FC (1996) Phytoremediation of soils contaminated with organic pollutants. In *Advances in Agronomy*, Sparks DL, ed., Academic Press, San Diego, CA, Vol. 56, 55–114.

Damalas CA (2009) Understanding benefits and risks of pesticide use. *Sci Res Essay* 4(10):945–949.

Davey ME, O'Toole GA (2000) Microbial biofilms: from ecology to molecular genetics. *Microbiol Molec Biol Rev* 64(4):847–867.

Davis LC, Banks MK, Schwab AP, Narayanan M, Erickson LE, Sikdar SK, Irvine RL (1998) *Plant Based Bioremediation, Bioremediation: Principles and Practice, Vol. 2, Biodegradation Technology Development*. Technomic Publishing Co., Lancaster, 183–219.

Davis LC, Erickson LE, Narayanan M, Zhang Q, McCutcheon SC, Schnoor JL (2003) Modeling and design of phytoremediation. In *Phytoremediation: Transformation and Control of Contaminants*, McCutcheon SC and Schnoor JL, eds. Wiley, New York, 663–694.

Davis LC, Lupher D, Hu J, Erickson LE (1999) Transport of trichloroethylene through living plant tissues. In *Proceedings of the 1999 Conference on Hazardous Waste Research*, Erickson LE, Rankin M, eds., Kansas State University, Manhattan, 205–209.

DeSchrijver A, DeMot R (1999) Degradation of pesticides by actinomycetes. *Crit Rev Microbiol* 25:85–119.

Diez MC, Mora ML, Videla S (1999) Adsorption of phenol and color from BKME using synthetic allophanic compounds. *Water Res* 33(1):125–130.

Dixon B (1996) Bioremediation is here to stay. *ASM News* 62:527–528.

Dorigo U, Leboulanger C, Bérard A, Bouchez A, Humbert JF, Montuelle B (2007) Lotic biofilm community structure and pesticide tolerance along a contamination gradient in a vineyard área. *Aquat Microb Ecol* 50:91–102.

Dosnon-Olette R, Couderchet M, Oturan MA, Oturan N, Eullaffroy P (2011) Potential use of Lemna minor for the phytoremediation of isoproturon and glyphosate. *Int J Phytoreme* 13(6):601–612.

Dua M, Singh A, Sethunathan N, Johri A (2002) Biotechnology and bioremediation: successes and limitations. *Appl Microbiol Biotech* 59(2–3):143–152.

Dubey RC (2004) *A Text Book of Biotechnology*, 3rd Edition, S. Chand and Company Ltd., New Delhi, 365–375.

Dzantor KE (2007) Phytoremediation: the state of rhizosphere engineering for accelerated rhizodegradation of xenobiotic contaminants. *J Chem Technol Biotech* 82:228–232.

Eapen S, Singh S, D'Souza SF (2007) Advances in development of transgenic plants for remediation of xenobiotic pollutants. *Biotechnol Adv* 25:442–451.

Enrica G (1994) The role of microorganisms in environmental decontamination, contaminants in the environments, a multidisciplinary assessment of risk to man and other organisms. *ED Renzoni Arister* 25:235–246.

European Environment Agency (1999) *Environmental Assessment Report No. 3, Groundwater Quality and Quantity in Europe*. European Environment Agency, Copenhagen.

Federal Remediation Technology Roundtable (FRTR) (2006) Remediation technologies screening matrix and reference guide. Version 4.0, pp 1–10. https://hwbdocuments.env. nm.gov/Los%20Alamos%20National%20Labs/TA%2011/3654.pdf.

Felsot AS, Mitchell JK, Bicki TJ, Frank JF (1992) Experimental design for testing landfarming of pesticide-contaminated soil excavated from agrochemical facilities. *Pesticide Waste Manag* 22:244–261.

Fragoeiro S (2005) Use of fungi in bioremediation of pesticides [dissertation]. Cranfield University, Bedford.

Fragoeiro S, Magan N (2008) Impact of *Trametes versicolor* and *Phanerochaete crysosporium* on differential breakdown of pesticide mixtures in soil microcosms at two water potentials and associated respiration and enzyme activity. *Int Biodeterior Biodegrad* 62:376–383.

Frenich AG, Rodriguez MJG, Vidal JLM, Arrebola FJ, Torres MEH (2005) A study of the disappearance of pesticides during composting using a gas chromatography-tandem mass spectrometry technique. *Pest Manag Sci* 61:458–466.

Frutos FJV, Escolano O, García F, Babín M, Fernández MD (2010) Bioventing remediation and ecotoxicity evaluation of phenanthrene-contaminated soil. *J Hazard Mater* 183(1–3):806–813.

Fu GP, Cui Z, Huang T, Li SP (2004) Expression, purification, and characterization of a novel methyl parathion hydrolase. *Protein Expr Purif* 36:170–176.

Fulekar MH (2009) Bioremediation of fenvalerate by *Pseudomonas aeruginosa* in a scale up bioreactor. *Roman Biotechnol Lett* 14(6):4900–4905.

Gamon M, Saez E, Gil J, Boluda R (2003) Direct and indirect exogenous contamination by pesticides of rice farming soils in a Mediterranean wetland. *Arch Environ Contam Toxicol* 44:141–151.

Germaine KJ, Liu X, Cabellos GG, Hogan JP, Ryan D, Dowling DN (2006) Bacterial endophyte-enhanced phytore-mediation of the organochlorine herbicide 2,4- dichlorophenoxyacetic acid. *FEMS Microbiol Ecol* 57:302–310.

Gouma S (2009) Biodegradation of mixtures of pesticides by bacteria and white rot fungi. Ph.D. Thesis, School of Health Cranfield University, p 416.

Ha J, Engler CR, Lee SJ (2008) Determination of diffusion coefficients and diffusion characteristics for chlorferon and diethylthiophosphate in Ca-alginate gel beads. *Biotechnol Bioeng* 100(4):698–706.

Ha J, Engler CR, Wild J (2009) Biodegradation of coumaphos, chlorferon, and diethylthiophosphate using bacteria immobilized in Ca-alginate gel beads. *Bioresour Technol* 100:1138–1142.

Heath E, Šcancar J, Zuliani T, Milacic R (2010) A complex investigation of the extent of pollution in sediments of the Sava River: Part 2: Persistent organic pollutants. *Environ Moni Assess* 163:277–293.

Hickey WJ, Fuster DJ, Lámar RT (1994) Transformation of atrazine in soil by *Phanerochaete chrysosporium. Soil Biol Biochem* 26:1665–1671.

Hoagland RE, Zablotowicz RM, Locke MA (1994) Propanil metabolism by rhizosphere microflora. In *Bioremediation through Rhizosphere Technology*, Anderson TA, Coats JR, eds., ACS Symposium Series 563. ACS, Washington, DC, 160–183.

Hong MS, Farmayn WF, Dorth IJ, Chiang CY, Mcmillan SK, Schnoor JL (2001) Phytoremediation of MTEB from groundwater plume. *Environ Sci Tech* 35:1231–1239.

Husain Q, Husain M, Kulshrestha Y (2009) Remediation and treatment of organopollutants mediated by peroxidases: a review. *Crit Rev Biotech* 29(2):94–119.

Hussain S, Siddique T, Arshad M, Saleem M (2009) Bioremediation and phytoremediation of pesticides: recent advances. *Crit Rev Environ Sci Technol* 39(10):843–907.

Iqbal M, Edyvean RGJ (2004) Biosorption of lead, copper and zinc ions on loofa immobilized biomass of *Phanerochaete chrysosporium. Min Eng* 17:217–223.

Jung H, Sohn KD, Neppolian B, Choi H (2008) Effect of soil organic matter (SOM) and soil texture on the fatality of indigenous microorganisms in integrated ozonation and biodegradation. *J Hazard Mater* 150:809–817.

Kadian N, Gupta A, Satya S, Mehta RK, Malik A (2008) Biodegradation of herbicide (atrazine) in contaminated soil using various bioprocessed materials. *Biores Technol* 99(11):4642–4647.

Kao CM, Chen CY, Chen SC, Chien HY, Chen YL (2008) Application of *in situ* biosparging to remediate a petroleum-hydrocarbon spill site: field and microbial evaluation. *Chemosphere* 70(8):1492–1499.

Karthikeyan R, Lawrence CD, Erickson LE, Al-Khatib K, Kulakow PA, Barnes PL, Hutchinson SL, Nurzhanova AA (2004) Potential for plant-based remediation of pesticide-contaminated soil and water using nontarget plants such as trees, shrubs and grasses. *Crit Rev Plant Sci* 23:91–101.

Kawahigashi H (2009) Transgenic plants for phytoremediation of herbicides. *Curr Opin Biotech* 20:225–230.

Kennedy DW, Aust SD, Bumpus JA (1990) Comparative biodegradation of alkyl halide insecticides by the white rot fungus *Phanerochaete chrysosporium* (BKM-F-1767). *Appl Environ Microbiol* 56:2347–2353.

Kennedy JF, Cabral JMS (1983) Organic synthesis using immobilized enzymes. In *Solid Phase Biochemistry*, Schouten WH, ed., Wiley Pub., New York.

Khan NU, Varma B, Imrana N, Shetty PK (2011) Phytoremediation using an indigenous crop plant (wheat): the uptake of methyl parathion and metabolism of p-nitrophenol. *Ind J Sci Technol* 4(12):1661–1667.

Koffas M, Roberge C, Lee K (1999) Stephanopoulos G. Metabolic engineering. *Annu Rev Biomed Eng* 1:535–557.

Kookana RS, Rogers SL (1995) Effects of pulp mill effluent disposal on soil. *Rev Environ Contam Toxicol* 142:13–64.

Korade DL, Fulekar MH (2009) Rhizosphere remediation of chlorpyrifos in mycorrhizospheric soil using ryegrass. *J Hazard Mater* 72:1344–1350.

Kruger EL, Anderson TA, Coats JR (1997a) Phytoremediation of soil and water contaminants. ACS Symposium Series 664, Washington, DC.

Kruger EL, Anhalt JC, Sorenson D, Nelson B, Chouhy AL, Anderson TA, Coats JR (1997b) Atrazine degradation in pesticide-contaminated soils: phytoremediation potential. In *Phytoremediation of Soil and Water Contaminants*, ACS Symposium Series; American Chemical Society: Washington, DC, 1997. 54–64.

Lambert JM, Yang T, Thomson NR, Barker JF (2009) Pulsed biosparging of a residual fuel source emplaced at CFB borden. *Inter J Soil Sedi Water* 2(3):6.

Larkin M, Kulakov L, Allen C (2005) Biodegradation and *Rhodococcus*-masters of catabolic versatility. *Curr Opin Biotech* 16:282–290.

Lee TH, Byun IG, Kim YO, Hwang IS, Park TJ (2006) Monitoring biodegradation of diesel fuel in bioventing processes using in situ respiration rate. *Water Sci Tech* 53(4–5):263–272.

Lemmon CR, Pylypiw HM (1992) Degradation of diazinon, chlorpyrifos, isofenphos and pendimethalin in grass and compost. *Bull Environ Contam Toxicol* 48(3):409–415.

Liu YY, Xiong Y (2001) Purification and characterization of a dimethoate-degrading enzyme of *Aspergillus niger*ZHY256 isolated from sewage. *Appl Environ Microbiol* 67:3746–3749.

Lorenzo DV (2008) Systems biology approaches to bioremediation. *Curr Opinion Biotech* 19:579–589.

Lowrance RR, Sharpe JK, Sheridan JM (1986) Long-term sediment deposition in the riparian zone of a coastal plain watershed. *J Soil Water Conser* 41:266–271.

Lowrance RR, Todd RC, Fail J, Hendrickson O, Leonard RA, Asmussen LE (1984) Riparian forests as nutrient filters in agricultural watersheds. *BioScience* 34:374–377.

Lusta KA, Starostina NG, Fikhte BA (1990) Immobilization of microorganisms: cytophysiological aspects. In: *Proceedings of an International Symposium: Physiology of Immmobilized Cells*, De Bont JAM, Visser J, Mattiasson B, Tramper J, eds., Elsevier, Amsterdam.

Ma X, Burken JG (2003) TCE diffusion to the atmosphere in phytoremediation applications. *Environ Sci Technol* 37:2534–2539.

Ma Y, Prasad MNV, Rajkumar M, Freitas H (2011) Plant growth promoting rhizobacteria and endophytes accelerate phytoremediation of metalliferous soils. *Biotech Adv* 29:248–258.

Magan N, Fragoeiro S, Bastos C (2010) Environmental factors and bioremediation of xenobiotics using white rot fungi. *Mycobiology* 38(4):238–248.

Maphosa F, Lieten SH, Dinkla I, Stams AJ, Smidt H, Fennell DE (2012) Ecogenomics of microbial communities in bioremediation of chlorinated contaminated sites. *Front Microbiol* 3:1–14.

Martin M, Mengs G, Plaza E, Garbi C, Sánchez A, Gibello A, Gutierrez F, Ferrer E (2000) Propachlor removal by Pseudomonas strain GCH1 in a immobilized-cell system. *Appl Environ Microbiol* 66(3):1190–1194.

Mazmanci M, Unyayara (2005) Decolourisation of reactive black 5 by *Funalia trogii* immobilized on *Loofa cylindrical* sponge. *Process Biochem* 40:337–342.

McGuinness M, Dowling D (2009) Plant-associated bacterial degradation of toxic organic compounds in soil. *Int J Environ Res Public Health* 6:2226–2247.

McIntyre T (2003) Phytoremediation of heavy metals from soils. *Adv Biochem Eng Biotechnol* 78:97–123.

Megharaj M, Ramakrishnan B, Venkateswarlu K, Sethunathan N, Naidu R (2011) Bioremediation approaches for organic pollutants: a critical perspective. *Environ Int* 37:1362–1375.

Mendez MO, Maier RM (2008) Phytostabilization of mine tailings in arid and semiarid environments—an emerging remediation technology. *Environ Health Perspec* 116(3): 278–283.

Mervat SM (2009) Degradation of methomyl by the novel bacterial strain *Stenotrophomonas maltophilia* M1. *Elect J Biotech* 12(4):1–6.

Mohamed AT, El Hussein AA, El Siddig MA, Osman AG (2011) Degradation of oxyfluo-rfen herbicide by soil microorganisms: biodegradation of herbicides. *Biotechnology* 10:274–279.

Mohammed MS (2009) Degradation of methomyl by the novel bacterial strain *Stenotrophomonas maltophilia* M1. *Euro J Biotechnol* 12:1–6.

Mougin C, Laugero C, Asther M, Dubroca J, Frasse P, Asther M (1994) Biotransformation of the herbicide atrazine by the white rot fungus *Phanerochaete chrysosporium*. *Appl Environ Microbiol* 60:705–708.

Mukherjee I, Kumar A (2011) Phytoextraction of endosulfan a remediation technique. *Bull Environ Contam Toxicol* 88(2):250–254.

Mulchandani A, Kaneva I, Chen W (1999) Detoxification of organophosphate pesticides by immobilized *Escherichia coli* expressing organophosphorus hydrolase on cell surface. *Biotechnol Bioeng* 63:216–223.

Muller WP, Korte F (1975) Microbial degradation of benzo-(a)-pyrene, monolinuron and diel-drin in waste composting. *Chemosphere* 4(3):195–198.

Muller WP, Korte F (1976) Ecological chemical evaluation of waste treatment procedures. In *Environmental Quality and Safety: Global Aspects of Chemistry, Toxicology and Technology as Applied to the Environment*, Coulston F, Korte F, eds., Academic Press, New York, 215–236.

Nair DR, Burken JG, Licht LA, Schnoor JL (1993) Mineralization and uptake of triazine pes-ticide in soil-plant systems. *J Environ Eng* 119:842–854.

Narayanan M, Russell NK, Davis LC, Erickson LE (1999) Fate and transport of trichloroeth-ylene in a chamber with alfalfa plants. *Int J Phytorem* 1:387–411.

Nawaz K, Hussain K, Choudary N, Majeed A, Ilyas U, Ghani A, Lin F, Ali K, Afghan S, Raza G, Lashari MI (2011) Eco-friendly role of biodegradation against agricultural pesticides hazards. *Afri J Microbiol Res* 5(3):177–183.

Nester EW, Denise G, Anderson C, Evans RJ, Pearsall NN, Nester MT (2001) *Microbiology: A Human Perspective*, 3rd edition, McGraw-Hill, New York.

Newman LA, Reynolds CM (2004) Phytodegradation of organic compounds. *Curr Opin Biotechnol* 15:225–230.

Nicolella CM, van Loosdrecht, Heijnen JJ (2000) Review article wastewater treatment with particulate biofilm reactors. *J Biotechnol* 80:1–33.

O'Laughlin J, Belt GH (1995) Functional approaches to riparian buffer strip design. *J Forest* 93:29–32.

Oshiro K, Kakuta T, Sakai T, Hirota H, Hoshino T, Uchiyama T (1996) Biodegradation of organophosphorus insecticides by bacteria isolated from turf green soil. *J Ferment Bioeng* 82:299–305.

Oudot J, Dutrieux E (1989) Hydrocarbon weathering and biodegradation in a tropical estuarine ecosystem. *Mar Environ Res* 27:195–213.

Pankow JF, Thompson NR, Johnson RL, Baehr AL, Zogorski JS (1997) The urban atmosphere as a non-point source for the transport of MTBE and other volatile organic compounds (VOCs) to shallow groundwater. *Environ Sci Technol* 31:2821–2828.

Paraíba LC, Cerdeira AL, Da Silva EF, Martins JS, Coutinho HLA (2003) Evaluation of soil temperature effect on herbicide leaching potential into groundwater in the Brazilian cerrado. *Chemosphere* 53(9):1087–1095.

Paraiba LC, Spadotto CA (2002) Soil temperature effect in calculating attenuation and retardation factors. *Chemosphere* 48:905–912.

Perkovich BS, Anderson TA, Kruger EL, Coats JR (1995) Enhanced mineralization of [14C]-atrazine in *Kochia scoparia* rhizospheric soil from a pesticide-contaminated site. *Pesti Sci* 46:391–396.

Peterjohn WT, Correll DL (1984) Nutrient dynamics in an agricultural watershed: observation of a riparian forest. *Ecology* 65:1466–1475.

Petruska J (1985) A benchtop system for evaluation of pesticide disposal by composting. *Nucl Chem Waste Manag* 5(3):177–182.

Pieper DH, Reineke W (2000) Engineering bacteria for bioremediation. *Curr Opin Biotechnol* 11:262–270.

Pinay G, Decamps H (1988) The role of riparian woods in regulating nitrogen fluxes between the alluvial aquifer and surface water: a conceptual model, regulated rivers. *Res Manag* 2:507–516.

Pletsch M, de Araujo BS, Charlwood BV (1999) Novel biotechnological approaches in environmental remediation research. *Biotechnol Adv* 17(8):679–687.

Pointing SB (2001) Feasibility of bioremediation by white-rot fungi. *Appl Microbiol Biotechnol* 57:20–33.

Prasad MNV, Freitas H, Fraenzle S, Wuenschmann S, Markert B (2010) Knowledge explosion in phytotechnologies for environmental solutions. *Environ Poll* 158:18–23.

Prescott LM, Harley JP, Klein DA (2002) Microbiology. *Fundamen Appl Microbiol* 2:1012–1014.

Quintero JC, Lu-Chau T, Moreira MT, Feijoo G, Lema JM (2007) Bioremediation of HCH present in soil by the white-rot fungus *Bjerkandera adusta* in a slurry batch bioreactor. *Int Biodeter Biodeg* 60:319–326.

Qureshi A, Mohan M, Kanade GS, Kapley A, Purohit HJ (2009) *In situ* bioremediation of organochlorine-pesticide-contaminated microcosm soil and evaluation by gene probe. *Pest Manag Sci* 65:798–804.

Racke KD, Frink CR (1989) Fate of organic contaminants during sewage sludge composting. *Bull Environ Contam Toxicol* 42(4):533.

Rajendhran J, Gunasekaran P (2008) Strategies for accessing soil metagenome for desired applications. *Biotechnol Adv* 26:576–590.

Rajkumar M, Ae N, Prasad MNV, Freitas H (2010) Potential of siderophore producing bacteria for improving heavy-metal phytoextraction. *Trends Biotech* 28(3):142–149.

Rajkumar M, Prasad MNV, Freitas H, Ae N (2009) Biotechnological applications of serpentine soil bacteria for phytoremediation of trace metals. *Crit Rev Biotech* 29(2):120–130.

Rao N, Grethlein HE, Reddy CA (1995) Mineralization of atrazine during composting with untreated and pretreated lignocellulosic substrates. *Compost Sci Util* 3(3):38–46.

Rayu S, Karpouzas DG, Singh BK (2012) Emerging technologies in bioremediation: constraints and opportunities. *Biodegradation* 23:917–926.

Richins D, Kaneva I, Mulchandani A, Chen W (1997) Biodegradation of organophosphorus pesticides by surface-expressed organophosphorus hydrolase. *Nat Biotechnol* 15:984–987.

Richins R, Mulchandani A, Chen W (2000) Expression, immobilization, and enzymatic characterization of cellulose-binding domain-organophosphorus hydrolase fusion enzymes. *Biotechnol Bioeng* 69:591–596.

Riser-Roberts E (1998) *Bioremediation of Petroleum Contaminated Site*, CRC Press, Boca Raton, FL.

Ritmann BE, Jacson DE, Storck SL (1988) Potential for treatment of hazardous organic chemicals with biological process. *Biotreat Sys* 3:15–64.

Rockne K, Reddy K (2003) *Bioremediation of Contaminated Sites*, University of Illinois, Chicago.

Rubilar O, Feijoo G, Diez MC, Lu-Chau TA, Moreira MT, Lema JM (2007) Biodegradation of pentachlorophenol in soil slurry cultures by *Bjerkandera adusta* and *Anthracophyllum discolour. Indust Engin Chem Res* 4:6744–6751.

Samanta SK, Singh OV, Jain RK (2002) Polycyclic aromatic hydrocarbons environmental pollution and bioreme-diation. *Trends Biotechnol* 20:243–248.

Seech A, Bolanos-Shaw K, Hill D, Molin J (2008) *In situ* bioremediation of pesticides in soil and groundwater. *Remedia Winter* 19:87–98.

Sessitsch A, Reiter B, Pfeifer U, Wilhelm E (2002) Cultivation independent population analysis of bacterial endophytes in three potato varieties based on eubacterial and *Actinomycetes*-specific PCR of 16S rRNA genes. *FEMS Microbiol Ecol* 39:3–32.

Shabir G, Afzal M, Anwar F, Tahseen R, Khalid ZM (2008) Biodegradation of kerosene in soil by a mixed bacterial culture under different nutrient conditions. *Inter Biodeter Biodegra* 61:161–166.

Shalaby SEM, Abdou GY (2010) The influence of soil microorganisms and bio- or -organic fertilizers on dissipation of some pesticides in soil and potato tube. *J Plant Protec Res* 50(1):86–92.

Shanahan P (2004) *Bioremediation, Waste Containment and Remediation Technology*. Spring 2004, Massachusetts Institute of Technology, MIT Open Course Ware.

Shann JR, Boyle JJ (1994) Influence of plant species on *in situ* rhizosphere degradation. In *Bioremediation Through Rhizosphere Technology*, ACS Symposium Series, Anderson TA, Coats JR, eds., American Chemical Society, Washington, DC, Vol. 563, 70–81.

Shen YJ, Lu P, Mei H, Yu HJ, Hong Q, Li SP (2010) Isolation of a methyl parathion-degrading strain Stenotrophomonas sp. SMSP-1 and cloning of the ophc2 gene. ***Biodegrad Dordrec*** 21(5):785–792.

Shimizu H (2002) Metabolic engineering–integrating methodologies of molecular breeding and bioprocess systems engineering. *J Biosci Bioeng* 94:563–573.

Shukla KP, Singh NK, Sharma S (2010) Bioremediation: developments, current practices and perspectives. *Gen Eng Biotechnol J* 3:1–20.

Singh BK, Walker A (2006) Microbial degradation of organophosphorus compounds. *FEMS Microbiol Rev* 30(3):428–471.

Singh DK (2008) Biodegradation and bioremediation of pesticide in soil: concept, method and recent developments. *Ind J Microbiol* 48(1):35–40.

Singh DP, Khattar JI, Nadda J, Singh Y, Garg A, Kaur N, Gulati A (2011a) Chlorpyriphos degradation by the cyanobacterium *Synechocystis* sp. strain PUPCCC 64. *Environ Sci Poll Res Int* 18(8):1351–1359.

Singh N, Megharaj M, Kookana R, Naidu R, Sethunathan N (2004) Atrazine and simazine degradation in pennisetum rhizosphere. *Chemosphere* 56:257–263.

Singh S, Sherkhane PD, Kale SP, Eapen S (2011b) Expression of a human cytochrome P4502E1 in *Nicotiana tabacum* enhances tolerance and remediation of g-hexachlorocyclohexane. *New Biotechnol* 28:423–429.

Strobel KL, McGowan S, Bauer RD, Griebler C, Liu J, Ford RM (2011) Chemotaxis increases vertical migration and apparent transverse dispersion of bacteria in a bench-scale microcosm. *Biotechnol Bioeng* 108(9):2070–2077.

Strong PJ, Burgess JE (2008) Treatment methods for wine-related ad distillery wastewaters: a review. *Bioreme J* 12:70–87.

Surekha RM, Lakshmi PKL, Suvarnalatha D, Jaya M, Aruna S, Jyothi K, Narasimha G, Venkateswarlu K (2008) Isolation and characterization of a chlorpyrifos degrading bacterium from agricultural soil and its growth response. *Afr J Microbiol Res* 2:26–31.

Tang CY, Criddle QS, Fu CS, Leckie JO (2007) Effect of flux (transmembrane pressure) and membranes properties on fouling and rejection of reverse osmosis and nanofiltration membranes treating perfluorooctane sulfonate containing waste water. *Environ Sci Technol* 41:2008–2014.

Tao L, Yang H (2011) Fluroxypyr biodegradation in soils by multiple factors. *Environ Monit Assess* 175:227–238.

Taylor AW, Spencer WF (1990) Volatilization and vapor transport processes. In *Pesticides in the Soil Environment Processes, Impacts and Modeling*, Cheng HH, ed. Soil Sci Soc Am Inc, Madison, WI, 214–269.

Tortella GR, Diez MC, Duran N (2005) Fungal diversity and use in decomposition of environmental pollutants. *Crit Rev Microbiol* 31:197–212.

Trapp S, Karlson U (2001) Aspects of phytoremediation of organic pollutants. *J Soils Sedi* 1:37–43.

United States Environmental Protection Agency (2006) A Citizen's Guide to Bioremediation, pp. 1–2. https://semspub.epa.gov/work/HQ/158703.pdf.

Utmazian MN, Wenzel WW (2006) *Phytoextraction of Metal Polluted Soils in Latin America*. Environmental Applications of Poplar and Willow Working Party. 18-20 May 2006, Northern Ireland, pp. 1–7. https://www.fao.org/forestry/11114-07881fab8de72b-c1ae18a2f90c9367d2f.pdf

Van Deuren J, Lloyd T, Chhetry S, Raycharn L, Peck J (2002) Remediation technologies screening matrix and reference guide. Federal Remediation Technologies Roundtable 4.

Van Dyk JS, Brett P (2011) Review on the use of enzymes for the detection of organochlorine, organophosphate and carbamate pesticides in the environment. *Chemosphere* 82:291–307.

Vandervoort C (1997) Fate of selected pesticides applied to turfgrass: effects of composting on residues. *Bull Environ Contam Toxicol* 58(1):38–45.

Vidali M (2001) Bioremediation, an overview. *Pure App Chem* 73(7):1163–1172.

Villena GK, Gutiérrez-Correa M (2003) Biopelículas de *Aspergullus niger* para la producción de celulasas algunos aspectos estructurales y fisiológicos. *Rev Peru Biol* 10(1):78–87.

Wanatabe K, Kodama Y, Harayama S (2001) Design and evaluation of PCR primers to amplify bacterial 16S ribosomal DNA fragments used for community fingerprinting. *J Microbiol Methods* 44:253–262.

Wang XX, Chi Z, Ru SG, Chi ZM (2012 Genetic surface-display of methyl parathion hydrolase on Yarrowia lipolytica for removal of methyl parathion in water. *Biodegradation* 23:763–774.

World Health Organization (1990) Report on fenvalerate, environmental health criteria. International Program on Chemical Safety.

Wu NF, Deng MJ, Liang GY, Chu XY, Yao B, Fan YL (2004) Cloning and expression of ophc2, a new organophosphorus hydrolase gene. *Chin Sci Bull* 49:1245–1249.

Wu T, Crapper M (2009) Simulation of biopile processes using a hydraulics approach. *J Hazard Mater* 171(1–3):1103–1111.

Yadav JS, Loper JC (2000) Cytochrome P450 oxidoreductase gene and its differentially terminated cDNAs from the white rot fungus *Phanerochaete chrysosporium*. *Curr Genet* 37:65–73.

Yan Q-X, Hong Q, Han P, Dong X-J, Shen YJ. Li SP (2007) Isolation and characterization of a carbofuran-degrading strain *Novosphingobium* sp. FND-3. *FEMS Microbiol Lett* 271:207–213.

Yañez-Ocampo G, Penninckx M, Jiménez-Tobon GA, Sánchez-Salinas E, Ortiz-Hernández ML (2009) Removal of two organophosphate pesticides employing a bacteria consortium immobilized in either alginate or tezontle. *J Hazard Mater* 168:1554–1561.

Yates SR, McConnell LL, Hapeman CJ, Papiernik SK, Gao S, Trabue SL (2011) Managing agricultural emissions to the atmosphere: state of the science, fate and mitigation and identifying research gaps. *J Environ Qual* 40:1347–1358.

Yoon JM, Oliver DJ, Shanks JV (2008) Phytotransformation of 2, 4-dinitrotoluene in *Arabidopsis thaliana*: toxicity, fate and gene expression studies *in vitro*. *Biotech Progress* 19:1524–1531.

Yu H, Yan X, Shen W, Hong Q, Zhang J, Shen Y, Li S (2009) Expression of methyl parathion hydrolase in *Pichia pastoris*. *Curr Microbiol* 59:573–578.

Zablotowicz RM, Hoagland RE, Locke MA, Anderson TA, Coats JR (1994) Glutathione S-transferace activity in rhizosphere bacteria and the potential for herbicide detoxification. In *Bioremediation through Rhizosphere Technology*, ACS, Washington, DC, ACS Symposium Series, Vol. 563, 184–198.

Zavoda J, Cutright T, Szpak J, Fallon E (2001) Uptake, selectivity and inhibition of hydroponic treatment of contaminants. *J Environ Eng* 127:502–508.

Zhang Q, Davis LC, Erickson LE (2001) Transport of methyl tert-butyl ether (MTBE) through alfalfa plants. *Environ Sci Technol* 35:725–731.

Zhang R, Cui Z, Jiang J, Gu X, Li S (2005) Diversity of organophosphorus pesticides degrading bacteria in a polluted soil and conservation of their organophosphorus hydrolase genes. *Can J Microbiol* 5:337–343.

Zhao B, Poh LC (2008) Insights into environmental bioremediation by microorganisms through functional genomics and proteomics. *Proteomics* 8:74–881.

Zhongli C, Shunpeng L, Guoping F (2001) Isolation of methyl parathion-degrading strain m6 and cloning of the methyl parathion hydrolase gene. *Appl Environ Microbiol* 67(10):4922–4925.

Zhou J, He Z, Van Nostrand JD, Wu L, Deng Y (2010) Applying Geo Chip analysis to disparate microbial communities. *Microbe* 5:60–65.

20 Microbial remediation of agricultural soils contaminated with agrochemicals

M. Madakka and G. Mary Sandeepa

CONTENTS

20.1 INTRODUCTION

During 'Green revolution (GR) period' (1966–1985), the agriculture innovations had transformed the agriculture practice and productivity. A remarkable growth in world crop production has occurred in the past six decades. It is mainly because of increase in the utilization of fertilizers, pesticides, new plant varieties and development in technologies like irrigation. In agriculture this development feeds more or less six billion people in the world. World population is rising day by day and the change in diets also demand the 70% of food production rates (FAO, 2009). According to Oerke (2006) 35% of global food production yield is decreasing because of pests. The fast improvement of the agrochemical field was occurred after 1945 many pesticides like

DOI: 10.1201/9781003394600-23

herbicides, fungicides, insecticides and other many chemicals were introduced to control pests. From 1990 to 2018, the average quantity of pests has increased from 1.55 kg·ha^{-1} to 2.63 kg·ha^{-1} globally.

Pesticides are either chemical or biological compounds utilized for killing or to eliminate pests. The pesticides are classified into many groups which depends on the targeted pests, namely larvicides, fungicides, bactericides, insecticides and rodenticides. In 19th and 20th centuries, pyrethrins (plant extracts) were used as herbicides, fungicides and insecticides. In 1930s, the inorganic chemicals like sulfur and arsenic containing compounds were used for the protection of crops. The arsenic act as insecticides and the sulfur compounds were toxic to fungus. At the time of Second World War beginning, many pesticides like aldrin, dichloro diphenyl trichloroethane (DDT) and dieldrin were used as insecticides and 2-methyl-4-chlorophenoxyacetic acid (MCPA), and 2,4-dichlorophenoxyacetic acid (2,4-D) were used as herbicides (Matthews, 2018).

20.2   AGROCHEMICALS FATE IN AGRICULTURAL SOIL

When these agrochemicals used for long time they will stay in sediments and soil and sediments. If the pesticide enters in the soil environment, various factors will influence it. They will enter into food chain or may undergo to various degradations naturally. These are volatilization, photo-decomposition, adsorption, leaching, adsorption and many microbial activities. Different pesticides react in different manner and often many of these process occurred simultaneously. Soil quality and texture has great impact on microbial activity, leaching and adsorption, whereas climatic conditions will influence the activity, volatility and photo-decomposition. Adsorption is a phenomena in which a component mixture is accumulated at an interface. The pesticide leaching ability of a pesticide will depend upon (1) evapotranspiration, (2) adsorption to particles of soil and (3) the compound entry into solution. Moreover, the pesticide follows or leaches. When dried soil surface is there the pesticide will drained into to ground water, if evapotranspiration ratio is high it leads to more water utilization by plant thus more pesticides also transported into the plant via roots. Depending on the fate we can categorized the pesticides which again depend on three process i.e. transport, pesticide transfer and transformation of that pesticide. In the transport process, the pesticides enter into the environments from its application site and spread throughout the surface water system. This process is considered as controlling one which involves different factors in environment and soil distribution. Lastly, the transformation is chemical and biological process which degrades the POPs in to simpler compounds. Generally the fate of pesticides in surface water system (surface) depends on the chemical, structural and biological attributes of pesticides.

Pesticides or some parts of pesticides remain in the soil this accumulation in soil will affects the living microbes in that soil environment. When humans are exposed to these contaminated water and food or directly inhaled the polluted air from agriculture and household use. The pesticide may enter the human's body through oral, respiratory, eye and dermal pathways (Kim et al., 2017). The toxicity depends on dosage, exposure time, structure and electronic properties (Hamadache et al., 2016; Heard, 2017). The pesticides toxicity depends on the molecule structure, electronic properties, dosage and number of exposure times (Hamadache et al., 2016; Heard, 2017).

So by above reasons the pesticide residual concentration must be reduced in the soil and efficient remediation methods must be applied. The efficient eco-friendly process is bioremediation; it is used as alternative technique to very toxic approach and more expensive techniques. In biodegradation the pesticides are degraded by microbial activity of microbes. The microbes used are bacteria (Doolotkeldieva et al., 2018), or fungi (Erguven, 2018) degrade pesticides into simple compounds like carbon di oxide, water, minerals and oxides. Enzymes play an essential role in these reactions and can act as catalysts (Senko, 2017). The bioremediation is important for pesticide-polluted soil because the pesticides have ubiquitous distribution, high persistence and their effect on human's health. The research on effective bioremediation reveals that we can optimize the microbe's activity by addition of nutrients. But it is not completely successful; because of less solubility they are not available to microbes; it is a major problem of biodegradation in polluted soils. Organophosphates have medium water solubility, thus organochlorine pesticides are very less soluble than carbamates esters of N-methyl carbamic acids (Zheng and Wong, 2010).

20.3 PESTICIDE'S BIOAVAILABILITY FOR MICROORGANISMS

In the biodegradation of pesticide-polluted soil, the main constraint is bioavailability. The definition of bioavailability is the amount of pesticides that can be taken up by microbes readily. Pesticide bioavailability to microorganisms affects the biodegradation in a number of ways; they are mentioned as follows: (1) pesticides in low concentration make microbes to initiate catabolic genes of biodegradation that lowers the process. (2) At less pollutants levels and nutrient conditions, microbes may degrade the contaminant but growth rate decreases leading to decrease in the microbe's pesticide uptake (Odukkathil and Vasudevan, 2013). Many studies showed that the pesticides bounded to soil are unavailable for microbial degradation. Another reason is low bioavailability for the persistence of various pesticides using. The low bioavailability is due to imbalance distribution of pesticides and microbes spatially and the diffusion of substrate is rendered by soil matrix.

The bioremediation process will depend on the microbial metabolic potential to transform or degrade the pollutants, which will depend on both bioavailability and accessibility (Antizar-Ladislao, 2010).

20.3.1 BIOSURFACTANTS

Many biosurfactants are studied for bioremediation are rhamnolipids by *Pseudomonas aeruginosa*, sophorolipids and surfactine. Lab studies confirmed that when rhamnolipids are used they will act more effectively in the oil polluted effluents than tweet or triton commercial surfactants. Rhamnolipids can remove greater than 90% of the styrene which is adsorbed in a mixture of soil and sand. Scientist identified that BSP3 (*Burkholderia cenocepacia*) strain produces one type of biosurfactant which belongs to glucolipid. This shows improvement in solubilization of pesticides like methylparathion, ethylparathion and trifluraline. *Bacillus subtilis* SK320 from

TABLE 20.1

Some examples of bioremediation technology developed recently

Pesticide	Bioremediation technique used	Microbes involved	Efficiency of bioremediation	Scale/contamination	References
DDTs (insecticide)	Natural attenuation	Endogenous flora	after 7 weeks 23% is biodegraded	Laboratory	Ortíz et al. (2013)
	Biostimulation (phenol, hexane and toluene)		After 7 weeks > 56% biodegraded		
Pentachlorophenol (herbicide)	Biostimulation	Endogenous flora	97 % degradedafter 6 days	Laboratory (soil from paddy field)	Chen et al. (2012)
Fenpropathrin (insecticide)	Bioaugmentation	Bacillus sp. DG-02	93.3% was biodegraded aafter72 hours	Laboratory (solution)	Chen et al. (2014)
2,4-Dichloro phenoxy acetic(herbicide)	Bioaugmentation	Novosphingobium, strain DY4	After 3 and 7 days 50–95% biodegraded respectively	Laboratory (soil from paddy field)	Dai et al. (2015)
Chlorpynfos(herbiside)	Bioaugmentation	Bacillus cereus, strain Ct3	After 7 days 88% biodegradation occur	Laboratory (Agricultural soil)	Farhan et al. (2014)
Atrazine (herbicide	Bioaugmentation	Strain A6 (Acinetobacter)	30% Biodegraded 6 days later	Laboratory (Agricultural soil)	Singh and Cameotra (2014)
	Bioaugmentation, Bioavailability enhancer	Strain A6, rhamnolipids and triton X-100	80% Biodegraded 6 days later		
Chlorpyrifos (herbicide)	Bioaugmentation	CS2 strain	55% Biodegraded 6 days later	Laboratory (Agricultural soil)	Singh et al. (2016)
	Bioaugmentation, Bioavailability enhancer	CS2 strain, rhamnolipid	88.3% Biodegraded 6 days later		

endosulfan-polluted soils produces lipoproteins (biosurfactant) which is mediated by enzymes.

20.3.2 TECHNOLOGIES INVOLVED IN BIOREMEDIATION

Technologies of bioremediation are categorized into (1) in situ bioremediation or (2) ex situ bioremediation. In in situ, bioremediation-polluted material is treated in the land, and in ex situ bioremediation, the polluted material is treated elsewhere (Chowdary et al., 2012). Some examples of bioremediation technology developed are shown in Table 20.1. It is important to note that bioremediation processes may be either aerobic (Wiegel and Wu, 2000) or anaerobic (Komancová et al., 2003).

20.4 MICROBIAL DEGRADATION MECHANISMS

In bioremediation process, pesticides will be degraded or mineralized by microorganisms; they use these pollutants as their nutrients or nutrient source for their energy. The enzymes such as peroxidases, oxygenases and hydrolases played an essential role in these mechanisms

The pesticides are degraded in three stages.

- Stage 1: Pesticides will be converted into less toxic and more water-soluble products by oxidation or hydrolysis and reduction.
- Stage 2: The stage-1 products are again transformed into amino acids having lower toxicity and higher solubility and sugars.
- Stage 3: Transformation of the stage-2 metabolites into secondary conjugates with less toxicity.

When bioremediation activity is planned the time (taken for degradation) is considered as a relevant parameter to be assessed. Bioremediation efficiency mainly depends on pollutant or pesticide concentration at the first and last stage of the process, water content, microbial activity, pesticide leaching and temperature availability in the soil.

20.4.1 MICROORGANISMS USED IN BIOREMEDIATION

Usually more than one microorganism is used during pesticide degradation. Extensively bacterial species are being utilized for bioremediation purposes (Sylvia et al., 2005). Many bacterial types are involved in active bioremediation process: (1) gamma proteobacteria (*Aerobacter, Pseudomonas, Plesiomonas, Acinetobacter, Aerobacter, Plesiomonas, Moraxella*), (2) beta proteobacteria (Neisseria, *Burkholderia, Neisseria*), (3) alpha proteobacteria (*Sphingomonas*), (4) *Flavobacteria* (*Flavobacterium*) and (5) *Actinobacteria* (*Micrococcus*) and *Flavobacteria* (*Flavobacterium*) (Mamta and Khursheed, 2015). Endo sulfan is actively degraded by five bacterial genera which include *Acinetobacter, Klebsiella, Flavobacterium Alcaligenes, Flavobacterium* and *Bacillus*. In this degradation process, endosulfan lactone, endosulfan diol and endosulfan ether were produced but had lesser toxicity than the original one (i.e. endosulfan).

20.5 APPLICATION OF MICROBIAL REMEDIATION

The in situ process is aerobic and it is carrying out in the polluted site. For this oxygen should be provided to the soil.

The important in situ techniques are as follows.

20.5.1 NATURAL ATTENUATION

In this process, contaminants are degraded naturally by indigenous soil microbes present there. These are (1) biological degradation, (2) volatilization, (3) dispersion, (4) sorption, (5) dilution and (6) radioactive decay of the pollutant. For example, researchers explained that two metabolites of insecticide endosulfan are endosulfan sulfate and endosulfan diol, both are degraded by microbes present in the polluted soils.

20.5.2 BIOSTIMULATION

- In this technique the kind and amounts of nutrients are used to initiate and improve the indigenous microbes growth are optimized. The nutrients like oxygen, nitrogen carbon dioxide and phosphorous are very important for microorganism's life and give them energy, improving their growth.

In this method, mainly N_2 and PO_4 are added because they initiate and improve the bioremediation and microbial diversity. Scientists have studied the DDD, DDT and DDE degradation which stimulates the microorganism's growth by addition of surfactant. Throughout the bioremediation process, the supplied nutrient concentration should be controlled, because more or less amounts of stimulants can decrease the activity and diversity of microbes. The biodegradation of tebuconazole is done in the soil by using the biostimulation technique. The tebuconazole acts on microbial flora and the enzymes negatively; for this reason, in soil concentration of tebuconazole should be decreased. When pesticide is increased the microbial population will decreased. In their experimental tests, they have tested the two types biostimulation substances (bird droppings and compost) effects on the bioremediation process. The results reveal that both substances had a positive effect on development and enzyme activity of the microorganism's oil. The tebuconazole was degraded more intensely in the bird dropping's fertilized soil when compared to compost.

20.5.3 BIOAUGMENTATION

In this process, the addition of single strains or microbial consortia to the soil will be done so. The microorganisms having particular metabolic capabilities induce and improved the biodegradation processes.

In soils, the concentration of pesticides is considered as an important one which conditions the remediation, because high pesticide's concentration will suppress the various metabolic activities of soil microbes. Doolotkeldieva et al. (2018) have one work on bacterial remediation in soils of pesticide-polluted dumping areas. In a preliminary study, they discovered that many bacterial types were present in that polluted soils. They again studied the aldrin remediation. According to their results the bacterial strains such as *Pseudomonas fluorescens* and *Bacillus polymyxa* with specific genes like cytochrome P450 can degrade the aldrin with in a very short time. In polluted soil, the concentrations of pesticides may vary at different depths of soil since the pesticides leach into the soil subsurface and will be adsorb on to the soil particles, making them less bioavailable. The studies and evaluation on bioaugmentation process was done by Odukkathil and Vasudevan [95], in their experiment they took a glass column having 4,500 cm³ volume. The results showed that the concentration of pesticide in the central soil is low; this is because of higher microbial activity which favors the degradation, whereas bottom soil has higher concentration, because pesticides drift downwardly during the water seepage.

20.5.4   BIOVENTING

Through unsaturated soil zones the oxygen is fed to initiate the indigenous microorganisms growth which have the capability of degrading the contaminants.

20.5.5   BIOSPARGING

Air is injected with pressure into the soils (saturated) to improve the amount of oxygen and induce the growth of microbes for the pollutant's degradation. These processes are less cost and efficient methods. The important application of these methods is that the contaminated soil won't be mobilized. Vice versa, in ex situ remediation techniques, the polluted soil is removed from contaminated areas and transferred to other area where we can clean. The important techniques are given in Table 20.1.

20.5.6   BIOREACTORS

The polluted soil is treated with effluents to get a slurry form and improve the reactions of microbes which can remove the pollutants.

20.5.7   COMPOSTING

The polluted soil will be mixed or combined with amendments to improve the pesticide degradation aerobically. Landfarming technique and biopiles technique are the two techniques involved in this method. In on-site methods, the soil is taken out and processed in the area nearer to the contaminated area. The landfarming treatment is an example can also reduce the operation cost when compared with ex situ methods. In all bioremediation techniques, the parameters like oxygen, nutrients, pH, water content and temperature should be controlled to improve the removal efficiency.

20.6   CONCLUSION

The utilization of pesticides and other products used for plant production were increased heavily after second world war in developed as well as in developing countries but these are considered as very potent to different extends and impact humans and the ecosystem. More of them have less degradability and remain for a longer time. Soil biodegradation carried out exploiting either indigenous or specific microbes (fungi or bacteria) or degradation enzymatically. While in literature at lab scale soil biodegradation is available but only few findings on land scale activities.

Unfortunately this is importantly due to local authorities imposing a soil cleanup, very poor cooperation between research laboratories and many other companies involved in the bioremediation sector in pollution contaminated soil.

It will become beneficial when cooperation becomes united highly to disseminate the result and experiences.

REFERENCES

Antizar-Ladislao B (2010) Bioremediation: working with bacteria. *Elements* 6:389–394.

Chen M, Shih K, Hu M, Li F, Lic C, Wu W, Tong H (2012) Biostimulation of indigenous microbial communities for anaerobic transformation of pentachlorophenol in paddy soils of Southern China. *J Agric Food Chem* 60:2967–2975.

Chen S, Chang S, Deng Y, An S, Dong YH., Zhou J, Hu M, Zhong G, Zhang LH (2014) Fenpropathrin Biodegradations pathways in Bacilus sp. Dg-02 and its potential for bire-mediation of pyrethroid – Contaminated soils. *J Agric Food Chem* 62:2147–2157.

Chowdary S, Bala NN, Dhauria P (2012) Bioremediatin - A natural way for cleaner environ-ment. *Int Pharmaceut Chem Boil Sci* 2:600–611.

Doolotkeldieva T, Konurbaeva M, Bobusheva S (2018) Microbial communities in pesticide-contaminated soils in Kyrgyzstan and bioremediation possibilities. *Environ Sci Pollut Res* 25:31848–31862. doi: 10.1007/s11356-017-0048-5. [PMC free article] [PubMed] [CrossRef] [Google Scholar]

Erguven GO (2018) Comparison of some soil fungi in bioremediation of herbicide acetochlor under agitated culture media. *Bull Environ Contam Toxicol* 100:570–575. doi: 10.1007/s00128-018-2280-1. [PubMed] [CrossRef] [Google Scholar]

FAO (2009) *Feeding in the world in 20150. World agricultural summit on Food security 16-19 Nov 2009*. Food and Agriculture Organization of the United Nations, Rome.

Farhan M, Ali-Butt Z, Khan AU, Wahid A, Ahmad M, Ahmad F, Kanwal A (2014) Enhanced biodegradation of chlorpyrifos by agricultural soil isolate. *Asian J Chem* 26:3013–3017.

Hamadache M, Benkortbi O, Hanini S, Amrane A, Khaouane L, Si Moussa C (2016) A quantita-tive structure activity relationship for acute oral toxicity of pesticides on rats: Validation, domain of application and prediction. *J Hazard Mater* 303:28–40. doi: 10.1016/j.jhazmat.2015.09.021. [PubMed] [CrossRef] [Google Scholar]

Heard MS, Baas J, Dorne JL, Lahive E, Robinson AG, Rortais A, Spurgeon DJ, Svendsen C, Hesketh H (2017) Comparative toxicity of pesticides and environmental contaminants in bees: Are honey bees a useful proxy for wild bee species? *Sci Total Environ* 578:357–365. doi: 10.1016/j.scitotenv.2016.10.180. [PubMed] [CrossRef] [Google Scholar]

Kim KH, Kabir E, Jahan SA (2017) Exposure to pesticides and the associated human health effects. *Sci Total Environ* 575:525–535. doi: 10.1016/j.scitotenv.2016.09.009.

Komancová M, Jurčová I, Kochánková L et al. (2003) Metabolic pathways of polychlorinated biphenyls degradation by Pseudomonas sp. 2. *Chemosphere* 50:537–543.

Mamta RJR, Khursheed AW (2015) Bioremediation of pesticides under the influence of bac-teria and fungi Chapter 3. In: *Handbook of Research on Uncovering New Methods for Ecosystem Management through Bioremediation*, IGI Global Publishers. pp. 51–72.

Matthews GA (2008) *A History of Pesticides*. CABI, Boston, MA. [Google Scholar]

Odukkathil G, Vasudevan M (2013) Toxicity and bioremediation of pesticides in agricultural soil. *Rev Environ Sci Biotechnol* 12:421–444.

Oerke E (2006) Crop losses to pests. *The Journal of Agricultural Science,* 144(1), 31–43. doi:10.1017/S0021859605005708

Ortiz-Hernandez L, Sanchez-Salinas E, Dantán-González E, Castrejón-Godínez, M (2013) Pesticide biodegradation: Mechanisms, genetics and strategies to enhance the process. In: Rolando Chamy (ed.), *Biodegradation - Life of Science,* IntechOpen.

Senko O, Maslova O, Efremenko E (2017) Optimization of the use of His_6-OPH-based enzymatic biocatalysts for the destruction of chlorpyrifos in soil. *Int J Environ Res Public Health* 14:1438. doi: 10.3390/ijerph14121438. [PMC free article] [PubMed] [CrossRef] [Google Scholar]

Singh AK, Cameotra SS (2014). Influence of microbial and synthetic surfactant on the biodegradation of atrazine. *Environ Sci Pollut Res Int.* 21(3): 2088–2097. doi: 10.1007/s11356-013-2127-6. Epub 2013 Sep 12. PMID: 24026208.

Singh P, Saini HS, Raj M (2016) Rhamnolipid mediated enhanced degradation of chlorpyrifos by bacterial consortium in soil-water system. *Ecotox Environ Safe* 134:156–162.

Sylvia DM, Fuhrmann JF, Hartel PG, Zuberer DA (2005) Principles and applications of soil microbiology. In: Pearson Education Inc; NJ. https://microbewiki.kenyon.edu/in.

Wiegel J, Wu Q (2000) Microbial reductive dehalogenation of polychlorinated biphenyls. *FEMS Microbiol Ecol* 32:1–15.

Zheng G, Wong JW (2010) Application of microemulsion to remediate organochlorine pesticides contaminated soils. *Proc Annu Int Conf Soils Sediment Water Energy* 15:21–35.

Index

Note: **Bold** page numbers refer to tables and *italic* page numbers refer to figures.

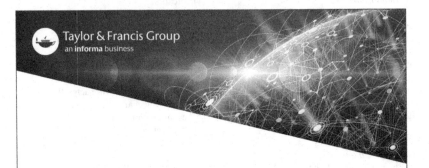